THE MILITARY ORDERS, VOLUME 5

The Military Orders, Volume 5
Politics and Power

Edited by

PETER W. EDBURY
Cardiff, UK

editorial committee
Malcolm Barber, Helen J. Nicholson, Jonathan Phillips,
Denys Pringle, William Purkis

Routledge
Taylor & Francis Group

LONDON AND NEW YORK

First Published 2012 by Ashgate Publisher

2 Park Square, Milton Park, Abingdon, Oxfordshire OX14 4RN
52 Vanderbilt Avenue, New York, NY 10017

Routledge is an imprint of the Taylor & Francis Group, an informa business

First issued in paperback 2018

British Library Cataloguing in Publication Data
The Military Orders. Vol. 5, Politics and Power.
    1. Knights of Malta – Congresses. 2. Templars – Congresses. 3. Military religious orders
    – Europe – History – Congresses. 4. Hospitalers – History – Congresses.
    I. Edbury, P. W. (Peter W.)
    271.7'91-dc23

Library of Congress Cataloging-in-Publication Data
LCCN: 94017896

ISBN: 978-1-4094-2100-9 (hbk)
ISBN: 978-1-138-11709-9 (pbk)

Contents

PART 7: TEMPLAR MYTHOLOGY

List of Illustrations

3.6   Plan of the citadel area showing the plan of the later
       chapter house in the circle (graphics: G. Buzás)   57
3.7   Theoretical reconstruction of the western façade of the
       later chapter house (graphics: G. Buzás)   58
3.8   Profiles of the arch and the window frame of the
       early chapter house (graphics: G. Buzás)   60
3.9   Reconstructed window of the early chapter house
       (graphics: G. Buzás)   60
3.10   Fragment of the central arch supporting the ceiling of the
       early chapter house (photo: G. Buzás)   62
3.11   Reconstructed interior of the early chapter house
       (graphics: G. Buzás)   62
3.12   Console fragment of the early chapter house (photo: G. Buzás)   64
3.13   Plan of citadel area showing the plan of early chapter house
       in the circle (graphics: G. Buzás)   64

4.1   The excavation area of S3 and I2 with the remains of domestic
       structures at the end of the rescue excavation of the Syro-Hungarian
       Archaeological Mission in 2007 (photo: B. Major)   67
4.2   The percentual distribution of animal remains
       (graphics: I. Kováts)   70
4.3   Domestic sheep (Ovis aries L.1758) humerus fragments in
       posterior view (photo: I. Kováts)   70
4.4   Chop marks on domestic sheep (Ovis aries L.1758)
       vertebrae fragments (photo: I. Kováts)   71

5.1   Acre: plan of the city showing the suggested locations of
       the churches mentioned (compiled by Denys Pringle, drawn
       by Ian Dennis)   76

13.1   Grand Master d'Amboise's squadron and the relief of Rhodes
       (© Museo Textil i d'Idumentaria, Barcelona)   149

14.1   Frontispiece of the *Statuta Hospital Hierusalem* (1588)
       (© National Library of Malta)   160
14.2   On the common treasury, *Statuta Hospitalis Hierusalem*
       (© National Library of Malta)   161

16.1. The military orders' properties in Wales, indicating the places
       mentioned in the text (graphics: N² Productions)   194
16.2. Hospitaller houses in England and Wales which lodged
       travellers in 1338, according to Philip de Thame's report
       (graphics: Helen J. Nicholson and N² Productions)   203

Tables

List of Abbreviations

AASS	*Acta Sanctorum Bollandiana*
AHN OO.MM	Archivo Histórico Nacional, Madrid, Ordenes Militares
AOL	*Archives de l'Orient Latin*
ASVen	Archivio di Stato, Venice
BL	British Library
BN	Bibliothèque Nationale de France
CH	*Cartulaire général de l'Ordre des Hospitaliers de Saint-Jean de Jerusalem, 1100–1310*, ed. J. Delaville le Roulx, 4 vols (Paris, 1894–1906)
Cont WT	*La Continuation de Guillaume de Tyr (1184–1197)*, ed. M.R. Morgan, Documents relatifs à l'histoire des croisades, 14 (Paris, 1982)
CT	*Cartulaire général de l'Ordre du Temple 1119?–1150. Recueil des chartes et des bulles relatives à l'ordre du Temple*, ed. Marquis d'Albon (Paris, 1913)
Eracles	*L'Estoire de Eracles Empereur et la Conqueste de la Terre d'Outremer*, in *RHC Occ*, 1.2 (Paris, 1859)
HC	*A History of the Crusades*, gen. ed. K.M. Setton, 2nd edn, 6 vols (Madison, 1968–89)
Malta, Cod.	Archives of the Order of St John, National Library of Malta, Valletta
MGH SS	*Monumenta Germaniae Historica. Scriptores*
MO 1	*The Military Orders: Fighting for the Faith and Caring for the Sick*, ed. M. Barber (Aldershot, 1994)
MO 2	*The Military Orders*, vol. 2: *Welfare and Warfare*, ed. H. Nicholson (Aldershot, 1998)
MO 3	*The Military Orders*, vol. 3: *History and Heritage*, ed. Victor Mallia-Milanes (Aldershot, 2008)
MO 4	*The Military Orders*, vol. 4: *On Land and By Sea*, ed. Judi Upton-Ward (Aldershot, 2008)
NLM	National Library of Malta, Valletta
PL	*Patrologia Latina*
PPTS	Palestine Pilgrims' Text Society
PUTJ	*Papsturkunden für Templer und Johanniter*, ed. R. Hiestand, 2 vols (Gottingen, 1972–84)
QuStDO	Quellen und Studien zur Geschichte des Deutschen Ordens
RHC	*Recueil des Historiens des Croisades*

Arm	*Documents arméniens*
Occ	*Historiens occidentaux*
Or	*Historiens orientaux*
RHGF	*Recueil des Historiens des Gaules et de la France*
RIS	*Rerum Italicarum Scriptores*
ROL	*Revue de l'Orient Latin*
RRH	Regesta Regni Hierosolymitani *and* Additamentum (*Ad*), ed. R. Röhricht (Innsbruck, 1893–1904)
RS	Rolls Series
RSJ	*The Rule of the Spanish Military Order of St James, 1170–1493*, ed. E. Gallego Blanco (Leiden, 1971)
RT	*La Règle du Temple*, ed. H. de Curzon (Paris, 1886)
SDO	*Die Statuten des Deutschen Ordens nach den ältesten Handschriften*, ed. M. Perlbach (Halle, 1980)
SRP	*Scriptores Rerum Prussicarum*, ed. T. Hirsch et al. (Leipzig, 1861)
WT	Guillaume de Tyr, *Chronique*, ed. R.B.C. Huygens, Corpus Christianorum, Continuatio Mediaevalis, 63, 63A (Turnhout, 1986)

List of Contributors

Nadia Bagnarini graduated from the University of Rome la Sapienza. Her thesis, under the supervision of Antonio Cadei, focused on medieval art in the preceptory of Santa Maria in Carbonara in Viterbo. She is now studying for a PhD at the Riccardo Francovich School at the University of Siena. Her primary interest is in Templar architecture in central Italy and is expanding her field to include Teutonic and Hospitaller architecture in Lazio.

Elena Bellomo teaches in the Dipartimento di Scienze dell'Antichità, del Medioevo e geografico-ambientali of the Università degli Studi di Genova. Her publications include several articles on the crusades and the military orders and the books *A servizio di Dio e del Santo Sepolcro. Caffaro e l'Oriente latino* (2003), and *The Templar Order in North-west Italy (1142-c. 1330)* (2008).

Ilya Berkovich wrote his MA thesis on 'The Battle of Forbie: A Historical and Topographical Study' under the supervision of Professor Benjamin Z. Kedar at the Hebrew University of Jerusalem. He is currently a PhD candidate at Peterhouse, Cambridge.

Pierre Bonneaud studied medieval history at the École des Hautes Études en Sciences Sociales in Paris. He researches the military orders in the Crown of Aragon and the Hospitallers of Rhodes in the late Middle Ages. His publications include *Le prieuré de Catalogne, le couvent de Rhodes et la couronne d'Aragon* (2004), *Els Hospitalers catalans a la fi de l'Edat mitjana, L'Ordre de l'Hospital a Catalunya i a la Mediterrània, 1396–1472* (2008) and several articles.

Karl Borchardt is Professor of Medieval and Regional History at the University of Würzburg and works at the Monumenta Germaniae Historica in Munich on the thirteenth-century letter-collections named after Petrus de Vinea. His PhD on ecclesiastical institutions in Rothenburg ob der Tauber prior to the Reformation (1988) included research on Hospitaller commanderies. He has published several articles on the Hospitallers and co-edited a forthcoming publication of Hospitaller documents concerning Cyprus 1409–1459.

Nicolas Buchheit teaches history in Strasbourg. He is currently preparing a PhD thesis at the University of Strasbourg on the commanderies of the Order of St John and their networks in Lower Alsace during the thirteenth and fourteenth centuries.

Emanuel Buttigieg is a lecturer at the University of Malta, where he is also a member of the Board of Studies for the MA in Hospitaller Studies. He read history at the universities of Malta and Cambridge (Peterhouse), and has recently completed his PhD on 'The Hospitallers c.1580–c.1700, with reference to Nobility, Faith and Masculinity'.

Gergely Buzás is an archaeologist and art historian attached to the King Matthias Museum, Visegrád, Hungary. His special field of interest is medieval planning and construction techniques and he is the head of the largest medieval-site reconstruction programme in Hungary.

Nicholas Coureas is a senior researcher at the Cyprus Research Centre in Nicosia. He is the author of several books on the history of Lusignan Cyprus, among them *The Latin Church in Cyprus, 1195–1312* (1997), *The Assizes of the Lusignan Kingdom of Cyprus* (2002) and, most recently, *The Latin Church in Cyprus, 1313–1378* (2010). He has also published various articles on the commercial, religious, cultural and strategic aspects of the history of Lusignan Cyprus.

Robert Dauber is a law graduate. He has written books on the history of the Hospitallers with special reference to their navy (1989), their connections with the Holy Roman Empire (1996–99) and on biographies of individual Hospitallers (2007). He is now preparing the first-ever comprehensive study of the navy of the Knights of Rhodes.

Renger de Bruin studied history at Utrecht University, where he was awarded a doctorate in 1986 for a thesis on the effects of the French Revolution in Utrecht. He has worked at the universities of Utrecht, Leiden and Greifswald. He has been Curator of Utrecht History at the Centraal Museum in Utrecht since 1994, and was appointed Professor of Utrecht Studies at Utrecht University in 2001.

Peter Edbury is Professor of Medieval History at Cardiff University. His publications include *William of Tyre: Historian of the Latin East* (with J.G. Rowe) (1988) and *The Kingdom of Cyprus and the Crusades* (1991) as well as new editions of the legal treatises by John of Ibelin (2003) and Philip of Novara (2009). He is currently preparing a new edition of both the *Chronicle of Ernoul* and the *Old French Continuations of William of Tyre*.

Sven Ekdahl is Research Professor of Medieval History at the Instytut Polsko-Skandynawski (Polish–Scandinavian Research Institute), Copenhagen, honorary PhD at the University of Vilnius, and Foreign Member of the Lithuanian Academy of Sciences. He has published extensively on the history of the Teutonic Order in Prussia, and has also treated Polish, Baltic and Scandinavian themes.

Luis Adão da Fonseca is Professor of Medieval History at the University of Oporto and Vice-rector at the Lusíada University (Oporto). He is a member of the Portuguese Academy of History and the Maritime Portuguese Academy, director of the series *Militarium Ordinum Analecta* (Oporto) and Editor-in-chief of the *e-Journal of Portuguese History* (Oporto University and Brown University).

Alan Forey, who is now retired, taught at the universities of Oxford, St Andrews and Durham. His publications include *The Military Orders from the Twelfth to the Early Fourteenth Centuries* (1992).

Éva Galambos is an assistant lecturer at the Hungarian Academy of Fine Arts, Budapest. She is a specialist in conservation and restoration with particular expertise in laboratory analyses of paints and colour pigments.

Cristian Guzzo is a specialist on the military orders in southern Italy. He has published a monograph on the Templars in the *Regnum Siciliae* and essays in leading journals including *Sacra Militia* and *Studia Monastica*. He directs *Deus Vult*, a new international review dedicated to military orders.

Zsolt Hunyadi is Associate Professor at the Institute of History at the University of Szeged (Hungary). He has published extensively on the religious orders and the crusades in central Europe. Among his research to have been published recently are *The Hospitallers in the Medieval Kingdom of Hungary c.1150–1387* (2010) and *The World of the Crusades* (2011).

Philippe Josserand is maître de conférences at the University of Nantes. He is the author of *Église et pouvoir dans la péninsule Ibérique. Les ordres militaires dans le royaume de Castille (1252–1369)* (2004) and *Regards croisés sur la guerre sainte. Guerre, religion et idéologie dans l'espace méditerranéen latin (XIe–XIIIe siècle)* (2006). He is a co-editor of *Prier et combattre. Dictionnaire européen des ordres militaires au Moyen Âge* (2009).

István Kováts works as an archaeologist and archaeozoologist at the King Matthias Museum, Visegrád, Hungary. He is a specialist on animal exploitation and animal keeping during the medieval and post-medieval period. He has directed excavations in several areas of the late medieval town and citadel of Visegrád.

Anthony Luttrell studied at Oxford, Madrid, Rome and Pisa, and taught at Swarthmore College, Philadelphia, and the universities of Edinburgh and Malta; he was also Assistant Director and Librarian of the British School at Rome. His major research interest has been in the Hospitallers on Rhodes.

Balázs Major is an Arabist, archaeologist and historian. He is the Hungarian director of the Syro-Hungarian Archaeological Mission (SHAM) responsible for the research programme at al-Marqab, Syria. He teaches in the Department of Arabic and Islamic Studies at the Catholic University of Hungary.

Victor Mallia-Milanes is Professor of Early Modern History at the University of Malta, where he is a former head of the History Department and dean of the Faculty of Arts. He has published extensively on this history of Venice, the Order of St John, and Malta in the early modern period. His most recent works include *In the Service of the Venetian Republic: Massimiliano Buzzaccarini Gonzaga's Letters to Venice's Magistracy of Trade 1754–1776* (2008).

Joel Silva Ferreira Mata is an associate professor at the Lusíada University, Oporto, and member of CEPESE (Centro de Estudos da População, Economia e Sociedade). His publications include *A Comunidade Feminina da Ordem de Santiago: A Comenda de Santos em Finais do Século XV e no Século XVI. Um Estudo Religioso, Económico e Social*, Militarium Ordinum Analecta. Fontes para o Estudo das Ordens Religioso-Militares 9 (2007).

Manuel Lamas Silveira de Mendonça holds a BA in Historical Sciences and a PhD in Medieval History from the Universidade Lusíada. He is a researcher of CEPESE (Centro de Estudos da População, Economia e Sociedade). His research interests are centered on the economic and social perspectives of the Portuguese military orders.

Helen J. Nicholson is a reader in history at Cardiff University and has published on the military orders, crusades and various related subjects. Her most recent book is *The Knights Templar on Trial: The Trial of the Templars in the British Isles, 1308–1311* (2009). She is currently completing an edition of the Templar trial proceedings in the British Isles.

Fernanda Olival is Assistant Professor at the University of Évora, research member of CIDEHUS (Centro Interdisciplinar de História, Culturas et Sociedades) and the coordinator of a project on the Commissaries of the Military Orders and the Holy Office in Portugal funded by the Portuguese Foundation for Science and Technology.

Gregory O'Malley studied at the universities of London and Cambridge and is a former research fellow of Emmanuel College, Cambridge. His book on the *Knights Hospitaller of the English Langue, 1460–1565* was published in 2005. He has also published several articles on the 'British' and Irish Hospitallers and on links between the British Isles and the Levant in the fifteenth and sixteenth centuries.

Simon Phillips was educated at Reading and Southampton. Between 2006 and 2009 he was a research fellow at the University of Winchester. Currently he is a research associate at the University of Cyprus. His interests are on the military orders and late medieval ecclesiastical history. His publications include *The Prior of the Knights Hospitaller in Late Medieval England* (2009).

Karol Polejowski completed his PhD at University of Gdansk (Poland), Institute of History. He has published extensively on the history of the Teutonic Order in France in the Middle Ages. Currently his research interests are concerned with the activities of the de Brienne family in the thirteenth and fourteenth centuries, both in the Mediterranean region and in France.

Denys Pringle is a professor at the School of History, Archaeology and Religion, Cardiff University. He has published widely on the medieval archaeology of Italy, North Africa, the Middle East and Scotland. Among his books are *The Red Tower (al-Burj al-Ahmar)* (1986), *Fortification and Settlement in Crusader Palestine* (2000), *Belmont Castle* (with R.P. Harper) (2000) and *The Churches of the Crusader Kingdom of Jerusalem: A Corpus*, 4 vols (1993–2009).

Mariarosario Salerno holds degrees from the universities of Calabria, Bari and Basilicata. Her doctoral thesis is entitled 'Gli Ospedalieri di San Giovanni di Gerusalemme tra Vicino Oriente e Mezzogiorno d'Italia'. She is a researcher in medieval history at the University of Calabria where she teaches the economic and social history of the Middle Ages.

Ana Cláudia Silveira is a PhD student at Universidade Nova de Lisboa researching the history of the town of Setúbal, a property of the Order of Santiago. Her research interests include the military orders, their relationship with local institutions, and their role in local administration.

Rombert Stapel studied medieval history at Leiden. He is currently studying for a PhD at Leiden University and the Fryske Akademy in Leeuwarden, preparing a study on the *Jüngere Hochmeisterchronik* that will lead to a new edition of this source based on the Middle Dutch manuscripts.

Maria Starnawska is Professor of Medieval History at the University of Podlasie in Siedlce, Poland. She is the author of several works on the history on the military orders and of the cult of the relics of the saints in medieval Poland.

Kristjan Toomaspoeg holds a PhD from the University of Paris X-Nanterre, and his thesis was published as *Les Teutoniques en Sicile (1197–1492)* (2003). He now works as *ricercatore* and *professore aggregato* at the University of Salento (Lecce), specializing in the history of the military orders in Italy, relations between northern and southern Europe in the Middle Ages, and various aspects of the history of the Kingdom of Sicily.

Theresa Vann is the Joseph S. Micallef Curator, Malta Study Center, Hill Museum and Manuscript Library, Collegeville, Minnesota.

António Pestana de Vasconcelos graduated in historical sciences at the University Portucalense (Oporto) in 1991. He obtained a Master degree in 1995 and in January 2009 presented his PhD dissertation to the Faculty of Letters, Oporto. He is a researcher of CEPESE (Centro de Estudos da População, Economia e Sociedade) and specializes in the history of the nobility and the military orders.

John Walker studied medieval history at St Andrews University and is now a lecturer in the History Department at Hull University. His research interests include the history of Yorkshire during the Middle Ages and the after-history of the Templars.

Paul Webster is an early career academic in the School of History, Archaeology and Religion at Cardiff University. He completed his doctoral thesis, 'King John's Piety, c.1199–c.1216', at Peterhouse, Cambridge, in 2007. He is currently preparing a monograph developing this thesis, and related articles focusing on the reign of King John.

Juliette Wood is a professional folklorist. After completing her doctoral research at the University of Pennsylvania she specialized in Celtic literature at Oxford University. Her research interests include medieval folk narrative and the modern revivals of magic and Celticism. She is currently Associate Lecturer in the School of Welsh, Cardiff University, and a director of The Folklore Society.

Editor's Preface

This volume contains papers from the fifth conference on the military orders which was held on 3–6 September 2009 at the Glamorgan Building, Cardiff University. The conference took place under the joint auspices of the London Centre for the Study of the Crusades, the Military Religious Orders and the Latin East, and the Cardiff Centre for the Crusades. We welcomed scholars from at least seventeen countries who between them presented over sixty papers. It has only been possible to publish a selection of the papers here; however, I should like to take this opportunity to thank the rather large number whose contributions sadly had to be declined, for helping make the event such a success.

As editor I should like to express my gratitude to my editorial committee, Malcolm Barber, Helen J. Nicholson, Jonathan Phillips, Denys Pringle and William Purkis, for their support and sage advice throughout. Thanks are also due to John Smedley, Beatrice Beaup, Kirsten Weissenberg and their colleagues at Ashgate Publishing for their help and patience.

Like all successful conferences, careful planning was essential, and here we are particularly indebted to my Cardiff colleagues, Helen Nicholson and Denys Pringle. A special word of thanks is also due to Mark Redknap and the National Museum Wales for hosting the opening session, laying on a reception and mounting a conference exhibition that ran from September 2009 until April 2010 in the Origins Gallery of the National Museum in Cathays Park, Cardiff. I should also like to thank the Seven Pillars of Wisdom Trust for a subvention towards the publication costs; Cadw: Welsh Historic Monuments for a grant of £1,000 to assist two young researchers from overseas to attend the conference; Beacon for Wales for a subsidy for local people to attend the conference; the prior of the Order of St John in Wales, Mr D.H. Thomas CBE, GCStJ, DL, for opening the conference; and Mr Geoffrey Phillipps CStJ of Slebech Hall, Pembrokeshire for his welcome at the conference outing which visited Slebech, until the Reformation the site of the principal commandery of the Order of St John in Wales.

Finally, a word of thanks to all those who helped with the organisation and running of events, in particular Claire Rees and Annie Brown; Steve Donaldson and Aled Cooke, technicians; Samantha Emmott and Susan Hayward-Lewis of Cardiff University Conference Office; Andy Smith and Ros Davies, security staff; our caterers; and our students, Betty Binysh, Philip Handyside and Kevin Lewis.

Peter Edbury

Introduction

Jonathan Riley-Smith

The fifth of the British four-yearly conferences on the military orders could not make use of its home in the old Hospitaller priory buildings in Clerkenwell because of building work. Cardiff University provided its first-class conference facilities, and thanks are due to the Vice-chancellor and the academic staff, who, under the leadership of Helen Nicholson, Denys Pringle and Peter Edbury, worked so hard to make the conference an outstanding success. Peter Edbury generously took the responsibility of editing this volume.

I have drawn attention elsewhere to the phenomenal increase in the number of scholars working on the military orders in the last few decades. When I started research in 1960 I do not suppose that worldwide there were more than could be counted on the fingers of two hands, whereas no fewer than 240 contributed to the *Dictionnaire*, edited by Nicole Bériou and Philippe Josserand, that was published in 2009. I used to feel that this renaissance needed explanation, but perhaps I was wrong to be puzzled and the fact that most established historians of the Middle Ages and the modern period regarded the orders as esoteric and of tangential significance was neither here nor there. The topic is intrinsically very important. The orders were major institutions in medieval and early modern times, and their representatives, disposing of huge estates, were influential throughout Europe. There are, moreover, large caches of unpublished and unexploited source material. The real significance of military order history is being demonstrated by the number of young scholars flooding into the field.

The thirty-eight papers published in this volume provide only a rough guide to the direction of research, because they are but a selection of those read at the conference. For example, the presence in this volume of only four relating to archaeology and three to art and architecture could lead an outsider to underestimate the contribution archaeologists and art historians are making to the subject, but in other respects the trends in research I have noted in previous introductions to conference proceedings are maintained. Interest in the Order of the Hospital of St John predominates. The Hospital is represented by seventeen papers, and it plays a part in four others that cover more general topics. This is to be expected when one considers its relative importance and its longevity. Its nearest competitors are the Iberian Orders, which are represented by seven papers. Spain and Portugal have research centres of international importance, and Portugal's strong showing in this volume reflects the achievements of a school of historians gathered around Luis Adão da Fonseca. The Order of the Knights Templar trails some way behind with five papers, two of which deal with the Templar myth. The Teutonic Order is considered in four papers and the tiny English Order of St Thomas in one.

Interest in the late medieval period, in which Anthony Luttrell worked almost alone for so long, continues to grow. There are almost as many contributions concentrating on the two centuries after 1300 as there are those that deal with topics from the twelfth and thirteenth centuries, although one should add that the reports published here on the excavations at the Hospitaller castle at al-Marqab (Margat) demonstrate that exciting discoveries can still be made in a field that has been more worked over than others. If the growing concentration on the Rhodian period (fourteenth–sixteenth centuries) of Hospitaller history is one of the most striking features of the last fifty years, Malta (sixteenth–eighteenth centuries) still waits in the wings. It is represented here by five papers, although its intrinsic interest, the wealth of source material and the appearance on the scene of some outstanding young historians will certainly bear fruit in the next few decades. The post-Malta period (discussed in one paper) has still to fire the imagination of most historians, but it is one in which the two surviving military orders of the Church, the Hospital of St John and the Teutonic Order, developed in striking ways and the non-Catholic Orders that claimed a share in their history came to the fore.

One of the most distinguished historians of the military orders has questioned whether research on crusading could continue at its present pace and intensity. He may well be proved right with respect to general crusade studies. Periods of rapid development in historiography are often followed by decades of consolidation, during which the insights and discoveries are absorbed. But there is no sign of abatement when one turns to the religio-military orders. We can look forward to many years of enthusiasm and intellectual advance.

Chapter 1

The Military–Religious Orders: A Medieval 'School for Administrators'?

Karl Borchardt

For the twelfth and thirteenth centuries, the crucial and formative period of the crusades and the military–religious orders, almost all relevant sources are known, even though not all of them are properly edited and interpreted. If someone were to discover and present a source that is entirely new, that would almost cause a sensation. On the other hand, given the vast literature about the crusades and the military–religious orders – and many of the participants in this conference are continuously adding to these publications – it is not easy to find new questions or interpretations from the twelfth- and thirteenth-century sources. No honest scholar can therefore claim to be able to present entirely new ideas about the military–religious orders between 1100 and 1300. And yet, even for this well-studied period, there are important problems that may deserve further research.

The present paper focuses on what one could call the constitutional or administrative history of the military–religious orders. Of course past scholars have not neglected this problem,[1] but much of their work has been bedevilled by their legal training and thinking.[2] This often results in anachronistic terminology based on later or even modern concepts, and a lack of understanding of the historical situation that conditioned or influenced the development of the orders. Nevertheless, the constitutional and the administrative practices of the military–religious orders have been studied and continue to be studied by a close reading of the relevant sources. There are rules, statutes, and other normative texts of greater or lesser legal authority, which were collected in manuscript form. Specialists discuss the dates and the relations between the manuscripts, for instance the vexed question whether the French or the Latin version of particular Hospitaller or Templar texts deserves precedence. An example is provided by Christian Vogel's recent PhD thesis on the subject of Templar legislation, 'Das Recht der Templer',

---

[1]   J. Riley-Smith, *Templars and Hospitallers as Professed Religious in the Holy Land* (Notre Dame, IN, 2010), pp. 2–3 deplores the fact that the religious importance of these orders has been neglected in comparison with their economic, social and political role.

[2]   The best general discussion of the status of the members of the military–religious orders is G. Constable, 'The Place of the Crusader in Medieval Society', *Viator*, 29 (1989), 377–403, now revised in his *Crusaders and Crusading in the Twelfth Century* (Farnham and Burlington, VT, 2008), pp. 143–82.

supervised by Rudolf Hiestand and published in 2007, where the author dates the corpus of Templar *retrais*, legal decisions by Master and Convent collected in various manuscripts, to the 1190s, half a century later than we had thought.[3] Furthermore, there is legal and constitutional terminology in contemporary charters and, to a lesser extent, in contemporary chronicles that can be studied for the history of administration or the history of communication if we want to use this more popular, but somewhat unspecific, paradigm among present-day historians.

Yet communication theory can be as misleading for historians as the older legalistic approach, popular since the nineteenth century as 'Verfassungs-' or 'Rechtsgeschichte'. There are two problems with such studies on the constitutional history of military–religious orders. Firstly, we must not interpret the sources between 1100 and 1300 in the light of later medieval and early modern constitutional history. An officer in charge of a house in the twelfth century should not be called a 'preceptor' or 'commander' too readily, because it is by no means certain that this officer was appointed and acted in the same way as his namesakes in later centuries. Secondly, it is even more important that we must interpret the sources within their contemporary historical background, since constitutional and administrative innovations were usually introduced to solve new problems created by political, social, economic or other changes. Judith Bronstein, for example, has recently shown that the Hospitallers intensified regional and local administration in response to the post-1187 challenge to organize a steadier flow of support from Europe to the Holy Land.[4] Of course we can and should compare the great military–religious orders active in all European countries and regions with other more or less centralized religious orders, especially the big ones such as Cluniacs, Cistercians or Premonstratensians. But learning from Judith Bronstein's and similar studies, we should always try to explain constitutional life and administrative routine as being shaped by specific historical situations. The general topic of this conference, 'Politics and Power', suggests what is needed. Prosopographical data as collected by Jochen Burgtorf in his PhD thesis on the leading officers of the Templars and Hospitallers between 1100 and 1300 are essential.[5] But we also need to understand

---

[3] C. Vogel, *Das Recht der Templer: Ausgewählte Aspekte des Templerrechts unter besonderer Berücksichtigung der Statutenhandschriften aus Paris, Rom, Baltimore und Barcelona*, Vita regularis: Ordnungen und Deutungen religiösen Lebens im Mittelalter, Abhandlungen, 33 (Berlin, 2007), pp. 104–8, 345–6. Hitherto the *retrais* (*RT*, pp. 77–278) were thought to have been collected by the Master Bertrand de Blanquefort 1156–69 or even earlier in the 1130s and 1140s. J. Burgtorf, *The Central Convent of Hospitallers and Templars: History, Organization, and Personnel (1099/1120–1310)* (Leiden, 2008), p. 180 and passim prefers the date c.1165. The question deserves further study.

[4] J. Bronstein, *The Hospitallers and the Holy Land: Financing the Latin East, 1187–1274* (Woodbridge, 2005).

[5] Burgtorf, *Central Convent*. The importance of prosopography for the history of administration has already been stressed by A. Demurger, 'L'apport de la prosopographie à l'étude des mécanismes des pouvoirs, XIIIe–XVe siècles', in F. Autrand (ed.), *Prosopographie et genèse de l'état moderne* (Paris, 1986), pp. 289–301.

the political and social background. The military–religious orders in the Holy Land responded to varying challenges in ever-changing historical situations, after the failure of the Second Crusade, after the loss of Jerusalem in 1187 and the failure to recover it through the Third Crusade, and after the subjugation of the quasi-independent French nobility by the Capetian monarchy in the middle of the thirteenth century that proved to be the end of the traditional crusades, just to hint at some of these varying circumstances.

These basic facts about the origins and early development of the crusades and the great military–religious orders are well known and here need only a brief summary. In the period under consideration, the crusades were military expeditions privileged by the pope with the prestigious aim of recovering or defending Jerusalem and the Holy Land, although this concept was soon transferred to other theatres of war with the infidel and with the enemies of the church. The role of the reformed late eleventh-century papacy was essential for the emergence of the crusades, as Urban II and his successors took up an idea once propounded by Gregory VII that the papacy should direct Christian *milites* to the noble and meritorious goal of fighting the enemies of the faith in the East. As many emperors and kings were opposed to the papal programme at that time, the crusades provided Urban II and his successors with an important instrument in establishing their claims as leaders of western Christianity. Although the military–religious orders were not technically crusaders – their members were forbidden to take any vows other than chastity, poverty and obedience in their order – they rose to prominence between 1100 and 1300 as a form of standing militia supporting Jerusalem and the Holy Land. Both the Templars and the Hospitallers had their twelfth-century central headquarters in Jerusalem. They were founded and endowed to support Jerusalem. The Hospitallers and the Order of the Holy Sepulchre were recognized by Paschal II in 1113 and 1114 respectively to provide pilgrims to Jerusalem with alms and shelter. The Templars were founded in probably 1119/20 to protect pilgrims in the vicinity of Jerusalem against Muslim raids. Papal privileges followed and contributed to the rapid growth of these orders, although one should not go as far as Hans Prutz and claim that the Templars and Hospitallers were merely or even primarily instruments of the popes,[6] a claim that is sometimes repeated for the Dominicans and the Franciscans founded in the thirteenth century. The popes might protect these new orders against bishops and parish priests, and in return the popes might enjoy personal services or financial support, but the constitutional development of the military–religious orders was always shaped primarily by their own needs and desires.

---

[6]   H. Prutz, *Die Geistlichen Ritterorden: Ihre Stellung zur kirchlichen, politischen, gesellschaftlichen und wirtschaftlichen Entwicklung des Mittelalters* (Berlin, 1908), pp. 143–4. L. García-Guicharro Ramos, 'Exemption in the Temple, the Hospital and the Teutonic Order: Shortcomings of an Institutional Approach', *MO* 2, pp. 289–93 also warns not to overestimate the papal privileges.

The Jerusalem-based Hospitallers and Templars were in a special situation, since they had to collect alms in the West, transfer men, material and monies to the East, and soon began to acquire properties in the West in order to support their eastern obligations. In this respect they had similarities with the Augustinian canons of the Holy Sepulchre. But the Hospitallers and Templars both fulfilled much more important tasks for the Latin East, and the pressure to develop new constitutional and administrative devices was much stronger for them. This is probably the reason why the military–religious orders differ significantly from other new religious communities or orders of the twelfth century. General histories of religious orders, among them those by the German historian Gert Melville and his colleagues and pupils published in the series 'Vita regularis' since 1996 – there are now almost forty volumes – usually concentrate on Cluniacs, Cistercians or Premonstratensians and tend to mention the military–religious orders only in passing as if Hospitallers and Templars merely imitated these other monastic communities.[7] The main purpose of this paper is to question this view.

Compared to Cluniacs, Cistercians or Premonstratensians, the major innovation we have to explain is the prominence of the master and his eastern headquarters. The abbots and monasteries of Cluny, Cîteaux or Prémontré enjoyed prestige among their respective communities. But with the partial exception of Cluny that had some dependent houses governed by priors who were appointed, controlled and sometimes recalled by the grand abbot of Cluny,[8] the new *ordines* – a term developed for Cîteaux early in the twelfth century – were communities of separate houses with their own brethren, their own possessions and their own abbots. The Templars, the Hospitallers and, later on, the other military–religious orders were radically different, because they grew around a prominent eastern headquarters. There the master lived and administered the order's affairs together with the convent, the community of brethren living in this headquarters.[9] From around

---

[7]   The exception that proves the rule is of course J. Sarnowsky, *Macht und Herrschaft im Johanniterorden des 15. Jahrhunderts: Verfassung und Verwaltung der Johanniter auf Rhodos (1421–1522)*, Vita regularis: Ordnungen und Deutungen religiösen Lebens im Mittelalter, 14 (Münster, Westfalen, 2001). Most other studies of this series, especially those on the twelfth and thirteenth centuries, do not mention the military–religious orders although many of them are devoted to a comparative approach concerning monastic and religious orders.

[8]   D.W. Poeck, *Cluniacensis Ecclesia: Der cluniacensische Klosterverband (10.–12. Jahrhundert)*, Münstersche Mittelalter-Schriften, 71 (Munich, 1998); P. Racinet, *Crises et renouveaux: Les monastères clunisiens à la fin du Moyen Âge (XIIIe–XVIe siècles): De la Flandre au Berry et comparaisons méridionales* (Arras, 1997); *Les maisons de l'ordre de Cluny au Moyen Âge: evolution et permanence d'un ancien ordre bénédictin au nord de Paris*, Bibliothèque de la Revue d'histoire ecclésiastique, 76 (Brussels, 1990).

[9]   Rule of Saint Benedict c. 2; Regel der Templer c. 66: *Ceterum magister, qui baculum et virgam manu tenere debet, baculum videlicet, quo aliorum virium imbecillitates sustentet, virgam vero, qua vitia delinquentium zelo rectitudinis feriat ...*, Vogel, pp. 256–7; M.-L. Bulst-Thiele, *Sacre domus militiae Templi Hierosolimitani magistri:*

1140 onwards the Templar master styled himself *Dei gratia* and was addressed as *venerabilis* like any bishop or abbot;[10] and the Hospitaller master soon followed this example.[11] Yet we should not press this too far, as also subordinate officers such as the Hospitaller prior of St Gilles and the Templar master in Spain were being called *Dei gratia* at the same time,[12] and as the quasi-monarchical position of the master was checked and balanced by the brethren present in the headquarters.[13] Important decisions had to be achieved with the consent of the chapter. The conventual chapters gradually evolved into chapters general where, for example, subordinate officers for the administration of possessions outside the headquarters were installed.

The important point, however, is that all these subordinate officers both in the convent and outside were, at least in theory, only lieutenants of the master.[14] That is why these officers were called *baillis*, a term adopted from the Norman, Angevin and Capetian royal or princely administrations. Like the similar term 'commander', whose precise origins are not yet clear, this usage emphasizes the fact that the officers were only lieutenants of the master and could be dismissed at any time.[15] As in royal or princely administrations, the officers were responsible to the master, were obliged to send subsidies to the headquarters, to hand in reports and submit accounts, and they were subject to visitations. Any important decision such as the alienation of lands needed, at least in theory, permission from the central headquarters. And, at least in theory, new brethren could only be admitted by licence from the central headquarters. Of course the master, or the master together with the convent, might delegate such powers to regional or local officers. Yet we should note here the basic difference in comparison with other twelfth-century religious communities or orders where the single house or monastery always retained the right to decide such matters for itself.

In this sense the Templars and the Hospitallers – and not the Cluniacs, Cistercians or Premonstratensians – were the first truly centralized religious orders,[16] and that was undoubtedly an important twelfth-century innovation. The debate about the origins of the military–religious orders usually has its focus on

---

Untersuchungen zur Geschichte des Templerordens 1118/19–1314, Abhandlungen der Akademie der Wissenschaften in Göttingen, Philologisch-historische Klasse, 3. Folge, 86 (Göttingen, 1974), p. 9; M. Barber, *The New Knighthood: A History of the Order of the Temple* (Cambridge and New York, 1994), p. 17.

[10] *CT*, no. 145, p. 102 and no. 205, p. 143. Vogel, p. 257; Burgtorf, pp. 181–2.

[11] *CH*, no. 177 dated around 1150, but not no. 189 dated 1151. In 1140 *venerandus magister*: *CH*, no. 136.

[12] *CH*, no. 214 (Saint-Gilles), no. 220 (Spain).

[13] *RT*, §98; Baltimore fol. 31v: ... *et li Maistres si doit estre obedient a son covent.*

[14] *RT*, §530: ... *tous les autres baillies qui tienent en lor provinces luec de Maistre, por ce que il sont en luec de Maistre.*

[15] Vogel, p. 256; Burgtorf, pp. 179–80.

[16] J. Riley-Smith, 'The Origins of the Commandery in the Temple and the Hospital', in A. Luttrell and L. Pressouyre (eds), *La Commanderie, Institution des ordres militaires*

their military activities. The recognition of bloodshed not only as dire necessity but as meritorious Christian duty is indeed a fascinating phenomenon in the context of the crusades and the military–religious orders: Simonetta Cerrini called it 'La révolution des Templiers'.[17] But it is perhaps of equal importance to explain the comparatively high degree of early administrative centralization. The special prestige of the Jerusalem headquarters may have been helpful in achieving this, as many donation charters refer to the spiritual rewards expected by those who supported Jerusalem. It was, however, not entirely new for support to be given to religious institutions with more or less distant possessions that required some form of central administration, especially, but by no means exclusively, those religious communities that ran hospitals.

In 1090 Urban II had recognized the monastic community of Vallombrosa in Tuscany where the abbot of the central convent would appoint provosts for the dependent houses that he visited in person on a regular basis.[18] As a former monk of Cluny, Urban II was well acquainted with similar procedures by the grand abbot of Cluny. Other rich and prestigious abbeys such as Montecassino, Fruttuaria or St-Victor in Marseilles followed similar practices.[19] When Urban II returned from Clermont to Italy in 1096, he consecrated and privileged the church of Sta Croce in Mortara west of Milan, headed by a certain Airaldus who later became bishop of Genoa and himself consecrated two churches in Genoa, San Michele a Cameri and San Theodoro di Fassolo; all three churches were devoted to the support of pilgrims and became centres of small religious communities with dependent houses.[20] Around 1080 a similar institution had already been founded at Altopascio near Lucca on the pilgrim routes to Rome; this community was given

---

dans l'Occident médiéval (Paris, 2002), pp. 9–18 at p. 9 calls them the 'earliest true religious orders'.

[17]   S. Cerrini, La révolution des Templiers: une histoire perdue du XIIe siècle (Paris, 2007).

[18]   PL 151, col. 322–3, and the vita of the founder Johannes Gualbertus by Atto, PL 146, col. 667–706 at col. 680: '… per singula loca sollicitus pater praepositos ordinavit. Quos … per se ipsum visitare curabat et honeste corrigere moresque fratrum ad meliora corrigere.'

[19]   P. Amargier, Un âge d'or du monachisme: Saint-Victor de Marseille (990–1090) (Marseilles, 1990). General survey with references to the sources, R. Molitor, Aus der Rechtsgeschichte benediktinischer Verbände: Untersuchungen und Skizzen, 1 (Münster, Westfalen, 1928), pp. 86–102; J. Wollasch, Mönchtum des Mittelalters zwischen Kirche und Welt, Münstersche Mittelalter-Schriften, 7 (Munich, 1973), pp. 136–86.

[20]   C. Andenna, Mortariensis ecclesia: una congregazione di canonici regolari in Italia settentrionale tra XI e XII secolo, Vita regularis: Ordnungen und Deutungen religiösen Lebens im Mittelalter, Abhandlungen, 32 (Berlin and Münster, Westfalen, 2007), pp. 203–42.

possessions in many parts of Italy, and in 1239 Gregory IX endowed it with the Hospitaller rule, although it retained separate *consuetudines*.[21]

The great Jerusalem-centred orders naturally enjoyed special prestige and received special favours from the faithful. Accordingly, the Templars and the Hospitallers set the example, because they supported the Holy Land and, at least from the 1140s onwards, were present in all parts of western Christendom. It was their new and special task to organize support for the Holy Land that profoundly affected their constitutional development. This challenge became even more pressing when in the course of time the crusades themselves dwindled. In response, the military–religious orders were encouraged to develop a much greater degree of centralization than any of the other new twelfth-century orders, including the Cistercians and Premonstratensians, both of whom based their communities on separate monasteries with their own abbot, their own brethren and their own possessions. As early as the years between 1140 and 1155 the military–religious or other charitable orders may have influenced the Grande Chartreuse when it instituted chapters general at which the priors of the single houses had to report to, and might be deposed by, the prior of the Grande Chartreuse himself who thus came to be acknowledged as supreme master of the whole community. The commanding position accorded the order's supreme head was contrary to the practice at Cîteaux,[22] but followed the example set by the grand abbot of Cluny and especially by the masters of the Hospital and Temple in Jerusalem.

From the very beginnings, and in contrast to Cluny, each military–religious order was an entity in itself and not a community of juridically separate preceptories or commanderies. Both Templars and Hospitallers did not as a rule have single houses under a preceptor or commander in the West before the thirteenth century when efficient local administration of estates became a necessity as alms and donations fell away. Most early western donation charters name as recipients the *fratres*, the hospital or the militia in Jerusalem and not some regional or local officer or

---

[21]   *Les Registres de Gregoire IX*, ed. L. Auvray (Paris, 1896–1955), no. 4799. A. Spicciani, *L'ospedale lucchese di Altopascio: storia economica e finanziaria nei secoli XI–XII* (Pisa, 2006); A. Santangelo, *Sulla lingua della Regola dei frati di San Jacopo d'Altopascio* (Florence, 1983); recently A. Meyer, 'Organisierter Bettel und andere Finanzgeschäfte des Hospitals von Altopascio im 13. Jahrhundert (mit Textedition)', in G. Drossbach (ed.), *Hospitäler in Mittelalter und Früher Neuzeit. Frankreich, Deutschland, Italien. Eine vergleichende Geschichte = Hôpitaux au Moyen âge et au temps modernes. France, Allemagne, Italie. Une historie comparée*, Pariser historische Studien, 75 (Munich, 2007), pp. 55–105; idem, 'Altopascio, Lucca e la questua organizzata nel XIII secolo', in A. Esposito and A. Rehberg (eds), *Gli ordini ospedalieri tra centro e periferia: Giornata di studi, Roma, Istituto Storico Germanico, 16 giugno 2005*, Ricerche del'Istituto storico germanico di Roma, 3 (Rome, 2007), pp. 195–209.

[22]   F. Cygler, *Das Generalkapitel im hohen Mittelalter: Cisterzienser, Prämonstratenser, Kartäuser und Cluniazenser,* Vita regularis: Ordnungen und Deutungen religiösen Lebens im Mittelalter, 12 (Münster, Westfalen, 2002), pp. 222–34.

house in the West.[23] This is the reason why, contrary to a widespread assumption, we cannot explain the constitutional development of the military–religious orders in terms of following the example of the Cistercians. That is somewhat surprising because Bernard of Clairvaux was instrumental and influential in the foundation and recognition of the Templars, because the Cistercians continued to promote military activities against Muslim or heathen enemies of the faith even outside the Holy Land in Spain and on the Baltic, and because in Spain military–religious orders such as Calatrava were originally Cistercian affiliates.[24] But we must not overemphasize the role of Cîteaux. For the military–religious orders the model of a religious community around a hospital with sometimes distant possessions intended for its maintenance was apparently of greater importance. This is true not only for the Hospitallers but also for the Templars, although they did not run a hospital at Jerusalem. But contemporaries considered their task of fighting to protect pilgrims as equally charitable, and this made the development of a remarkably centralized constitution both possible and necessary for them.

When the Teutonic Order emerged in the Holy Land during the 1190s it copied both the Templars and the Hospitallers, and sometimes also the famous hospital of Santo Spirito in Sassia at Rome.[25] Contrary to the assertions of Indrikis Sterns, the most recent editor of the Teutonic Order's statutes, there was no influence from the Cistercian *Carta caritatis*.[26] The Teutonic Order originally used the liturgy of the Holy Sepulchre, as did the Templars and the Hospitallers. But since many brethren did not know this liturgy properly, the Order received a papal licence to adopt the more popular Dominican breviary[27] and was permitted to adapt it to its own

---

[23]   Riley-Smith, 'Origins of the Commandery', pp. 9–18.

[24]   A. Forey, *The Military Orders from the Twelfth to the Early Fourteenth Centuries* (London, 1992), pp. 170–73.

[25]   *SDO*, p. 57 n. 1–2, p. 58 n. 2–3, p. 59 n. 2. See now H. Houben, 'Regola, statuti e consuetudini dell'Ordine Teutonico: *status quaestionis*', in C. Andenna and G. Melville (eds), *Regulae Consuetudines Statuta: Studi sulle fonti normative degli ordini religiosi nei secoli centrali del Medioevo*, Atti del I e del II Seminario internazionale di studio del Centro italo-tedesco di storia comparata degli ordini religiosi (Bari/Noci/Lecce, 26–27 ottobre 2002 / Castiglione delle Stiviere, 23–24 maggio 2003), Vita regularis: Ordnungen und Deutungen religiösen Lebens im Mittelalter, 25 (Münster, Westfalen, 2005), pp. 375–85. For Santo Spirito in Sassia see now G. Drossbach, *Christliche caritas als Rechtsinstitut: Hospital und Orden von Santo Spirito in Sassia (1198–1378)*, Kirchen- und Staatskirchenrecht, 2 (Paderborn, 2004).

[26]   I. Sterns, 'The Statutes of the Teutonic Knights: A Study of Religious Chivalry', PhD thesis, Univ. of Pennsylvania, 1969, p. 99; idem, 'Crime and Punishment among the Teutonic Knights', *Speculum*, 57 (1982), 84–111. See the review by U. Arnold, 'Die Statuten des Deutschen Ordens: Neue amerikanische Forschungsergebnisse', *Mitteilungen des Instituts für Österreichische Geschichtsforschung*, 83 (1975), 144–53.

[27]   'Ceterum, quia divinum officium secundum ordinem Sancti Sepulchri pro eo, quod a pluribus ex iisdem fratribus clericis ignoratur, vix absque scandalo, sicut accepimus, in vestro potest ordine observari, quod illud secundum ordinem fratrum Predicatorum

purposes.[28] So even the smallest and latest of the three great military–religious orders in the Holy Land was not dependent upon the Cistercian model.

Alongside the higher degree of centralization, there is a second difference between the military–religious orders and the Cistercians. To administer and organize their possessions outside the headquarters they soon began to create territorial districts, whereas the Cistercians followed the filiation principle to maintain unity and to ensure efficient administration. The abbot of Cîteaux had the right to visit only those monasteries that had been founded directly from Cîteaux; the other Cistercian monasteries had their own father-abbots. By the 1130s the houses of Prémontré and of Arrouaise imitated this Cistercian model of supervision through their respective father-abbots.[29] When from the 1150s the communities grew and the number of abbeys increased, the Premonstratensians complemented this with the visitation by two abbots of the same province called *circatores* who were charged with inspecting their neighbouring abbeys and reporting to the annual chapter at Prémontré; thus by around 1200 the Premonstratensians had developed a fully-fledged system of *circarie*.[30] Neither the Hospitallers nor the Templars ever practised supervision by the head of the founding house. On the contrary, from their very beginnings they administered their possessions according to geographical districts, as collecting alms for Jerusalem was their main task and

---

amodo in vestris ubique domibus celebretur, vobis concedimus facultatem': Tabulae ordinis Teutonici: Ex tabulario regii Berolinensis codice potissimum, ed. E. Strehlke (Berlin, 1869; repr. Toronto, 1975), no. 471, p. 357.

[28]   Permission by Alexander IV in 1257 to change the breviary '*ad quamdam formam secundum Deum religioni vestre congruam et salubrem*': Strehlke, no. 536, p. 378.

[29]   The Vita B of Norbert of Xanten, *PL* 170, cols. 1253–344 at cols. 1330–31 §101, expressly declares, '*... deinceps, ad similitudinem ex exemplum Cisterciensium, annuatim se reversurus ...*'. 'Les premiers statuts de l'ordre de Prémontré: Le clm 17174 (XIIe siècle)', ed. R. van Waefelghem, *Analectes de l'ordre de Prémontré*, 9 (1913), 35: '*Iterum statutum est, quod semel in anno gratia sese visitandi, ordinis reparandi, confirmande pacis, conservande karitatis, abbates omnes ad colloquium pariter conveniant in loco competenti quem communi consilio providerint, ubi in sinistris corrigendis domno abbati Premonstrate ecclesie que mater est aliarum, sanctoque illi conventui reverenter singuli humiliterque obediant et clamati veniam petant ...*'. J. Oberste, *Visitation und Ordensorganisation: Formen sozialer Normierung, Kontrolle und Kommunikation bei Cisterziensern, Prämonstratensern und Cluniazensern (12.– frühes 14. Jahrhundert)*, Vita regularis: Ordnungen und Deutungen religiösen Lebens im Mittelalter, 2 (Münster, Westfalen, 1996), pp. 175–7.

[30]   Statutes from around 1150, Dist. 4 c. 7: *Les Statuts de Prémontré au milieu du XIIe siècle*, eds P.F. Lefévre and W.M. Grauwen, Bibliotheca Analectorum Praemonstratensium, 12 (Averbode, 1978), pp. 47–8: '*Quia vero pro multitudine abbaciarum et remocione locorum patres abbates filias abbacias quandoque, sicut statutum est, visitare non possunt, provisum est, ut per diversas provincias abbacie, que sibi vicine sunt, de abbatibus earundem ecclesiarum singulis annis duos habeant circatores, qui singulos abbacias visitent, et si qua ibi corrigenda invenerint, aut per se corrigant aut ad patres in annuo colloquio diligenter inquisita referant.*'

was done according to diocese in order to avoid quarrels. That was probably the reason why both the Hospitallers and the Templars organized their administration according to territorial principles, especially from the 1150s onwards when, after the failure of the Second Crusade, their fund raising proved to be vital for the steady support of the Franks in the Levant. Some recent publications on religious orders by Gert Melville and his pupils, usually in the same series 'Vita regularis' which includes Christian Vogel's *Das Recht der Templer* and Jürgen Sarnowsky's *Macht und Herrschaft im Johanniterorden*,[31] chose to ignore the military–religious orders, and, in consequence, miss, I think, an important factor that may explain the growing popularity of the territorial principle in administration.

This territorial principle was strengthened in 1215 by a decree of the Fourth Lateran Council that all monasteries should be subject to control according to dioceses and provinces. Every third year exempt and non-exempt abbots and priors of the monasteries of each province were to come together and discuss matters following the example set by the chapter general of Cîteaux. Presiding at these chapters would be four abbots, two of them Cistercians.[32] The prominent role of Cîteaux in this decree of 1215 should not blur the fact that it was precisely the Cistercians that were the one great religious order that never adopted the territorial principle. In the thirteenth century the Dominicans and the Franciscans had provinces.[33] Even Cluny began to develop provinces after 1200.[34] The secular church had used dioceses and provinces from antiquity, but among the religious orders it was apparently the Templars and the Hospitallers who had proved the usefulness of geographical districts.

Apart from the rather strict centralization around the Jerusalem headquarters and the territorial districts, a third innovation attributable to the Templars and

---

[31]   See above nn. 3 and 7.

[32]   Extra 3.35.7, and 3.35.8 as an extension by Honorius III: *Constitutiones concilii quarti Lateranensis una cum commentariis glossatorum*, ed. A. García y García, Monumenta Iuris Canonici A, 2 (Città del Vaticano, 1981), pp. 60–61; U. Berlière, 'Les chapitres généraux de l'Ordre de St. Benoît', *Revue bénédictine*, 18 (1901), 364–98 and 19 (1902), 38–75, 268–78, 374–411; Oberste, *Visitation*, pp. 52–5.

[33]   The Dominicans followed the example of Prémontré. G.R. Galbraith, *The Constitution of the Dominican Order, 1216 to 1360* (Manchester, 1925), pp. 175–9, despite the reservations by A.H. Thomas, *De oudste constituties van de Dominicanen: voorgeschiedeneis, tekst, bronnen, ontstaan en ontwikkeling (1215–1237), met uitgave van de tekst*, Bibliothèque de la Revue d'histoire ecclésiastique, 42 (Louvain, 1965), pp. 153–7, 384. It remains a question whether the later provinces of other orders were influenced by the Premonstratensian *circarie* or more directly by the military–religious orders.

[34]   Francia, Lyons, Auvergne, Poitou, Provence, Gascogne, Spain, Lombardy, Alamania-Lorraine, England. At the same time Cluny began to have regular chapters general: Oberste, *Visitation*, pp. 283–364; Cygler, *Generalkapitel*, pp. 315–470; idem, 'Le chapitre général de Cluny (XIIe–XIVe siècle): État de la question', in J. Hoareau-Dodinau and P. Texier (eds), *Anthropologies juridiques. Mélanges Pierre Braun*, Cahiers de l'Institut d'Anthropologie Juridique (Limoges, 1998), pp. 213–35.

Hospitallers would appear to be the limited duration of office. Their regional and local officers were theoretically lieutenants of the master and were not, as a rule, appointed for life. There may, however, have been exceptions, and in any case in practice the removal of a regional or local officer was always tied up with problems of politics and power. In the Hospital it was apparently supposed that priors from the West were to report to the master in the East every five years,[35] and for the Templars there are hints that this may have occurred every four years.[36] Unfortunately there is not enough evidence for the twelfth and thirteenth centuries to describe in detail how often chapters general met and how officers were appointed or replaced. The great distance between the Holy Land and the European houses must have made communication difficult and expensive. If there were rules for European officers to travel to the East every four or five years, it seems unlikely that they were strictly applied and obeyed. But it seems clear that the western officers of both the Hospitallers and the Templars held their posts for only limited periods of time. That contrasted with the practice of other monastic or religious orders where the heads of the houses in particular would stay in office for life. Limited duration of office was not entirely new, as for instance the *prepositi* dependent upon the abbot of Vallombrosa were in a similar situation. All Cistercian abbots, however, and not only the one at Cîteaux, were elected for life, as was customary for all Benedictine abbots. The same was true of all Premonstratensian abbots, not only the abbot of Prémontré, and of all Carthusian priors and not just the prior of the Grande Chartreuse. Yet in the Temple and the Hospital it was solely the master, the supreme head of the whole order, who according to the constitution held office for life. This was undoubtedly a third important innovation of the military–religious orders during the twelfth century.

It was a step in a process that ended with limited duration of office for the supreme head of other orders in the later Middle Ages, as for example in the case of the Celestinians. At the beginning of the thirteenth century the annual chapter general of the Dominicans had the duty to correct, and, if necessary, the power to depose the master general or to call for his resignation, but in practice this did not prevent a respected and uncontroversial master general to be confirmed and hold office for life.[37] The Franciscans were even more radical. Their minister general regularly held his office for only three years, although he could be re-elected and in fact frequently was re-elected by the chapter general.[38] This was a step beyond that taken by the military–religious orders, and it may be explained

---

[35] *CH*, nos. 4310, 4462; J. Riley-Smith, *The Knights of St. John in Jerusalem and Cyprus c.1050–1310* (London, 1967), p. 361.

[36] A. Forey, *The Templars in the Corona de Aragón* (London, 1973), p. 313, and, sceptical, Vogel, p. 308 and n. 1596.

[37] Galbraith, pp. 93, 133–9; W.A. Hinnebusch, *The History of the Dominican Order: Origins and Growth to 1500* (Staten Island, 1965–73), 1, p. 223.

[38] J.R.H. Moorman, *A History of the Franciscan Order from its Origins to the Year 1517* (Oxford, 1968), p. 106.

by the tradition in towns of having *consules* or similar officials for one year only. The resurgence of Roman Law during the twelfth century brought about a renaissance of the ideal of limited tenure of office, especially among citizens, and it should therefore not come as a surprise that the mendicant orders were the first to introduce limited duration of office for their supreme heads. From the early thirteenth century onwards most new orders that, contrary to the decree of the Fourth Lateran Council of 1215, continued to be founded and recognized by the popes, had as their supreme head a person with a limited period of office; that was true even when the head was an abbot following the rule of St Benedict, as in the case of the Celestinians who took their name from Pope Celestine V (1294). In the fourteenth century the Celestinians and similar later orders were even to introduce statutes against the re-election of their supreme abbot, a fact that illustrates a drastic change concerning the ideas about ecclesiastical office.[39] This change may have started with the Hospitaller and Templar subordinate officers of the twelfth century.

In sum the military–religious orders of the Holy Land had an important and hitherto perhaps underestimated influence on constitutional practices among other religious orders. Templars and Hospitallers paved the way towards an unprecedented degree of centralization under the master. They helped to spread the territorial principle for control of dependent houses, as contrasted with the filiation principle of the Cistercians. And they spread the concept of only limited tenure of office, as contrasted with the lifelong elected Cistercian or Premonstratensian abbots. The role of the Templars and Hospitallers from the twelfth century onwards can thus be compared to the role of the Dominicans and Franciscans from the thirteenth century onwards. Whereas the mendicant orders were particularly close to the towns, the military–religious orders were influential among the nobility. It is almost commonplace to praise the mendicant orders and their constitutional principles as championing representative democracy.[40] Such present-day ideologies apart, the military–religious orders probably exerted a similar influence upon important circles within medieval society: we could call them a 'school for administrators'.[41] In medieval Europe this may have been even more important than the early modern 'school for ambassadors', as David Allen called the Maltese Hospitallers at the second of our conferences.[42] Since

---

[39]   K. Borchardt, *Die Cölestiner: Eine Mönchsgemeinschaft des späteren Mittelalters*, Historische Studien Ebering 488 (Husum, 2006).

[40]   B. Tierney, 'Freedom and the Medieval Church', in R.W. Davis (ed.), *The Origins of Modern Freedom in the West: The Making of Modern Freedom* (Stanford, CA, 1995), pp. 64–100 at p. 83, mentions the Dominicans as spreading representative government from 1216 onwards, following the Fourth Lateran Council of 1215.

[41]   Somewhat anachronistically E.E.A. Staehle, *Geschichte der Johanniter und Malteser* (Gnas, 2002) considers the Hospitallers to have been some medieval school for managers.

[42]   D.F. Allen, 'The Order of St John as a "School for Ambassadors" in Counter-Reformation Europe', *MO* 2, pp. 363–79.

the military–religious orders had to organize support for the Holy Land, they may be regarded primarily as administrative organizations run by religious people. Unlike the Cistercians and Premonstratensians, they were not communities of otherwise independent religious houses that shared certain monastic ideals and acknowledged a common *regula*, common *instituta* and *consuetudines*. It is true that in the same way as other communities, the Templars, the Hospitallers and their fellow military–religious orders needed a recognizable identity that was maintained by common codes of behaviour for dress, food, liturgy and so on. Yet it was not monastic uniformity but primarily administrative effectiveness that they tried to achieve. The fact that brethren lived outside the headquarters in castles, hospitals or manor houses, had, for the military–religious orders, come about as a consequence of their administrative task of organizing support for the Holy Land, whereas among Cistercians, Premonstratensians and others this fact followed from a desire to unite a plurality of legally independent religious houses through the same way of life. It was their specific economic purpose in supporting the Holy Land that was the main characteristic of the military–religious orders, and this explains their comparatively high degree of centralization. Apparently it was this purpose and degree of centralization that served as a model for the Dominicans and the Franciscans.[43] For their task of sending respected and convincing preachers whose style of life imitated the apostles throughout Christendom, the friars apparently borrowed administrative principles from the military–religious orders.[44]

The constitutional ideals of the military–religious orders, the appointment of officers for limited periods of time only, the regular sending of surpluses to the central headquarters, the rendering of written accounts,[45] and the visitations popularized patterns of behaviour among the nobility that were relevant not only for ecclesiastical but also for temporal administrations. Relations between the military–religious orders and royal administration in France, England or Sicily or of other temporal lords deserve further research. This question would concern prosopography, the use of Hospitallers or Templars as rectors in the papal states[46] and as officers by other temporal lords. It also concerns the administrative techniques

---

[43]   Significantly Galbraith, p. 177, calls the Dominicans an army with a mobile headquarters.

[44]   Even the Franciscans, who called regional officers *custodes* and cared for the *pauperes Christi*, a terminology reminiscent of the *pauvres du Christ* and of the Master as *custos pauperum Christi* in the Hospital.

[45]   A Hospitaller example is found in a manuscript extant in Prague with weekly accounts of Saint-Gilles from before the fall of Acre in 1291; the author of the present paper is studying it together with Damien Carraz and Jiří Mitáček. For the Teutonic Order see *Visitationen im Deutschen Orden im Mittelalter*, eds M. Biskup and I. Janosz-Biskupowa, 1, pp. 1236–449 and 2, pp. 1450–519, QuStDO 50 (= Veröffentlichungen der Internationalen Historischen Kommission zur Erforschung des Deutschen Ordens, 10) (Marburg, 2002–2004).

[46]   Three examples: *Das Kammerregister Papst Martins IV. (Reg. Vat. 42)*, ed. G. Rudolph with T. Frenz, Littera antiqua, 14 (Città del Vaticano, 2007), nos. 82–7, 291, nos. 128, 183, 190 and nos. 271–2.

and mentalities themselves, the formulae used for appointments, judgements, accounts, visitations and so on. By 1202 the Templar Fr Aymard had developed a sophisticated accounting system for Philip Augustus that laid the ground for the central administration of French royal revenues into the later Middle Ages.[47] There is a German tradition of linking the Teutonic *Ordensstaat* in late-medieval Prussia with the Hohenzollern monarchy of the eighteenth and nineteenth centuries, as both were hailed for allegedly unbiased administration of justice and economical use of resources.[48] Prussian ideology apart, all medieval military–religious orders must indeed have been a school for experienced administrators. These administrators were usually nobles, although twelfth- and thirteenth-century recruitment into the military–religious orders did include people from important families in towns. At that time there was not yet a clear-cut dichotomy between nobility and bourgeoisie, and not only priests or sergeants but also many *milites* of the military–religious orders may have had an urban background.[49] The same may be true for the royal or princely bureaucracies established during the twelfth and thirteenth centuries.

Studies concerned with the influence of the mendicant orders on medieval towns should therefore be complemented by studies on the influence of the military–religious orders on princely or royal bureaucracies. Accounts, inventories and registers were used by both ecclesiastical and secular administrations, and studies might compare these sources. In addition a comparative history of terms such as *bailli* / *baiulivus*, *comandor* / *commendator* or *preceptor*[50] might prove rewarding. Finally, we should note that not only the friars but also the military–religious orders developed some kind of control to avoid mismanagement by the master, the head of the administration. The convent and its chapters that gradually evolved into chapters general were given a role in decision making. The *proceres*, *probi homines*, *proudommes* of the order had to be heard by the master. This development may have helped to popularize the idea among the nobility that kings or princes should govern with the consent of their *proceres*, their estates, following

---

47   I. de la Torre Muñoz de Morales, 'The London and Paris Temples: A Comparative Analysis of their Financial Services for the Kings during the Thirteenth Centuries', *MO* 4, pp. 121–7 at p. 122.

48   M. Biskup and G. Labuda, *Dzieje Zakonu krzyżackiego w Prusach: Gospodarka – Spoleczństwo – Państwo – Ideologie* (Gdansk, 1988), trans. J. Heyde and U. Kodur, *Die Geschichte des Deutschen Ordens in Preußen: Wirtschaft, Gesellschaft, Staat, Ideologie*, Klio in Polen, 4 (Osnabrück, 2000); H. Boockmann, *Der Deutsche Orden: Zwölf Kapitel aus seiner Geschichte* (Munich, 1981), trans. R. Traba and W. Lipnik, *Zakon Krzyżacki: dwanaście rozdziałów jego historii*, Klio w Niemczech, 3 (Warsaw, 1998).

49   See for example K. Borchardt, 'Die deutschen Johanniter zwischen Ministerialität und Meliorat, Ritteradel und Patriziat', in S. Schmitt and S. Klapp (eds), *Städtische Gesellschaft und Kirche im Spätmittelalter, Kolloquium Dhaun 2004*, Geschichtliche Landeskunde, 62 (Stuttgart, 2008), pp. 67–74.

50   Cf. the *preceptor monasterii* at Gembloux around 950: J.F. Niermeyer, *Mediae Latinitatis Lexicon minus*, revised edn by J.W.J. Burgers (Leiden and Boston, 2002), 2, p. 1079.

the famous maxim *Quod omnes tangit, ab omnibus debet approbari.*[51] It is not by chance that Michael Mitterauer in his book *Warum Europa? Mittelalterliche Grundlagen eines Sonderwegs* mentions the great centralized religious orders that came into being from the eleventh century onwards among the factors that may explain western success because they taught administrative efficiency.[52]

One last point should be mentioned in this context. The administrative ideals of the military–religious orders were one thing, historical reality another. Numerous studies by Anthony Luttrell and others have shown for the fourteenth and fifteenth centuries that the rule, the statutes and other norms were time and again flouted, because political circumstances or the personal relations of the officers involved proved to be stronger.[53] We can safely assume that this was not basically different between 1100 and 1300,[54] although lack of written sources often makes it impossible to prove such disdain for the rule, the statutes and similar norms. On

---

[51]   Cod. 5.29.5.2; Dist. 63 dictum post c.25; Extra 1.23.7: Tierney, 'Freedom and the Medieval Church', pp. 86–8; Y. M.-J. Congar, 'Quod omnes tangit ...', *Revue historique du droit français et étranger*, 36 (1958), 210–59; G. Post, *Studies in Medieval Legal Thought* (Princeton, NJ, 1964), pp. 91–220.

[52]   M. Mitterauer, *Warum Europa? Mittelalterliche Grundlagen eines Sonderwegs*, 3rd edn (Munich, 2003), pp. 188–98 at p. 196: 'Wie die Papstkirche haben auch die universalen Ordensgemeinschaften der westlichen Christenheit durch die Dichte ihrer überregionalen Organisation sozialräumliche Verhältnisse nachhaltig bestimmt. Wie im Vergleich von Religionsgemeinschaften [sc. Islam, Buddhism, Eastern Christians, K.B.] angedeutet werden konnte, ist ein derartig hoher Organisationsgrad keineswegs die Regel – im Gegenteil: die seltene Ausnahme. Die Intensität der überregionalen Organisation von Papstkirche und universalen Ordensgemeinschaften hat in der westlichen Christenheit sehr gute Voraussetzungen für Prozesse der sozialen und kulturellen Integration bzw. Penetration geschaffen.' R. Stark, *The Victory of Reason: How Christianity Led to Freedom, Capitalism, and Western Success* (New York, 2005), however, somewhat neglects the administrative achievements of the religious orders, although he emphasizes the importance of Christian theology for the rise of western capitalism.

[53]   A. Luttrell, *Studies on the Hospitallers after 1306: Rhodes and the West*, Variorum Collected Studies Series (Aldershot and Burlington, VT, 2007), especially XIV, XV, and XVIII–XX; *The Hospitaller State on Rhodes and its Western Provinces, 1306–1462*, Variorum Collected Studies Series (Aldershot and Brookfield, VT, 1999), especially IV, V, XI–XV; *The Hospitallers of Rhodes and their Mediterranean World*, Variorum Collected Studies Series (Aldershot and Brookfield, VT, 1992), especially III, IV, IX, XI and XV–XVI; *Latin Greece, the Hospitallers and the Crusades 1291–1440*, Variorum Collected Studies Series (Aldershot and Brookfield, VT, 1982), especially I, XIV–XVI; *The Hospitallers in Cyprus, Rhodes, Greece and the West 1291–1440*, Variorum Collected Studies Series (Aldershot, 1978), especially II, XI–XIV and XXIII.

[54]   A. Forey, 'Constitutional Conflict and Change in the Hospital of St. John during the Twelfth and Thirteenth Centuries', *The Journal of Ecclesiastical History*, 33 (1982), 15–29 (repr. *Military Orders and Crusades*, Variorum Collected Studies Series (Aldershot and Brookfield, VT, 1994), X). See also T. Füser, *Mönche im Konflikt: Zum Spannungsfeld von Norm, Devianz und Sanktion bei den Cisterziensern und Cluniazensern (12. bis frühes 14.*

the other hand, we should not doubt that the rule, the statutes and other norms existed among the military–religious orders, that deviance was seen as a problem, and that therefore such administrative ideals began to educate a significant part of the European nobility.[55]

This paper is intended to emphasize the importance of the military–religious orders as a 'school for administrators', and to encourage research about this topic. Historians of religious orders in general, if they want to do more than just to tell the story of their own order or of their favourite order, usually focus on the Cistercians and the mendicants. The Cistercians and the various mendicant orders remain important enough, but to neglect the importance of the military–religious orders may be the consequence of some anti-feudal ideology that deliberately tries to diminish the importance of the nobility in European history. The marginality of the military–religious orders in mainstream historical research, about which Julien Théry and Alain Demurger recently complained,[56] certainly has an ideological background that unfortunately outlives the demise of the self-styled socialist regimes of the twentieth century. The theories of Karl Marx and Friedrich Engels about feudalism and capitalism are still influential especially among Continental European historians and may ultimately be responsible for the undeserved neglect of Templars, Hospitallers and other military–religious orders in many modern studies about religious orders and medieval society, the only exception being perhaps the financial skills of the Templars that are frequently praised.[57]

Therefore the present, very general paper is at pains to emphasize the contemporary context of the military–religious orders and their innovative role in medieval societies. The recent publications by Gert Melville and his pupils on religious orders that were consulted for this purpose are somewhat boring to read, even for those who know German, because their language is cumbersome and sometimes virtually unintelligible.[58] Basic questions such as the use of paper

---

Jahrhundert), Vita regularis: Ordnungen und Deutungen religiösen Lebens im Mittelater, 9 (Münster, Westfalen, 2000).

[55]   Recruitment was of course not restrained on a well-defined class of *nobiles* and deserves further studies: A.J. Forey, 'Novitiate and Introduction in the Military–Religious Orders during the Twelfth and Thirteenth Centuries', *Speculum*, 61 (1986), 1–17 (repr. *Military Orders*, III), and idem, 'Recruitment to the Military Orders', *Viator*, 17 (1986), 141–71 (repr. *Military Orders*, II).

[56]   *Les ordres religieux–militaires dans le Midi (XIIe–XIVe siècle)*, ed. J. Théry, Cahiers de Fanjeaux, 41 (Toulouse, 2006), Introduction p. 7; A. Demurger, 'L'étude des ordres religieux–militaires en France: la fin de la marginalité?', *Cahiers de recherches médiévales*, 15 (2008), 167–73 at p. 173.

[57]   See now I. de la Torre Muñoz de Morales, *Los Templarios y el origen de la banca* (Madrid, 2004).

[58]   Concerning Oberste, *Visitation* see H. Seibert in *Deutsches Archiv*, 53 (1997), 723–4 at p. 724: 'Trotz ihres nicht immer leicht lesbaren Stils, mancher entbehrlicher Längen und des Fehlens jeglicher Register gehört die Arbeit zu den wichtigsten Beiträgen vergleichender Ordensgeschichtsforschung der letzten Jahre.'

for accounts or the survival chances of possible earlier media such as wax tablets are seldom discussed,[59] although the introduction of a relatively cheap medium to write on must have been essential for the development of administration from the thirteenth century onwards. Reports and accounts existed earlier, but registers and inventories began to be prescribed by statutes and other legislation only at the end of the thirteenth and the early fourteenth centuries when paper was available in sufficient quantities.[60] Moreover, the series 'Vita regularis' and similar publications amply use sociological terminologies and theories[61] that at best explain the working of institutions but contribute next to nothing for a better understanding of historical change; 'Politics and Power' as in this conference are much more helpful categories.

Alan Forey rightly noted that the military–religious orders departed from contemporary practice in the matter of organization and foreshadowed to some extent forms of organization adopted by the friars in the thirteenth century.[62] This is indeed an important and promising problem for useful further research. And despite the aversion that readers may nurture, there is something in Gert Melville's 'Vita regularis' series and similar German publications on religious orders and medieval society that can be useful for further inquiries and therefore merits closer attention by the specialists here assembled. To draw attention to this

---

[59]   E. Goez, *Pragmatische Schriftlichkeit und Archivpflege der Zisterzienser: Ordenszentralismus und regionale Vielfalt, namentlich in Franken und Altbayern (1098–1525)*, Vita regularis: Ordnungen und Deutungen religiösen Lebens im Mittelalter, 17 (Münster, Westfalen, 2003), especially pp. 185–269; J. Oberste, '"Ut domorum status certior habeatur ...", Cluniazensischer Reformalltag und administratives Schriftgut im 13. und frühen 14. Jahrhundert', *Archiv für Kulturgeschichte*, 76 (1994), 51–76; idem, 'Normierung und Pragmatik des Schriftgebrauchs im cisterziensischen Visitationsverfahren bis zum beginnenden 14. Jahrhundert', *Historisches Jahrbuch*, 114 (1994), 312–48; idem, 'Institutionalisierte Kommunikation im Verwaltungsalltag religiöser Orden des hohen Mittelalters', in G. Melville (ed.), *De ordine vitae: Zu Normvorstellungen, Organisationsformen und Schriftgebrauch im mittelalterlichen Ordenswesen*, Vita regularis: Ordnungen und Deutungen religiösen Lebens im Mittelalter, 1 (Münster, Westfalen, 1996), pp. 59–99.

[60]   J. Oberste, F. Cygler and G. Melville, 'Aspekte zur Verbindung von Organisation und Schriftlichkeit im Ordenswesen: Ein Vergleich zwischen den Cisterziensern und Cluniazensern im 12./13. Jahrhundert', in C.M. Kasper and K. Schreiner (eds), *Viva vox et ratio scripta: Mündliche und schriftliche Kommunikationsvorstellungen im Mönchtum des Mittelalters*, Vita regularis: Ordnungen und Deutungen religiösen Lebens im Mittelalter, 5 (Münster, Westfalen, 1997), pp. 205–80, especially pp. 276–7 on a papal order of 1289 for Cluniac houses to keep *registra* and on Cluniac statutes from 1291 to 1314, characteristically without any discussion of parchment or paper.

[61]   Oberste, *Visitation*, pp. 395–6 with reference to Max Weber, Niklas Luhmann, Norbert Elias, Jürgen Habermas and Michael Foucault. *Institutionalität und Symbolisierung: Verstetigungen kultureller Ordnungsmuster in Vergangenheit und Gegenwart*, ed. G. Melville (Cologne, 2000).

[62]   Forey, *The Military Orders*, p. 3.

series is therefore a secondary purpose of this paper, apart from presenting the military–religious orders as a medieval 'school for administrators'. The twelfth-century Hospitallers and Templars in the Holy Land may have been responsible for the spread of three major constitutional innovations, strict centralization under a headquarters, territorial administrative districts and limited duration of office for subordinate posts. They set an example for the mendicant orders of the thirteenth century. An interesting question for further comparative studies not only about military–religious orders[63] or orders in general but about both ecclesiastical and secular constitutional history remains the relation of military–religious orders with, or their influence on, techniques applied by people from the same, usually noble social background in royal or princely administrations of France, England, Sicily and elsewhere.

---

[63]   The comparison with other religious orders and with secular administrations is somewhat neglected by two recent general histories of the military religious orders in general – Forey, *The Military Orders* and A. Demurger, *Chevaliers du Christ: Les ordres religieux–militaires au Moyen-Âge (XIe–XVIe siècle)* (Paris, 2002) (trans. W. Kaiser as *Die Ritter des Herrn: Geschichte der geistlichen Ritterorden* (Munich, 2003)) –, and even more so by recent studies about single military religious orders such as K. Militzer, *Von Akkon nach Marienburg: Verfassung, Verwaltung und Sozialstruktur des Deutschen Ordens*, QuStDO, 56 (= Veröffentlichungen der Internationalen Kommission zur Erforschung des Deutschen Ordens, 9) (Marburg, 1999), or H. Nicholson, *The Knights Hospitaller* (Woodbridge, 2001) and idem, *The Knights Templar* (Stroud, 2001).

PART 1
The Latin East

Chapter 2

Archaeological and Fresco Research in the Castle Chapel at al-Marqab: A Preliminary Report on the Results of the First Seasons

Balázs Major and Éva Galambos

Qal'at al-Marqab is among the most imposing and most important crusader castles in the Levant and possesses one of the largest medieval chapels built on a military site in the Latin East. Despite its widely acknowledged significance, it has been the subject of only a few, mostly historical studies with, until now, no archaeological work having taken place there. The Syro-Hungarian Archaeological Mission (SHAM) began a full-scale research project on the site in October 2007 with a multi-disciplinary team made up of historians, archaeologists and archaeozoologists, art historians, experts in geology and geophysics, architects and conservators of objects and frescos, together with laboratory experts, with the aim of getting a better understanding of the site and its evolution through almost a millennium.[1] Among the dozens of research zones inside the castle and its suburb, special attention has been devoted to the castle chapel that proved to contain far more scientific data and artistic treasures than anticipated (Fig. 2.1).

I. Historical Background[2]

The large castle, noted by a number of European travellers but only described in detail for the first time by Deschamps,[3] is relatively recent by comparison with

---

[1]   The Syro-Hungarian Archaeological Mission is a joint research project headed by the Directorate General of Antiquities and Museums (DGAM) and the Catholic University of Hungary with architects Edmond el-Ajji and Marwan Hasan as the directors from the Syrian side and Balázs Major as the director from the Hungarian side. The research project involves a number of other Hungarian institutions and universities, among which the Hungarian Academy of Fine Arts, the National Museum of Hungary, and the Directorate of the Museums of Baranya County are the most important partners in the research into the chapel.

[2]   For a more detailed historical sketch see B. Major, *Historical Background: The Master Plan of al-Marqab Citadel*, Project Defence System on the Mediterranean Coast – Euromed Heritage II Project (Spain, 2008), pp. 162–74.

[3]   P. Deschamps, *Les châteaux des croisés en Terre Sainte, III. La défense du comté de Tripoli et de la principauté d'Antioche* (Paris, 1973), pp. 259–85.

Figure 2.1   The chapel of al-Marqab from the north-west (photo: B. Major)

Figure 2.3   Interior of the chapel from the south-west corner (photo: B. Major)

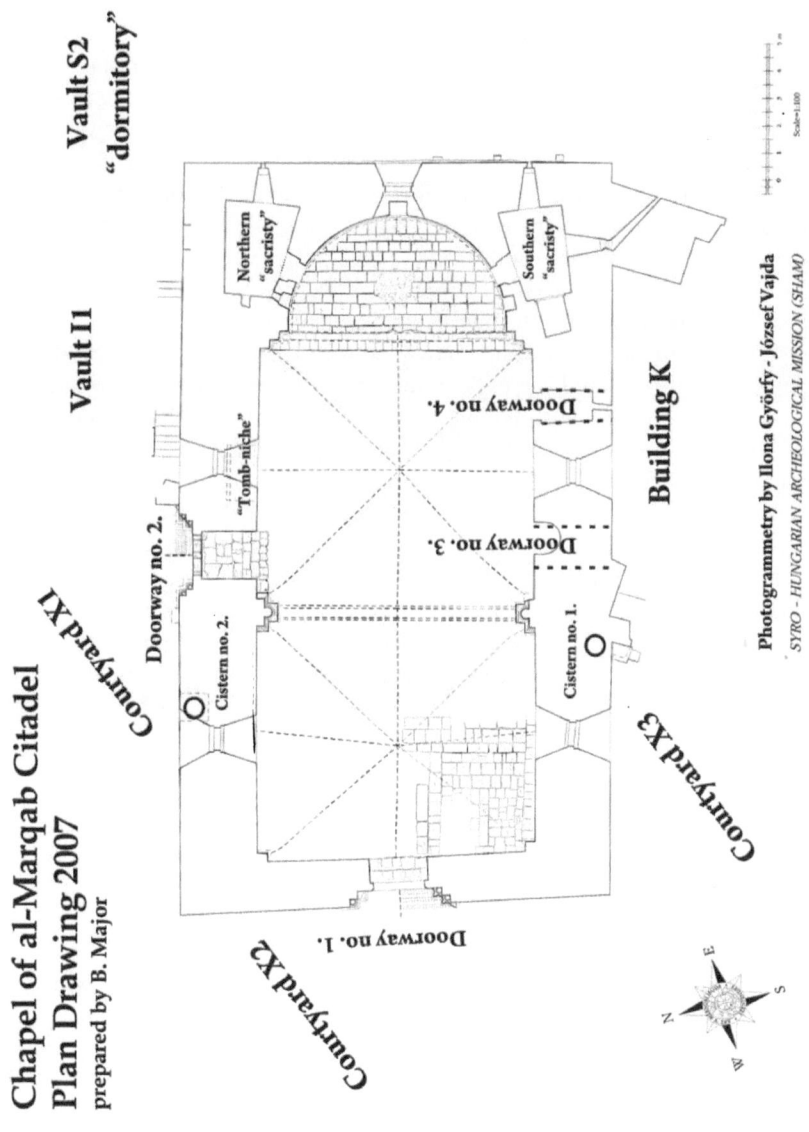

Chapel of al-Marqab Citadel
Plan Drawing 2007
prepared by B. Major

Vault II

Vault S2
"dormitory"

Northern "sacristy"

Southern "sacristy"

Courtyard X1

Doorway no. 2.

"Tomb-niche"

Cistern no. 2.

Doorway no. 4.

Doorway no. 3.

Building K

Cistern no. 1.

Courtyard X3

Courtyard X2

Doorway no. 1.

Photogrammetry by Ilona Györfy - József Vajda
SYRO - HUNGARIAN ARCHEOLOGICAL MISSION (SHAM)

Scale=1:100

Figure 2.2   Chapel of al-Marqab Citadel (graphics: B. Major)

other castles in the Syrian coastlands. According to the Arabic sources, it was established by the local population around the year 1062,[4] and, in view of its strategic location overlooking the coastal route and its mountainous hinterland, was named al-Marqab, the 'look-out post'. After a brief period of Byzantine occupation that began in 1104,[5] it was first taken by crusaders from Antioch in 1117–18 and settled with Franks and Armenians.[6] The castle seems to have been acquired by the Muslims in the 1130s,[7] and then regained by a certain Renaud de Mazoir in the year 1140.[8] *Margat*, as the crusaders called it, became the main castle of this most influential baronial family in the principality of Antioch. As such it must have been the 'praesidium … inexpugnabile et munitissimum' that served as the centre of baronial resistance led by Renaud de Mazoir against Prince Bohemond III, and to which the patriarch of Antioch, a supporter of the revolt, also transferred his seat for a period.[9]

After more than two decades of donating properties piecemeal to either the Templars or the Hospitallers, the family decided to sell its entire possessions including al-Marqab and the town of Bānyās (the crusader Valenia) below it. The competition was won by the Hospitallers, staunchly supported by Bishop Anterius of Valenia,[10] and the charter signalling their purchase of the Mazoir properties was issued on 1 February 1187.[11] The Hospitallers immediately started a complete remodelling of the whole site including not only the citadel but also the suburban area. Construction work can have hardly begun when, just a few months after the purchase, the crusaders lost the battle of Hattīn and, within a short period, most of their former territories in the south. Saladin's victorious armies struck at the northern crusader states the following year, but the campaign in July 1188 avoided laying siege to the major fortified sites belonging to the military orders with the exception of the poorly guarded Tartūs. In fact the Muslim army had great difficulty crossing along the narrow coastal defile beneath al-Marqab where it was

---

[4]   Yāqūt, *Mu'jam al-buldān*, ed. Farīd 'Abd al-'Azīz al-Khubadī (Beirut, n.d.), 5, p. 127.

[5]   Anna Comnena, *The Alexiad*, trans. E.R.A. Sewter (London, 1969), p. 365.

[6]   Ibn 'Abdazzāhir, *Tashrīf*, ed. Murād Kāmil (Cairo, 1961), p. 85.

[7]   Deschamps, pp. 260–61.

[8]   Cafari Caschifelone, 'De liberatione civitatum orientis liber', *RHC Occ*, 5, pp. 66–7.

[9]   WT, p. 1014 (s.a. 1180); *A History of Deeds Done beyond the Sea*, trans. E.A. Babcock and A.C. Krey (New York, 1943), 2, p. 454.

[10]   J. Burgtorf, 'Die Herrschaft der Johanniter in Margat im Heiligen Land', in R. Czaja und J. Sarnowsky (eds), *Die Ritterorden als Träger der Herrschaft: Territorien, Grundbesitz und Kirche*, Ordines Militares, Colloquia Torunensia Historica, 14 (Toruń, 2007), p. 30.

[11]   *CH*, 1, no. 809; H.E. Mayer, *Varia Antiochena. Studien zum Kreuzfahrerfürstentum Antiochia im 12. und frühen 13. Jahrhundert* (Hannover, 1993), p. 176. The conventional dating placed the donation a year earlier, e.g. Deschamps, p. 263.

subjected to an artillery barrage from a Sicilian fleet in the castle's port.[12] After passing behind hastily erected wooden palisades, Saladin's troops destroyed the undefended episcopal town of Valenia but then left the area for good. This brief invasion cannot have had much effect on the grandiose construction project, as recent archaeoseismic research conducted by the SHAM has shown that nearly all the major elements of the Hospitaller phase of al-Marqab were standing by the time of the 1202 earthquake.[13] This in itself proves the great importance the Order attached to al-Marqab, an importance that was further demonstrated by the fact that, contrary to its rule, the famous Chapter General of 1204–1206 was held there.[14] After the destruction of Valenia, the bishop moved his residence to al-Marqab,[15] and the site seems to have served as the seat of the bishopric[16] until it fell to the Muslims on 25 May 1285.[17] Although much of its importance came to an end with the expulsion of the crusaders, al-Marqab still had a garrison in Mamluk and Ottoman times and also served as the administrative centre of the region. Its suburb and part of the castle area was inhabited until 1958 when the DGAM relocated the population.

II. The General Layout of the Chapel

The chapel building in the citadel of al-Marqab is one of the largest such structures still standing in a crusader castle (Fig. 2.2). It has a nave measuring 23.6 × 18.6m, which is divided into two groin-vaulted bays with a pointed transverse arch dividing them and springing from compound pilasters (Fig. 2.3). The nave has two still functioning original doorways, one opening on the western façade and one on the north side (respectively doorways 1 and 2 in Fig. 2.2). The nave terminates in a semi-circular apse 7.6m wide covered by a pointed-arched semi-dome. The apse is flanked by two barrel-vaulted rooms of a rather irregular rectangular plan, the so-called sacristies. The chapel was lit by six lancet windows: one over the

---

[12]   Ibn al-Athīr, *al-Kāmil fi'l-ta'rīkh*, ed. 'Umar 'Abd al-Salām Tadmurī (Beirut, 1999), 10, p. 49; Abū Shāma, *al-Rawdatayn*, ed. Ibrāhīm al-Zaybaq (Beirut, 1997), 4, pp. 18–19.

[13]   M. Kázmér and B. Major, 'Distinguishing Damages from Two Earthquakes – Archaeoseizmology of a Crusader Castle (al-Marqab citadel, Syria)', *The Geological Society of America Special Paper*, 471 (2010), 185–98.

[14]   J. Burgtorf, 'The Military Orders in the Crusader Principality of Antioch', in K. Ciggaar and M. Metcalf (eds), *Antioch (696–1268)*, Orientalia Lovaniensia Analecta 147 (Leuven, 2006), pp. 217–46.

[15]   Wilbrand von Oldenburg, 'Itinerarium Terrae Sanctae', ed. S. de Sandoli, *Itinera Hierosolymitana Crucesignatorum (saec. XII–XIII)* (Jerusalem, 1983), 3, p. 212.

[16]   E.G. Rey, *Les colonies franques de Syrie aux XIIme et XIIIme siècles* (Paris, 1883), p. 335.

[17]   Ibn 'Abdazzāhir, pp. 77–81.

doorway of the western façade, one in the apse, and two pairs facing each other on the northern and southern façades respectively.

III. Investigation of the Chapel Building

III.1. Some observations on the construction

The results of the geophysical survey executed by Gábor Bertók (Directorate of Baranya County Museums, Hungary) with a geo-radar (Mala X3M Integrated Radar Control Unit) in the fieldwork season of 2007 indicate that the chapel was not the first building to have been constructed on this site. The radar picture showed the contours of what seems to have been the corner of a huge rectangular building approximately 70cm deep under the stone pavement of the nave of the chapel. Test soundings opened in the summer season of 2010 within the chapel verified the geo-radar information and revealed the extensive remains of buildings with more than one plastered floor level inside. The corner of the building appearing on the geo-radar picture turned out to be a multi-period structure, the southern and eastern walls of which were partially excavated. Excavations immediately to the north of the chapel indicated that the eastern wall of this building continued outside the area of the later chapel. If the wall fragments discovered indeed belonged to the eastern wall of this same building, then it must have been a considerable rectangular building with walls exceeding 90cm in width. The thickness of the wall is insufficient to allow for the possibility that this structure could have been a donjon, but the eminent position of the building on the highest point of the southern plateau, the size of the structure, and the remains of a simple pilaster in its wall demonstrate its prominence and suggest that it could well have been the great hall of the pre-Hospitaller castle. It is tempting to link the construction of such a vast structure to the influential Mazoir family, and that would strengthen the likelihood that this family, whose heads must otherwise have spent considerable time at the court of the princes of Antioch, did indeed use al-Marqab as their seat. It would also strengthen the possibility that the unnamed Mazoir castle that served as a refuge during the baronial resistance to Bohemond III was none other than al-Marqab. Excavations have made it clear that whatever buildings existed here were swept away after 1187 as part of the Hospitallers' redesigning of the castle, and this area was levelled in order to create ample space for the chapel.

It has been long established that the chapel was among the first structures to have been built by the Hospitallers,[18] and, according to the recent wall-texture research and periodization survey of the SHAM, there is only one building attributable to the Hospitallers that clearly preceded its construction. This is the huge pointed barrel-vaulted hall (vault S2) standing to the north-east of the chapel, which, given its position and the huge latrine tower attached to it, could have easily been the

---

[18]   Deschamps, p. 227.

knights' original dormitory. The chapel can be understood as the product of a single grandiose building campaign, although it shows traces of at least one modification during the construction work. Gergely Buzás (National Museum of Hungary) has concluded that the conspicuous change in the groin vault over the bays is the result of modifying the plan from a semi-circular groin vault to a pointed one, perhaps attributable to a change of architect. Despite the magnificent impression that the finished chapel conveys, the photogrammetry plans and section drawings produced by the SHAM reveal that it was not a faultless structure. We find most of the anomalies in the western bay with the north-western and south-western corners of the chapel far from being orthogonal, and the vault plus the frame of the western lancet window of the southern wall having a serious distortion.

The two pastophoria or sacristy rooms on each side of the apse in the wall thickness also show serious differences, but that is mainly due to their apparently different functions. The southern one, with a deep cupboard in its western wall, seems to have been the vestry. The northern sacristy has a neatly executed small window inserted into the western end of its southern wall. Excavations conducted in the sacristies in April 2009 have proved that the original crusader-period floor level was much lower than today.[19] In the case of the northern sacristy it was about 20cm below the stone threshold of the doorway. This means that the little window was well placed to provide a view of the assumed location of the altar for someone sitting on a chair and using this room as his private oratory.

The building programme of the chapel also included the construction of a pointed arched niche 1.1m deep in the interior to the east of the northern doorway of the chapel. A closer inspection of the fabric revealed that the niche was in all possibility intended for the housing of a relic or for the tomb of an important person.

The chapel was constructed in the first phases of the Hospitaller building activities and was subject to a number of later alterations. It is apparent that, besides the western and northern monumental doorways, the chapel had another narrow doorway opening on the eastern half of the southern wall of the eastern bay. It was later closed on its southern side, thereby forming a rectangular niche in the chapel (doorway 3 in Fig. 2.1). Wall-texture research of the SHAM has found the frame of another doorway to the east of doorway no. 3, again located in the eastern bay.[20] It stood opposite to the northern doorway of the chapel and had the same dimensions, doorframe and segmental arch (doorway 4 in Fig. 2.1). Both these southern doors to the chapel were blocked when, in a later, but still crusader-period construction phase, a new two-storey wing was attached to the eastern half of the chapel's southern wall (Building K). In the post-crusader period, when the chapel was turned into a mosque, the relatively soft crusader infilling of doorway 4 was partially removed and a large ashlar *mihrab* niche inserted. Another Muslim

---

[19]   Excavated by Balázs Major (Catholic University of Hungary), Adrián Berta (University of Szeged, Hungary), and Mayssam Youssif (DGAM, Syria).
[20]   Research conducted by Professor Kornélia Forrai (Hungarian Academy of Fine Arts).

addition to the chapel was a small square minaret platform partially constructed from *spolia* of gothic ribs from the hall belonging to the residential area over the gate tower of the citadel.

III.2. Fresco research

The first frescos in the chapel at al-Marqab were found accidentally in the northern sacristy in the spring of 1978 and were cleaned, bordered and in many places retouched by experts from the DGAM. The only previous study of them is by Jaroslav Folda, who was assisted in the field in the spring of 1979 by fresco conservator Pamela French. He summarized the results of his research, including a superb iconographic analysis of the northern sacristy paintings, in 1982.[21] In his paper he stated that the large painting on the slightly pointed barrel-vaulted ceiling over the northern sacristy was a representation of Pentecost, while the fragmentary remains of the fresco on the western wall once depicted a Nativity scene. He noted the presence of a row of saints on both the northern and southern walls of which only very limited fragments survived. Folda's researches extended into the body of the chapel itself, where his soundings traced lines from the underpainting of a geometric pattern in the dado area of the apse with a grid in which lozenge-shaped forms containing equal-armed crosses alternated with those containing vertical stripes. His conclusion was that the painting programme had begun in the northern sacristy and had gradually extended into the apse area where it only reached the underpainting phase and was never finished. His superficial examinations of the walls of the southern sacristy and the nave found no fine plaster similar to the apse or any sign of painted surface. In this he saw a sudden interruption in the painting programme which he connected to the shock caused by the passage of Saladin's armies in 1188 after which the Hospitallers turned their attention to military matters never to resume the painting project.[22]

In view of the importance of the chapel and the previously discovered frescos, the SHAM launched a fresco research project in al-Marqab with experts and laboratory support from the Hungarian Academy of Fine Arts in Budapest.[23] This

---

[21]   J. Folda, 'Crusader Frescoes at Crac des Chevaliers and Marqab Castle', *Dumbarton Oaks Papers*, 36 (1982), 196–210.

[22]   Ibid., pp. 208–9; J. Folda, *Crusader Art in the Holy Land, from the Third Crusade to the Fall of Acre* (Cambridge, 2005), p. 34.

[23]   The documentation and laboratory research for the project is headed by Éva Galambos (Hungarian Academy of Fine Arts) with the participation of the fresco conservation experts Professor Kornélia Forrai (Hungarian Academy of Fine Arts), Anna Selmeczi, Péter Gedeon and Sister Gabriella Sári (Sisters of Social Service). The Restoration Workshop of the DGAM is represented by conservator Nada Sarkis, who was assisted in 2008 by Du'ā Danadshī and Nibāl al-Kafrī. Initial iconographical research into the recently discovered frescos is by Balázs Major and the programme of 3D computer reconstruction of the buildings and their painted surfaces is by Gergely Buzás (National Museum of Hungary).

project, which encompasses the entire castle area, has resulted in the discovery of huge medieval fresco panels inside the chapel and established that a large part of the chapel was painted with high quality frescos with some areas redesigned more than once. Furthermore, it was not only the chapel that was painted. A brief survey of the results achieved hitherto indicates that further fieldwork might result in a much better understanding of the artistic and technical aspects of crusader painting.

III.2.1. The nave (southern wall of eastern bay) The first set of hitherto unknown frescos was discovered in 2007 after removing the two to three layers of later plaster from the southern wall of the eastern bay adjacent to the apse area.[24] The medieval fresco panel once covered the whole area between the southern lancet window of the eastern bay and the pilasters of the apse conch starting from a height of 3m above the pavement of the chapel (Fig. 2.4). Its original area is estimated to have been 10.5m², but with the disappearance of most of its western half we have only about 6m² left. Although there are traces of painted surfaces continuing on to the eastern part of the ashlar framing of the lancet window, it seems that the plastered surfaces of the bay to the west of the lancet did not receive any painting. The fresco panel represents the torments in hell with twenty-seven figures on three levels above each other, the height of each register being around 0.9m (Fig. 2.5).

The uppermost level depicts a dead forest with truncated trees and people hanging on them in a way that seems to represent their sin. In the case of several of the surviving eleven figures, the type of sin is decipherable with tolerable certainty, helped by a number of snakes who are providing additional torment on the very organ with which the sinner committed his offence. The fifth person hung by his tongue with a snake biting it might have been a liar, scold or curser, or a combination of all three. His neighbour to the right with eyes covered and eaten by snakes could be a warning of the fate awaiting traitors.[25] Figure number eight, who is hanging by his hair on the frame of the picture and is stretched by rounded millstones tied to his legs, could depict vanity. The eleventh and last surviving person in the upper row of hell has obviously committed his sin with his genital organ as this is the part that is attacked by a snake. Figures two and seven, both hanging upside down with millstones attached to their necks and snakes around the lower part of their bodies, may have erred in a somewhat similar way to number eleven. These two, hanging in an absolutely identical position, are a warning that the final decoding of the picture is far from complete. Harder to identify are the sins of the first figure, hanging by his hair with snakes around the loin, neck and arm. Interpretation is not helped by the sometimes fragmentary nature of certain areas, as in the case of figure number three who appears to have been spiked on a

---

[24]   Research was conducted by Anna Selmeczi assisted by Zsolt Szécsi. Cleaning and stabilizing in the following seasons was done by Anna Selmeczi and Péter Gedeon.

[25]   For a possible parallel from a later period, see H. Maddocks, '"Me thowthe as I slepte that I was a pilgrime": Text and Illustration in Deguilleville's "Pilgrimages" in the State Library of Victoria', *The La Trobe Journal*, 51–2 (1993), plate 23 at p. 74.

Figure 2.4   The fresco panel depicting hell (photo: E. Galambos)

branch of the tree. It is also not entirely clear what has happened to number nine, who is hanging with his neck broken, perhaps the result of being pulled down by an indistinct demon figure beneath him.

The middle level of the fresco panel, with thirteen extant figures, is made up of two main subjects. The eastern half of the picture is occupied by a huge spoked wheel with elongated triangular blades jutting out from it, on which six people with their arms tied are impaled. They are also hit by a demon in the easternmost edge of the picture holding a huge hammer-like instrument. In the middle of the western half of the surviving fresco one can see a person with a mitre, evidently a Latin bishop, sitting in a huge cauldron with flames shooting out around it. He is surrounded by four demons, two sitting on either side of him, and two flying over him. These latter two are pouring something over him from jugs. The two sitting demons (nos. 20 and 23) both have huge horns, and it is clear that they have more than one face. The demon to the right (no. 24), for example, clearly possesses three faces, one on his head, one jutting out from his breast, and a third one with eyes closed on his loins. Although badly damaged, it is clear from a closer inspection at the site that demon no. 19 flying to the left of the bishop has two bat-like wings attached to his greenish hairy body and at first glance the outline of his head

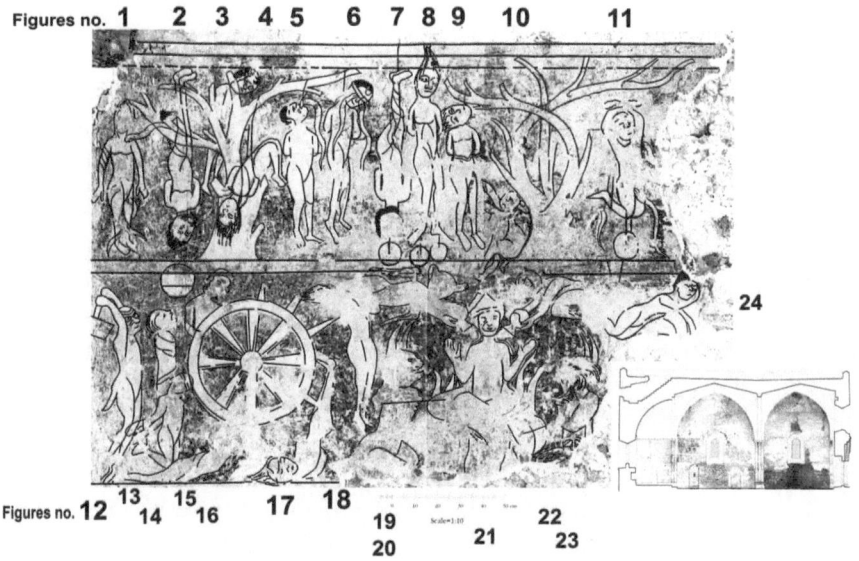

Figure 2.5   The two upper registers of the fresco depicting hell in the chapel of
al-Marqab (graphics: B. Major)

resembles a cow or bull. The other flying demon (no. 22) seems to have a face
resembling a monkey. The third row of the hell panel is heavily damaged, but
the outlines of three figures are nevertheless discernable. The one in the middle
is definitely a sinner apparently impaled on a spike. He is flanked on the left by
a demon with a long tail hitting him with a bent sword-like instrument, while
another demon on his right is pouring something over him from a jug identical to
those held by the flying demons above.

The iconography and the style of the fresco found on the southern wall of the
eastern bay clearly show that it was the work of a European painter. This is quite
rare in crusader art, which was heavily influenced, and in many cases executed,
by eastern artists who for the most part followed eastern models and styles. The
depiction of hell in the chapel at al-Marqab has its closest parallels in Europe
in depictions like the one preserved in the twelfth-century *Hortus Deliciarum*
by Herrad von Hohenburg.[26] This example occupies a complete folio page and
depicts hell with several levels showing a large number of people being tormented
by being hung in different ways, surrounded by demons. In the left corner of the
picture Lucifer himself appears possessing more than one face.

What is anomalous in the fresco at al-Marqab is the prominence accorded the
bishop. It is not unusual to have evildoing ecclesiastics depicted in hell scenes, but

---

[26]   R. Green et al., *The Hortus Deliciarum of Herrad of Hohenburg* (London and
Leiden, 1979), 2, p. 439.

they are always in groups. The lone figure of a bishop at al-Marqab, originally in the very centre of the fresco panel with no less than four devils in attendance, may indicate that he might perhaps represent more than simply the general sins of the secular church. Of course it is almost impossible to make detailed evaluations of the iconography of the fresco as more than one third of it is missing, but one has the impression that this highly individualized bishop could have been a message for an actual person. One is tempted to toy with the idea that the Hospitallers might have had the bishop painted as a representation of their former friend Anterius. According to the documentary sources, the bishop of Valenia, who had to relocate his residence to al-Marqab after the destruction of his seat in 1188, fell out with the Hospitallers very quickly. Quarrels broke out between the two parties over the rights to tithes in the diocese.[27] So could the image of the bishop being cooked in a cauldron have been a Hospitaller mockery of this 'greedy bishop' of Valenia? It would not have been the first time the Hospitallers had expressed their disrespect to hostile ecclesiastical leaders, among them the patriarch of Jerusalem – and in ruder forms.[28]

The surviving fresco panel is painted on the first layer of plaster on the chapel wall. The plaster layer comprises lime-mortar mixed with tiny basalt pebbles probably originating from coastal gravel. The plaster was covered with a very thin layer of lime, onto which the main lines of the composition were scratched. This served as the base for the layer of underpaint sketch for the fresco, painted with red lines containing iron oxide. Elaborate though this underpaint was, the final scenes differed from it in a number of places. The execution of the fresco continued with the painting of several layers of the background colour. The bottom layers were usually dark brownish-red, while the upper layers were painted with green earth. An analysis of the pigments was made with Polarized Light Microscopy. The colours employed on the fresco are mainly earth pigments: red iron oxide, yellow iron oxide, ochre, and green earth. This was measured by x-ray diffraction, which proved that it was of a celadonite type. The tree trunks comprised a single yellow colour, but the figures' body colour was built up with several layers of pigments. The basic colour was a middle-tone pink, which was tempered with lighter and darker tones: for example, in the shaded parts brownish and greenish colours were used. The body colours were mixed from red containing sedimentary chalk and iron oxide. The last element to be added to the painting was the cinnabar colour of the mouths and the black contours of the figures. Being painted with secco technique, little remains from the upper contour layers, although there is enough to allow an assessment of the original painting style. This is all the more important as the painting seems to be essentially European, not only in its iconography but also in the technique employed. It was also confirmed by on-site research that certain spots of paint seeped into the damaged areas of the plaster layer. That could mean that the plaster was already damaged by the time the first painted layers were

---

[27]   Burgtorf, 'Die Herrschaft der Johanniter', pp. 37–8.
[28]   WT, pp. 812–14; trans. Babcock and Krey, 2, pp. 239–41.

applied, or that the damage was later and subsequently repaired. Further study of the damage might enable us to identify its causes (e.g. earthquake), and that could provide precious information about the date of construction.

III.2.2. The nave (northern wall of the eastern bay) In tandem with the uncovering of the southern wall, investigations started on the surfaces of the eastern bay of the northern wall as well. Opposite to the hell scene, a fresco panel roughly similar in size and general layout is emerging, but with a totally different subject. Exploration is still far from complete, as the plastered surface of this wall has been badly damaged by constant water seepage from the direction of the chapel's freestanding outer northern wall. Besides weakening the painted layers, the moisture has also resulted in the build-up of thick layers of incrustation, the complete removal of which is time consuming and requires great care. At the moment, the lower part of the fresco reveals the most about its subject. As in the case of the hell fresco, this panel is divided into horizontal strips of which only the two lowest have been partly freed from later plaster. In the two horizontal registers naked people are marching towards the altar in serried ranks with hands folded in prayer. Most of what remains of them is the red underpainting lines delineating their bodies. By contrast with the dark brown and green colours of the hell scene, the background colour of this northern fresco panel is bright yellow. All this indicates that the main subject painted on the northern wall is a representation of heaven with the blessed in the two bottom rows. Above these two registers the red underpainting contours of a huge drapery or a cloak (possibly belonging to a huge figure) were found in the western half of the panel close to the lancet window. Soundings and test windows opened into the upper half of the plastered surfaces show that the fresco panel on the northern wall extended more than 1m higher than that on the southern wall of the eastern bay. Like the fresco panel on the southern wall, the heaven panel did not continue on the other side of the lancet window. Close inspection has revealed traces of repairs as in the case of the hell fresco. Although there is much work still to be done, it is clear that the fresco panel of heaven was painted with the same technique and in an identical style as the panel depicting hell. It was done by the same European workshop.

The northern wall of the eastern bay also contains the 'tomb-niche', positioned below the lancet window. The research of the SHAM has revealed that the niche, enclosed by a simple pointed arch, was more ornate than it appears now. It had a moulding running all along the outer façade of the arch, two fragments of which were found in rather unexpected places. One was discovered in a putlog hole of the southern wall of the eastern bay at a height of 8m above the ground. This putlog hole evidently dates from the period of the chapel's construction and was obviously blocked by the Muslims when they re-plastered the chapel. The second fragment was found during the excavations in the kitchen area to the south-west of the chapel, also in post-crusader layers. This latter piece of moulding preserved traces of yellow painting, and it is probable that not only the moulding of the niche was painted.

Another decorated surface was traced between the 'tomb-niche' and the apse pilasters. Its basic motif consisted of double circles scratched into the wet plaster and the area between them was later painted red. The interior of these circles was then given a simple decoration of palmettes painted with the same red colour. In all likelihood this formed part of a painted drapery, but as no other similar decorative motifs have yet been found, it is not impossible that this painted drapery was restricted to this particular area beside the 'tomb-niche'.

It is also interesting that, in the plastered layer above the painted palmette drapery, there is the graffiti of a ship; this would seem to be crusader, with some clear parallels to those in the south-eastern tower at Crac des Chevaliers. If so, that would indicate that the painted drapery surface had already passed out of use before the end of the crusader period. A definite date for the plaster containing the graffiti is all the more important as traces of the same plastered layer were found in the upper registers of the nave as well; that would indicate that at least one major replastering of the chapel took place in the crusader period and that this new plaster seems not to have been painted. However, this hypothesis needs further proof.

III.2.3. Other areas of the nave Preliminary soundings made by the SHAM in the western bay of the nave have detected a thick painted horizontal line in the area of the lancet windows. As it is on the first plastered layer it would appear that it was intended as the border for further fresco panels, but no traces of figural paintings have yet been found. Our present state of knowledge seems to indicate that only the eastern half of the eastern bay was painted with the strongly didactic representations of heaven and hell, and the rest remained devoid of figural decoration.

Research on the compound pilasters supporting the transverse arch that separates the two bays has revealed two distinct painted layers. The first layer of painting contained light pink and red colours that are likely to have been employed to give the surfaces a marble appearance. The basic colour of the second layer was yellow, on which were painted red lines imitating ashlar borders. On this layer large reddish areas were also painted with ashlar imitation in black. This could have been a later development, possibly a third repainting phase also aiming some sort of marble effect on the pilasters.

It was not only the interior of the chapel nave that received painted decoration. Thorough inspection of the doorways revealed remains of red and yellow pigments on the ashlars of the northern portal's outer façade. The recently discovered southern doorway of the chapel, opposite to the northern portal, also preserves traces of yellow paint on its jambs. Recent findings point to the possibility that most of the neatly carved architectural elements of the chapel's doorways were painted, the joints of the ashlars being articulated with dark red stripes.

III.2.4. The apse Of all painted areas of the chapel, it is the apse that is the most difficult to investigate and periodize. This partly stems from the fact that, contrary to the conclusions of earlier research,[29] the painting of the apse appears not to have been completed uninterruptedly; it is evident that there were later repaintings and modifications executed in separate phases.

According to the latest results, the earliest decoration in the apse is to be found around the doorway of the northern sacristy. It was a pattern imitating the ashlar frame and architrave of the doorway, with its rectangular units painted in alternating dark red and pale yellow. The architrave was painted to resemble a flat arch with joggled voussoirs (Fig. 2.6) This decorative motif of imitation joggled voussoirs was widespread in crusader architecture, featuring on the single block architraves of many sites; the closest surviving example to al-Marqab is found carved into the architrave of a postern gate in the Templar castle of 'Arīma. The border lines of the one at al-Marqab were painted with a lighter red colour, and in the western corner considerable remnants of a painted surface imitating marble have also come to light. The plaster around the doorway bears close resemblance to the mortar binding of the ashlar stones of the chapel and no resemblance to the later plaster layers employed in the chapel. Scanty remains of vertical and rectangular lines painted with dark red and detected above the ashlars of the dado hint at the possibility that the dado area of the apse beneath the lancet window received ashlar contours either directly on their surface or over a very thin layer of lime.

In the layer overlaying the plastered frame of the doorway in the northern half of the apse was the geometric pattern found by Folda. It was made up of lozenge-shaped forms containing equal-armed crosses alternating with those containing vertical lines. They are described as having been executed in a pale red pigment only,[30] although a thorough re-examination and new soundings of the apse have proved that the geometric pattern did receive colouring, and can be considered as having being completed. The crosses were painted white with the lozenges around them in yellow. The vertical stripes received alternating white and orange colouring (Fig. 2.7). Perhaps it was at this time that the little window niche immediately to the west of the northern sacristy received its yellow paint with the red lines.

Folda reported that his examination of the southern half of the nave walls revealed the geometric grid-cross pattern reappearing close to the entrance of the southern sacristy. The SHAM examination has found further remains of the lozenge-shaped grid, but its size differs from that of the lozenges on the northern walls. Another point of difference is that, although the cross pattern is clearly visible, there is no trace of the vertical stripes. Instead, thorough examination and raking-light photography has discovered that the seemingly empty lozenges contained a heraldic eagle with black colouring. The head of the eagle was scratched into the plaster with the help of a stencil and then coloured with black pigments. The cross pattern was bordered with red with white crosses on yellow background. The arms of the crosses were not closed by vertical and horizontal lines, but extended to the corners of the lozenge frame (Fig. 2.8) This painted

---

[29] Folda, 'Crusader Frescoes', pp. 208–9; idem, *Crusader Art in the Holy Land*, p. 34.
[30] Folda, 'Crusader Frescoes', p. 198.

Figure 2.6   Reconstruction of the ashlar painting around the doorway to the
               northern sacristy (graphics: E. Galambos)

Figure 2.7   Partial reconstruction of the decorative pattern containing 'crosses
               and vertical stripes' (graphics: E. Galambos)

layer also underwent some alteration at a later stage, as one of the eagle heads has traces of red overpainting. Further work is needed to define the relation of the two painted patterns in the southern and northern halves of the apse. Much additional data is also expected from the ongoing laboratory analyses (e.g. by Polarized Light Microscopy) of the pigments. Another field of research might also produce valuable information. It is not impossible that the two different patterns with their heraldic decoration (a cross with vertical stripes and a cross with black eagle) were actually the coat of arms of actual Hospitaller officers or of some other high-ranking individual affiliated to al-Marqab. If it were possible to identify them and connect them with particular persons or families, it would certainly help dating.

Recent research of the SHAM revealed that the window-zone of the apse was also covered with paintings. The summer season of 2010 saw the complete uncovering of the section of the apse south of the lancet window. The very fragmentary fresco remains include what appear to be a banqueting table and the fragment of a female figure beside which a male person is striking at the hallowed head of a saint (Fig. 2.9). This is evidently a depiction of the beheading of St John the Baptist, the patron saint of the Order. The female figure surviving from below the waist would thus be Salome. The other fresco fragment found after the removal of the later plastered layers is found immediately to the west of the beheading scene. It depicts a blessing hand emerging from the clouds flanked by two angels who are spreading incense over a younger and an older person. The very fragmentary nature of this second fresco necessitates more iconographical studies to decipher its original content, although a scene from Abraham's sacrifice of Isaac or a depiction of a miracle connected to the relics of St John are both possible. Exploration windows cut through later plaster layers in the northern half of the window-zone of the apse indicate the presence of further frescos yet to be uncovered. The detail found recently shows the underpainted contours of a male figure leaning with a double-handled jar in his hand under a neatly decorated pointed arch (Fig. 2.10). What the precise subject of this was is hard to define, but, in the light of the figure, a scene from the wedding at Cana would not be impossible. However fragmentary the fresco fragments of the window-zone frescos might be, one thing is certain: the style of the painting and its execution technique are very similar to the heaven and hell frescos, so they are also clearly European made.

Although no traces of painting were found on the interior surfaces of the half dome covering the apse, high-resolution photography detected traces of painting on the triumphal arch preceding the apse. The surviving fragment has a dark blue background colour with a white lozenge-pattern and dots of lime imitating pearls. Its style is closest to that of the paintings in the northern sacristy.

III.2.5. The northern sacristy The work of the SHAM in the northern sacristy is mainly focused on the conservation of the frescos studied in great detail by Folda and the cleaning of their environment to stop further damage. For this reason the debris accumulated in the small guardroom above the sacristy, which was behaving like a sponge transmitting rainwater into the vault of the frescos, was removed in the

Figure 2.8   Reconstruction of 'eagle and cross' pattern around the entrance of
               the southern sacristy (graphics: E. Galambos)

summer of 2009 after detailed documentation.[31] Another serious source of danger is
the cement pointing that was employed during the early conservation work and that
is now also being removed from between the stones of the side walls. It seems that
much damage was caused by numerous types of fungi and other bacteria for which
water seepage, the salty environment, and the old conservation liquids employed at
the end of the seventies created an ideal environment.[32] Laboratory analyses and the
testing of materials on site have aimed to eliminate the bacteria and also to remove
the conservation liquid and the careless retouching of 1979. That also necessitated
the removal of the bacteria-contaminated debris on the floor of the northern sacristy,
and this in turn resulted in determining the original floor levels and the discovery
of a large number of well preserved but tiny fresco fragments that in all probability
came from the Pentecost scene.

A detailed analysis of the plastered layers preserved on the walls of the room,
especially in relation to the recently discovered painted layers, was also necessary.
This research found that the Pentecost scene in the northern sacristy room was
painted on a third plastered layer (Fig. 2.11), with the first one being identical to the
plaster of the nave under the fresco panels. The first layer of plaster contains a high

[31]   Excavation directed by Father Tony 'Eid (Antonine Maronite Order, Lebanon).
[32]   Laboratory analyses of the bacteria were done by Dr Gyula Vágvölgyi, Associate
Professor and Head of the Department of Microbiology at the University of Szeged
(Hungary).

Figure 2.9   The beheading of St John the Baptist with Salome on the left in the apse window-zone (photo: B. Major)

Figure 2.10   A figure with a jar from the decoration in the window-zone of the apse (photo: B. Major)

quantity of volcanic basalt ingredients. On the base of petrographic analysis we can conclude that this plaster is identical to that in the chapel on which most frescos were painted. Traces of yellow pigment over its surface make it possible that the northern sacristy room was originally yellow in colour. The two plastered layers above it contain less basalt particles, but more lime and biogen limestone, plus straw or cavings. The surface of the intermediate layer is polluted, which means that it was exposed for a long time. That indicates that a considerable amount of time elapsed between the construction of the first plastered layer and the application of the third. This third layer is much whiter than the earlier ones because of its high lime content. All this indicates that, contrary to the former periodization, the fresco programme of the northern sacristy or oratory was last to be painted. If the bishop of Valenia was really using the castle chapel as his cathedral church and not another building, perhaps in the suburb, the northern sacristy would have served as an ideal private oratory for him. A relatively later date for the luxurious decoration, which employed lapis lazuli colourants, would be in keeping with the period after the Hospitallers had won a decisive victory over the secular church in the diocese of Valenia and were electing the bishop from among their own clergy.[33]

A possible parallel and *terminus postquem* for the frescos of this sacristy might be provided by the guard tower beneath the castle overlooking the same coastal defile through which Saladin had had to struggle. Paul Deschamps was first to notice the existence of a fresco fragment on the vault of the northern loophole in the eastern wall of the first-floor room.[34] The painting, covering less than one square metre, had been applied on an ultramarine background with brown, yellow and white colours. At the centre of the fresco is Christ seated on a throne, but only part of his halo and a piece of the throne have survived. Two other halo fragments on the fresco and a brown figure that looks like the head of an ox are also visible on the remains. It is not impossible that the painting depicted Christ and the evangelists, and this decorated loophole facing east was an oratory for the guards in the tower. The style and the materials employed are very similar to the frescos in the oratory (northern sacristy) of the chapel. A *terminus postquem* for this tower and consequently the fresco in the loophole are given by the chronicle of Ibn Wāsil, which states that during the Muslim siege of al-Marqab in 1204–1205 the army of Aleppo succeeded in destroying the coastal tower.[35]

III.2.6. Tentative Periodization Summary of the Chapel's Decoration In all probability the chapel was among the first structures to be constructed by the Hospitallers after they bought al-Marqab in 1187. While the chapel building seems to have been the result of a single construction project, its decoration – according

---

[33]   On this process see J. Riley-Smith, *The Knights of St. John is Jerusalem and Cyprus c. 1050–1310* (London, 1967), p. 413.

[34]   Deschamps, p. 285 n. 5.

[35]   Ibn Wāsil, *Mufarrij al-kurūb fī akhbār Banī Ayyūb*, eds Jamāl al-Dīn al-Shayyāl and Hasanayn Muhammad Rabī' (Cairo, 1954–75), 3, p. 165.

Figure 2.11   Detail of the Pentecost panel in the northern sacristy showing that
the painting is on the third plastered layer (photo: E. Galambos)

to the research of the SHAM – is clearly the result of several consecutive painting
programmes (see Fig. 2.12).

The first decorative elements in the chapel could have been the ashlar contours
painted with dark red lines over a very thin layer of lime applied directly over the
ashlar architectural elements: mostly on the doorframes, the dado area of the apse
and the pilasters. In all probability the red and pale yellow ashlar imitation framing
the entrance of the northern sacristy belongs to this period. This would underline
the special importance attached to the northern sacristy from the earliest times and
make its use as private oratory more probable. The alternating darker and lighter
ashlars of the triumphal arch, still unpainted, would have perfectly matched the
overall simple decorative style of this early phase.

In a next phase the apse received a thin plaster layer whitewashed, into which
the eagle-cross pattern was scratched then painted on the southern surface of the
apse. The same scratching of contour lines could be observed on the palmette-
pattern beside the 'tomb-niche', which might indicate that this latter feature was
of roughly the same date. Comparison of the plasters indicates that this might have
also been the time when the fresco panels of heaven and hell were painted on the
side walls of the eastern bay. These latter painted surfaces also show traces of
contour scratching of the registers and of some of the figures. It is almost certain
that the figural fresco decoration of the apse is contemporary with those in the bay.
This large-scale painting activity could also have seen the pilasters and door frames

Figure 2.12   Computer graphic of the chapel interior with recently discovered
               painted surfaces reconstructed (graphics: G. Buzás)

being covered with a very thin layer of plaster, basically for levelling the surfaces,
over which a yellow colour with red contour lines of ashlar imitation was applied.

Traces of a third repainting activity over many areas of the chapel interior have
begun to emerge, based both on on-site observations and on the initial laboratory
analyses of the yellow pigments. The frescos of the eastern bay show traces of
repair, while the 'eagle and cross' decorative pattern has also preserved evidence
a number of overpaintings done with a yellow paint with high lead content. The
yellow colour identified with EDS and Polarized Light Microscopy analyses occurs
on many surfaces and always in the areas of repair in the upper pigment layers. This
paint was employed in the 'cross and stripes' decorative pattern on the northern
surfaces of the apse's dado area. It is likely that by this time the 'cross and eagle'
pattern was already overpainted, although more research is needed to verify this
theory.

It is not impossible that at least one large-scale replastering of the nave took
place during the Crusader period, the scant traces of which have been found on many
places both on the lower parts of the northern wall and over the heaven panel. It is
clear that this layer was already damaged before the first definite layer of Muslim
plaster was applied over it, and that would indicate a relatively long timespan in the
life of the crusader chapel, when neither the hell nor the heaven fresco was seen.
This overplastering could have been the result of a serious damage to the frescos
(for example the earthquake of 1202) or the fading of the upper pigments, repairs
to which was judged impossible without the original European-trained masters.

The plaster layers of the northern sacristy remain difficult to relate to the layers
in the nave with the exception of the first. However, the fresco there is certainly
one of the last crusader period decorative works in the chapel. Besides the heavily

orientalizing style, the use of the precious ultramarine pigment also sets it apart from the rest of the chapel. It would be tempting to have it painted by a local master long after the European painters of the nave were gone. If it has any relation to the fresco fragment found in one of the loopholes of the coastal guard tower, then it is likely to have been painted after 1204–1205.

Although there are still many questions to be answered, one issue concerning the possible dating of the chapel and its decoration is now clear. The former assumption that Saladin's northern campaign had an effect on the decoration of the chapel can be dismissed. First of all, the more precise dating of the charter recording the purchase puts the Hospitaller takeover one year later than originally supposed, and that would leave less than eighteen months for the construction of the chapel before Saladin's arrival. This, in view of other construction activities in the castle, is very unlikely. Moreover, as the SHAM research indicates, the decoration programme in the chapel was a multi-phase project possibly lasting for several years or even decades.

Table 2.1   Concise table of plastered layers

	Nave (eastern bay)	Apse	Pilasters	Northern Sacristy
Period 1–2	First plastered layer with high basalt ingredients, lime coating	Pointing of ashlars, Decoration around entrance of northern sacristy	Pointing of ashlars, traces of red contour lines	First plastered layer with high basalt ingredients
	Lime coating and frescos	First plastered layer, geometric pattern	Yellow background colour, red marble effect	Intermediary plastered layer with high lime ratio
Period 2–3	1. thin replastering with high lime ratio			Plaster with high lime ratio, fresco (Pentecost, Nativity)
	2. replastering with high lime ratio	Lime plaster		
Muslim period				
	3. thick replastering with high lime ratio	Lime plaster		replastering with high lime ratio

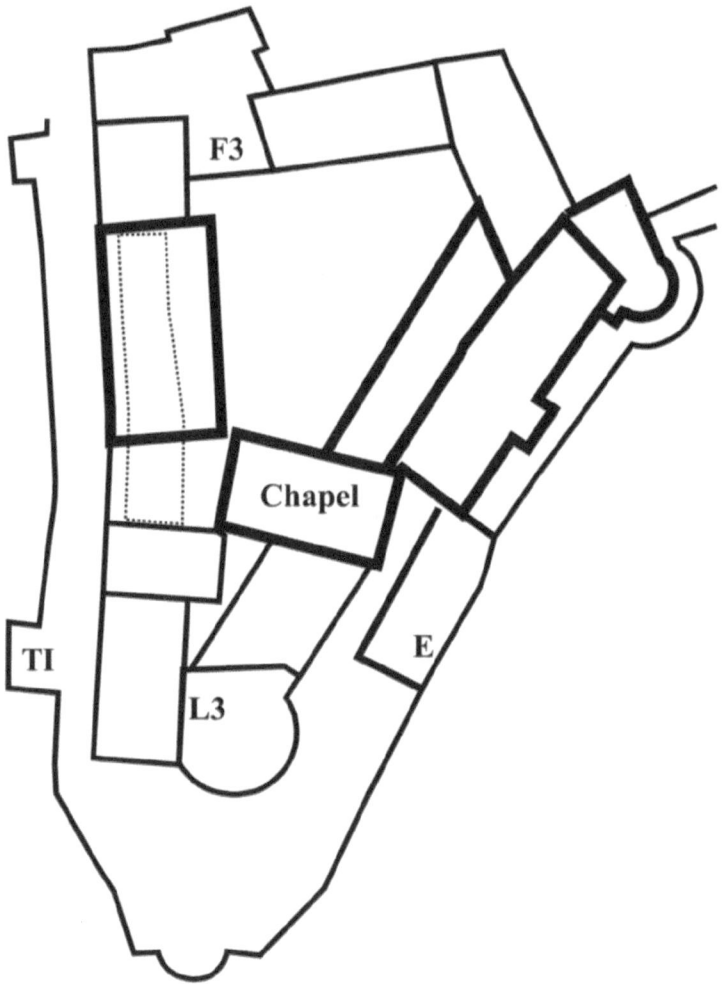

Figure 2.13   al-Marqab Citadel (graphics: B. Major)

III.2.7. Other Painted Surfaces in al-Marqab The large quantity of frescos in its chapel and also the relatively numerous examples of painted surfaces in other, non-ecclesiastical buildings make al-Marqab is a unique fortified site from the crusader period (Fig. 2.13). There are two areas where ashlar imitations painted with dark brown or black lines on a white plastered surface are preserved. One is the vault over the southern forecourt of the inner gate tower complex of the citadel (F3 on plan); the second is the vault of the early outer gate tower (TI) close to the chapel. Traces of red paint were detected in the southernmost window niche of vault E in the outer circuit of walls to the east of the chapel. Clear remains of a fragmentary decorative pattern painted with orange colour were found by the SHAM in the alcove of the apartment inserted into the haunch of the vault of the donjon (L3). Recent research has indicated that a decorative painting tradition at the site might have preceded the Hospitaller acquisition of al-Marqab. Plaster fragments with colour pigments on them were retrieved in significant quantities from soundings to the north of the chapel. In the course of investigations into the texture of the outer defensive walls in the south-west, two recycled ashlars were found with decorative painted plaster still clinging to them. The wall in which they were re-employed as filling material it likely to date from the earliest Hospitaller construction phase or could be even older.

The surveys and excavations of the year 2010 proved that the western slope of the mountain below the castle comprised a huge fortified suburb. The rescue excavations conducted in the summer of 2010 found the traces of a crusader-period chapel with an adjacent medieval cemetery. Plastered fragments retrieved from the debris of the chapel show that it once had very fine plaster with richly painted decoration. Some fragments make it clear that the chapel was re-plastered and painted at least once during its history.

Summary

The initial results of the archaeological and fresco research programme of the Syro-Hungarian Archaeological Mission have shown that the famous fortified site of al-Marqab is truly unique in terms of its fresco decoration. The findings are also giving rise to the hope that a thorough investigation supported by detailed laboratory analyses of the samples might enable us to make a much clearer reconstruction of the fresco painting programme and its subjects in the near future that might contribute considerably to our knowledge of crusader art history.

Chapter 3

The Two Hospitaller Chapter Houses at al-Marqab: A Study in Architectural Reconstruction

Gergely Buzás

The research project launched by the Syro-Hungarian Archaeological Mission (SHAM) in 2007 at al-Marqab has as one of its most important aims the exploration and identification of the functions of the individual buildings on the site. Special attention has been given to finding the most probable location of the chapter house, a basic element of every major Hospitaller castle. Earlier proposals placed the chapter house in the cross-vaulted room on the first floor of annexe J adjoining the southern side of the inner gate tower complex of the citadel.[1] This hall is labelled J2 on the standard plan of the castle. Written sources for the history of crusade-period al-Marqab, not to mention specific buildings, are scarce. However, at least one important event had to take place in the chapter hall: the famous Chapter General held sometime between 1204 and 1206.[2] This important meeting, attended by dozens of Hospitaller officials from all over the Holy Land and Europe, necessitated the existence of a large hall, and the room immediately beside the gate tower would have been far too small.

For one reason or another all the existing structures within the castle can be discounted as possible sites for the chapter house, and we have directed our attention to the area of a demolished medieval building immediately to the south of the inner gate tower complex and its annexe on the western façade of the citadel (Fig. 3.1). Here a cursory examination showed the clear outline of a large building measuring 18.6 × 37.2 m with thick walls and a walled-up doorway on its southern façade opening into the courtyard in front of the chapel. Another clear indicator for the importance of this building was a neatly carved, almost completely intact medieval window on the edge of the western façade of the former hall overlooking

---

[1] P. Deschamps, *Les châteaux des croisés en Terre Sainte, III. La défense du comté de Tripoli et de la principauté d'Antioche* (Paris, 1973), pp. 279–80; J. Mesqui, 'Qal'at al-Marqab, le château de Margat. Description archéologique', http://www.castellorient.fr/0-Accueil/indexfran.htm (accessed 30 April 2011).

[2] J. Burgtorf, 'The Military Orders in the Crusader Principality of Antioch', in K. Ciggaar and M. Metcalf (eds), *Antioch (696–1268)*, Orientalia Lovaniensia Analecta 147 (Leuven, 2006) p. 236.

Figure 3.1   The area of the chapter houses before the excavations (photo: G. Buzás)

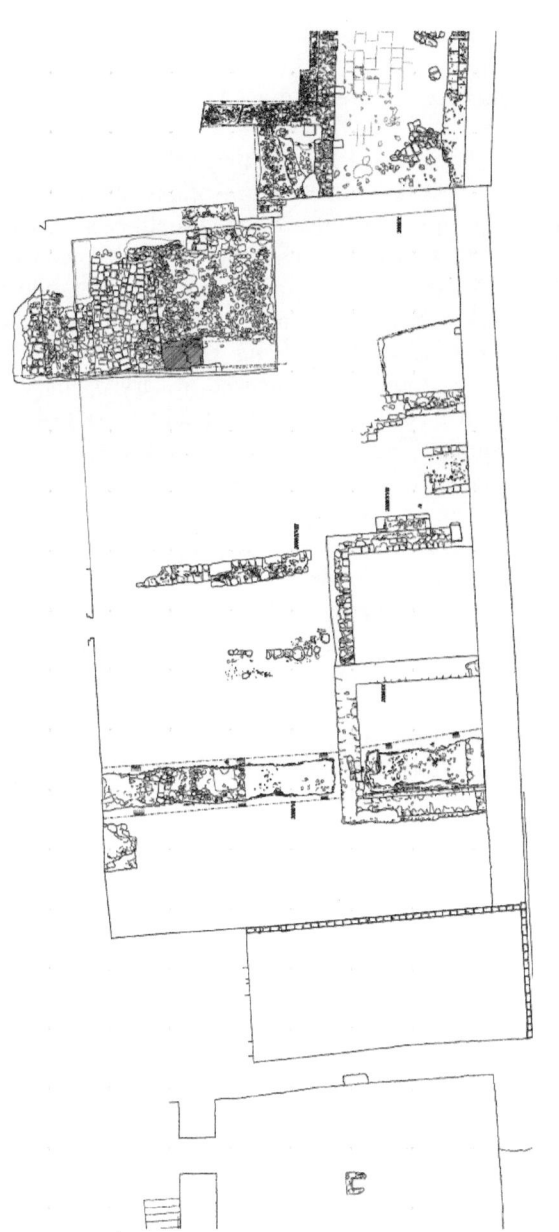

Figure 3.2   Photogrammetry plan of the area of the chapter houses showing structural remains in the excavation trenches (measured by J. Vajda and I. Győrfy) (graphics: G. Buzás)

in the sea. The presence of a huge vaulted cellar running under the area also indicated the presence of an upper structure, which was later destroyed, and an Ottoman palace, much of which is still standing, built on the site.

Archaeological Excavations in the Area of the Supposed Chapter House

Given the apparent importance of the area, the SHAM made the clarification of this building's history one of their first tasks in the excavations that started in 2007. The first excavation trench was opened in the presumed south-east corner of the building, and during 2008 and 2009 further trenches were opened to examine the inner parts of the site[3] (Fig, 3.2). The 2007 excavation immediately produced results that reinforced our hypothesis about the former existence of a building here. We discovered the south-west corner of the building, as well as the bases of two square pillars inside it (Fig. 3.3). Both bases consisted of carved limestone blocks. One of the pillars was built directly next to the southern wall, 4.90m west from the newly discovered corner, whereas the other stood free, at about the same distance from the previous pillar to the north. With two ashlar courses, the southern pillar survived to a greater height. The lower row of stones – the base – is wider, but not moulded, and measures 1.00 × 1.00m. The upper row formed part of the shaft, which measures 85 × 85cm. In the debris of the site, peppered with large stones, two carved stone elements of a pointed groin vault were unearthed. We also found the floor level of the room. A 70cm-wide stone bench running along the eastern wall was also uncovered. In front of the latter, there was a 2.30m-wide floor made of carved stones and raised one step above the general floor level of the hall. The general floor itself was made of middle-sized stones set in thick mortar. A test sounding excavated into this floor proved that it actually forms part of the cellar vault under the building and that it also extends beyond the top of the cellar's eastern wall. On the surface of this stone and mortar floor, which extends under the carved-stone raised floor, we detected the remnants of a demolished wall running north to south. In the test trench the inner surface of this stone wall also became visible, revealing that its stones survive one ashlar course in height. This approximately one-metre-wide wall did not run all the way to the southern wall. It ended close to the southern wall and was subsequently partly demolished by the 50cm-wide foundation ditch of the later wall. The mortar of the base of the older wall was clearly different from that of the later building. The remnants of these walls thus prove that this building had two distinct periods of construction. In the first period the cellar was built, and above it a building with a 12.5m inner width – somewhat more than the width of the cellar. The eastern and southern walls of

---

[3]   Excavations have been conducted in the past three years by Gergely Buzás (National Museum of Hungary), Eszter Bechtold (Historical Museum of Budapest), Balázs Major (Catholic University of Hungary), Zsolt Petkes, Zsolt Vágner and Mayssam Youssif (DGAM).

Figure 3.3   The excavation trench in the south-east corner of the later chapter
house (photo: G. Buzás)

this building were demolished to floor level, and a new large hall with pillars was
erected in its place, on a site now widened by 3.3m towards the east and 0.5m
towards the south.

The pillar bases and vaulting elements excavated in 2007 suggested a structure
consisting of three aisles with five sections, each consisting of three cross-vaulted
bays on a square ground plan. On the basis of this working hypothesis, a new
excavation trench was laid across near the northern end of the supposed hall where
we expected to find the fourth pair of pillars (counting from the south), and the
base of the east pillar of this pair was indeed uncovered. The west pillar in this area
was destroyed by a wall built over it in the Ottoman era. This discovery proved
that the second hall was a building with regular internal architectural features.

The SHAM excavations thus resulted in the recognition of two hall structures
on an exposed area of the castle, both of which are thought to be much larger than
any other building previously identified as the chapter house.

Theoretical Reconstruction of the Second Medieval Building (the Second Chapter House)

The reconstruction of this later, pillared hall is based on the results of the 2007–
2008 excavations and on the existing remains of the wall on the northern side.
The northern section of the eastern wall of this hall survives incorporated into the

Ottoman buildings. In this wall, near the north-east corner, there is the southern part of a medieval opening. The floor level of this opening is only 20cm higher than that of the Gothic-style room J2 adjoining the great hall from the north, but located one storey higher. (The rib vault of this Gothic room was once supported by two columns.) The previously mentioned opening itself is located 5.5m above the floor level of the great hall. The opening does not have a stone frame and never did have one. It was probably 1.8m wide and was topped by an arch whose springing was located at the height of 2.8m. The size and the structure of this opening suggest that it was not a window but an entrance. The height of its floor level can only be explained if we assume the existence of a gallery at the northern end of the great hall, a gallery that was also connected to the adjoining Gothic room J2 with the two columns. The existence of this gallery is also supported by the fact that traces of a walled-up door can be identified in the western area of the south wall of the hall J2 (the one that it shares with the great hall).

Because of the position of this door and the previously mentioned arched opening, one could theoretically also think that the great hall had two storeys, and that these openings led to its upper level, while only its 5.5m-high lower level would have been vaulted and had three aisles. This tempting hypothesis is, however, refuted by the fact that no springing of a vault (or traces thereof) can be observed on the inner surface of the eastern wall (this can be very well studied on its northern section that survives except for the northernmost bay which is incorporated into the eighteenth-century building), even though this medieval wall surface is intact. The excavations revealed that there were no pillars standing along the eastern wall, so if there had been vaulting it could have started only from a springing incorporated into the wall itself. Therefore, the absence of a springing in this wall section excludes the possibility of imagining this hall as a two-storey structure.

The arched opening in the east wall probably led to this gallery from a vestibule. This is suggested by the presence of a stone corbel located on the outside of the east wall of the great hall and just below the floor level of the arched opening. It must have supported the wooden floor of the vestibule. Two stone corbels that once held a wooden ceiling and are located on the southern façade of the gate hall that leads to the courtyard of the inner castle may be related to the vestibule, since the other side of the wooden ceiling held by these corbels could not but lean on the northern wall of the vestibule. On the site of the vestibule there is now the terrace of an eighteenth-century building from the Ottoman era. The function of the vestibule was apparently that it housed the stairs that led up to the arched opening of the gallery.

The remains of the arched opening provide important information not only about the destroyed vestibule and gallery but also about the interior of the hall itself. As the arched opening is located right in the corner of the hall, the vaulting of the latter cannot have started from lower than the springing height of the arch of the opening. This suggests a height of 8.3m for the springing of the vaulting of the hall, and a height of approximately 12m for its total inner height. The diameter of the cross vaults is 7m, which establishes the height of the vault as 3.5m if we hypothesize the use of round arches. The carved stone elements of the vault excavated in the chapter hall prove that precisely this kind of vaulting was used for the hall. If to this we add the thickness of the vaulting, we come to a total height of c. 12.5m, which is identical to the height of the original ceiling level of the hall J2 adjoining the second chapter hall from the north (Fig. 3.4). This ceiling of J2 was later replaced by a thirteenth-century rib vault, but its blocked putlog holes are still clearly discernible.

Another Ottoman-period building preserves a short section of the western façade of the second great hall, including a window (Fig. 3.5). The façade is faced with ashlars carved from rough limestone, and is divided by two moulded string courses. A large niche topped by a pointed arch and containing a window is located above the lower string course. The upper string course runs at the height of the springing of the arch of the niche, and turns around it above, following its arched shape. The recessed surface of the niche is also faced with limestone ashlars and pierced by the rectangular window opening, which is not moulded. The interior niche of the window is topped by a segmental arch that is lower than the outer one. The window must have been located under the hypothesized northern gallery. It is likely that this hall, which had a great inner height, also once had a second series of windows located higher up, but this is impossible to ascertain as the west façade has survived only to the level of the top of the arch of the exterior niche around the window. The bottom of the surviving window was located about one metre above the floor of the great hall. At the same height along the southern end of the western wall, two stone corbels were found. These details suggest that there could have been some sort of a wooden platform running alongside this section of the western wall, and it is not impossible that it was an elevated pulpit.

The long bench along the eastern wall and the probable wooden platform can be taken as further proof for identifying the building as the chapter house. Benches and seats running along walls are typical of monastic chapter houses. The presence of a wooden pulpit or balcony could have been used for readings or giving speeches on this side of the hall. The interpretation of the hall as a chapter house is also supported by its central location, its vicinity to the chapel, and the fact that this was far the largest room of the castle (Fig. 3.6[).

Despite the enormous size, the carefully executed limestone ashlar revetment of the exterior, and the complex interior with vaults, three aisles and a gallery, the chapter house had a rather austere appearance. The pillars supporting the groin-vaulting had a simple prism shape on a square ground plan with no moulding at their bases. Neither were the windows ornamented with moulded frames, moulded embrasures, or columns. The original appearance of the western façade may have resembled the courtyard façade of the chapter house of the Templar castle at Tartūs: the similarities include the double string course and the double series of windows (the existence of which is, however, hypothetical at al-Marqab). The lower row of the large semi-circular arched windows in the chapter house in Tartus may, however, have been more elaborately worked than those in al-Marqab

Figure 3.4   Theoretical reconstruction of the interior of the later chapter house
              (graphics: G. Buzás)

Figure 3.5   The sole surviving window of the later chapter house (photo: G. Buzás)

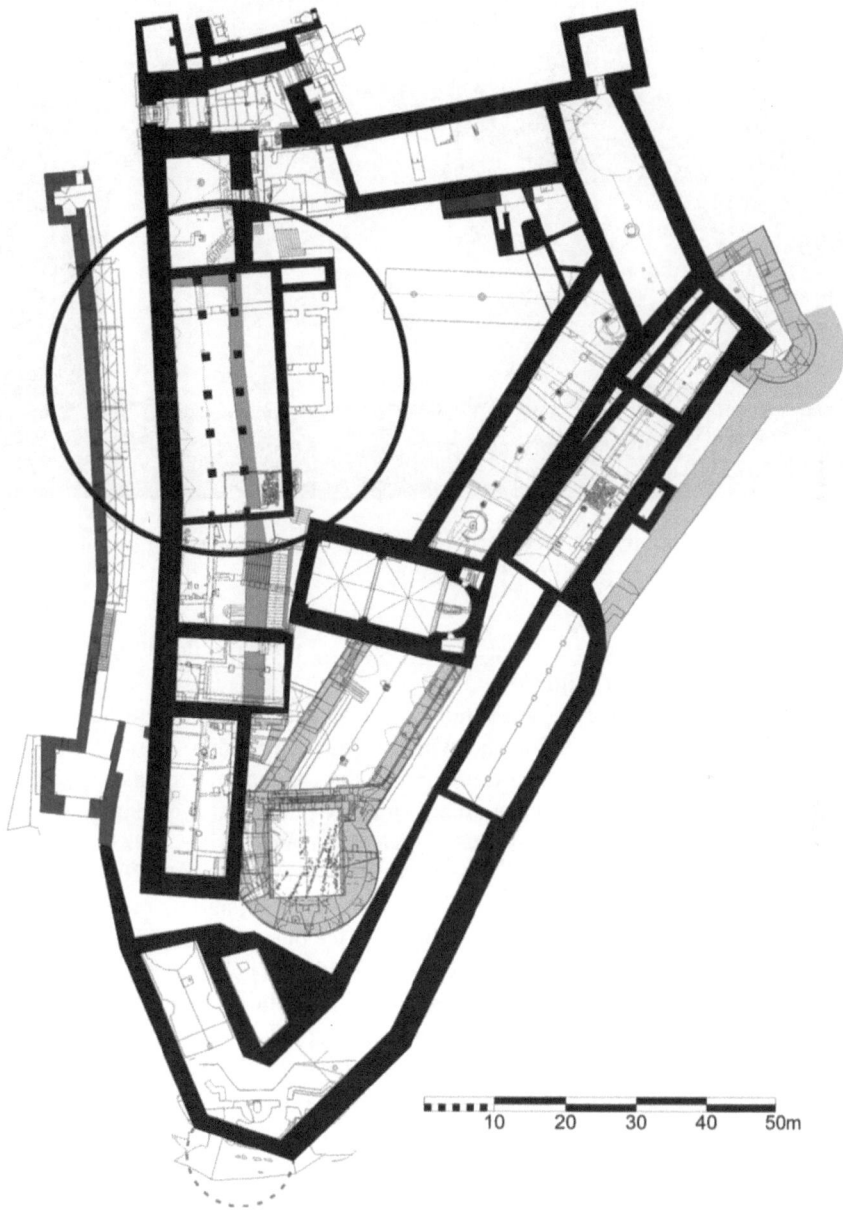

Figure 3.6   Plan of the citadel area showing the plan of the later chapter house
              in the circle (photo: G. Buzás)

Figure 3.7   Theoretical reconstruction of the western façade of the later chapter
               house (graphics: G. Buzás)

(Fig. 3.7). That, at least, is what is suggested by their moulded embrasures that
survive in a walled-up condition.

The same austere style, executed to a high standard, and the use of carved
limestone ashlars for the door and window frames are also characteristic of the
later wings built in the courtyard in front of the east and north-east ranges of the
inner castle at al-Marqab. These architectural features differ conspicuously from
the more ornamental and sophisticated style that characterizes hall J2, with its rib
vault and two columns, and which probably dates from around 1220. On a stylistic
basis the chapter house seems to pre-date this structure, and its construction may
be related to the programme of rebuilding after the earthquake in 1202[4] and to
the Chapter General held in the castle around 1204–1206. Reconstruction in
anticipation of a forthcoming Chapter General would also provide a reasonable
explanation for the hasty nature of the work executed.

The description by the English traveller Richard Pococke strongly suggests
that the chapter house was still standing at the time of his visit that had begun in
the year 1737. He saw to the west of the chapel 'a large saloon arched over, and
supported by pillars in a very magnificent manner, which might be a refectory for
the priests.'[5] As the chapter house is just to the north-west of the chapel and was

---

[4]   H.E. Mayer, 'Two Unpublished Letters of the Syrian Earthquake of 1202', in S.A.
Hanna (ed.), *Medieval and Middle Eastern Studies in Honour of A.S. Atiya* (Leiden, 1972),
p. 303.

[5]   R. Pococke, *A Description of the East, and Some Other Countries: Observations
on Palestinae or the Holy Land, Syria, Mesopotamia, Cyprus, and Candia* (London,

supported by a number of impressive pillars, it could well be that this was the building to which Pococke referred. If his description does refer to the chapter house, then his visit was literally at the twenty-fourth hour as the inscription on an Ottoman building erected in the north-eastern corner of the site of the chapter house dates its construction to 1738/39. This date would preclude any possibility that it was the famous earthquake of 1752 that destroyed the chapter house.[6] The only other building in the castle that possesses pillars (only two) is the Gothic-style room J2, but it is very unlikely that Pococke's reference was to this rather small structure. Both archaeological excavations and another inscription prove that, by the first third of the nineteenth century, the whole area of the former chapter house was occupied by an Ottoman courtyard house of which only those portions standing on the northern parts of the chapter house survive.

Theoretical Reconstruction of the First Medieval Building (First Chapter House)

As previously mentioned, the archaeological excavations also shed light on the fact that the thirteenth-century chapter house discussed above must have been preceded by a somewhat smaller building with a width of approximately fifteen metres. Since the walls of this earlier building were demolished to ground level, there is no hope of finding any of its structural stone elements on the site of the chapter house. Fortunately, however, we came across a number of carved stones dating from about the end of the twelfth century scattered all over the area of the citadel and the suburb. These stones very likely originate from the earlier chapter house. They form two closely related groups, both of which consist of stones carved of rough limestone and characterized by rich mouldings.

One is a group of 38.5cm-thick ribs that have a radius of curvature of only 60cm. They are moulded in the same way on both sides, with hollow and convex mouldings (Fig. 3.8). A radius this small suggests that these must be fragments of window frames. On the basis of their mouldings one can only think of double windows in which no panes were inserted. This type of window is typical of cloisters and chapter houses. One fragment from this group was later built into a partition wall on the upper level of the gate hall of the inner castle, which was erected during the crusader era but later than the late twelfth-century construction of the gate hall. This indicates that the fragment originated from a building demolished by the Hospitallers themselves in the thirteenth century.

1743–45), 2, p. 200.

[6]   J. Plassard and B. Kogoj, 'Catalogue des séismes ressentis au Liban', *Annales Mémoire de l'Observatoire de Ksara*, 4 (1962), 8; M.R. Sbeinati, R. Darawcheh and M. Mouty, 'The Historical Earthquakes of Syria: An Analysis of Large and Moderate Earthquakes from 1365 B.C. to 1900 A.D.', *Annals of Geophysics*, 48 (2005), 396.

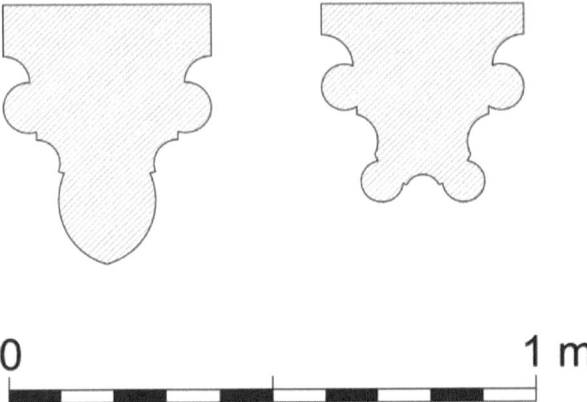

Figure 3.8   Profiles of the arch and the window frame of the early chapter house
              (graphics: G. Buzás)

Figure 3.9   Reconstructed window of the early chapter house (graphics: G. Buzás)

A further fragment that can perhaps be associated with this group of window frame fragments is a tall Attic base that once belonged to small mullion column carved of rough limestone. This came to light in 2009 in the course of excavations in the northernmost corner of the castle hill. The plinth size of this column – which usually equals that of the abacus – corresponds to the width of the lower part of the moulding of the arch. It is precisely this part of the moulding that usually rests on the mullion in the case of double windows (Fig. 3.9).

The other group of carvings consists of 39.5cm-thick rib fragments whose mouldings are related to those of the window fragments. The upper parts of the mouldings are identical to those of the windows, but the lower parts consist not of two convex mouldings but of a single large bowtel moulding with a plum-shaped cross section, without a fillet. An even more important difference is that the radius of curvature of these ribs is much larger: approximately five metres. Among the fragments belonging to this type there is also an element once belonging to the springing of an arch, in which two such ribs meet with their mouldings turned the opposite way (Fig. 3.10). This proves that these are not fragments of a vault but of an arcade. On the basis of the element of the springing one can calculate the form of the springing of the arcade that narrowed down to the bowtel moulding only, in an extremely daring way, at the base of the springing. Unfortunately, no keystones have survived and so the width of the arches cannot be calculated, but if we employ the usual proportions of the pointed arches used in the castle in the late twelfth century, we come to a width of 8m, which would establish the height of the arches as 5.5m. Yet this arcade, despite its height and large span, was only 39.5cm thick, and the base of the springing was even thinner, about 18.5cm. Such a graceful structure could in no way support a vault, only a wooden ceiling (Fig. 3.11). One can imagine this structure as being similar to the early thirteenth-century arcade that still stands in the great hall of the château at Blois.[7] Here columns aligned along the longitudinal axis of the hall support a series of arches that in turn hold up two arched wooden ceilings arranged side by side. The height of the earlier chapter house in al-Marqab was at least ten metres or even higher.

On the basis of its style, material and size, we may also link a large engaged capital of which only the bottom 45cm-high part survives to this structure (Fig. 3.12). This capital and another one like it must have supported each end of the arcade where it joined the walls. This must have been a Gothic capital ornamented with sedge leaves ending in buds, with six-lobed palmette-like leaves between the sedge leaves. The form of the bottom of this capital reveals that it once stood on a pilaster or – even more likely – on a corbel with a square ground plan. The surviving part of the capital is 37cm wide, so its original width roughly corresponds to the width of 39.5cm of the early Gothic ribs.

In conclusion we can state that the early Gothic group of carvings from the end of the twelfth century seems to have belonged to a large room whose wooden

---

[7]   J. Mesqui, *Châteaux et enceintes de la France médiévale. Da la défense à la résidence* (Paris, 1993), 2, pp. 78–9.

Figure 3.10   Fragment of the central arch supporting the ceiling of the early
               chapter house (photo: G. Buzás)

Figure 3.11   Reconstructed interior of the early chapter house (graphics: G. Buzás)

ceiling was supported by an arcade resting on columns and (probably) corbels and was lit through elaborate double windows. In the light of our present knowledge, such a large twelfth-century room with an airy interior and constructed to such high standards could only be the castle chapter house. This identification is also supported by what is known of the early thirteenth-century demolition of the room and the secondary use of some elements from this group of carvings in the thirteenth century.

We can reconstruct the interior of the earlier chapter house as a hall with three columns and two aisles, with the columns aligned on top of the pointed arched barrel vault of the cellar below. Four air vents from the cellar opened into the floor of the hall between the columns (Fig. 3.13). The ground plan of the interior of the hall measured 12.5 × 34.3m, and its height was probably 12.5m. As far as the size of the two chapter houses in al-Marqab is concerned, the two ground plans of 12.5 × 34.3m and 15.8 × 34.3m are comparable to the middle-sized French great halls that measure approximately 10–15 × 30–35 m. With its 18 × 30m size, the great hall at Blois – which resembles the first chapter house of al-Marqab not only in size but in structural form as well – also belongs to this group. The arcade that supported the ceiling must have been a graceful and daring structure, without any support from the sides. This may have actually caused its collapse. The 1202 earthquake resulted in east–west movements[8] – as one can observe on the basis of the cracks in the donjon – so the arcade in the chapter house would have been affected by it in the most sensitive way.

Summary

The analysis of the archaeological excavations and the surviving stone carvings found at al-Marqab have made it possible to identify an earlier and a later chapter house and attempt a theoretical reconstruction of these very important but entirely destroyed medieval interiors. The history and the reconstruction of these halls help us understand the construction history of the castle better and allow us to appreciate the importance of the role it played in the history of the Knights Hospitallers in the Holy Land.

---

[8]   M. Kázmér and B. Major, 'Distinguishing Damages from Two Earthquakes: Archaeoscizmology of a Crusader Castle (al-Marqab citadel, Syria)', *The Geological Society of America Special Paper*, 471 (2010), 185–98.

Figure 3.12   Console fragment of the early chapter house (photo: G. Buzás)

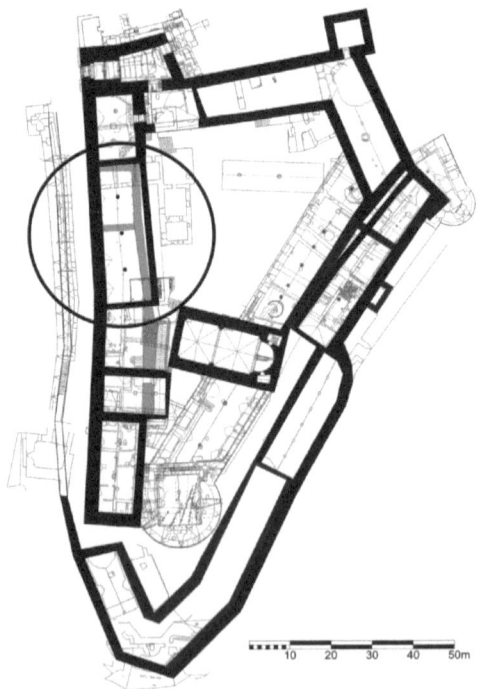

Figure 3.13   Plan of citadel area showing the plan of early chapter house in the
               circle (graphics: G. Buzás)

Chapter 4

Meat Consumption and Animal Keeping in the Citadel at al-Marqab: A Preliminary Report[1]

István Kováts

Introduction

Extensive fieldwork surveys and associated archaeological documentation have been carried out in the coastal zone of Syria since the year 2000 by the Syro-Hungarian Archaeological Mission (SHAM) under the direction of Dr Balázs Major. The area of investigation first attracted attention in the West at the end of the eleventh century, when European crusaders arrived in great numbers. Much of the coastal zone of Syria was subsequently controlled by the Hospitallers and Templars, who built the largest castles – notably Crac des Chevaliers, al-Marqab, Tartūs and Safita – to defend their territories.

In 2007 the SHAM launched a new multi-disciplinary research project at al-Marqab, the crusader *Margat*, the most important fortress in the area which extends over 5.7 hectares of an eroded volcanic cone. This joint Syrian–Hungarian enterprise is directed by Edmond el-Ajji (DGAM) on the Syrian side and by Balázs Major (Catholic University of Hungary) on the Hungarian. Significant discoveries were made during the excavation seasons in 2007, 2008 and 2009,[2] and these shed new light on the military architecture, material culture and daily life at this site during the crusader period. Environmental and scientific research[3] has also been conducted within the framework of the Archaeological Mission. Archaeozoological studies, aimed at the identification, analysis and interpretation

---

[1] This text of this paper was reviewed and translated by László Bartosiewicz (Institute of Archaeological Sciences, Eötvös Loránd University, Budapest), whose help is acknowledged here.

[2] See the papers by Balázs Major and Éva Galambos and by Gergely Buzás in this volume.

[3] É. Galambos, 'A pokol pigmentjei', *Élet és Tudomány* (13.03.2009), 401–3; M. Kázmér, 'Earthquake Damages in al-Marqab Citadel, Syria', in Á. Török and B. Vásárhelyi (eds), *Mérnökgeológia – Kőzetmechanika* (Budapest, 2008), pp. 159–64; M. Kázmér, 'Földrengés-sebek Margatban', *Élet és Tudomány* (11.03.2009), 337–9.

of animal remains from both a natural and a cultural point of view have formed an important part of these investigations.

In addition to their significance as archaeological artefacts, animal remains recovered from archaeological sites can offer information on diet, the environment, and the roles played by different animals during the period in question. Numerous observations can be made concerning the non-dietary use of animals, e.g. in relation to craft activities. It is important to remember that, from a human point of view, it was the larger creatures of the so-called macrofauna, including domesticated animals that were of most direct relevance. Most of their remains ended up in archaeological deposits as a result of conscious human activity, and they therefore give only a selective representation of the fauna existing on the site.[4]

A great quantity of animal bones has been brought to light in and around the domestic quarters, kitchens and latrines at al-Marqab during the course of the excavations. The evaluation of this rich material is still in progress, and this report provides only a brief summary of archaeozoological finds from a particular section of the excavated area. It is important to note that the proportions of material tabulated below relate to the animal remains embedded in a single fill recovered from just one well-defined area of the castle: they do not represent material from all periods or from all parts of the site.

The Site

The large inner courtyard at al-Marqab is surrounded by some of the most important buildings of the castle. The eastern side of the courtyard is defined by two vaulted buildings running parallel to each other. The outer building is actually made up of two storeys, the lower one possibly being a pre-Hospitaller construction, and the upper one, with a substantial latrine tower attached to it, having been built in all probability as a large dormitory for the Hospitallers. The next main construction programme resulted in the erection of a new corner tower at the northern end of the supposed dormitory and the addition of a new vault attached to the dormitory on the west. It seems that residential buildings with wooden roofs were, from the earliest time, located on top of the dormitory hall and were later several times expanded. These installations included dwellings, a kitchen, baths, latrines and a sewage system that was constantly being improved during the thirteenth century. After a period of destruction, possibly the Mamluk siege and takeover of 1285, a further renovation of this eastern wing was carried out in the later part of the century. However, this area seems not to have been used in the Ottoman period which began in the early sixteenth century. The rich assemblage of artefacts brought to light by our excavations comprise the utilitarian

---

[4]   C. Renfrew and P. Bahn, *Régészet. Elmélet, módszer, gyakorlat* (Budapest, 1999), p. 234.

Figure 4.1   The excavation area of S3 and I2 with the remains of domestic
structures at the end of the rescue excavation of the Syro-Hungarian
Archaeological Mission in 2007 (photo: B. Major)

objects – pottery, coins, iron tools – dating from the occupation by the Order of St
John and then the Mamluks[5] (Fig. 4.1).

The Quantity and Identity of the Animal Remains

The site yielded a total of 1,541 animal bone fragments. These were recovered
in a generally well-preserved state. Only 243 fragments could not be identified
beyond the general small/large animal category. The 'large mammal' category
included cattle, horse and red deer, while species of the size of a pig, sheep or
larger dog were classified as 'small mammals'. In the case of bird and fish remains,
only preliminary identifications could be made. The number of completely non-
identifiable bone fragments totalled fifty-two.

The relatively good preservation of animal bones is related to the collection of
finds by hand: small pieces of large bone as well as the remains of the so-called

[5]   Excavation of the area in the autumn of 2007 was conducted by Balázs Major
(Catholic University of Hungary), Zsolt Vágner, Zsolt Petkes, Gergely Buzás (National
Museum of Hungary) and 'Afīf Maarūf (DGAM). Special thanks are due to Gergely
Buzás and Balázs Major, who provided the architectural description and archaeological
stratigraphy of the investigated area.

microfauna can only be properly recovered using fine methods of recovery.[6] The lack of screening and/or water-sieving favours the selective recovery of clearly visible, large and well-preserved bones. As the excavation on the S3 was a rescue excavation, there was no opportunity for the time-consuming water-sieving.[7]

The overwhelming majority of identifiable bones originated from domestic mammals. Bone fragments of sheep and/or goat were especially numerous (80.4 per cent of the assemblage), showing the unambiguous dominance of meat from these animals in the diet for the period and area of the castle under review. It may be presumed that most of the bones representing non-identifiable 'small mammals' also came from these two small ruminant species. In Table 4.1 sheep and goat are included under the term *Caprinae*, as heavily fragmented bones do not usually allow for a distinction between these two species. A distinction was possible, however, when diagnostic elements of the skull (sutures, cross-section of the horn core) and distal limb segments, especially metapodia, were encountered.

A reasonable amount of bones of cattle were also found, while other domestic animals were represented by only sporadically occurring remains. In the case of birds, the presence of the domestic hen could be unambiguously observed (4.58 per cent of all identifiable remains). Twenty-two fragments of avian bone could not be further identified in the absence of diagnostic morphological features. The taxonomic composition and percentual proportions of the assemblage are listed in Table 4.1.

Table 4.1   The taxonomic distribution of animal remains recovered from the studied wing of the building

Species	No.	%
Cattle *Bos taurus* L. 1758	83	6.82
Sheep *Ovis aries* L. 1758	54	4.44
Goat *Capra hircus* L. 1758	18	1.48
Sheep/Goat Caprinae Gray 1852	978	80.40
Pig *Sus domesticus* Erxl.1777	15	1.23
Horse *Equus caballus* L. 1758	4	0.30
Dog *Canis familiaris* L. 1758	1	0.08
Domestic Hen *Gallus domesticus* L. 1758	59	4.85
Domestic Animals	**1212**	**99.70**
Red Deer *Cervus elaphus* L. 1758	1	0.08
Deer family *Cervidae*	3	0.22
Wild Animals	**4**	**0.30**

---

[6]   L. Bartosiewicz, *Régenvolt háziállatok. Bevezetés a régészeti állattanba* (Budapest, 2006), p. 76.

[7]   At this site finds were collected by hand except from the latrines, where samples were screened and water-sieved. These methods of fine recovery yielded impressive quantities of small vertebrate remains including those of rodents, birds and small fish.

Species	No.	%
Identifiable	*1216*	*100.00*
Mammals (Large size) *Mammalia indet.*	55	
Mammals (Small size) *Mammalia indet.*	188	
Birds *Aves sp.*	22	
Fish *Pisces sp.*	3	
Shell *Bivalvia*	5	
Non-identifiable bone fragment *Ossa indet.*	**52**	
Total	**1541**	

The relative frequency of identifiable remains illustrates the great importance of sheep and goat meat. Their proportion is notably large by comparison with other economically important species such as cattle or pig.

Brief Description by Animal Species

*Sheep and goat (*Ovis aries *L.1758 and* Capra hircus *L.1758, Caprinae subfamily)*

Due to their sheer numbers, the remains of sheep and goat carry most information. It is likely that the chief form of exploitation was their butchery for meat. This is also illustrated by the observation that most of the caprine remains recovered in this part of the site represent good quality meat, from body regions designated as categories A (vertebrae, scapula and pelvis, proximal limb segments) and B (ribs and lower limb bones) in the system developed by Hans-Peter Uerpmann[8] (Fig. 4.2) The remains of meat-purpose animals are usually poorly preserved, partly due to intentional butchery. During the Middle Ages powerful metal blades were available whose cut marks may be recognized on even fragmented material. Evidence of perpendicular and oblique hacking, inflicted using larger tools, are visible on caprine vertebrae (Fig. 4.3). These may have been left by cleavers or even large knives, the marks perhaps attributed to primary butchering when the carcass is disarticulated for sale and or shipment. Cut surfaces are usually smooth, with clearly defined edges that are indicative of the use of sharp blades. Cut marks made by smaller blades such as those of ordinary knives may be interpreted with greater probability as evidence of secondary butchering, i.e. skinning, further carcass partitioning in the kitchen, food preparation and consumption. Such cut marks include consistently occurring, fine perpendicular cuts across both ends of several caprine rib segments.

---

[8]   H.P. Uerpmann, 'Animal Bone Finds and Economic Archaeology: A Critical Study of "Osteo-archaeology" Method', *World Archaeology*, 4 (1973), 307–22.

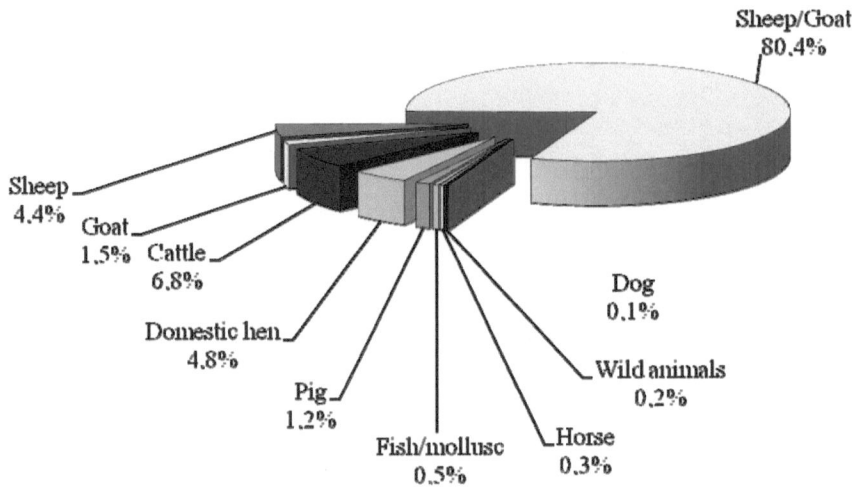

Figure 4.2   Domestic sheep (Ovis aries L.1758) humerus fragments in posterior view (photo: I. Kováts)

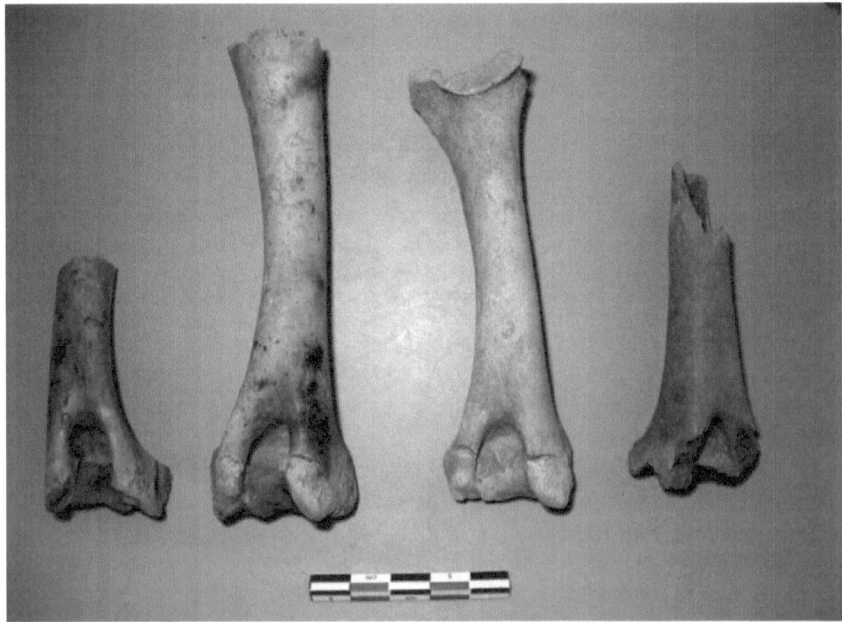

Figure 4.3   Domestic sheep (Ovis aries L.1758) humerus fragments in posterior view (photo: I. Kováts)

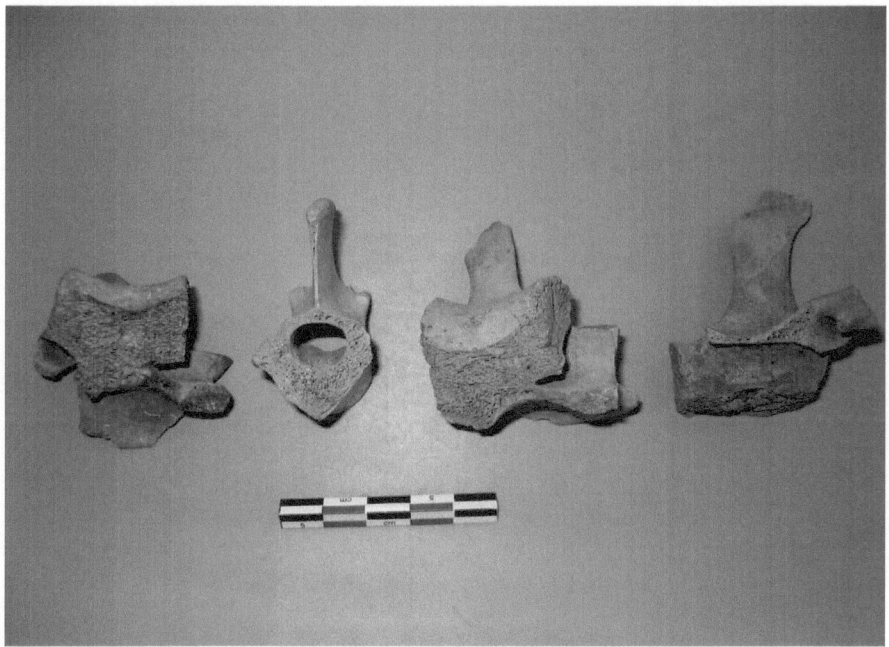

Figure 4.4   Chop   marks   on   domestic   sheep   (Ovis   aries   L.1758)
vertebrae fragments (photo: I. Kováts)

In addition to primary exploitation for meat and fat, however, female sheep
and goats may also have been milked throughout their lives, and the use of wool
should also be kept in mind.[9]

*Cattle (*Bos taurus *L.1758)*

Cattle bones made up only 6.82 per cent of the identifiable remains, including small
fragments, many of which would have resulted from post-mortem fragmentation.[10]
This proportion may be considered quite low, considering that cattle may be
exploited in a number of ways including milking and draught work. The small
number of heavily fragmented cattle bones does not allow for far-reaching
interpretations. Practically no bones survived in a complete or measurable state.
The best preserved skeletal elements include dry limb bones poor in meat, such as
phalanges and a few metapodium fragments. The absence of complete limb bones
and of horn cores among the food remains should be noted, and it was not possible
to distinguish the remains of cows, bulls and oxen.

---

[9]   Bartosiewicz, *Régenvolt háziállatok*, pp. 96–100.
[10]   L. Bartosiewicz, 'Faunal Material from Two Halstatt Period Settlements in
Slovenia', *Arheoloski Vestnik*, 42 (1991), 199–205.

The general appearance of medieval cattle in the eastern Mediterranean region can be imagined on the basis of the animals shown in Byzantine mosaics and illuminated manuscripts. These iconographic sources usually show short-horned, *brachyceros* types, and some look like light beasts of small withers in height and gracile skeletal makeup, a type of cattle widespread in Syria and Turkey to this day.[11]

*Pig (*Sus domesticus *Erxl.1777)*

A total of fifteen pig bones could be identified in the assemblage. This number is far too small for any conclusions to be drawn beyond the mere consumption of pork. This does, however, raise a rather important aspect of interpreting this assemblage, as pig bones would only be expected in occupation layers from the period of European control. In this case archaeozoology can provide useful identification data for periodization.

*Horse (*Equus caballus *L.1758)*

Horses contributed minimally (0.4 per cent) to the animal bone assemblage. None of the remains survives in a measurable state. The small number of bones in itself shows that these finds are unlikely to have derived from food consumed in the eastern wing of the building. In 732 in Christian Europe, Pope Gregory III had, by papal edict, banned the consumption of horse meat. Although the prophet Muhammad refrained from eating horses, he did not ban this custom, although later Islamic groups frowned on it.[12]

*Dog (*Canis familiaris *L.1758)*

This species is represented only by a single poorly preserved mandible fragment. The near complete absence of canine remains, of course, may be explained by the fact that, as these animals played no role in the diet, their bones would only be mixed with food refuse in extraordinary circumstances. However, the presence of dogs in the castle is shown by marks caused by dogs gnawing the bones of other animals recovered at this site.

---

[11]   Z. Kádár and A. Tóth, *Az egyszarvú és egyéb állatfajták Bizáncban* (Budapest, 2000), pp. 41–2.

[12]   L. Bartosiewicz, 'Gondolatok a "lovas nomád" hagyományról' – 'Thoughts about the "Equestrian Nomadic" Tradition', in L. Bartosiewicz, E. Gál and I. Kováts (eds), *Csontvázak a szekrényből. Válogatott tanulmányok a magyar archaeozoológusok Visegrádi Találkozóinak Anyagából. 2002–2009. Skeletons from the Cupboard: Selected Studies from the Visegrád Meetings of Hungarian Archeozoologists 2002–2009* (Budapest, 2009), pp. 73–83.

*Red deer and other cervids (*Cervus elaphus *L. 1758, Cervidae)*

The remains of hunted animals are almost completely missing from the material under discussion here. The few bones from game are fragmented and only one of them can be attributed to a red deer. In the other cases not even the size of the animal can be readily appraised: these cervid remains may equally originate from small red deer, fallow deer (*Dama dama* L. 1758) and large roe deer (*Capreolus capreolus* L. 1758). The remarkable absence of bones from game animals among the food refuse is of interest in itself, as the Hospitaller statutes explicitly banned both hunting and falconry.[13]

*Domestic hen (*Gallus domesticus *L.1758)*

Among the bird remains, the bones of the domestic hen could be clearly identified. These bones are small and fragile. On some of the long bones cut marks left by either food preparation or during defleshing at the table could be recognized. Although the primary form of exploitation for hens must have been meat, the egg shells found in large quantities in various parts of the castle also show the importance of eggs in the diet.

Bone Manufacturing

Marks left by occasional bone manufacturing and use could be detected on only a few skeletal elements from caprines and cattle. Typically holes had been drilled or cut at either articular end of metapodium or radius fragments. A heavily worn diaphysis fragment from a sheep or goat is indicative of its use in some form of smoothing activity. None of the modifications, however, seems to result from any planned and patterned bone manufacturing. Most of the raw material had been extracted from the food refuse.

Summary

It may be stated that the overwhelming majority of animal bones recovered from the studied section of the site represent food refuse originating from domestic animals. The archaeozoological remains are dominated by the bones of sheep and goat. Hardly any evidence of wild mammals could be identified. Similarly, fish seems to be under-represented in the material, although its absence may be related

---

[13]   'Concerning hunting and hawking: hunting and hawking are forbidden both this side the sea and beyond' (statutes of Hugh Revel (1258–77), no. 41). E.J. King, *The Rule Statutes and Customs of the Hospitallers 1099–1310* (London, 1934), p. 64. I am most grateful to Balázs Major for drawing this reference to my attention.

to inadequate methods of recovery in the course of a rescue excavation. Although manufactured bone was found in the form of ad hoc implements, no evidence of systematic bone working could be identified. The results briefly summarized in this preliminary report are, by definition, far from final. Reconstructing medieval animal exploitation at al-Marqab with regard to diet, food preferences and meat consumption in general can only be fully accomplished after the complete analysis of the finds. In addition to the mammal bones reviewed in this report, it will also be necessary to study bird and fish remains found at the site in depth. Analyses of the microfauna (small vertebrates such as rodents) will further contribute to a reconstruction of the natural environment of the fortress in the medieval centuries.

Addendum

Ongoing excavations resulting in the recovery of further substantial quantities of bones and their identification seem to indicate that our impression of the ratio of the types of animals consumed and the overall picture of animal husbandry may change considerably. Excavation trench 2010/XXXVI opened in the central courtyard of the castle yielded enormous quantities of pig bones (*Sus domesticus* Erxl.1777), and that would point to the fact that, although the proportion of pigs is still lower than the other animals kept for meat, pork consumption was markedly present in al-Marqab during the crusader period. Pigs were mainly kept for their meat, and that is supported by the fact that their remains were mostly found in the context of kitchen waste.

Chapter 5

The Order of St Thomas of Canterbury in Acre

Denys Pringle

The Order of St Thomas of Acre had its beginnings in a chapel and cemetery established outside the walls of Acre during the siege that resulted in the city's recapture by the Franks in 1191.[1] According to Ralph of Diceto, dean of London, its founder was his own chaplain, William, who also became the first prior.[2] Ralph's account is repeated in a shortened form by Roger of Wendover[3] and by Matthew Paris.[4] The Annals of Dunstable Priory, however, attribute the foundation to Hubert Walter, bishop of Salisbury and later archbishop of Canterbury,[5] while Matthew Paris himself later credited King Richard I with both pledging to establish a chapel to St Thomas in Acre while he was still at sea and with putting the promise into effect when he was at Jaffa in 1192.[6] As Alan Forey has pointed out, it is possible that all three had a hand in the foundation, though Richard's was to be the part more often recalled by posterity.[7]

After the capture of the city, the house evidently moved inside the walls (see Fig. 5.1). On 10 February 1192, King Guy confirmed the German hospital's possession of some land in Acre on which their houses and hospital had been built. The description of this indicates that the German house lay just inside the east wall of the city, on the north side of the street leading to the gate of St Nicolas.[8] On the east and south its boundary extended

---

[1]   A. Forey, 'The Military Order of St. Thomas of Acre', *English Historical Review*, 92 (1977), 481–503, repr. *Military Orders and Crusades*, Variorum Collected Studies Series (Aldershot, 1994), xii; D. Pringle, *The Churches of the Crusader Kingdom of Jerusalem: A Corpus* (Cambridge, 1993–2009), 4, pp. 161–4.

[2]   *Opera Historica*, ed. W. Stubbs, RS 68 (London, 1876), 2, pp. 80–81, trans. T.A. Archer, *The Crusade of Richard I, 1189–92* (London, 1912), pp. 114–15.

[3]   *Flores Historiarum*, ed. H.G. Hewlett, RS 84 (London, 1890), 1, pp. 178–9.

[4]   *Chronica Majora*, ed. H.R. Luard, RS 57 (London, 1872–84), 2, p. 360.

[5]   'Annales Prioratus de Dunstaplia', in *Annales Monastici*, ed. H.R. Luard, RS 36 (London, 1864–69), 3, p. 126.

[6]   *Historia Anglorum*, ed. F. Madden, RS 44 (London, 1866–69), 2, pp. 14, 38.

[7]   Forey, 'Military Order of St. Thomas', pp. 481–2.

[8]   On the German hospital, see Pringle, *Churches*, 4, pp. 131–6, no. 425.

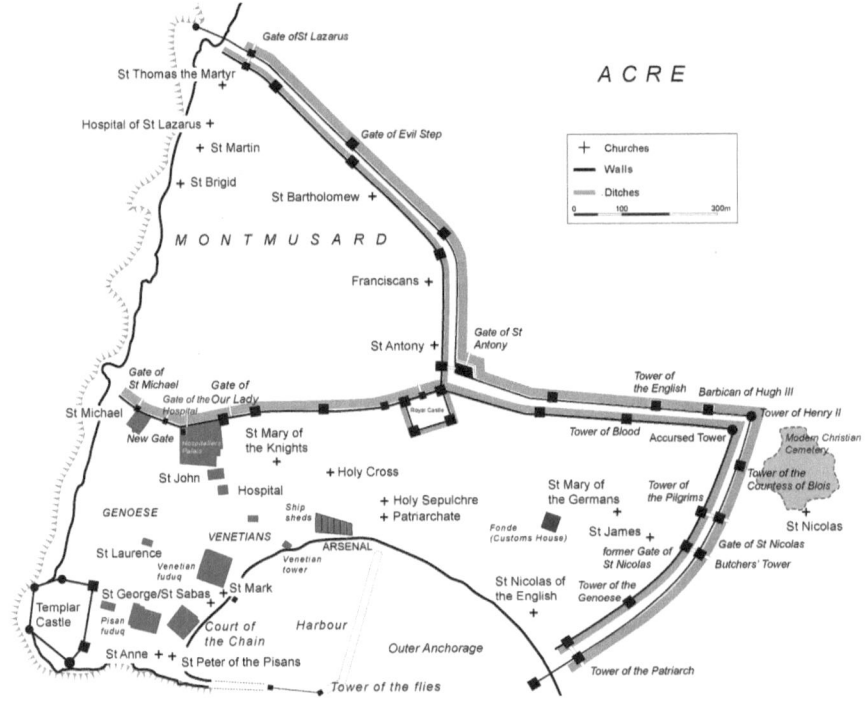

Figure 5.1   Acre: plan of the city showing the suggested locations of the churches mentioned (compiled by Denys Pringle, drawn by Ian Dennis)

from the two stairs of the gate-tower (*turris perforata*), such that the steps remain outside your [i.e. the Germans'] enclosure towards the tower, along the boundary between you and the brothers of St Thomas, up to the public street that extends to the gate of St Nicolas.[9]

This suggests that the property of the brothers of St Thomas lay immediately inside the gate of St Nicolas, on the north side of the street between the city wall and the hospital of the Germans. In February 1193, however, Henry of Champagne and Queen Isabella granted the German hospital a vault next to the gate of St Nicolas.[10] This sounds suspiciously like the property, or a part of it, that the year before had been in the hands of the brothers of St Thomas. Even if that were the case, however, it remains likely that the hospital of St Thomas was located somewhere close by.

———

   [9]   *Tabulae ordinis Theutonici:*, ed. E. Strehlke (Berlin, 1869), repr. with preface by H.E. Mayer (Toronto, 1975), pp. 23–4, no. 27; *RRH*, no. 701.
   [10]   Strehlke, p. 25, no. 29; *RRH*, no. 710.

It appears that, in the somewhat chaotic period of Frankish resettlement in Acre immediately after the fall of the city, when in the general free-for-all King Philip II Augustus of France and King Richard I of England had to intervene on behalf of burgesses whose property had been occupied by newcomers,[11] the new English house of St Thomas acquired a number of churches and other properties in the city that had been left vacant by their previous owners. The *Liber Censuum* of the Church of Rome, compiled in 1192, for example, lists the hospital of St Thomas the Martyr along with St George of the Portico (*de Sisto*).[12] This was evidently a combined house, for on 10 January 1208, Pope Innocent III wrote to the prior and brothers of St Thomas and St George *de Xisto* confirming their privileges. These included the churches of St Mary, St Peter and St Nicolas of the Field of the English (*de Campo Anglorum*), the hospital of St Thomas, and other possessions.[13] The identity of these churches has always been something of a puzzle. The purpose of this paper is to propose some identifications and to suggest how the Order of St Thomas may have managed its property portfolio in Acre in the following years.[14]

There is good reason to believe that the church of St George of the Portico was the church later known as St Sabas. This was located in the quarter of the Chain, near the harbour, and it was a house belonging to it in the same area that became the centre of the dispute that resulted in the 'War of St Sabas' between the Venetians and the Genoese between 1256 and 1258.[15] Now in his confirmation of 1208, Pope Innocent III also upheld the claim made by the prior of the combined house to a privilege that had been accorded to Simon, prior of St George of the Portico (*Sancti Georgii de Xisto*) during the pontificate of Alexander III (1159–81), allowing him the use of an episcopal ring, pastoral staff and mitre.[16] This right was subsequently contested by the bishop of Acre, whose objection was sustained in a letter sent by Innocent III on 25 October 1212. This letter also makes it clear that St George's was subject to the church of St Thomas and governed by the same prior.[17] It is uncertain why Prior Simon had been accorded such a privilege in the

---

[11]   *Eracles*, pp. 175–6; *Gesta Regis Henrici II et Ricardi I*, ed. W. Stubbs, RS 49 (London, 1867), 2, p. 181.

[12]   *Le Liber Censuum de l'Église romaine*, eds P. Fabre, L. Duchesne and G. Mollat (Paris, 1889–1952), 1, p. 238.

[13]   *Regesta, PL*, 215, cols. 1525–6, no. 209.

[14]   Further details on all these churches may now be found in Pringle, *Churches*, 4, pp. 79–80, no. 407 (St George), pp. 136–7, no. 426 (St Mary), pp. 155–5, no. 439 (St Nicolas), pp. 157–8, no. 441 (St Peter), pp. 161–4, no. 447 (St Thomas).

[15]   D. Jacoby, 'Crusader Acre in the Thirteenth Century: Urban Layout and Topography, *Studi medievali*, 3s., 20 (1979), 1–45 at pp. 15–19, 28; idem, 'Three Notes on Crusader Acre', *Zeitschrift des Deutschen Palästina-Vereins*, 109 (1993), 83–96 at pp. 83–8; cf. J. Prawer, *Histoire du royaume latin de Jérusalem*, 2nd edn (Paris, 1975), 2, pp. 359–73.

[16]   *Regesta, PL*, 215, cols. 1525–6, no. 209.

[17]   Ibid., 217, cols. 706–7, no. 185; R. Hiestand, *Vorarbeiten zum Oriens Pontificius, 3: Papsturkunden für Kirchen im Heiligen Lande*, Abhandlungen der Akademie der

first place; but one possible explanation is that St George's had originally been an Orthodox house and before 1187 its prior had been acting effectively as bishop to the Orthodox community in the city.

Support for the idea that St George's had been an Orthodox house is also provided by the Life of the Serbian monk St Sava, son of Stephen Nemanja, who passed through Acre on his way to Jerusalem in 1229. On his way home the following year, he lodged in Acre in a house belonging to the Latin church of St Nicolas. Subsequently he purchased the Latin monastery of St George, and gave it to the Orthodox monastery of St Sabas, near Jerusalem.[18] In 1234, Sava revisited Palestine and spent some time in the Orthodox house that had by then been established in Acre in the monastery that he had bought.[19] Now the church of St Nicolas in which Sava had stayed on his second visit to Acre seems unlikely to have been the cemetery chapel of St Nicolas situated outside the city walls.[20] More probably it was the other church of St Nicolas, which was mentioned as belonging to the Order of St Thomas in 1208. The location of that church and of the Field of the English in which it evidently stood are uncertain, though an inventory of Genoese properties in Tyre and Acre compiled in May 1250 does refer to a church of St Nicolas in the eastern part of the walled city, adjoining a Genoese garden close to the sea.[21] This is also relatively close to the part of the city in which the hospital had first settled. It would therefore seem probable that it was from the hospital of St Thomas, the owners of the church of St Nicolas, that Sava purchased the monastery of St George, which subsequently became St Sabas.

No further details can be provided of the church of St Peter that the Order also possessed in 1208, save to say that it was evidently not the Pisan church of St Peter, which lay in their quarter in the southern part of the city.[22] By process of elimination, however, the church of St Mary may perhaps be identifiable with St Mary of the Knights, in which the canons of Nazareth were installed by 1256.[23] The archbishop and chapter of Nazareth had been resident in Acre from 1198,

---

Wissenschaften in Göttingen, Phil.-hist. Klasse, 3s., 136, p. 394, no. 199; Forey, 'Military Order of St. Thomas', pp. 485–7.

[18]   On the monastery of St Sabas, near Jerusalem, see Pringle, *Churches*, 2, pp. 258–66, nos. 216–17.

[19]   Domentijan, *Vita S. Savae*, ed. Dj. Daničić, *Život svetoga Simeuna i svetoga Save* (Belgrade, 1865), p. 302; Jacoby, 'Three Notes', pp. 83–4; idem, 'New Venetian Evidence on Crusader Acre', in P. Edbury and J. Phillips (eds), *The Experience of Crusading*, 2: *Defining the Kingdom* (Cambridge, 2003), pp. 240–56 at p. 246.

[20]   *Pace* Jacoby, 'Three Notes', p. 84. On that church and cemetery, see Pringle, *Churches*, 4, pp. 151–5, no. 438.

[21]   'Quatre titres des propriétés des gênois à Acre et à Tyre', ed. C. Desimoni, *AOL*, 2.2 (1884), 213–30 at pp. 222–4, no. 3; *RRH*, no. 1187; cf. Desimoni, p. 219, no. 2.

[22]   Pringle, *Churches*, 4, pp. 156–7, no. 440.

[23]   *Les Registres d'Alexandre IV*, eds C. Bourel de la Roncière *et al.* (Paris, 1902–1959), no. 1300, cf. no. 1330; B. Hamilton, *The Latin Church in the Crusader States: The Secular Church* (London, 1980), p. 95.

if not before, but we do not know precisely where. Following a brief return to Nazareth between 1241 and 1255, they finally returned to Acre, selling the remaining temporalities of the diocese to the Hospitallers in 1259.[24] If the knights of the church's designation were indeed the knights of St Thomas, it could be hypothesized that the archbishop would have acquired it from them after the militarization of the canons, which occurred in 1227–28.

Although they received various donations in England, the canons of St Thomas remained poorly endowed. A letter of Pope Gregory IX explains that this was because King Richard had died before he could make proper provision for them.[25] While in Acre in 1227–28, however, Peter des Roches, bishop of Winchester, noting the unsuitability of the house's location and the general laxity of the canons, reformed them as a military order with a rule based on that of the Teutonic Order. He also transferred their house from the old city to a new location beside the sea in Montmusard, a developing suburb on the northern side of the city, which had been provided with a double wall by 1211–12.[26] In doing this he was advised by the patriarch and the magnates of the kingdom and he himself donated some 500 marks to facilitate the transformation. The former prior resigned and returned to England, where he fell dead immediately after celebrating mass in Westminster Abbey.[27] The new arrangement was confirmed in February 1236 by Pope Gregory IX, who in March of the same year allowed the brothers to use as their insignia a half-red and half-white equal-armed cross to distinguish them from the Templars. The knights also wore a black cloak and the clergy and *conversi* one of russet cameline cloth.[28]

According to the Annals of Dunstable, as well as moving their house to Montmusard, where the air was considered to be purer, Peter des Roches also changed the canons 'into the Order of the Spanish Sword (*in ordinem Spatae Hispaniensis*)'.[29] The latter may be identified as the Spanish military confraternity of St James, which later had a role in assisting the master of the Hospital in the defence of the city.[30] The annalist was evidently confused, however, for there was

---

[24]   Pringle, *Churches*, 2, pp. 66–8 and 4, pp. 136–7, no. 426.

[25]   *Les Registres de Grégoire IX*, ed. L. Auvray (Paris, 1896–1995), no. 2944.

[26]   D. Jacoby, 'Montmusard, Suburb of Crusader Acre: The First Stage of its Development', in B.Z. Kedar, H.E. Mayer and R.C. Smail (eds), *Outremer: Studies in the History of the Crusading Kingdom, presented to Joshua Prawer* (Jerusalem, 1982), pp. 205–17.

[27]   'Annales Prioratus de Dunstaplia', p. 126; Matthew Paris, *Chronica Majora*, 2, p. 490.

[28]   *Les Registres de Grégoire IX*, nos. 2944, 3005; Forey, 'Military Order of St. Thomas', pp. 487–8; idem, *The Military Orders from the Twelfth to the Early Fourteenth Centuries* (Basingstoke–London, 1992), pp. 21–2; B. Dichter, *The Orders and Churches of Crusader Acre* (Acre, 1979), p. 111, fig. 64.

[29]   'Annales Prioratus de Dunstaplia', p. 126; *RRH*, no. 1386n.

[30]   *Excidii Acconis Gestorum Collectio*, ed. R.B.C. Huygens, *The Fall of Acre 1291*, Corpus Christianorum, Continuatio mediaeualis 202 (Turnhout 2004), p. 62 (and notes by

no institutional connection between this confraternity and the Order of St Thomas. It is possible, however, that the source of his confusion was the transfer of the former church of the canons of St Thomas to the confraternity of St James, after the former had moved to Montmusard. Following this interpretation, the church in question would have been that of St James, which belonged to the abbey of Mount Tabor and passed into the hands of the Hospital of St John when that Order acquired Mount Tabor in 1254; at the same time the two priors of St James became *confratres* of the Hospital and all its members were required to swear allegiance to the grand master each year.[31] The church probably lay in the eastern part of the city and could have formed the nucleus for the hospital of St Thomas when the canons moved inside the walls in 1191.[32] The fact that they did not own the building would also be consistent with what is known from other sources concerning their impecunious state.

A charter of November 1240 locates the new house of the Order of St Thomas in Montmusard beside the sea; to the east of it was a public street and to the south a property that was in the process of being granted by the Templars to the brothers of St Lazarus. The charter refers to this property as lying in the English quarter (*in vico Anglicorum*) and the house of an Englishman named Nicolas stood south of it.[33] Matthew Paris's maps of Acre also describe Montmusard as 'mostly inhabited by English'. One of them shows the 'house of St Thomas the Martyr' beside the sea to the south of the house of St Lazarus,[34] while another illustrates the church in the same position but without identifying it.[35] According to the *Pardouns de Acre* (1258–64), pilgrims visiting *Seint Thomas* would receive fifteen years' remission of purgatory, but only seven on Tuesdays.[36] This text places the house in the sequence – St Brigid, St Martin of the Bretons, St Lazarus, St Thomas, St Bartholomew, St Antony –, suggesting that it lay north of St Lazarus, as the charter of 1240 seems to indicate, rather than to its south.[37]

A. Forey on pp. 14 and 62); Ludolphus de Sudheim, *de Itinere Terre Sancte*, ed. G.A. Neumann, *AOL*, 2.2 (1884), 305–77 at p. 340, *fratres de gladio*; Pringle, *Churches*, 4, p. 82.

[31]   *CH*, 2, nos. 2666, 2689; *RRH Ad*, nos. 1214a, 1216a.

[32]   Pringle, *Churches*, 4, pp. 81–2, no. 409.

[33]   'Fragment d'un cartulaire de l'ordre de Saint-Lazare en Terre Sainte', ed. A. de Marsy, *AOL*, 2.2 (1884), 121–58 at pp. 155–7, no. 39; *RRH*, no. 1096.

[34]   Corpus Christi, Cambridge, Ms. 26, fols iii^v–iv^r.

[35]   BL London, Lansdowne Ms. 253, fols 230v–231r; cf. 'Itinéraire de Londres à Jérusalem', in *Itinéraires à Jérusalem et descriptions de la Terre Sainte*, eds H. Michelant and H. Raynaud, Publications de le Société de l'Orient Latin, série géographique 3 (Geneva, 1882), pp. 123–40 at p. 135–6.

[36]   'Pelrinages et Pardouns de Acre', in *Itinéraires à Jérusalem*, pp. 227–36 (at p. 236).

[37]   On these churches see Pringle, *Churches*, 4, pp. 71–2, nos. 397–8 (St Antony, St Bartholomew, St Brigid), pp. 121–3, no. 418 (St Lazarus), pp. 129–30, no. 421 (St Martin).

In April 1245, the Order is recorded in possession of a piece of land lying west of *Casal Blanc* (unlocated) and Kuwaikat (*Coquet*) in the plain of Acre.[38] In March 1255, there is also reference to a prior named Alan.[39] On 30 October 1256, Pope Alexander IV granted the brothers similar rights and privileges regarding such matters as the receipt of alms, the retention of tithes, and the burial of members of the Order that were already held by the Templars and Hospitallers, though these did not apparently extend to exemption from episcopal jurisdiction.[40] In 1257, Matthew Paris records the master of St Thomas arriving at St Albans via Rome, bringing with him a letter from the Holy Land containing inter alia accounts of the divine destruction of Muḥammad's 'temple' and tomb in Mecca, a flood in Baghdad and the incursions of the Tartars.[41]

In 1261, Bishop Florence of Acre wrote to King Henry III of England, drawing the king's attention to the relative neglect from which the house had suffered since its foundation by his predecessor and its need of financial assistance if it was to reap the benefit of the reforms that had since then been implemented. The bishop also rather pointedly contrasted the relative liberality with which Henry was accustomed to give alms to the Templars and Hospitallers.[42] It seems, however, that donations to the house in Acre remained at a modest level. In August 1266, Odo, count of Nevers, left the house a coat and surcoat of Persian cloth in his will;[43] and, in October 1267, Sir Hugh de Nevill bequeathed it his grey palfrey and a suit of armour.[44] In the 1270s, King Edward I wrote to Hugh III of Cyprus commending the brothers and their master, Ralph of Donmbe.[45] On 15 September 1279, Robert de Cardolio and his fellow brothers of the hospital and church of St Thomas of Acre, during the continued absence of the master in England, wrote to Edward I to thank him for his letter promising assistance and noted that the paucity of income deriving from their estates in Cyprus and Syria had resulted in the church of St Thomas, which they had begun building in Acre long before, still

[38]   *CH*, 2, no. 2353; *RRH*, no. 1135.

[39]   *CH*, 2, no. 2722; *RRH*, no. 1228.

[40]   *Registres d'Alexandre IV*, no. 1553; Forey, 'Military Order of St. Thomas', p. 491; idem, *The Military Orders*, p. 113; Dichter, *Orders and Churches*, pp. 111–12, fig. 65.

[41]   *Chronica Majora*, 5, pp. 630–31 and 6, pp. 348–50, add. no. 183; S. Lloyd, *English Society and the Crusade, 1216–1307* (Oxford, 1988), p. 254.

[42]   *Diplomatic Documents preserved in the Public Record Office*, 1: *1101–1272*, ed. P. Chaplais (London, 1964), pp. 241–2, no. 343; *List of Ancient Correspondence of the Chancery and Exchequer*, PRO Lists and Indexes, 15 (London, 1902, revised edn 1969), 47.27; Lloyd, *English Society and the Crusade*, p. 240 n.183, p. 249.

[43]   'Inventaire et comptes de la succession d'Eudes, comte de Nevers (Acre 1266)', ed. M. Chazaud, *Mémoires de la Société Nationale des Antiquaires de France*, 4s., 11 (1871), 164–206 at p. 198.

[44]   M.S. Giuseppe, 'On the Testament of Sir Hugh de Nevill, written at Acre, 1267', *Archaeologia*, 56 (1899), 351–70 at p. 352.

[45]   L. de Mas Latrie, *Histoire de l'île de Chypre sous le règne de la maison de Lusignan*, 2 (Paris, 1852, repr. Famagusta, 1970), pp. 81–2; *RRH*, no. 1386.

remaining unfinished; they also requested that the master, who had been absent for a long time, should be sent back to them.[46]

In the final years before Acre fell to the Mamluks, moves were afoot to amalgamate the Order with the Templars, though this option did not command universal support within the Order. Another undated letter to the English king from around this time suggests that the Order had by then effectively mortgaged its house in Acre to the Templars, to whom it was paying a fixed annual rent.[47] The knights of St Thomas appear to have played little part in the defence of Acre against the Mamluks.[48] After the fall of the city, some members of the Order sought refuge in Cyprus, where they held the church of St Nicolas in Nicosia.[49] The master of the Order was still residing near Limassol as late as 1344.[50] By 1379, however, the Order's headquarters appears to have moved definitively to London.[51]

---

[46]   *List of Ancient Correspondence*, 20.99; de Mas Latrie, *Histoire*, 2, pp. 82–3; *RRH*, no. 1432; Forey, 'Military Order of St. Thomas', pp. 492–3; idem, *Military Orders*, p. 124; Lloyd, *English Society and the Crusade*, p. 250; N. Coureas, *The Latin Church in Cyprus, 1195–1312* (Aldershot, 1997), p. 179.

[47]   Forey, 'Military Order of St. Thomas', pp. 493–5.

[48]   Forey, 'Military Order of St. Thomas', pp. 488–9.

[49]   Coureas, *Latin Church in Cyprus*, p. 180.

[50]   'Itinerarium cuiusdam anglici (1344–45)', in G. Golubovich, *Biblioteca bio-bibliografica della Terra Santa e dell'Oriente francescano* (Florence, 1906–1927), 4, pp. 427–60 (at p. 446), trans. E. Hoade, *Western Pilgrims*, Studium Biblicum Franciscanum, Collectio major, 18 (Jerusalem, 1952), pp. 47–76 at pp. 58–9; cf. Coureas, *Latin Church in Cyprus*, pp. 179–80.

[51]   Forey, 'Military Order of St. Thomas', pp. 496–7.

Chapter 6

Templars, Franks, Syrians and the Double Pact of 1244

Ilya Berkovich[1]

Despite the disunity prevailing in their camp, the defeat of part of their forces by the Egyptians, and their troops' apparently pointless wandering up and down the coast, the Barons' Crusade of 1239–41 ended with some substantial achievements. The castle of Beaufort was regained; its former master, the sultan of Damascus al-Ṣāliḥ Ismāʿīl, had to besiege his own garrison that had mutinied, disgusted that such a strong fortress was to be transferred to the Christians. Eastern Galilee was repossessed and the massive refortification of Saphet allowed the Franks to bring a substantial area with over 200 villages under their control. More to the south, a new castle was built over the ruins of Ascalon, adding some fifty kilometres to the Frankish littoral.

Moreover, the series of agreements with the warring Ayyūbid princes of Egypt and Syria that made these gains possible in the first place, promised additional territorial expansion. Just like al-Ṣāliḥ Ismāʿīl, the sultans of Egypt and Kerak, al-Ṣāliḥ Ayyūb and al-Naṣir Dāwūd, were willing to grant concessions, being particularly generous when it came to territories controlled by their opponents. This allowed the Franks to lay claim to all the lands between Ascalon and Jerusalem, together with the hinterland of Gaza.

While their former territorial holdings were slowly but steadily being restored, the politics of the Frankish world remained fissile. The Latins were already divided between the supporters and opponents of Emperor Frederick II, who was regent of Jerusalem for Conrad, his son by Queen Isabel II. The Barons' Crusade had reinforced those differences by leaving the Franks split over the question whether the Kingdom of Jerusalem should favour an agreement with Egypt or Damascus. Frederick's truce with al-Kamil and his subsequent good relations with Egypt meant that the imperial camp, which by 1241 included not only the imperial *bailli* but also the Hospitallers and the Teutonic Knights, favoured the former. The baronial party and the Templars, on the other hand, preferred a renewal of the

[1]   I would like to thank Michael Chernin, Erez Cohen and Professor Benjamin Kedar for their generous help. I am also grateful to Professors Reuven Amitai and Martin van Creveld and Dr Yuval Noah Harari for commenting on earlier versions of this paper.

truce with Damascus that, for much of the twelfth century, had been a tacit ally and sometimes even a client of the Frankish kingdom.[2]

According to the standard accounts, in the aftermath of the Barons' Crusade, once the imperial party had been finally defeated with the expulsion of the Filangieri brothers from Tyre, the politics of the kingdom came to be dominated by the baronial party and the Templars. The winter of 1243–44 saw the renewal of the Franco-Damascene alliance, joined also by Kerak and Homs, under its ruler, Sultan al-Manṣūr Ibrāhīm. While the ultimate aim of this treaty was the joint conquest and partition of Egypt, the Franks were able to reap some immediate advantages: all their gains from the treaties of 1240–41 were confirmed, and, more importantly, the Christians were granted the sole possession of Jerusalem. Even the Temple Mount was surrendered, despite the fact that, according to previous agreements, it was reserved for Muslim worship. In preparation for the upcoming war, Al-Manṣūr Ibrāhīm travelled to Acre in person where he was received with great honour, while the army of Damascus moved south and occupied Gaza. Faced by this threatening turn of events, al-Ṣāliḥ Ayyūb was left with no alternative but to summon the Khwarizmians. Their arrival in July 1244 led to the final loss of Jerusalem and the military catastrophe at Forbie from which the kingdom was never to recover.[3]

A somewhat different explanation for the events of 1244 is given by Peter Jackson, who has provided the most detailed study of the aftermath of the Barons' Crusade. Jackson has shown, for instance, that the Khwarizmian invasion in the summer of that year had little to do with the situation in the Holy Land itself. After the crushing defeat inflicted on the Seljuk sultan of Rum the previous year, in May 1244 the Mongols were poised to start a new campaign in northern Syria. The Khwarizmian arrival less than two months later was not so much an invasion as an escape from this new threat.[4] Nevertheless, Jackson's account of the events on

---

[2]   Here I largely follow S. Painter, 'The Crusade of Theobald of Champagne and Richard of Cornwall, 1239–1241', *HC*, 2, pp. 463–85; M. Balard, 'La croisade de Thibaud IV de Champagne (1239–1240)', in Y. Bellenger and D. Quéruel (eds), *Les Champenois et la croisade* (Paris, 1989), pp. 85–95; M. Lower, *The Barons' Crusade: A Call to Arms and its Consequences* (Philadelphia, 2005), pp. 158–77; P. Jackson, 'The Crusades of 1239–41 and their Aftermath', *Bulletin of the School of Oriental and African Studies*, 50 (1987), 35–48.

[3]   S. Runciman, *A History of the Crusades* (Cambridge, 1951–54), 3, pp. 223–4; J. Prawer, *Histoire du Royaume Latin de Jérusalem*, trans. G. Nahom (Paris, 1969–70), 2, pp. 307–10; H.E. Mayer, *Geschichte der Kreuzzüge*, 7th edn (Stuttgart, 1989), p. 227; J. Richard, *Histoire des Croisades* (Paris, 1996), pp. 338–40; C. Tyerman, *God's War: A New History of the Crusades* (Cambridge, MA, 2006), pp. 770–71. It is largely repeated in specialized studies such as R.S. Humphreys, *From Saladin to the Mongols: The Ayyubids of Damascus, 1193–1260* (Albany, 1977), pp. 272–4; J.S.C. Riley-Smith, *The Knights of St. John in Jerusalem and Cyprus c. 1050–1310* (London, 1967), pp. 178–81; M. Barber, *The New Knighthood: A History of the Order of the Temple* (Cambridge, 1994), pp. 142–7.

[4]   Jackson, *Crusades*, pp. 55–6. The links between the Khwarizminans and al-Ṣāliḥ Ayyūb went back to late 1230s when he was prince of Ḥiṣn Kayfā in Djazīra. M.L. Bulst-

the Frankish side in 1243–44 is largely similar to the one described above.[5] Nor is it as detailed as his analysis of the events of 1239–43, but there is a good reason for this. Both Frankish and Arabic primary sources are generally weak when it comes to describing their adversaries' political affairs. Jackson comments that sometimes authors even fail to mention important details concerning their own side. Jackson's study, however, was the first to make full use of the last section of the *Siyar al-abāʾ al-baṭārika*, a history of the Coptic patriarchs of Egypt. Written by a knowledgeable contemporary, this source reports in some detail the Muslims' dealings with the Franks – something that the more famous Arabic chroniclers of the era, Sibṭ Ibn al-Djawzī and Ibn Wāṣil, had seldom done. Unfortunately its author breaks off his account in 1243, shortly before the new Franco-Damascene alliance was concluded.[6]

The aim of this paper is to look again at the events of the year immediately preceding Forbie. When exactly was the treaty made? Was it the only treaty made that year? What role did the orders play in the pact? In answering these questions, it should be possible to offer a more concise reconstruction of the events of 1244 and also to reassess the relative strength of the Second Frankish Kingdom of Jerusalem and the role it played in the power-politics of its Muslim neighbours.

In June 1243, the Templar–Damascene campaign against the Egyptians was abandoned after al-Ṣāliḥ Ismāʿīl suddenly retreated with his army.[7] This was followed by a short period of reconciliation between Damascus and Egypt. On 11 September 1243 al-Ṣāliḥ Ismāʿīl acknowledged al-Ṣāliḥ Ayyūb as his overlord by ordering the Friday prayers to be recited in the name of the Egyptian Sultan.[8] This helps to establish the earliest possible date of the agreement that transferred the whole of Jerusalem to the Christians, something that could not have happened until relations between the Ayyūbids deteriorated once more. A definite *terminus ante quem* for this treaty is the visit of Ibn Wāṣil to Jerusalem towards the end of 641 A.H. (8 June 1244), when he witnessed the Christian liturgy being performed in the Dome of the Rock.[9]

The most detailed description of this treaty is found in a letter from master of the Temple, Armand of Périgord, to Robert Sandford, the Order's preceptor in

---

Thiele, 'Zur Geschichte der Ritterorden und des Königreichs Jerusalem im 13. Jahrhundert bis zur Schlacht bei La Forbie am 17. okt. 1244', *Deutsches Archiv für Erforschung des Mittelalters*, 22 (1966), 219.

[5]   Jackson, 'Crusades', pp. 55–6.

[6]   Ibid., pp. 32–5.

[7]   *History of the Patriarchs of the Egyptian Church: Known as the History of the Holy Church*, eds and trans. A. Khater and O.H.S. Khs-Burmester (Cairo, 1943–74), 4/2, p. 142. According to Jackson it was probably due to the news of the Mongol success arriving from the north. See 'Crusades', pp. 54–5.

[8]   Abū Shāma, *Tarājim rijāl al-qarnayn al-sādis waʾl-sābiʿ al-maʿrūf biʾl-dhayl ʿalā al-rawḍatayn*, ed. M. al-Kawtharī (Cairo, 1947), p. 173.

[9]   Ibn Wāṣil, *Mufarridj al-kurūb fī akhbār Banī Ayyūb*, ed. Dj. al-Shayyāl et al. (Cairo, 1954–77), 5, pp. 333–4.

England. Matthew Paris, whose *Chronica Majora* preserves this source, places it immediately after events that took place in early February 1244. The first two dates mentioned by Matthew after recording Armand's message are 3 July and 1 March of that same year.[10] It is, however, unlikely that anything more specific can be inferred from Matthew other than the fact that Armand's message arrived in England sometime in 1244. Jackson, however, discounted the Frankish annals that place the treaty in 1244, and, citing Armand's statement that fifty-six years had elapsed since Jerusalem was last exclusively in Christian hands, concluded that the letter was sent in 1243. Jackson's inference was that the treaty transferring the whole of the city to the Christians was made in that year. Armand's letter, therefore, either refers to a treaty made after the break between Damascus and Egypt in late 1243 or even to the truce concluded between al-Ṣāliḥ Ismāʿīl and the Templars in anticipation of the campaign against Gaza the previous June. Further evidence adduced by Jackson to support the second possibility is a letter from the pope to the Latin patriarch of Jerusalem, Robert of Nantes, relating to the reconstruction of the city walls and dated 5 August 1243.[11]

Jackson is definitely correct that in the summer of 1243 Jerusalem was under Frankish control. After changing hands a number of times during the Barons' Crusade, it was finally retaken by the Christians in the summer of 1241, and there is no evidence that it was lost again before the Khwarizmian invasion.[12] However, Innocent's letter makes no reference to the Temple Mount[13] that, according to all previous understandings, was to remain in Muslim hands. Secondly, the *Annales de Terre Sainte* clearly distinguishes between the truces made in 1243 and 1244: the first leading to the abortive campaign against the Egyptians, while the second resulted in the Christian recovery of the Temple Mount. The 'Templar of Tyre' also dates the treaty that returned Jerusalem to the Christians to 1244.[14]

Moreover, assuming that the figure given by Armand is not merely an approximation, the fifty-sixth year since the fall of Jerusalem to Saladin ended on October 2 1243, and so the return of the Temple Mount to the Christians had to take place after that date. Finally, the letter of Master of the Hospital, William of Châteauneuf, sent to the West after the battle of Forbie, says that the treaty returning the Temple Mount to the Christians was concluded at the beginning of the previous summer.[15] Obviously, this could not be the summer of 1243, as argued by Jackson, but that of 1244.

---

[10]   Matthew Paris, *Chronica Majora*, ed. H.R. Luard, RS 57 (London, 1872–83), 4, pp. 288–91 for the letter and pp. 287, 294, 296 for the dates.

[11]   Jackson, 'Crusades', p. 55 n. 168.

[12]   Al-Nuwayrī, *Nihdyat al-arab fi funun al-adab*, quoted in Jackson, 'Crusades', p. 49.

[13]   *MGH Epistolae Saeculi XIII e Regestis Pontificum Romanorum Selectae*, ed. C. Rodenberg (Berlin, 1883–94), 2, no. 6.

[14]   'Annales de Terre Sainte', ed. R. Röhricht, *AOL*, 2 (1884), p. 441; *Gestes des Chiprois*, ed. G. Raynaud (Geneva, 1887), p. 146.

[15]   *Chronica Majora*, 4, p. 307.

It can be established, therefore, that the Temple Mount was not returned to the Franks as part of the Templar–Damascene truce in the summer of 1243. The treaty that allowed the Franks to regain full control of Jerusalem could not have been concluded before the autumn of 1243, and the balance of evidence is strongly in favour of the transfer of Temple Mount to the Franks taking place in 1244, perhaps even as late as the spring of that year. For the purposes of this present discussion, 1244 will henceforth be the accepted date, if only to distinguish the agreement that surrendered the whole of Jerusalem to the Christians from the short Templar–Damascene truce that led to the abortive campaign against Gaza in June 1243.

While modern studies usually acknowledge that the truces of 1241–43 and even 1244 were the result of Templar policy, it is generally assumed that the agreement that returned Jerusalem to the Christians was made by the entire kingdom. The general histories cited above mostly refer to the Christian side as 'Franks' without distinguishing their political parties.[16] But how far did the situation in late 1243 to early 1244 differ from that prevailing two years or even one year before? We know that in 1241–43 it was only the Templars who were allied with Damascus. But powerful as the Order was, especially if one adds its allies from the baronial faction, it was not the only substantial force in Frankish politics.

Upholding the truce made with Egypt by Richard of Cornwall, the Hospitallers did not participate in the failed Templar–Damascene attempt against Gaza in the summer of 1241. This fact is underscored by the *Siyar*, which describes the arrival of the Templar envoys to Egypt later that year, noting that they were representing only those Franks who were at war with al-Ṣāliḥ Ayyūb, as the Christians holding Ascalon had already made peace with him.[17] The campaigns of 1242 and 1243 were also conducted by the Templars, accompanied by Geoffrey of Sergines.[18] A Templar victory over the Egyptians in 1242 is also reported by Matthew Paris.[19] One should ask therefore why the Hospitallers, who had hitherto kept out of the war with Egypt, would suddenly join the treaty in 1244, adopting a political course they had consistently opposed.

Moreover, the orders were drawn into a new round in the struggle between the baronial and the imperial parties.[20] An unsuccessful attempt by the imperial *bailli* to take over Acre in October 1241 had led to an open confrontation, with the

---

[16]   Above n. 3.

[17]   *History of the Patriarchs*, 4/2, p. 113. See also Jackson, 'Crusades', p. 49 n. 126 for the correct understanding of the later passage. The refortified Ascalon was handed by Richard Cornwall to an imperial representative. *Eracles*, p. 421; *Gestes*, p. 123; *Chronica Majora*, 4, p. 168.

[18]   'Annales', p. 441.

[19]   *Chronica Majora*, 4, p. 197. According to Jackson ('Crusades', p. 51), Matthew was referring sceptically to what appeared to be an inflated report following the Egyptian retreat from Gaza in late spring 1242.

[20]   P. Jackson, 'The End of the Hohenstaufen Rule in Syria', *Bulletin of the Institute of Historical Research*, 56 (1986), 33–4.

Hospitaller convent in the city being blockaded by the Ibelins and the Templars for some six months.[21] The situation in 1244 could have been different if one assumes that, after the fall of Tyre and the expulsion of the Filangieri brothers, the orders were reconciled, or, at least, the Hospitallers submitted to the winning party, thereby allowing the Templars to conduct their policies in the name of the entire kingdom.[22] But is there any proof this actually happened?

The primary sources are somewhat ambiguous as to whether Tyre surrendered to the barons on 10 July 1242 or 1243, with the modern scholarship leaning more towards the former date.[23] Whichever option is correct, however, there is good evidence to show that the Hospitallers firmly maintained their imperial sympathies. In mid-1243, for instance, Ascalon was transferred to the Order. The Hospitallers ensured that the grant was confirmed not only by Frederick but also by his son, with an additional diploma, this time from Conrad only, dated 30 March 1244.[24] The Hospitaller adherence to the emperor's cause can also be inferred from Frederick's letters of November 1244 and February 1245 in which he refers to the Hospitallers and the Teutonic Knights as acting and concluding truces in his name.[25]

The position of the Teutonic Knights is also illuminating in regard to the relations between the military orders. Until 1239, the Order had maintained a largely neutral position, being able to satisfy the emperor, the pope and the barons. However, after the death of Hermann von Salza, the new master, Conrad of Thuringia, changed tactics, and the Order joined the imperial camp. This immediately led to clashes with the baronial faction. Either in 1241 or in 1243, the house of the Teutonic Knights in Acre was attacked by the Templars. Furthermore, the Order lost a major legal case pertaining to its landholdings in the Galilee, the plaintiff, apparently, taking advantage of its political vulnerability. The case was settled on 7 July 1244; only four days later the Khwarizmians descended on Jerusalem.[26]

There is thus good reason to believe that the fall of Tyre did not bring about reconciliation between the orders. Both the Hospitallers and the Teutonic Knights appear strongly entrenched in their legitimist positions both in 1243 and also in

---

[21]   'Annales', 440–41; *Gestes*, 124–7, Riley-Smith, pp. 179–80, Barber, p. 142.

[22]   For an argument along these lines, see Riley-Smith, p. 180. Bulst-Thiele (p. 220) even claims that the treaty with the Syrians was concluded unanimously to oppose the Khwarizmian threat.

[23]   Jackson, 'End', pp. 22–6; D. Jacoby, 'The Kingdom of Jerusalem and the Collapse of the Hohenstaufen Rule in the Levant', *Dumbarton Oaks Papers*, 40 (1986), pp. 83–101.

[24]   *Historia diplomatica Frederici secundi*, ed. J.L.A. Huillard-Bréholles (Paris, 1852–61), 6/1, p. 116 and 6/2 pp. 848–9. According to Jackson ('End', p. 27), the Hospitaller insistence on the confirmation from Conrad, now titular king of Jerusalem, both demonstrated the Order's loyalty to the emperor and formed a pre-emptive legal measure against the baronial party.

[25]   *Historia diplomatica*, 6/1, pp. 236–40; *Chronica Majora*, 4, p. 302.

[26]   N.E. Morton, *The Teutonic Knights in the Holy Land 1190–1291* (Woodbridge, 2009), pp. 85–95. For the investment of Jerusalem, see *Chronica de Mailros*, ed. J. Stevenson (Edinburgh, 1835), p. 159.

1244, well after the new agreement with Damascus was in place. That hardly supports the prevailing view that this treaty was made in the name of the entire Frankish kingdom.

It is far more likely that once again the Templars were the only Frankish party behind the treaty, and the sources support this view. The 'Templar of Tyre' records that the truce with the sultan of Damascus was concluded by the Templars when, together with Geoffrey of Sergines, they were encamped near Jaffa.[27] Armand of Périgord refers to the same treaty, also presenting it as an entirely Templar affair. According to Armand, after the defeat suffered at the hands of the Order, the sultan of Egypt agreed to new negotiations for a truce. However, when al-Ṣāliḥ Ayyūb refused to surrender some of the lands he had promised in previous agreements, the Templars began talks with Damascus and Kerak instead. This decision was 'habita super his deliberatione provida et diligenti consilio praelatorum et quorundam terrae baronum'. Armand even complains that 'quamplures in his et aliis terrae utilitatibus promovendis, propter odium et invidiam, nobis existunt contrarii et infesti'.[28]

William of Châteauneuf is vaguer, simply saying that the truce was concluded by the Christians. He continues using that term while describing the conditions of the treaty and the transfer of Jerusalem. Yet when he tells of the events taking place after the arrival of Patriarch Robert in the Holy Land, he changes to first-person plural. He continues using 'we' when referring to the action taken by the Christians during the Khwarizmian invasion and at the subsequent battle of Forbie. The fact that he will not use the same language while describing the treaty with the Syrians might indicate William's detachment. Nor does he mention the Muslim allies that participated in the campaign on the Frankish side. Moreover, in the beginning of his letter William refers to some of his previous messages that had reported the ills having befallen the Holy Land as a consequence of espousing of the Damascene cause against Egypt.[29] This last point echoes Frederick, who emphasized that the agreement which provoked the war with Egypt was made by the Templars in total disregard of the Hospitallers and other Franks loyal to himself.[30]

In the light of the conditions prevailing early in 1244 and on the basis of the available evidence, it is far more plausible that the agreement with Damascus, which transferred the whole of Jerusalem to Christian control, was made by the Templars alone. It is therefore sensible to regard it as yet another of the factional deals made by the various Frankish parties with their Muslim neighbours. With that in mind, it might be asked whether this new treaty posed such a great threat to Egypt that it necessitated summoning the Khwarizmians. The impression given by Ibn Wāṣil is that that was indeed the case. Travelling to Egypt in the early summer of 1244, he observed that al-Naṣir Dāwūd was encamped near Jerusalem, while a part of the

---

[27]   *Gestes*, p. 146.
[28]   *Chronica Majora*, 4, pp. 289–90.
[29]   *Chronica Majora*, 4, pp. 307–11.
[30]   Ibid., 4, p. 302; *Historia diplomatica*, vol. 6/1, pp. 239–40.

Damascene army has already taken up position in Gaza.[31] It was being faced with the imminent threat of a combined offensive aimed at the division of Egypt that is usually cited as the reason al-Ṣāliḥ Ayyūb summoned the Khwarizmians.

However, it is not at all certain that the partition of Egypt was actually discussed in the initial negotiations in 1244. While they do take note of the Christian gains from the Damascene treaty, no Frankish or western sources make reference to any plans concerning Egypt proper. This in itself is not sufficient proof that such schemes were not contemplated: the question is whether they could have been seriously considered under the circumstances at that time.

The Templar–Damascene military actions in 1241–43 were not particularly remarkable. Apart from the sack of Nablus in October 1242, the only achievement of Templar arms was the Egyptian retreat from Gaza earlier that same year, and that was more due to internal intrigue than defeat at the hands of the Christians.[32] This is not to say that the conquest of Egypt could not have been considered early in 1244. Bearing in mind, however, the performance of the Templars and their Syrian allies during the preceding three years, it is highly unlikely that it would actually have been attempted. It is more plausible that the deal envisaged a continuation of what was largely a defensive war against al-Ṣāliḥ Ayyūb.

Interestingly, the reports of Franco-Syrian intentions to invade Egypt come exclusively from Arabic writings.[33] The authors usually associate these plans with the visit of al-Manṣūr Ibrāhīm to Acre, where, according to most modern accounts, the initial alliance against Egypt was sealed. But when exactly did this visit take place? Joinville, who has a digression on the events of 1244 in his biography of St Louis, states that the sultan of Homs was invited to Acre after the Khwarizmian attack. According to al-Makīn, al-Manṣūr's visit and stay in the Templar palace took place directly before the beginning of the campaign leading to Forbie.[34] Although no other contemporary source clearly states that al-Manṣūr visited Acre in the autumn of 1244, its description in the Arab accounts is always placed either after the Khwarizmian invasion or just before the battle. Similarly, Frederick describes the sultan of Homs's sojourn with the Templars and then immediately proceeds to discuss his role in the battle.[35]

---

[31]   Ibn Wāṣil, 5, pp. 332–3.

[32]   *History of the Patriarchs*, 4/2, p. 118; Al-Makīn ibn al-'Amīd, *Kitāb al-Madjmū' al-Mubārak*, published as 'La Chronique des Ayyoubides d'al-Makin ibn al-'Amid', ed. C. Cahen, *Bulletin d'Études Orientales*, 15 (1955–57), 152–3. This did not prevent the Templars to try and capitalize on that success; see above, n. 19.

[33]   For instance: Ibn Wāṣil, 5, p. 337; Al-Makīn, p. 155.

[34]   John of Joinville, *Vie de Saint Louis*, ed. J. Monfrin (Paris, 2002), p. 460; Al-Makīn, p. 155.

[35]   Sibṭ ibn al-Djawzī, *Mir'āt al-zamān fi ta'rikh al-a'yān* (Hyderabad, 1951–52), 2, pp. 745–6; Ibn Wāṣil, 5, p. 337; Abu 'l-Fidā, *Mukhtaṣar ta'rīkh al-bashar* (Beirut, 1972), 2, p. 276; Ibn al-Dāwādārī, *Kanz al-durar wa-jāmi' al-ghurar*, published as *Die Chronik des Ibn ad-Dawādārī*, eds H.R. Römer *et al.* (Cairo and Freiburg, 1960–94), 7, pp. 353–4; *Chronica Majora*, 4, pp. 302–3.

According to the patriarch, the Syrian military help had not yet arrived on 21 September, but on 4 October the Christian army had already left the city to do battle with the Egyptians and the Khwarizmians.[36] Therefore, if al-Manṣūr's visit had indeed taken place shortly before the beginning of the campaign, as the above sources imply, it can be pinpointed between those dates.

So if the Arab sources are to be believed, the 1244 agreement was an offensive alliance against Egypt concluded in the late summer or early autumn of that year. Can their version of the treaty be reconciled with the agreement that the western sources describe as a truce concluded in the early months of 1244? In fact there is additional evidence to suggest that we are dealing with two distinct agreements instead of one. According to Joinville, al-Manṣūr Ibrāhīm was invited to Acre entirely on the initiative of the Christians. This point is confirmed in a letter from Patriarch Robert to the pope shortly before Forbie. The letter further informs us that one of the reasons for the decision to summon the sultans was that the Franks did not have enough troops to fight the invaders by themselves, the only reinforcements available being a hundred warriors who had come as pilgrims.[37] But if an active military partnership between the Latins and the Syrian Ayyūbids was already an established fact, why would the Franks summon the sultans only after realizing that the kingdom on its own was too weak? We might also expect that the existence of such an agreement would allow a Franco-Syrian army to take the field much earlier than October, almost three months after the Khwarizmian invasion.

Another important point demonstrated by the patriarch's letter is that in September 1244 all the Christians were acting in unison. This was certainly not the case earlier that year. Especially striking is the comparison with Armand's letter that openly admits the truce with Damascus and Kerak was made exclusively by the Templars, who discussed it only with those Franks who were close to their party. On the other hand, the patriarch underscores the summoning of the Syrian sultans was done following deliberations involving all the nobility, clerics and heads of the military orders.[38]

While the evidence adduced so far is largely circumstantial, we possess one primary source stating directly that 1244 witnessed two distinct agreements. Ibn al-Furāt, the author of what is perhaps the best compendium of previous authorities, tells us that in 641 A.H. (1243–44), after the unsuccessful negotiations between the Ayyūbids, al-Ṣāliḥ Ismā'īl and al-Naṣir Dā'ūd came to an agreement with the Franks and handed them Jerusalem. Then in 642 A.H., after the arrival of the Khwarizmians, the sultans of Damascus and Homs made a treaty against Egypt together with the Franks. Al-Manṣūr Ibrāhīm left Damascus at the head of the Syrian forces arriving in Acre where he lodged in the Templar palace. Joined by all the Franks, the united army proceeded

---

[36]   *Mailros*, p. 158; *Chronica Majora*, 4, p. 341.

[37]   *Mailros*, p. 158.

[38]   Ibid., p. 158; see also *Chronica Majora*, 4, p. 339.

southward against the Khwarizmians and the Egyptians.[39] If our only evidence came from the *Ta'rikh al-duwal*, it would be hard to prove the existence of two separate Franco-Muslim agreements in 1244. It is always possible that a later compiler might mistakenly refer to the same treaty as two separate ones, even if Ibn al-Furāt was less likely to make such an error than some of his colleagues. However, the circumstantial evidence considered so far vindicates Ibn al-Furāt's version of the events.[40]

The two agreements of 1244 show that in that year Franco-Muslim relations underwent two distinct phases, with different participants and changing aims. They also demonstrate the political changes wrought by the Khwarizmian invasion. The first pact, concluded in early 1244, was initiated by the Syrians. Joined only by one of the Frankish factions, it would put into the field merely a fraction of the available Christian forces. Its real aspirations, and even its acknowledged aims, were probably modest – a continuation of what was essentially a defensive war against Egypt. The second agreement was arranged in the late summer or early autumn. The initiative this time came from the Franks, with all their parties acting as signatories. Based on the understanding that far more could be achieved this time, when the united forces of the entire kingdom would join the war, the allies contemplated a far grander goal – the joint conquest and partition of Egypt.

The greater ambitions of this new alliance suggest that, when united, the Latins had at their disposal a force that could considerably alter the military balance between the Ayyūbid principalities. Evidence from the campaign of 1244 supports this point – the Christians made up two out of the three divisions of the allied army. Despite the support of some of the Franks, al-Manṣūr Ibrāhīm was overruled at the council of war on the day before Forbie, indicating that the Christians had the majority of the commanders there present.[41] However, earlier that year a treaty with the Templars alone sufficed to have the Temple Mount returned to the Christians. This was an unprecedented achievement that even the emperor had been unable to accomplish.

Ironically, the union between their political factions did not do the Franks any good. The army that fought at Forbie suffered from overconfidence, and the defeat that followed spelled the end of the kingdom as an independent military power. This, however, should not overshadow the vitality of the Second Kingdom of Jerusalem in the years preceding Forbie.[42] While the Latins certainly owed much to the conflicts between the various Ayyūbid rulers, this

---

[39]   Ibn al-Furāt, *Ayyubids, Mamluks, and Crusaders: Selections from the Ta'rikh al-duwal wa-l-muluk*, eds and trans. U. and M.C. Lyons (Cambridge, 1971), 1, pp. 1, 4–5.

[40]   C. Cahen, 'Ibn al-Furāt, Nāṣir', in *Encyclopedia of Islam*, 2nd edn, eds P.J. Bearman *et al.* (Leiden, 1960–2005), s.v.

[41]   Joinville, p. 462; *Eracles*, p. 249.

[42]   A recently published general history of the Crusades claims that Forbie demonstrates the Frankish 'peripheral' role in Near-Eastern politics: Tyerman, p. 771.

does not mean that their survival was entirely dependent on Muslim disunity. In fact, during much of that time the Latins were just as divided. The ability of the Templars to win major concessions shows that Frankish strength alone, even without crusader reinforcements, was, until 1244, a factor to be seriously reckoned with by the kingdom's Muslim neighbours.

Chapter 7

Royal and Papal Interference in the Dispatch of Supplies to the East by the Military Orders in the Later Thirteenth Century

Alan Forey

In the later thirteenth century military orders in the eastern Mediterranean were becoming increasingly dependent on their western provinces to provide support in the form of both regular responsions and additional contributions.[1] But the dispatch of supplies was hampered not only by financial problems within the orders but also, at times, by interference by outside powers, which either banned exports or appropriated responsions.

Restrictions on exports were commonly imposed by secular rulers in times of war or dearth, and more permanent limitations might be set in place as a means of raising money through the sale of licences, as happened in South Italy:[2] this last action did not prevent exports but merely increased their cost. Prohibitions on occasion sometimes covered not only goods but also money; and although, because of the common shortage of food and other supplies in the East, aid seems normally to have been sent in kind in the later thirteenth century, there were times when money was dispatched by the orders to the Holy Land or Cyprus.[3] As some supplies were transported through another country before being shipped across the Mediterranean, restrictions might affect not only the districts where they were imposed. Some Castilian and Portuguese responsions passed through Aragonese lands before being loaded onto ships in Catalan ports, although Cartagena and Alicante may also have

---

[1]   In the later thirteenth century supplies from Italy were being sent by military orders to Achaea as well as to the Holy Land and Cyprus: *I registri della cancelleria angioina*, ed. R. Filangieri (Naples, 1950–2006), 9, p. 30; 47, p. 285; *Actes relatifs à la principauté de Morée, 1289–1300*, eds C. Perrat and J. Longnon (Paris, 1967), p. 91, doc. 86.

[2]   *The Cambridge Economic History of Europe*, 3: *Economic Organization and Policies in the Middle Ages*, eds M.M. Postan, E.E. Rich and E. Miller (Cambridge, 1963), p. 402.

[3]   *CH*, 3, p. 555, doc. 4081; *Calendar of Close Rolls, 1302–1307* (London, 1908), pp. 137–8; A. Garcia i Sanz and M.-T. Ferrer i Mallol, *Assegurances i canvis marítims medievals a Barcelona* (Barcelona, 1983), 2, pp. 336–9, docs 26–9; Barcelona, Archivo de la Corona de Aragón (henceforth ACA), Cancillería real, Pergaminos, Jaime II 2071–3.

been used as departure points by Templars and Hospitallers in Castile.[4] Goods from Spain were also sometimes channelled through Marseille and Sicily.[5]

Yet numerous licences to export were issued in the later thirteenth century. Although the registers of the Angevin kings of Sicily have not survived, export licences for military orders are known to have been issued by Charles I in every year from 1269 until the early 1280s; and licences granted by Charles II have survived for most years in the 1290s.[6] The majority of the licences granted by rulers were for individual sailings, but more long-term concessions were at times also made. In 1262 James I of Aragon permitted the Hospitallers to load two ships every year free from customs dues.[7] In January 1295 Charles II allowed the Templars to export 2,000 *salmae* of wheat, 3,000 of barley, and 500 of legumes to Cyprus each year, and in the following month he made the further concession that they could transport 4,000 *salmae* of grain annually to Cyprus – provided that a quarter of it went to noble refugees from Acre. These long-term concessions were, however, to last only as long as it pleased the king.[8]

It is further clear that the needs created by famine and war did not always prevent shipments by military orders. Although in 1279/80 Charles I of Sicily had banned food exports because of dearth in the kingdom, and although the vicar of the patriarch of Jerusalem, writing from Acre in October 1280, informed the

---

[4]   ACA, Cancillería real, Registro (henceforth R.) 63, fol. 89; R. 66, fol. 57v; Pergaminos, Pedro II (III) 292; *Fueros y privilegios de Alfonso X el Sabio al reino de Murcia*, ed. J. Torres Fontes (Murcia, 1973), p. 111, doc. 100; P. Josserand, '*Et succurere Terre sancte pro posse*: les Templiers castillans et la défense de l'Orient latin au tournant des XIIIe et XIVe siècles', *Cahiers de recherches médiévales*, 15 (2008), 228.

[5]   ACA, Cancillería real, Cartas Reales Diplomáticas (henceforth CRD), Templarios 334; Filangieri, 6, p. 175.

[6]   References to licences issued by kings of South Italy and Sicily are to be found mainly in Filangieri; see also E.M. Jamison, 'Documents from the Angevin Registers of Naples: Charles I', *Papers of the British School at Rome*, 17 (1949), 87–180; *CH*, 3, pp. 204, 208–9, 231, 273, 338, 342, 381, 390, 414–15, docs 3350, 3360, 3401, 3466, 3599, 3609, 3690, 3717, 3758; A.A. Scotti, *Syllabus membranarum ad regiae Siciliae archivum pertinentium* (Naples, 1824–45), 1, p. 8; 2.1, pp. 71, 152–3, 213–14; 2.2, pp. 8, 22. K. Toomaspoeg, 'Le ravitaillement de la Terre Sainte. L'exemple des possessions des ordres militaires dans le royaume de Sicile au XIIIe siècle', in *L'expansion occidentale (XIe–XVe siècles). Formes et consequences* (Paris, 2003), p. 150, calculates the amounts known to have been sent from the kingdom of Sicily by military orders between 1269 and 1283; see also the listings in J.H. Pryor, '*In Subsidium Terrae Sanctae*: Exports of Foodstuffs and War Materials from the Kingdom of Sicily to the Kingdom of Jerusalem, 1265–1284', *Asian and African Studies*, 22 (1988), 144–5.

[7]   ACA, R. 12, fol. 61v.

[8]   L. de Mas Latrie, *Histoire de l'île de Chypre sous le règne des princes de la maison de Lusignan* (Paris, 1852–61), 2, pp. 91–2; N. Housley, 'Charles II of Naples and the Kingdom of Jerusalem', *Byzantion*, 54 (1984), 531, 533–5; see also M.L. Favreau-Lilie, 'The Military Orders and the Escape of the Christian Population from the Holy Land in 1291', *Journal of Medieval History*, 19 (1993), 225–6.

English King Edward I that supplies were not expected from Apulia or Sicily because of Charles of Anjou's wars,[9] several licences were issued in both years allowing the Hospitallers and Templars to export foodstuffs and animals from the kingdom of South Italy and Sicily to the Holy Land.[10] When Alfonso III of Aragon gave the Templars permission to export goods in 1290 he wrote to the grand master, William of Beaujeu, stating that

> Although these things are very necessary to us and our country because of the war in which we are involved, nevertheless for the honour of God, who protects and defends us and our people in our rights against our enemies in the said war, and because of our affection for you, we have freely granted permission.[11]

Support for the Holy Land was seen as a means of securing divine favour, and Alfonso also wanted at that time to obtain the appointment of a Catalan Templar as provincial master. Exports were in fact frequently allowed by the rulers of Aragon as well as those of Sicily or Naples even though in the later thirteenth century they were engaged in fairly constant warfare. This is apparent not only from surviving licences but also from references to the transporting of goods in agreements with merchants and in letters written by brethren about the dispatch of supplies to the East. In April 1282, when preparations were well advanced for the expedition to Tunis and Sicily that Pedro III of Aragon launched in June of that year, the Aragonese Templars were arranging for horses and other supplies to be dispatched in August from Barcelona to Acre;[12] and Templar correspondence reveals that the Aragonese provincial master, Berenguer of Cardona, was sailing for Cyprus – apparently with the supplies he had been gathering – in August 1300, only a few days after James II had summoned members of the Order for military service against Castile.[13]

Licences did, however, commonly impose certain limits. They frequently stated the amounts of food or animals that could be exported and often included a warning that these limits were not to be exceeded. But in some instances it is clear that the amounts allowed were those that military orders wanted to send to the East. In 1269, for example, Charles I stated that the prior of St Thomas of Acre wished

---

[9]   Filangieri, 44, pp. 619–20; *Foedera, conventiones, litterae et cuiuscumque generis acta publica*, ed. T. Rymer (1816–69), 1.2, pp. 586–7.

[10]   Filangieri, 21, pp. 36, 42; 23, pp. 39, 56; 44, pp. 619–20, 630–32; *CH*, 3, pp. 381, 390, docs 3690, 3717; Jamison, pp. 122, 125,132–4, nos. 114, 118, 155, 164; cf. *CH*, 4, p. 9, doc. 4536.

[11]   *Acta Aragonensia*, ed. H. Finke (Berlin, 1908–1922), 3, pp. 8–9, doc. 5.

[12]   ACA, Pergaminos, Pedro II (III), 292.

[13]   ACA, CRD, Templarios 68, 377; *Acta Aragonensia*, 1, pp. 78–9, doc. 55; A.J. Forey, *The Templars in the Corona de Aragón* (London, 1973), p. 412, doc. 41; idem, 'Letters of the Last Two Templar Masters', *Nottingham Medieval Studies*, 45 (2001), 153–4. On relations between Aragon and Castile in 1300, see A. Masiá de Ros, *Relación castellano-aragonesa desde Jaime II a Pedro el Ceremonioso* (Barcelona, 1994), 1, pp. 80–87.

to ship 600 *salmae* of wheat from Barletta to Acre, and he gave permission for this amount to be sent.[14] In some instances rulers were also prepared to allow the set limits to be exceeded. In March 1286 Alfonso III of Aragon allowed the Templars to bring twenty horses through the kingdom from Castile to be exported to the East: he stated that only these were to be shipped.[15] But in the following month he not only agreed that a further ten should be brought from Castile but also that another ten from Aragon could be dispatched.[16] Further, it should be remembered that licences were usually for individual sailings and did not set annual limits. In February 1270 the Hospitallers received from Charles I permission to send certain amounts of grain and animals to Acre, and a further concession was made by the king in August of that year.[17] The Hospitallers were clearly planning to send supplies at both of the normal sailing times that year.[18]

Yet licences might not only impose limits on amounts but also prohibit the dispatch of certain goods. Although at times warhorses were exported from the kingdom of South Italy and Sicily,[19] in 1276 Charles I allowed mules and rounceys to be sent to the East by the Hospitallers, but not warhorses (*equi ad arma*) or other prohibited goods.[20] The exclusion of *prohibita* occurs in many licences issued by the Angevin rulers.[21] War materials were probably among these: it may be noted that, although Aragonese kings at times permitted the export of arms and military equipment,[22] these items do not normally feature in the licences issued by the Angevin kings of Sicily and Naples.[23] Yet in 1293 James II of Aragon himself forbade the lieutenant of the castellan of Amposta to export horses or rounceys when he granted this Hospitaller a safe-conduct to go to Cyprus.[24]

---

[14]   Filangieri, 1, p. 292; see also 1, p. 293; 3, pp. 189, 239; 7, p. 62.

[15]   ACA, R. 63, fol. 89.

[16]   Ibid., R. 66, fols 57v, 58; see also fol. 62.

[17]   Filangieri, 3, p. 189; 7, p. 62; *CH*, 3, p. 231, doc. 3401; see also Filangieri, 45, p. 108; Mas Latrie, 2, pp. 97–8.

[18]   The Aragonese Templars, however, normally dispatched supplies only once a year: Forey, *Templars in the Corona de Aragón*, pp. 326–7.

[19]   Filangieri, 13, p. 34; 47, p. 149; *CH*, 4, p. 304, doc. 4855 bis; on the export of horses from the Sicilian kingdom by military orders, see J.H. Pryor, 'The Transportation of Horses by Sea during the Era of the Crusades: Eighth Century to 1285 A.D.', *The Mariner's Mirror*, 68 (1982), 110.

[20]   *CH*, 3, p. 338, doc. 3599.

[21]   See, for example, ibid., 3, pp. 414–15, 596, docs. 3758, 4163; Filangieri, 24, pp. 122–3; 45, p. 108; 47, pp. 148, 149, 285.

[22]   ACA, R. 66, fol. 61v; R. 81, fol. 81; *Acta Aragonensia*, 3, pp. 8–9, doc. 5.

[23]   In 1278 a Templar was allowed to take a bow, but this was apparently his own: Filangieri, 19, p. 193. There were also other restrictions commonly imposed by the rulers of South Italy and Sicily, but these did not hinder the transport of goods for the benefit of brothers in the East: these included time limits within which goods must be sent, and prohibitions on sending supplies to places other than the Holy Land or Cyprus.

[24]   *CH*, 3, p. 623, doc. 4223.

There were occasions, however, when licences could not be obtained at all or when orders had difficulty in securing them. In a letter to the English King Edward I in 1282 the Hospitaller Joseph of Cancy stated that Charles of Anjou was prohibiting the export of all foodstuffs from the kingdom of Sicily because of war.[25] In March 1290 Alfonso III of Aragon reminded the Hospitallers that, because of conflict with France and Castile, no horses, arms, grain or money were to be exported.[26] In 1297 Boniface VIII commented that because of wars the Hospitallers were not able to export horses from the Iberian Peninsula.[27] There were also occasions when military orders turned to the papacy for help in securing licences. In December 1291 Pope Nicholas IV asked Charles II to allow the Hospitallers to export grain from Provence, and in 1295 Boniface VIII requested the kings of England, France and Germany to permit the Templars to transport goods to Cyprus.[28] Such papal orders show that the Templars and Hospitallers were at times encountering, or at least anticipating that they would encounter, difficulties in securing royal assent. It may also be noted that for most of the 1280s very few licences are known to have been issued by Charles of Anjou and Charles II;[29] and in the opening decade of the fourteenth century the Hospitaller master Fulk of Villaret considered the problem to be sufficiently important to include a comment on it in the crusading proposal he submitted to Clement V.[30]

War and dearth were apparently not the only factors that hindered exports. Relations with a particular order may also have influenced decisions. Although in March 1290 Alfonso III of Aragon was not allowing Hospitaller exports to the Holy Land, in mid-April he permitted the Templars to dispatch horses, mules, wheat, barley, oats, arms and military equipment,[31] and it was towards the end of that month that the letter quoted above was sent to the Templar grand master. At the end of April, however, Alfonso wrote to the Hospitaller master, regretting that because of war he could not provide support for the Holy Land.[32] It was only in May, after he had allegedly learned of the loss of Tripoli, that he expressed readiness

---

[25] *Lettres de rois, reines et autres personnages des cours de France et d'Angleterre*, ed. J.J. Champollion-Figeac (Paris, 1839–47), 1, pp. 288–95.

[26] *CH*, 3, p. 555, doc. 4081.

[27] Ibid., 3, pp. 695–6, doc. 4334.

[28] Ibid., 3, p. 602, doc. 4177; *Foedera*, 1.2, p. 823; *Les registres de Boniface VIII*, eds G. Digard, M. Faucon, A. Thomas and R. Fawtier (Paris, 1884–1939), 1, p. 170, doc. 489; see also Filangieri, 35, pp. 108–9; *CH*, 4, pp. 132, 212–13, 214, docs. 4726, 4860, 4862.

[29] Filangieri, 27, p. 386 (1284); 29, p. 81 (1288).

[30] J. Petit, 'Mémoire de Foulques de Villaret sur la croisade', *Bibliothèque de l'Ecole des Chartes*, 60 (1899), 607; *CH*, 4, pp. 105–10, doc. 4681; *Projets de croisade (v.1290–v.1330)*, ed. J. Paviot (Paris, 2008), p. 194.

[31] ACA, R. 81, fol. 81; M. Fernández de Navarrete, 'Disertación histórica sobre la parte que tuvieron los españoles en las guerras de ultramar o de las cruzadas', *Memorias de la Real Academia de la Historia*, 5 (1817), 174–5, doc. 16.

[32] *CH*, 3, pp. 560–61, doc. 4090.

to allow the Hospitallers as well as the Templars to send goods to the East.[33] It is tempting to suggest that Alfonso's attitude was linked with his complaint in 1288 that some Hospitallers had assisted the French king in the latter's invasion of Aragon and that two leading Hospitallers, who were close to the Aragonese king, had been poorly treated by their Order: one of them, Boniface of Calamandrana, had been sent to Armenia, where his life had been endangered by the climate.[34]

The loss of the Holy Land apparently also made some western kings more reluctant to accede to petitions from military orders. When Boniface VIII wrote to the English, French and German kings in 1295 he stated that the Templars should be allowed to transport goods for the defence of Cyprus 'as they used to, with royal permission, when they resided in the kingdom of Jerusalem'.[35] The wording suggests that the fall of Acre led to a change in attitudes. This change became most marked in Aragon, where kings had experienced increasing difficulty in obtaining military support within the Peninsula from Templars and Hospitallers,[36] although the new royal policy does not become apparent until the beginning of the fourteenth century. In 1304 James II asked Benedict XI to agree that the orders in his kingdom should not send responsions to their superiors but devote all their revenues to the struggle against Granada,[37] and during negotiations about the fate of Templar property the Aragonese king repeatedly argued that Templar revenues should be used only in Spain: he alluded to the wording of the charter of 1143 that had granted Monzón and other strongholds to the Templars 'for the defence of the western church which is in Spain'.[38] In the following decades further attempts were made by Aragonese rulers either to prevent the Hospitallers from sending

[33]   ACA, R. 73, fol. 81v. Tripoli had fallen in April 1289.
[34]   *CH*, 3, pp. 518–19, doc. 4007; *Acta Aragonensia*, 3, pp. 3–4, doc. 2. The other was Raymond of Ribelles. On these brothers, see J. Burgtorf, 'A Mediterranean Career in the Late Thirteenth Century: The Hospitaller Grand Commander Boniface of Calamandrana', in K. Borchardt, N. Jaspert and H.J. Nicholson (eds), *The Hospitallers, the Mediterranean and Europe* (Aldershot, 2007), pp. 73–85; *The Central Convent of Hospitallers and Templars. History, Organization and Personnel (1099/1120–1310)* (Leiden, 2008), pp. 500–504, 630–34. In 1276 Charles I allowed the Teutonic Order to export twelve warhorses and twelve rounceys or mules to Acre, but permitted the Hospitallers to send only rounceys and mules: Filangieri, 13, p. 34; *CH*, 3, p. 338, doc. 3599.
[35]   *Foedera*, 1.2, p. 823; *Registres de Boniface VIII*, 1, p. 170, doc. 489.
[36]   A.J. Forey, 'The Military Orders and the Spanish Reconquest in the Twelfth and Thirteenth Centuries', *Traditio*, 40 (1984), 230–34.
[37]   *Acta Aragonensia*, 1, p. 158, doc. 108; V. Salavert y Roca, *Cerdeña y la expansión mediterránea de la Corona de Aragón, 1297–1314* (Madrid, 1956), 2, pp. 98–106, doc. 72.
[38]   A.J. Forey, *The Fall of the Templars in the Crown of Aragon* (Aldershot, 2001), pp. 157–8, 160, 163; *Colección de documentos inéditos del Archivo General de la Corona de Aragón*, eds P. de Bofarull y Mascaró *et al.* (Barcelona, 1847–1910), 4, pp. 93–9, doc. 43; *Col·lecció diplomàtica de la casa del Temple de Barberà, (945–1212)*, ed. J.M. Sans i Travé (Barcelona, 1997), pp. 110–14, doc. 35.

responsions or to limit the size of payments to the East.[39] The reason for changing attitudes in Aragon is expressed in a comment made by James II's envoy Vidal of Vilanova, who in 1305 advised the king to seek to have the military orders included in the tenth that James was negotiating with the pope, 'especially as they do not now undertake any expenditure in the Holy Land'.[40] Yet the Aragonese kings did not gain any permanent concessions on these points.

It was also after the loss of the Holy Land that responsions began to be appropriated, presumably – at least in some cases – for the reasons expressed by Vidal of Vilanova. For most of the thirteenth century kings and popes had not sought to commandeer the orders' responsions,[41] although taxes that they imposed on the orders made the dispatch of supplies to the East more difficult, and Aragonese kings had threatened confiscation of property for failure to provide military service.[42] After he had decreed the exaction of a tenth for Sicily, Nicholas IV in March 1290 specifically exempted the orders' *tertias* (responsions), even if their liability for payment on the other two-thirds of their revenues probably hampered the payment of responsions; and the exemption had to be repeated when it was not being observed.[43] But it was apparently only after the fall of Acre, and the Templars and Hospitallers had become established in Cyprus, that popes began to make direct demands on the orders' responsions. In August 1291, when he was planning to dispatch ships to the East, Nicholas IV instructed the heads of Templar and Hospitaller provinces in the West to send to him half of their responsions and of any other aid that they had formerly sent to the Holy Land as a contribution to the costs of the fleet that he was then fitting out.[44] When Boniface VIII was seeking a subsidy from the orders six years later, he mentioned that he had made no demands on the orders' responsions during the preceding two years, which suggests that, like Nicholas IV, he had previously been appropriating some of the

---

[39]   A.T. Luttrell, 'The Aragonese Crown and the Knights Hospitallers of Rhodes: 1291–1350', *English Historical Review*, 76 (1961), 9–10, 12–13. In 1339 Edward III asserted that the revenues of the English Hospitallers should be devoted to the defence of the realm, and not sent to the East: *Calendar of Close Rolls, 1339–1341* (London, 1901), pp. 256–7.

[40]   Salavert y Roca, 2, pp. 151–5, doc. 116.

[41]   Luttrell, p. 9, states that Aragonese kings sought to obtain responsions as early as 1277; but at that time Pedro III was merely appropriating some of the proceeds of the crusading tenth that had been deposited with the Templars and Hospitallers: ACA, R. 39, fol. 225–225v; *Documentos de Nicolás III (1277–1280) referentes a España*, ed. S. Domínguez Sánchez (León, 1999), pp. 282–5, doc. 80.

[42]   Forey, 'Military Orders and the Spanish Reconquest', p. 231.

[43]   E. Strehlke, *Tabulae ordinis theutonici* (Berlin, 1869), p. 426, docs. 665–6; *Les registres de Nicolas IV*, ed. E. Langlois (Paris, 1886–93), p. 614, docs 4204–6; *CH*, 3, p. 586, doc. 4147.

[44]   *CH*, 3, pp. 598–9, doc. 4168; *Registres de Nicolas IV*, p. 903, docs. 6796–9; on the fleet, see S. Schein, *Fideles crucis: The Papacy, the West, and the Recovery of the Holy Land, 1274–1314* (Oxford, 1991), pp. 77–8.

payments made by western provinces.[45] Yet this does not seem to have become a normal practice. In 1295, however, Boniface had himself criticized both Edward I of England and Dinis of Portugal, who were said to have seized Hospitaller responsions from their kingdoms.[46] The precise justifications for these royal actions are not known, but Edward's was presumably linked with the conflict with the French King Philip IV at that time over Gascony.

Military orders clearly did suffer at times from outside interference in the dispatch of supplies to the East, especially with regard to certain goods, and the collapse of the crusader states appears to have been an additional factor influencing the attitudes of kings and popes. Yet royal and papal interference seems to have had only an intermittent impact: for most years in the later thirteenth century there is some evidence of the dispatch of supplies, and appropriations of responsions after the loss of the Holy Land were apparently infrequent. Difficulties in the dispatch of responsions were probably more constantly occasioned by financial problems within the European provinces or priories of military orders.

---

[45]   *Registres de Boniface VIII*, 1, p. 914, doc. 2323.

[46]   *CH*, 3, p. 666, doc. 4283; *Registres de Boniface VIII*, 1, p. 168, doc. 479; *Codice diplomatico del sacro militare ordine gerosolimitano*, ed. S. Pauli (Lucca, 1733–37), 2, pp. 3–4.

Chapter 8

The Hospitallers and Charles I of Anjou: Political and Economic Relations between the Kingdom of Sicily and the Holy Land[1]

Cristian Guzzo

On 25 August 1270 King Louis IX of France died at the siege of Tunis, thereby putting an end to any hope that the settlers in the East may have had that they would ever receive effective support from western Europe. It was his brother, Charles of Anjou, who had convinced the French king to divert his expedition from the Holy Land to North Africa on the grounds that Mustansir, the emir of Tunis, seemed well disposed towards the Latins and ready to embrace Christianity.[2]

After defeating King Manfred in the battle of Benevento (26 February 1266), Charles had been crowned king of Sicily and thus fulfilled the ambitions of popes Urban IV and Clement IV to put an end to the power of the hated Hohenstaufen dynasty in Italy and place on the throne that had once belonged to Frederick II a monarch who would be interested in promoting and re-establishing southern Italy as a springboard for a new crusade to the East.[3] The former crusader and veteran of Louis IX's ill-fated crusade to Mansurah was an ambitious man who wanted to reconquer Constantinople and found the Mediterranean Empire that his Norman predecessors had never managed to create. The conquest of Tunis would have deprived the Ghibellines of a secure naval refuge, and, at the same time, guaranteed the Angevin monarchy's firm control over the eastern Mediterranean.[4]

When the Sultan Baybars learnt of Louis' project to take Tunis, he increased his pressure on the Christian fortresses in Syria, and between 7 April and 1 May 1271 he conquered two of the most important strongholds of the Knights of St John, Crac

---

[1] I would like to thank Elena Bellomo and Kristian Toomaspoeg for their valuable suggestions.

[2] E.G. Léonard, *Gli Angioini di Napoli*, trans. R. Liguori, 2nd edn (Varese, 1987), p. 126.

[3] N. Housley, *The Italian Crusades: The Papal–Angevin Alliance and the Crusades against Christian Lay Powers, 1254–1343* (Oxford, 1982), p. 68.

[4] S. Runciman, *A History of the Crusades* (Cambridge, 1951–54), 3, pp. 291–2; T.F. Madden, *The New Concise History of the Crusades* (Lanham, MD, 1999), p. 186; N. Housley, *Contesting the Crusades* (Oxford, 2006), pp. 73–4.

des of Chevaliers and the nearby castle of Akkar (Gibelacar).[5] It was now more necessary than ever before for the Hospital to summon its entire resources so as to overcome the serious situation in the East. Already, in April 1265, the Order had lost the town of Arsuf and with it ninety brothers who died in the fighting.[6] A further forty-five knights and the grand commander, Stephen of Meses, had been killed by the Muslims in October 1266 at ar-Ruwais (Carroblier) on the plain of Acre.[7]

Hospitaller finances in the Holy Land had been damaged by the truce agreed in 1267 between Baybars and the Hospitallers of al-Marqab; the Order was obliged to forego the tribute it received from some Muslim towns and districts, including the castles of the Assassins, whose tribute was estimated to be in the region of 1,200 gold dinars in addition to quantities of wheat and barley.[8] This agreement did not prevent the sultan from laying waste the area around al-Marqab and Crac de Chevaliers in 1270 and pasturing his army's horses on the growing crops in the vicinity.[9] The progressive reduction of the Order's resources in the East was so alarming that in 1268 the grand master, Hugh Revel, complained that he could only maintain 300 knights in Syria, as compared with 10,000 he alleged had once existed.[10]

Furthermore, the Hospitallers were badly damaged in the *Regnum Sicilie* during the Hohenstaufen era. As a direct result of the serious conflict between the Papacy and the Empire, the Order suffered losses, particularly in the Capitanata that produced the surplus necessary to maintain the *confrères* engaged in military operations in the East and the activities prescribed by the Rule.[11]

---

[5]   J. Cathcart King, 'The Taking of Le Krak des Chevaliers in 1271', *Antiquity*, 23 (1949), 83–92; N.A. Mirza, *Syrian Ismailism: The Ever Living Line of the Imamate, AD 1100–1260* (Richmond, 1997), p. 43; Madden, p. 186; K. Molin, *Unknown Crusader Castles* (London, 2001), p. 36.

[6]   J. Delaville Le Roux, *Les Hospitaliers en Terre Sainte et Chypre (1100–1310)* (Paris, 1904), p. 219; C.R. Humphery-Smith, *Hugh Revel: Master of the Hospital of St John of Jerusalem 1258–1277* (Chichester, 1994), p. 50; J. Waterson, *The Knights of Islam: The Wars of the Mamluks* (London, 2007), p. 103.

[7]   *Eracles*, p. 455; Delaville Le Roux, p. 219.

[8]   P.M. Holt, *Early Mamluk Diplomacy (1260–1290): Treaties of Baybars and Qalawun with Christian rulers* (Leiden, 1995), pp. 32ff.

[9]   J. Bronstein, *The Hospitallers and The Holy Land: Financing The Latin East, 1187–1274* (Woodbridge, 2005), p. 43.

[10]   Runciman, 3, p. 344.

[11]   For the conflict with Frederick II, C. Morris, 'The Case of the Missing Martyrs: Frederick II's War with the Church, 1239–1250', *Studies in Church History*, 30 (1993), 141–52. For the repercussions for Hospitaller interests, R. Iorio, 'Ospedalieri a Barletta e dintorni fra vescovi e papi, sovrani e sultani', *Studi Melitensi*, 2 (1994), 96ff. For Hospitaller presence in southern Italy during the Hohenstaufen period, G. de Troia, *Foggia e la Capitanata nel Quaternus Excadenciarum di Federico II di Svevia* (Foggia, 1994); A. Luttrell, 'Gli Ospedalieri nel Mezzogiorno', in G. Musca (ed.), *Il Mezzogiorno normanno-svevo e le crociate. Atti delle quattordicesime giornate normanno-sveve Bari 17–20*

It should also be emphasized that the Order had a constant need to recruit new fighters to replace those who died and or were injured in the Holy Land. Bearing in mind that between 1265 and 1266 the Hospitallers had lost 135 men, it may not be too fanciful to suppose that they sought to recruit new brothers from among the numerous pilgrims who arrived in Apulia from every part of Europe to visit the sanctuary of St Michael and then in some cases continue their journey to the Holy Land.[12]

The confiscation, at Frederick II's instigation, of many of the properties belonging to the Hospitallers would seem to have caused a drastic reduction in recruitment. Whereas the emperor's will required his heirs to return Templar properties, the Hospitallers did not benefit in the same way.[13] Nevertheless, in 1252 Conrad, the new king of the Romans, at the request of Rambaldus, the Hospitaller grand preceptor of Italy, Hungary and Austria, confirmed the Order's possession of all the assets it had owned during Frederick's reign and those of his predecessors. This provision presumably remained unrealized, for in January 1254 Pope Innocent IV censored Conrad for seizing the possessions of both the Templars and the Hospitallers and expelling the *praelatos et rectores*.[14]

The Order's position improved during the reign of Charles of Anjou, who in 1267, at the request of Pope Clement IV and with the approval of the grand master, Hugh Revel, was able to use the military resources of the former Hospitaller prior of France, Philip d'Egly, and other brothers in the fighting against the last Hohenstaufen partisans.[15] Philip was so zealous against Charles' enemies, that he placed most of the Order's resources in southern Italy, so important for the war in the East, as well as the income from his former French priory at the disposal of Angevins. The Hohenstaufen supporters retaliated by devastating Hospitaller property in Sicily.[16] Philip's conduct was condemned by Hugh Revel who found himself in the difficult position of having to please the king and the pope and, at

---

ottobre 2000 (Bari, 2002), pp. 289–300; K. Toomaspoeg, 'Le patrimoine des grands ordres militaires en Sicilie, 1145–1492', *Mèlanges de l'École Française de Rome. Moyen Âge*, 113 (2001), 313–41; idem, *Templari e Ospedalieri nella Sicilia Medievale* (Bari, 2003), pp. 57–68; R. Iorio, 'Uomini e sedi a Barletta di Ospedalieri e Templari come soggetti di organizzazione storica', in C.D. Fonseca and C. D'Angela (eds), *Barletta crocevia degli Ordini religioso-cavallereschi medioevali. Seminario di Studio, Barletta 16 giugno 1996* (Taranto, 1997), pp. 71–113.

[12]   For an example from c. 1191, G. Cherubini, *Pellegrini, pellegrinaggi Giubileo nel Medioevo* (Turin, 2000), p. 28 n. 41.

[13]   J. Riley-Smith, *The Knights of St. John in Jerusalem and Cyprus c.1050–1310* (London, 1967), p. 174.

[14]   B. Capasso, *Historia Diplomatica Regni Siciliae inde ab anno 1250 ad annum 1266* (Naples, 1874), p. 34 no. 64, pp. 60–61 no. 113.

[15]   For Philip, *I registri della cancelleria angioina*, ed. R. Filangieri (Naples, 1950–2006), 1, p. 118 no. 46. For the Hospitallers and the last Hohenstaufen supporters in southern Italy, H. Nicholson, *The Knights Hospitaller* (Woodbridge, 2001), p. 42.

[16]   Delaville Le Roux, p. 221.

the same time, protect the interests of the Hospital, whose condition in the Holy Land was dire.[17]

Despite the serious losses suffered in the kingdom of Sicily and the subsequent, albeit temporary, halt to the transfer of goods from southern Italy to the Holy Land, the Hospitallers were soon able to reap the benefits of their good relations with Charles. Although on the one hand Hugh Revel suffered the loss of some important resources, on the other hand he would have welcomed the fiscal and other privileges he obtained from the *regia curia*.[18] The king was grateful for the financial support he had from the Order of St John, and so he eased the path for the restitution of properties seized by Frederick II, many of which had been leased to private individuals and produced a good annual income for the crown.[19] Moreover, the king exempted the brothers from the payment of *jus exiture*, granting their ships the free use of the ports of the *Mezzogiorno* for repairs and the shipment of goods destined for their settlements in the East.[20] He also granted them exemption from the royal salt monopoly.[21] It is significant that in June 1271 the king ordered the *portolano* of Barletta, Risone of Marra, not to demand the payment of *ius ballistarum* from the local Hospitallers, on the grounds that they had been exempt from it since the time of Frederick II.[22] It is also recorded that on 6 July 1269, in the same year that King James I of Aragon decided to go to the East on a new crusade, Charles authorized Fr John de Vilers to send to the East five mules and horses, *iuxta mandatum maioris ultramarini Magistri*.[23] On 18 July of that year, the king, at the instance of Peter of Neocastro, Hospitaller prior of Barletta, authorized the Order to dispatch 2,000 *salme* of grain, 1,000 of barley, and 100 of legumes *de portibus Apulie* to Acre for the maintenance of the members of the Order, their households and the poor who were in the Order's care.[24] Meanwhile political tensions were increasing between the Knights of St John and Baybars, who in 1269 led his troops to al-Marqab before being forced to retreat to Damascus by bad weather.[25] In 1270 the grand master, Hugh Revel, wrote personally to Charles from Acre to ask permission for wheat, barley, and sixteen mules and horses to

---

[17]   Bronstein, pp. 93–4.

[18]   Delaville Le Roux, p. 227.

[19]   M. Salerno, *Gli Ospedalieri di San Giovanni di Gerusalemme nel Mezzogiorno d'Italia (secc. XII–XV)* (Taranto, 2001), p. 184. For Hospitaller properties at Foggia, de Troia, pp. 169–77.

[20]   Filangieri, 3, p. 189 no. 474; 15, p. 41 no. 166. P. Di Biase, *Puglia medievale e insediamenti scomparsi. La vicenda di Salpi* (Fasano, 1985), 146; D. Jacoby, 'Hospitaller Ships and Transportation across the Mediterranean', in K. Borchardt, N. Jaspert and H.J. Nicholson (eds), *The Hospitallers, the Mediterranean and Europe: Festschrift for Anthony Luttrell* (Aldershot, 2007), pp. 57–72.

[21]   Filangieri, 1, p. 292 nos. 428–9 and 27/1, p. 368 no. 792.

[22]   Ibid., 6, pp. 248–9 no. 1328.

[23]   Ibid., 3, p. 286 no. 2.

[24]   Ibid., 1, p. 286 no. 402.

[25]   Riley-Smith, p. 192.

be sent *in subsidium Terre Sancte* from the ports of Brindisi and Barletta. On 2 February 1271 the king excused Fr Ferrando Melardo from paying 30 gold ounces of the money owed to the custom officers in Messina in respect of the horses and mules *quos de Yspanie partibus detulit et ad ultramarinas partes intendit ducere*.[26]

The good diplomatic relations established with the court at Naples made it possible for some Hospitallers to hold important positions in the public administration and in the king's personal service. We know that in 1271 Fr Peter from the house at Barletta held the office of royal almoner, while Simon of Breban was a royal chaplain.[27] Fr Jacques de Taxi was very important, and his appointment as a political adviser to Charles was suggested by Hugh Revel in person, who, I believe, wanted to install in the Sicilian court a trustworthy agent who could find out in advance the king's political strategies.[28]

It would appear, however, that despite appearances Hugh Revel did not trust Charles. After all, the king was a man without scruples; he had not hesitated to dispose of the youthful Conradin of Hohenstaufen, having him beheaded in the Campo Moricino in Naples, and he had also turned the failed Tunis crusade to his own advantage, obtaining 210,000 gold ounces from the Emir Mustansir, the expulsion of the last Hohenstaufen supporters, and the doubling of the tribute paid from Tunis to the Neapolitan Crown.[29] It is therefore reasonable to assume that the grand master was not pleased when Pope Gregory X supported the rights of Maria of Antioch to the throne of Jerusalem against the legitimate incumbent, King Hugh III of Cyprus, and urged Maria to sell her rights to the crown to Charles of Anjou.

The last thing the Holy Land would have wanted was a king who was distant, both diplomatically and geographically, from the country he was to govern, especially at such a delicate moment for the fortunes of the Latin settlements in the East. So when Hugh III, exasperated by Templar opposition and by the continued intransigence of the barons and the Italian colonies, decided to abandon Acre and withdrew first to Tyre and then to Cyprus, Hugh Revel rushed to Tyre with the patriarch of Jerusalem and other leading figures to ask him to change his mind.[30] But Hugh III was inflexible, and on 7 June 1277 Count Roger of San Severino arrived at Acre with six galleys determined to assert the rights of his master Charles to the throne of Jerusalem. Charles in the meantime had purchased

---

[26] Filangieri, 3, p. 189 no. 474 and 6, p. 175 no. 907.

[27] Ibid., 5, p. 129 no. 121; Salerno, p. 185.

[28] Filangieri, 1, 286 no. 403; 6, p. 147 no. 747, p. 201 no. 1067; 7, p. 233 no. 142; 10, p. 30 nos. 108–9; 15, p. 57 no. 258. Toomaspoeg, *Templari e Ospedalieri*, p. 71.

[29] J. Richard, *The Crusades c. 1071–c. 1291*, trans. J. Birell (Cambridge, 1999), p. 432.

[30] *Eracles*, p. 474; Riley-Smith, p. 188; C. Guzzo, *Templari in Sicilia. La storia e le sue fonti tra Federico II e Roberto d'Angiò* (Genoa, 2003), p. 75.

Maria of Antioch's claim.[31] The Hospitallers offered no resistance and eventually accepted the new Angevin regime in the Holy Land.[32]

Hugh Revel died in 1277. The apparent submissiveness of the Order to Charles could have been the work of his successor, Nicholas Lorgne, who, following his predecessor's policy, had no wish to damage relations with the Angevin court but, at the same time, had no intention of ceasing hostilities against the Muslims; he himself was wedded to the idea of a Christian alliance with the Mongols as necessary to the eventual defeat of Baybars.[33] For his part, Charles hated the Mongols because of their friendship with the Byzantines, and he wanted to reach a rapid agreement with the Mamluk court so as to be able to pacify the East and to ensure the survival of Latin settlements that were now reduced to Acre, Tyre, Sidon, Tripoli, Jubail and Tortosa, with the isolated town of Lattakia and the castles of Atlit and al-Marqab.[34] His first priority, however, was the conquest of Byzantium, and the acquisition of the crown of Jerusalem could be seen as an attempt to complete the encirclement of Byzantine territory that he had begun in 1272 with the conquest of Albania, the acquisition of rights of succession in the principality of Achaia and the conclusion of friendship pacts with John Comnenus, duke of Patras, the Serbian king, Stephen Uroch, and Tsar Constantine Assen of Bulgaria.[35]

With Emperor Michael VIII's promise in 1274 at the Council of Lyon to secure the subjugation of the Greek Church to Rome, Charles had, temporarily, to abandon his designs on Constantinople, but he was more than ever disposed to make war on the Greeks.[36] The documents of the Angevin chancellery show how the Hospitaller houses had an important role in the programme of conquest of Byzantium undertaken by Charles. These houses were used by Charles for some of his financial operations as well. The king had to pay his army in Achaia and Romania with Venetian money. Accordingly he ordered his officials to buy this currency at the *pittagio* of the exchange in Barletta and then deposit the money in the local house of the Order.[37] On 1 April 1271 Charles ordered the prior of Barletta to hand over 1,000 ounces of gold to Angel of Marra, *magister rationalis* of the *curia*, to buy money from the republic of Venice. This Venetian currency

[31]   *Eracles*, pp. 475–8; *Cronaca del Templare di Tiro (1243–1314)*, ed. L. Minervini (Naples, 2000), p. 151; J. Dunbabin, *Charles I of Anjou: Power, Kingship and State-Making in Thirteenth-Century Europe* (London and New York, 1998), p. 96.

[32]   Riley-Smith, p. 189; Léonard, p. 155.

[33]   *Cronaca del Templare di Tiro*, pp. 152–3; Riley-Smith, p. 190.

[34]   Dunbabin, pp. 96–7; Runciman, 3, p. 348.

[35]   E. Fryde, *The Early Palaeologan Renaissance, 1261–c.1360* (Leiden, 2000), pp. 85–7; Dunbabin, p. 90; J.V.A. Fine, *The Late Medieval Balkans: A Critical Survey from Late Twelfth Century to the Ottoman Conquest* (Ann Arbor, MI, 1994), p. 170; Léonard, pp. 127–8.

[36]   J.M. Hussey, *The Orthodox Church in the Byzantine Empire* (Oxford, 1986), pp. 229ff.

[37]   Filangieri, 10, pp. 41–2 no. 141; F. Carabellese, *Carlo d'Angiò nei rapporti politici e commerciali con Venezia e l'Oriente* (Bari, 1911), p. 152.

would have been entrusted to the Hospitaller house at Barletta.[38] On 26 February 1273 Charles wrote to William of Barris, captain general of the troops located in Romania, to inform him of the arrival of Sergius Bove of Ravello with a ship loaded with 2,700 *salme* of wheat and 300 of barley. These assets would have been sold, and the proceeds delivered to the master of the Hospitaller house at Clarentza and used to pay the troops.[39]

Meanwhile in the Holy Land the Hospitallers continued to pursue an aggressive policy against the Muslims. On 1 July 1277 Baybars died, and Qalawun inherited his position. The Franks did not know how to take advantage of the situation, and Ilkhan Abaga and his vassal Leo III of Armenia called in vain for the formation of an alliance and a new crusade. Only the Knights of St John supported this strategy, while Charles of Anjou ordered a policy of coexistence with the new sultan.[40] He had no intention of opening a new military front in the East that would have forced him to divert men and resources from his planned war in Byzantium.

At the end of September 1280 a Mongol army crossed the Euphrates, occupied Aintab, Baghras and Darbsaq, and laid waste to Aleppo. Taking advantage of the general panic and without any regard for the policy of non-belligerence adopted by Roger of San Severino towards the Mamluk court, 200 Hospitallers from al-Marqab conducted a very fruitful raid around Maraclea, defeating a strong Muslim force that pursued them.[41] Shortly afterwards the sultan's troops retaliated by devastating the countryside around al-Marqab and attempting an assault on the castle itself. A contingent of 600 mounted Hospitallers made a sortie and managed to overwhelm the Muslims, gaining horses and equipment in the process.[42]

Meanwhile a Mongol ambassador arrived in Acre to inform the Franks that the Ilkhan wanted to send 100,000 men to Syria the following spring and to ask them to cooperate with men and materials. The Hospitallers forwarded this message to King Edward I of England, the only monarch still interested in a new crusade, but no answer was forthcoming from the Angevin government of Acre. Terrified by the Mongol threat, Qalawun sent messengers to Acre to propose a truce with the military orders that was signed on 3 May 1281.[43] Roger of San Severino could be satisfied, since this new agreement was in keeping with Charles' intentions.

Hospitaller survival in the Holy Land depended in large measure on the shiploads of food and other goods coming from their properties in southern Italy.[44] Grand Master Nicholas Lorgne accepted the truce agreed with the sultan because he did not want to break off diplomatic relations with the Angevins and perhaps

---

[38]   Filangieri, 10, p. 37 no. 126.

[39]   Ibid., 9, p. 166 no. 233.

[40]   Runciman, 3, pp. 348, 387.

[41]   *Cronaca del Templare di Tiro*, pp. 151–2. For the date, Delaville Le Roux, pp. 231–2 and n.; Riley-Smith, p. 194, n. 2.

[42]   *Cronaca del Templare di Tiro*, pp. 154–5.

[43]   Runciman, 3, pp. 390–91; Riley-Smith, p. 194.

[44]   Delaville Le Roux, p. 233.

because he had faith in a new crusade to be led by Edward of England that would unite all the Christian forces in the East against the common Muslim threat.

Charles constantly intervened in the territories of *Regnum Sicilie* in favour of the Knights of St John, defending them from the abuses of his officials and from the violence of private individuals who were envious of their privileges.[45] Opposition to his eastern policy would probably have affected the number of guarantees and privileges that the Hospitallers enjoyed, not only in the *Mezzogiorno*, but in the East as well. Moreover, the chronicler Ramon Muntaner ironically revealed that Charles loved to accumulate titles and styled himself 'Vicar General' of all land overseas and the supreme leader of all Christians in the East and of the orders of the Temple and the Hospital.[46]

But despite the authoritarianism of the king of Naples, Nicholas Lorgne did not hold back from forming an alliance with the Mongols and King Leo III of Armenia, so important was it to launch an offensive to reduce Mamluk power. The Hospitaller grand master could not officially send his Order's troops from Acre lest he alienate Roger of San Severino, who had to be given to understand that the Knights of St John were respecting the treaty with Qalawun and thus Charles's orders. All the same, Nicholas ordered reinforcement of the defences of al-Marqab, and, with the king of Naples's consent, had fresh consignments of food and horses arrive in Acre from southern Italy.[47] When in September 1281 two Mongol armies advanced into Syria, the Hospitallers at al-Marqab joined them, declaring that they were not to be in any way bound by the treaty signed by the Order at Acre.[48] The strategy of the Hospitallers seems clear. By strengthening al-Marqab still further, Nicholas Lorgne enabled his garrison there to withstand the retaliation that the sultan would have certainly inflicted as a result of their alliance with the Mongols and the Armenians. Meanwhile, the Knights of St John in Acre had officially respected the pact with the sultan, retained the favour of Roger of San Severino, passing off the actions of the commanders at al-Marqab as a local initiative.

Had the Mongol and Armenian coalition triumphed, the Knights of St John would have been lauded in the West as the saviours of the Holy Land and would have regained their political influence and prestige, in eclipse since the loss of their strongholds of Crac and Akkar. If the Order had put a stop to the run of Christian defeats, the sultan would not have dared attack Acre, fearing retaliation from King Charles whom he considered a powerful and formidable figure; instead he could have tried to vent his fury on the well-equipped castle of al-Marqab that would have been well able, as had already happened several times in the past, to resist Muslim attack.

---

[45]   Filangieri, 3, p. 288 nos 6, 8; Salerno, p. 66.

[46]   A. Demurger, *I Templari. Un Ordine cavalleresco cristiano nel Medioevo*, trans. E. Lana (Milan, 2006), p. 361.

[47]   Delaville Le Roux, p. 233 n. 2; Riley-Smith, p. 194.

[48]   Runciman, 3, p. 391.

On 30 October 1281 the Hospitallers of al-Marqab fought alongside the Mongols and the Armenians in the battle at Hims against the Mamluks. The Muslims were victorious, and Roger of San Severino went to meet the sultan and congratulate him on his victory, thereby emphasizing that the participation of the Hospitallers from al-Marqab in this battle should be seen as an action that was contrary to the intentions of the Angevin court.[49]

On the evening of 30 March 1282 the Sicilians rebelled against their Angevin masters and massacred all the French who were on the island. The revolt was fomented by the agents of the Byzantine emperor, Michael VIII Palaeologus, and more especially by those of Peter III of Aragon, who had inherited the claims to Sicily of Manfred of Hohenstaufen, thanks to his marriage to his daughter Constance. Charles was forced to recall the troops he had sent to Achaia to quell the revolt, and did not hesitate to involve the military orders in a conflict that was to undermine the integrity of his kingdom irreparably.[50] Between 23 December 1283 and the beginning of March 1284 the Hospitaller Peter of Moisac, prior of Sta Eufemia, played an active role in protecting Calabria on behalf of the Angevins, while on 29 April 1284, following a solemn council held in Melfi that imposed a tax on the military orders for the war against the Aragonese, the Hospitallers at Barletta and Capua were forced to send men and equipment to assist King Charles, or, alternatively, to produce a substantial subsidy.[51] These demands came at a time of deep crisis for the Order. The political destabilization of the Sicilian kingdom endangered its seaborne links with their settlements in the East and, at the same time, led to the inexorable erosion of Angevin power in Acre and the consequent revitalization of Qalawun's ambition to drive the Latins from the Holy Land forever. Once again, as in 1267, the financial resources that the Order had in the *Regno* were used to support the secular interests of the Angevin dynasty.

In June 1283 the economic problems facing the Hospitallers forced Nicholas Lorgne to renew the truce that Baybars had agreed ten years earlier with the Franks and that now was confirmed in a fresh treaty between the sultan and the new *bailli* in Acre, Odo Poilechien, who had replaced Roger of San Severino. Roger, in the meantime, had returned to Italy with all the troops available to him, leaving the Holy Land bereft men capable of bearing arms.[52]

---

[49] Ibid., 3, p. 392.

[50] *Cronaca del Templare di Tiro*, pp. 160–61; Léonard, pp. 176–7; Dunbabin, p. 109.

[51] M. Amari, *La guerra del vespro Siciliano*, 5th edn (Turin, 1852), p. 208 n. 1; C. Guzzo, 'Milites Templi Hierosolimitani in Regno Sicilie. Vecchi documenti, nuove acquisizioni', in C. Guzzo (ed.), *I Templari nell'Italia centro-meridionale. Storia ed Architettura* (Tuscania, 2008), p. 76.

[52] S. Runciman, *The Sicilian Vespers* (Cambridge, 1958), p. 277. For the truce, P.M. Holt, 'Qalawun's Treaty with the Latin Kingdom (682/1283): Negotiation and Abrogation', in U. Vermeulen and D. De Smet (eds), *Egypt and Syria in the Fatimid, Ayyubid and Mamluk Eras*, 2: *Proceedings of the 4th and 5th International Colloquium organized at the Katholieke Universiteit Leuven in May 1995, 1996*, Orientalia Lovaniensia Analecta 73 (Leuven, 1998), pp. 329ff.

The truce with the Hospitallers guaranteed the protection of their landed properties in the kingdom of Jerusalem, but Qalawun, who had not forgotten the Hospitaller participation at the Battle of Hims, prepared a surprise attack on the castle of al-Marqab.[53] Charles of Anjou died on the 17 January 1285, and his son Charles II was too busy fighting the Aragonese in Sicily to worry about the fate of the Holy Land. Certain of being able to operate completely undisturbed, the sultan appeared under the walls of al-Marqab on 17 April 1285, and on 25 May secured its surrender, permitting the surviving Hospitallers to retire in safety to Tortosa and then Tripoli.[54]

*   *   *

The fall of al-Marqab marked the end of Hospitaller power in the East. If the grand masters wanted friendly relations with the Neapolitan crown, they nevertheless followed a policy that conflicted with Angevin interests, as they held it necessary to maintain armed pressure on the Muslims as a means of curbing their political and military aggression. Whereas the Hospitallers owed Charles a big debt of gratitude for the many privileges he had given them, and while they also knew that their survival in the Holy Land depended in large part on supplies coming from southern Italy, they maintained a sort of diplomatic ambiguity that allowed them to counter discreetly, but at the same time vigorously, the political inertia of the Angevin monarchy, an inertia shared by the Templars. Instead they made the interests of the Latin East their first priority, unlike Charles who treated them as just one among his many interests.

The fall of al-Marqab shook the Christians of Acre from the torpor that had been forced on them by a *roi fainéant* whose failures in international policy had disastrous consequences for the Latin East. The death of Charles in 1285 allowed the Hospitallers temporarily to return as major players in the affairs of the Latin East, and they successfully supported the return of the Cypriot royal family to the throne of Jerusalem, a move that was endorsed, with some hesitation, by the Templars. But they were not destined to reap any benefit from the new political order, and within a few years the Muslims drove the Christians from their remaining strongholds.

---

[53]   Delaville Le Roux, pp. 234–5.
[54]   *Cronaca del Templare di Tiro*, pp. 166–7; Runciman, 3, p. 395.

Chapter 9

King James II of Cyprus and the Hospitallers: Evidence from the *Livre des Remembrances*

Nicholas Coureas

The Hospitallers constantly and unwaveringly supported Queen Charlotte during the civil war in Cyprus that took place between 1460 and 1464, and which ended when her illegitimate half-brother James succeeded in wresting the kingdom from her helped by a force of Mamluks. Yet documents from the years 1468–69 preserved in the *Livre des Remembrances*, the one extant register of acts recorded in the Cypriot royal chancery from James's reign, show that the Hospitaller Order and its members continued to possess estates and incomes in Cyprus and paid taxes to the crown in cash and kind. As king, James II himself corresponded with the Hospitaller grand master in Rhodes over the ransoming of Muslim captives, and his relationship with the Order was more nuanced than one might suppose in view of the support it was giving his half-sister, the lawful queen of Cyprus. In this paper the evidence from the documents will be examined, and an explanation offered for James's seemingly harmonious relations with the Hospitallers.

James's relations with the Hospitallers prior to the civil war had been cordial. In the summer of 1457 two Hospitallers killed a man named Sciarra, whose family and friends then accused Prince John of Coimbra, who had come to Cyprus as Charlotte's betrothed, of harbouring the murderers in his house. Sciarra's brother and his friends went to the house and in the fight that ensued outside two of the prince's servants and one of the other party were also killed, an event that allegedly occurred a few days before 21 June 1457.[1] Greatly saddened by this event, Prince John fell mortally ill and died on that date. Meanwhile the aggrieved Charlotte complained to James, then 17 years old, that Thomas the chamberlain of Cyprus, a foster-brother of Queen Helena Paleologina, had instigated the fight. James duly avenged his sister by having Thomas assassinated by two Sicilians in his pay on 13 July.[2] Towards the end of 1457 James, under a cloud because of Thomas's

---

[1] Tzortzes (M)poustrous (Georgios Bo(s)tr(y)enos e Boustronios), *Diegesis Kronikas Kyprou*, ed. G. Kehayioglou (Nicosia, 1997), pp. 4–5; George Boustronios, *A Narrative of the Chronicle of Cyprus 1456–1489*, trans. N. Coureas (Nicosia, 2005), §3.

[2] Kehayioglou (ed.), *Diegesis*, pp. 4–11; Boustronios, *Narrative*, §4; G. Grivaud, 'Une petite chronique chypriote du XVe siècle', in M. Balard, B.Z. Kedar and J. Riley-Smith

murder, went with a few companions to Rhodes, staying there for no less than five months as a guest of James de Milly, the Hospitaller grand master. After his return to Cyprus in March 1458 James asked Sir Anthony Silouan, the vicar of the Latin cathedral of the Holy Wisdom in Nicosia, to recommend him to King John II.[3] Sir Anthony is mentioned in a document of 1468 from the *Livre des Remembrances* as receiving a clerical assize of 60 bezants per annum from the Hospitaller lands in Cyprus,[4] and his connection with the Order may have gone back to the year 1458 or even earlier.

It was after the death of King John II on 26 July 1458 and the emergence of James's ambition to acquire the kingdom of Cyprus from his half-sister Charlotte, who was crowned on 15 October, that the Hospitallers turned against him. Following James's departure for Cairo in late 1458 and the arrival in Cyprus of Louis of Savoy in October 1459 to marry Queen Charlotte, both the Hospitallers and James solicited the support of the Mamluk sultan with regard to the throne of Cyprus. Eventually it was James who prevailed, and he returned to Cyprus on 18 September 1460 with a Mamluk army.[5] The Hospitaller envoy, John Dolfin the commander of Nisyros, was sent to Cairo in late 1459 in an attempt to win over the sultan, and he was still detained there in June 1461, when the Hospitaller grand master sent the Catalan merchant Bartholomew de Parete to Alexandria to deliver some letters to him; Bartholomew was then to proceed to Cairo to protest over Dolfin's detention and, if the occasion was opportune, ask the sultan to have Queen Charlotte restored to the throne.[6] The Hospitallers sent forces to garrison Kyrenia, held by Queen Charlotte's supporters. Yet a letter from James de Milly to Nicholas de Courogne, the commander of Treviso, and to John de Chailly, commander of Auxerre, instructing them to go to Cyprus and have Louis of Savoy escorted away from Cyprus should he so wish, also specified that they should seek to secure Louis' assent for a temporary accommodation with James so that their goods and incomes from Kolossi, then in the care of Brother William de Combort, would be secure. This is a clear indication that the Hospitallers wished to reach an accommodation with James, notwithstanding the assurances that the two commanders were instructed to give Louis of Savoy that the arrangement would not prejudice the resistance to James centred on Kyrenia. They were also to assure Louis that just as by custom the Hospitallers did not render homage to

---

(eds), *Gesta Dei per Francos: Crusade Studies in Honour of Jean Richard* (Aldershot, 2001), pp. 334, 338.

[3]   Kehayioglou (ed.), *Diegesis*, pp. 12–17, 34–7; Boustronios, *Narrative*, §§7, 9, 18.

[4]   *Le livre des remembrances de la secrète du royaume de Chypre (1468–1469)*, ed. J. Richard (Nicosia, 1983), no. 196.

[5]   Kehayioglou (ed.), *Diegesis*, pp. 72–7; Boustronios, *Narrative*, §§41–3; L. de Mas Latrie, *Histoire de l'île de Chypre sous le règne des princes de la maison des Lusignan* (Paris, 1852–61), 3, pp. 96–9; G. Hill, *A History of Cyprus* (Cambridge, 1940–52), 3, pp. 555–6.

[6]   de Mas Latrie, *Histoire*, 3, p. 86.

the kings of Cyprus for the Commandery of Kolossi, so now they would refuse to render homage to James. Furthermore, they would continue as always to guard it well so that James would not be able to take possession of it.[7] James de Milly, anxious to emphasize his declared support of Charlotte at a time when he was also making overtures to James, also wrote in November 1460 to the Hospitaller houses in western Europe, accusing James of apostasy to Islam and warning of the dangers resulting from a Muslim annexation of Cyprus, among which were the loss of Cypriot supplies for Rhodes such as grain.[8] Queen Charlotte, who had left for Rhodes in late 1460 or early 1461, was to stay there until 1474. Following the death in July 1473 of James, who had gained full control of Cyprus in 1464, the Hospitallers supported Charlotte's vain attempts to regain her throne, but on 4 July 1474 she left Rhodes for Italy, never to return.[9]

The Hospitallers gave overt support to Queen Charlotte on the international scene, but on Cyprus itself their policy was more nuanced. On 18 October 1460 the Grand Master James de Milly instructed Brother William de Combort, the lieutenant in command of the Cypriot Grand Commandery of Kolossi, not to place it in the possession of either James or the Mamluk sultan but to assure James that he would have the same obedience from the Order as had been given by custom to his predecessors. James clearly understood that the Hospitallers wished to have peaceful if not cordial relations with him, and on 3 March 1462 he gave William and his household of up to twenty men leave to either stay in Cyprus or travel to Rhodes at will.[10] Following his capture of Kyrenia in 1463 and of Famagusta in January 1464 James controlled the whole of Cyprus and could easily have meted out reprisals on the Hospitallers, constant in their support for Queen Charlotte. The documents found in the *Livre des Remembrances*, however, show that he had good relations with individual Hospitallers in Cyprus. A document of 6 March 1468 records James granting an annual income from the royal estates of 200 measures of corn and 300 measures of pulses to the Hospitaller brother Gomez d'Avila.[11] This brother is probably the Brother Gomez mentioned in the late fifteenth-century chronicle by George Boustronios, a partisan and contemporary of James, whose fiefs and *casalia* were granted after James's death to a certain John de Navarre on 21 January 1474.[12] The *Livre des Remembrances* mentions in a document of 14 June 1468 that the king had also granted Brother Gomez the fiefs of Vavatsinia and Trypi, the latter a dependency of the *casale* of Psimolophou, both having formerly

---

[7]   Kehayioglou (ed.), *Diegesis*, pp. 84–9; Boustronios, *Narrative*, §§50–51; de Mas Latrie, *Histoire*, 3, pp. 104–6.

[8]   Ibid., 3, pp. 108–13.

[9]   Kehayioglou (ed.), *Diegesis*, pp. 162–71; Boustronios, *Narrative*, §§113, 116–19, 123–4, 127; Hill, 3, pp. 599–601; A. Luttrell, 'Ta stratiotika tagmata', in Th. Papadopoullos (ed.), *Historia tes Kyprou*, 4, *Mesaionikon Basileion, Henetokratia* (Nicosia, 1995), p. 752.

[10]   de Mas Latrie, *Histoire*, 3, pp. 107–8, 164.

[11]   Richard (ed.), *Livre des Remembrances*, no. 4.

[12]   Kehayioglou (ed.), *Diegesis*, pp. 228–9; Boustronios, *Narrative*, §191.

belonged to the Latin patriarchate of Jerusalem. The document specifies that Brother Gomez was relieved of payment of all royal tithes and other assignations payable in cash or kind into the royal treasury.[13] The Hospitaller brother Gomez d'Avila was perhaps related to Peter d'Avila, the constable of Cyprus under King James, who was exiled to Venice in May 1474, and to Antonello d'Avila, the captain at Paphos replaced by the Venetian John Petinal in February 1474.[14]

On 31 August 1468 King James II instructed his councillors to exempt another Hospitaller brother, Sabat, from payment of the 'rate', a proportional royal tax on produce, due from him in 1468, to the sum of 100 ducats. The money was to be recovered from the royal camera following the camera's authorization for this to take place, but the king now instructed the camera to allow a certain Louis Spataro to continue receiving the 'rate' as he had been accustomed to. The officers of the camera were to grant Brother Sabat the 100 ducats due from him as a gift, albeit for the year 1468 only, and to enter them as a debit in the royal treasury or *secrète*, as was fitting. One month later, on 24 September 1468, the king formally granted Brother Sabat exemption from payment of the wheat and barley due from him by way of paying the 'rate' for the years 1467–68.[15] Like the above-mentioned Brother Gomez, the Hospitaller brother Sabat had also been a beneficiary of King James II's largesse. Florio Bustron, a relative of George Boustronios and author of a chronicle of Cyprus in the later sixteenth century, records that sometime after 1464 the king granted the *casalia* of Flasou and Omorphita to Sabat, whose first name is given as Francis in the manuscript in Genoa as used by the editor, but as Nicholas in the Paris manuscript. Florio Bustron also recounts how, shortly after the death of King James, Queen Catherine Corner, his Venetian widow, ordered Brother Sabat together with Nicholas de Morabit, who were both at Kyrenia, to take the oath of fealty; confusingly George Boustronios, who omits any mention of Sabat, states that Nicholas Morabit was the commander of the garrison at Paphos.[16]

A third Hospitaller who received *casalia* from King James II was the Italian Brother Marco Pasturana, who was also appointed the master of the royal hostel. A document of the *Livre des Remembrances* dated 10 October 1468 records how the king granted him exemption from the payments of the 'rate', amounting annually to 195 bezants payable in kind. Brother Marco was granted an exemption on half this sum, but payments already made were not refunded.[17] Brother Marco was already present in Cyprus in 1462, for on 11 September the Hospitaller grand

[13]   Richard (ed.), *Livre des Remembrances*, no. 164.

[14]   Kehayioglou (ed.), *Diegesis*, pp. 150–51, 222–5, 268–9, 310–11; Boustronios, *Narrative*, p. 61 and §§98, 182, 233, 277; Hill, 3, pp. 699–702.

[15]   Richard (ed.), *Livre des Remembrances*, nos. 50, 139.

[16]   Florio Bustron, 'Chronique de l'île de Chypre', ed. R. de Mas Latrie, in *Collection des documents inédits sur l'histoire de France: Mélanges historiques*, 5 (Paris, 1886), pp. 419, 433; Kehayioglou (ed.), *Diegesis*, pp. 152–5; Boustronios, *Narrative*, §102.

[17]   Richard (ed.), *Livre des Remembrances*, no. 66.

master Raymond Sacosta wrote to him ordering him to abandon forthwith the domain of the *casalia* of Phinikas and Anoyira and its incomes, which he had been holding unlawfully, and to hand them over to John Darlende, the commander of Valence. At some point he must have received from King James the *casalia* of Kophinou, Pergamos, Silikou and Kato Petra, mentioned in a late fifteenth-century Venetian document as forming part of the royal domain.[18]

Even when disputes arose with the Hospitallers under King James, these were resolved. In 1467 a Hospitaller galley arrested the pirate Michael of Malta and his booty. Michael may have been related to James of Malta, an enemy of the chronicler George Boustronios, who recounts that this James, a mercenary of King James who had been rewarded for his services by being made *chevetain* of Pendayia in the north-west of Cyprus, uncovered a conspiracy to assassinate the king by his nobles. One of the persons executed was the chronicler's own son Demetrios.[19] The Hospitallers' seizure of Michael of Malta and his ship was financially injurious to the king, who covered his losses by confiscating the goods and movables belonging to the Grand Commandery and the smaller commanderies of Cyprus until on 3 March 1468 the Hospitallers agreed to compensate him, whereupon the properties were returned.[20] Several documents of 1468 in the *Livre des Remembrances* record payments the king made for debts he owed to the Hospitaller Order, and they refer to sums that he had acquired from various Hospitaller estates, probably after their seizure of Michael of Malta.

Turning to these documents, one sees how on 3 March the king acknowledged owing John Rames, the Hospitaller Grand Commander of Cyprus, a total of 9,448 gold ducats, much of which originated from the revenues of Kolossi, the Hospitaller hostel in Nicosia and the estates of the Grand Commandery that the king had acquired. From this sum 5,000 ducats, representing the value of a ship from Ancona captured by the Sicilian Sor de Naves, a soldier in King James's service during the civil war, were subtracted. Ancona had been supplying the Genoese besieged in Famagusta by King James's forces during the civil war, as is shown by the instance of Francheschetto d'Alma's caravel that the Genoese had sent from Famagusta to Ancona to obtain wheat. This explains de Naves' action against the Anconitan vessel. The ducats were subtracted because the ship in question had subsequently been taken by the Hospitallers. The king undertook to repay the outstanding balance of 4,448 gold ducats by granting the Order 500 ducats in sugar from the royal *casale* of Kouklia, and, in payments over a three-year period, 1,500 ducats from the salt revenues of Salines near Larnaca, 1,650 ducats from the sugar produced at the royal estates of Kouklia, Akhelia, Lemba

[18]   de Mas Latrie, *Histoire*, 3, pp. 87, 512, 512 n. 1.
[19]   Kehayioglou (ed.), *Diegesis*, pp. 138–49; Boustronios, *Narrative*, §§93–5.
[20]   Hill, 3, p. 643.

and Emba, and 798 ducats from the revenues of the *casale* of Polemidhia in the diocese of Limassol.[21]

In a second document dated 4 March 1468 King James acknowledged owing the Hospitaller brother John Langlais, seneschal of the Order, 606 gold ducats on account of the sums he had acquired from the Hospitaller commanderies of Phinikas and Templos. The king undertook to repay the sum from the incomes originating from the *zambours* and molasses forming part of the sugar production from the royal estates of Kouklia, Akhelia, Lemba and Emba for the year 1468, as well as from the revenues of the *casale* of Tarsis for the year 1469. The sum would be paid in line with the accounts prepared by John Langlais's procurator, the Hospitaller Brother Anthony de Coronia, and Luke of Jerusalem, the *bailli* of the said commandery, who, despite his title, was representing the king, and to whom the king explicitly referred in the document as 'our good friend'. The fact that the *bailli* of a Hospitaller commandery was representing the king constitutes further evidence that the king had arrogated Hospitaller estates. One observes that George Boustronios, who recounted Luke of Jerusalem's death on 2 October 1474, mentions that he had been the *bailli* of the royal court. The document stated that 303 ducats would be repaid from the revenues of the royal estates of Kouklia, Akhelia, Lemba and Emba, and the remaining half from the revenues of the *casale* of Tarsis.[22] The death of John Rames sometime before 17 November 1468 did not impede repayments to the Hospitaller Order, for in a document of that date King James simply instructed his officers to make the payments to the grand master in Rhodes.[23]

The instances cited above of Hospitallers serving the king and of royal officials such as Luke of Jerusalem holding office in Hospitaller commanderies are not exhaustive. Mention should also be made of Sir John Stramballi, who was given along with his son-in-law an assignation of 126 bezants from the Hospitaller Order for one year in February 1468, something that he acknowledged in the document dated 8 February 1468. Sir John Stramballi, an official of the royal *secrète*, was referred to as 'de l'Opital' on account of this assignation, and elsewhere he is referred to as the treasurer of the *secrète* and as superintendent of sugar production, unless a namesake is involved.[24] George Boustronios records how King James betrothed the widow of Sir William Stramballi to the tailor Anthony Garcia Navarro, whom he himself had ennobled and granted the *casale* of Epikho, while Simon Strambali like Sir John was a royal administrator. This family, which remained important under the Venetians, included Diomede Strambali, who translated the Greek chronicle of Makhairas into Italian in the late fifteenth century. This family

---

[21]   Richard (ed.), *Livre des Remembrances*, no. 146; Hill, 3, pp. 642–3; Kehayioglou (ed.), *Diegesis*, pp. 122–5; Boustronios, *Narrative*, §81.

[22]   Richard (ed.), *Livre des Remembrances*, no. 147 and n. 3; Kehayioglou (ed.), *Diegesis*, pp. 176–7; Boustronios, *Narrative*, §142 and n. 281.

[23]   Richard (ed.), *Livre des Remembrances*, no. 84.

[24]   Ibid., no. 196.

migrated to Cephalonia in the Ionian Islands after the Ottoman conquest of Cyprus in 1570, retaining its prominence throughout the seventeenth century.[25]

The Hospitallers and King James cooperated on a number of other issues. On 7 April 1468 the Hospitaller grand master Giovanni Battista degli Orsini wrote to John Rames that he had learnt with great sadness from a letter from King James as well as from Brother John's own letters that a Hospitaller brother named Zante had committed a murder in Cyprus. At the king's demand and with a view to dispensing exemplary justice, the grand master instructed John Rames to have the miscreant tried and punished, just as he himself would have done, in conformity with the customs and usages of the Order.[26] Furthermore, on 5 October 1468 King James issued orders for the dispatch to Brother Peter Anthony, a mandatory of the Hospitaller grand master, of a consignment of white sugar of the first cooking, that is to say crystallized but as yet unrefined, originating from the royal estates of Kouklia and Akhelia and priced at 36.5 ducats per quintal so as to repay his debt to the Order's grand master degli Orsini, amounting to 693 Venetian ducats or 1,040 Rhodian ducats. The debt in question had been contracted as a ransom for seven former Muslim slaves, whom some years before the previous grand master, Raymond Sacosta, had purchased on King James's behalf on Rhodes and had then sent back to the Mamluk sultan in Egypt.[27] That the Hospitallers should be acting as intermediaries in the dealings of King James with his Mamluk suzerains indicates the high level of cooperation existing between them by this time. One notes in this context that on occasion Hospitaller grand masters acted of their own accord to ransom Muslims captured and sold into slavery. Hence in late 1459 or early 1460 the grand master James de Milly instructed his envoy to the Mamluk sultan, John Dolfin, then commander of Nisyros, to mention to the sultan how the grand master, having already ransomed numerous Muslim captives, would seek out others so as to secure their freedom.[28]

How are these instances of collaboration between King James and the Hospitallers, as well as the phenomenon of Hospitallers on Cyprus in the king's service, to be explained, given the Order's support for Queen Charlotte during the civil war of 1460–1464 and the queen's sojourn on Rhodes throughout the reign of King James? An explanation is furnished by the fact that King James desired not only power but also the legitimacy that only papal recognition could provide. Papal recognition would also bring economic benefits, such as the channelling of ecclesiastical taxes and incomes from the issue of indulgences towards the defence of Cyprus. The chronicler George Boustronios states that in 1471 King

---

[25]   Kehayioglou (ed.), *Diegesis*, pp. 226–9, 242–5; Boustronios, *Narrative*, §§189, 205 and n. 344; G.N. Moschopoulos, 'He kypriake oikogeneia Strambali sten Kephalonia (16os–17os ai)', in *Proceedings of the Second International Congress of Cypriot Studies* (Nicosia, 1985–87), 2, pp. 249–58.

[26]   de Mas Latrie, *Histoire*, 3, p. 91.

[27]   Richard (ed.), *Livre des Remembrances*, §63.

[28]   de Mas Latrie, *Histoire*, 3, p. 99 and n. 1.

James II sent Louis Perez Fabrigues, the Catalan who had become archbishop of
Nicosia that year, to Rome in order to persuade Pope Paul II to have him formally
crowned king of Cyprus, but the pope rejected this request because Queen
Charlotte was still alive.[29] In fact the chronicler mistakenly brought the incident
ten years forward. In 1461 James had sent Sir Philip Podocataro and Anthony
di Zucco, the Latin bishop of Limassol, as his ambassadors to Venice, Florence
and the pope. Only Florence recognized James as king, Venice remaining non-
committal and Pope Pius II rejecting James's claim to be king. Florio Bustron
states that Pope Pius II refused recognition because he was angered by James's
rejection of his proposal to marry his niece, but he wrongly gives 1471–72 as the
date of the mission, apparently oblivious to the fact that Pope Pius had died in
1464. In 1466, however, James, in an application supported by King Ferdinand of
Naples, obtained a measure of recognition from Pius's Venetian successor, Pope
Paul II. The pope acknowledged that James had effective dominion over Cyprus,
although he stopped short of formally recognizing him as king, and recognized
the legitimacy of William Goneme as archbishop of Nicosia, something the
Venetians had been soliciting since 5 June 1464, when the Venetian Senate wrote
requesting papal confirmation of his position. Having obtained papal recognition,
King James was betrothed by proxy to Catherine Corner, a Venetian noblewoman
whose family held considerable estates at Episkopi on Cyprus, on 10 June 1468.[30]
In view of these objectives and the importance of papal recognition, James would
not have wished to antagonize the Hospitallers, an Order of the Roman Church,
unduly: indeed, he would have been more than willing to cooperate with them.

Nemesis for the Hospitallers on Cyprus came after King James's death in July
1473. The Venetians warned the Hospitallers of Rhodes not to support Queen
Charlotte at this point, and the Hospitallers, mindful of Venice's naval power,
remained officially neutral. Even so, individual Hospitallers appear to have
continued to work against Venice, for as late as 1475 the Venetians received news
that the Hospitaller brother John de Canosa was arming four large ships in the
kingdom of Naples with a view to invading Cyprus, something that impelled the
Venetian government to lodge a strong protest before King Ferdinand of Naples,
who was greatly perturbed that the Venetians had got wind of this plan, allegedly
fomented by Rizzo di Marino. Furthermore, on a local level the Venetians were
angered by the involvement of Brother Nicholas Zaplana, who had acquired the
Grand Commandery of Kolossi in February 1471, in the plot hatched by former
mercenaries of King James II against his widow, Queen Catherine. One of the plot's
leaders was Nicholas's kinsman James Zaplana, whose valuables Nicholas had had

———

   [29]   Kehayioglou (ed.), *Diegesis*, pp. 148–9; Boustronios, *Narrative*, §96; Hill, 3,
p. 575; P.W. Edbury, 'Hoi teleutaioi Louzinianoi (1432–1489)', in Th. Papadopoullos (ed.),
Historia tes Kyprou, 4, *Mesaionikon Basileion, Henetokratia* (Nicosia, 1995), p. 228.
   [30]   Florio Bustron, 'Chronique', p. 432; Hill, 3, pp. 557–8, 575–9, 628–9, 631, 1159
(p. 631); G. Fedalto, *La Chiesa Latina in Oriente*, 2nd edn (Verona, 1981), 2, p. 175 n. 1;
Edbury, 'Hoi teleutaioi Louzignanoi', pp. 214–15, 231–3.

concealed in Kolossi after James had been forced to flee from Cyprus in January 1474 with his fellow conspirators. Once the Venetians had ascertained Nicholas's complicity he was deprived of office, and, under the Venetian Hospitallers, such as Marco Crispo, the commander of the priory of Verona and also Queen Catherine's uncle, who were appointed to succeed him, the Grand Commandery became gradually but firmly attached to the Venetian Corner family.[31]

---

[31]   Kehayioglou (ed.), *Diegesis*, pp. 218–31, 286–7; Boustronios, *Narrative*, §§176–7, 180, 182, 184–6, 192, 256; de Mas Latrie, *Histoire*, 3, pp. 93, 437; Hill, 3, p. 698 and n. 3; Edbury, 'Hoi teleutaioi Louzignanoi', pp. 250–51; Luttrell, 'Stratiotika tagmata', p. 753.

PART 2
Hospitaller Rhodes and Malta

Chapter 10

Smoke and Fire Signals at Rhodes: 1449

Anthony Luttrell[1]

Long before the Hospitallers established themselves on Rhodes between 1306 and 1310, smoke, fire and mirrors had been employed as means of communication.[2] The Hospitallers used carrier pigeons in Syria before 1291[3] and probably at Rhodes.[4] A version of an account by Ludolf de Sudheim, who was in the East in about 1336/1341, reported that, on the Hospital's island of *Carmellis*, evidently Kalymnos to the north of Kos, there was 'a castle with a very high tower from which they signal with mirrors – *speculantur* – when they see the Turks or other pirates coming from afar; if it is daytime, they make great smoke, and if instead it is night, they light a flaming fire so that from Rhodes and Kos and the islands of the Christians they can all hasten with arms so that the enemy is unable to prevail'.[5] In 1475 warnings of approaching Turkish fleets were to be sent, presumably by fire signals from Mount St Stephen just outside Rhodes town, to *Castellonovo*, that is, to the new castle at Kastellos on the north coast of Rhodes, and to the other islands as far as Kos.[6]

There was a considerable degree of intervisibility between castles and towers on Rhodes and between the castles and towers on Rhodes and those on the dependent islands; small coastal towers on or near the seashore were also used

---

[1]   The author is most grateful to Michael Heslop for much help and information.

[2]   E.g. P. Pattenden, 'The Byzantine Early Warning System', *Byzantion*, 53 (1983), 258–99; J. Rife, 'Leo's Peloponnesian Fire-Tower and the Byzantine Watch-Tower on Acrocorinth', in W. Caraher *et al.* (eds), *Archaeology and History in Roman, Medieval and Post-Medieval Greece: Studies on Method and Meaning in Honor of Timothy E. Gregory* (Aldershot, 2008), pp. 281–306; K. Molin, *Unknown Crusader Castles* (London, 2001), pp. 116–17, 157–61, 251–2.

[3]   S. Edgington, 'The Doves of War: The Part played by Carrier Pigeons in the Crusades', in M. Balard (eds), *Autour de la Première Croisade* (Paris, 1996), p. 173.

[4]   According to S. Spiteri, *Fortresses of the Knights* (Malta, 2001), p. 151, but without source.

[5]   Ludolphus de Sudheim, 'De Itinere Terre Sancte', ed. G. Neumann, *AOL*, 2 (1884), documents, p. 333; textual tradition documented in A. Luttrell, *The Town of Rhodes: 1306–1356* (Rhodes, 2003), pp. 214–16. See p. 217 for a different text giving not *carmellis* but *Castel Roys* or Kastellorizzo, which was too far from Rhodes for any intervisibility; that means that the date when the Hospital occupied Kastellorizzo cannot be established as 'c. 1340'.

[6]   Malta, Cod. 75, fol. 70 [modern fol. 78].

to watch out for Turks and other enemies.[7] By 1400 at latest there was a network of relatively strong castles on Rhodes into which the population could withdraw on the approach of danger.[8] There were detailed arrangements for the turcopoles or local guards and the *viglocomites* or watchmen on Rhodes itself,[9] but the communication system also involved advance warnings from the Order's outlying islands to the north that were within sight of Rhodes or of each other.[10]

Smoke and fire could transmit only very basic messages for which an agreed code of signals was necessary. An example of how the system functioned came in 1449 when three fire signals from Symi, which lay between Rhodes and the nearby mainland, were seen at Rhodes. The Hospitaller master and the council met on 18 June and issued instructions for the despatch of the Rhodian guard galley under the command of the grand bailiff of *Alamania*, Fr Johannes von Wittingen.[11] The organisation by the admirals' office of the manning of the galley by those Rhodian inhabitants who owed galley service as oarsmen under the system of *servitudo marina* must have taken a little time;[12] in 1447 the oarsmen were supposed to report within thirty hours of a summons to service.[13] There was normally only one galley available, which meant that it had to be deployed with caution. The captain's instructions were written in Italian; Wittingen understood only German,[14] but he was ordered to act with the advice of other brethren and of the crew, and he must have had interpreters of some kind.

---

[7]   N. Zarifis and D. Brokou, 'GIS and Space Analysis in the Study of the Hospitallers' Fortifications in the Dodecanese', in *Archaeological Informatics: Pushing the Envelope CAA 2001: Computer Applications and Quantitative Methods in Archaeology*, ed. G. Burenhult (Oxford, 2002), pp. 149–53; M. Heslop, 'The Search for the Defensive System of the Knights in Southern Rhodes', *MO* 4, pp. 189–200; idem, 'The Search for the Defensive System of the Knights in the Dodecanese (part I: Chalki, Symi, Nisyros and Tilos)', in H.J. Nicholson (ed.), *On the Margins of Crusading: The Military Orders, the Papacy and the Christian World* (Farnham, 2011), pp. 139–65; M. Losse, 'Kástro und Viglá: Burgen-Standorte auf Inseln der Südost-Ägäis', *Castrum Bene*, 9 (2006), 255–78; idem, 'Wacht- und Wohntürme aus der Zeit des Johanniter-Ordens (1307 bis 1522) auf der Ägäis-Insel Rhódos (Griechenland)', *Burgen und Schlösser*, 4 (2009), 245–60. See also Spiteri, pp. 106–218.

[8]   Text in A. Luttrell and E. Zachariadou, *Sources for Turkish History in the Hospitallers' Rhodian Archive 1389–1422* (Athens, 2008), pp. 99, 131.

[9]   Detailed study for the period after 1460 in G. O'Malley, *The Knights Hospitaller of the English Langue 1460–1565* (Oxford, 2008), pp. 304–13.

[10]   Spiteri, p. 151, writes, curiously, that in 1464 'there was no established method by which alarums and warning signals could be relayed along the length of the island'.

[11]   Malta, Cod. 361, fols 362–362v [374–374v].

[12]   On this *servitudo*, A. Luttrell, *The Hospitallers in Cyprus, Rhodes, Greece and the West: 1291–1440*, Variorum Collected Studies Series (London, 1978), IV.

[13]   J. Sarnowsky, *Macht und Herrschaft im Johanniterorden des 15. Jahrhunderts: Verfassung und Verwaltung der Johanniter auf Rhodos (1421–1522)* (Münster, 2001), pp. 646–7.

[14]   '... qui nisi linguam Alamanicam sciebat' and 'nesciebat aliud idioma nisi Alamanicum': Malta, Cod. 361, fols 313–314 [325–326] (10 February 1449).

The guard galley was to leave Rhodes at night and proceed with the utmost caution to Symi to discover there the reason for the three fire signals, and then, with the advice of the other brethren and sailors, to sail towards wherever it was suspected that the Turkish ships might be. If some of them were thought to be in the Gulf of Tracheia somewhere to the east of Symi,[15] or in some other gulf or out to sea, the captain was to follow them and, if there was no alternative, he was to burn them. There was also news at Rhodes that a Turkish ship had sailed towards Kastellorizzo, and if the captain learnt that the fires from Symi meant that that was the case, he was to chase the Turkish vessel. He was to return to Rhodes within ten or fifteen days, but before leaving Kastellorizzo he was to show himself openly to the Turks by day and then leave secretly at night. If the Turkish ships were reported to have sailed towards Kos and westwards, he was to follow and chase them and other Turkish ships he might happen to meet, but he was not to leave the area of the Rhodian islands. He was to send back any news of the situation and of his own movements by any ship that might be sailing to Rhodes.

The captain was to ensure that a proper guard was kept on all the Order's islands and that the smoke and fire signals were functioning according to the signal code, of which he was given a copy. Wittingen was to find out the reasons for any signals and to go to the closest island that had made fire or smoke to discover from what places the signals had come, from where they had first originated, and why they had been sent. If he met the 'galliot of Sicily', probably the galley purchased in Sicily in 1441,[16] and the brigantine of Kos, those vessels were to accompany him. If he went to Kos he was to stay away for thirty days and then to return to Rhodes; or, if the presence of Turkish shipping demanded a further stay, he could remain longer at his discretion. It was evidently important for the captain to communicate with Rhodes itself. If he saw or heard of fire or smoke from Rhodes he was immediately to return there with his galley, giving instructions on Symi that the guards there should always watch for signals from Mount St Stephen, the hill behind Rhodes town that was a key link in the system.[17] If the men on Symi saw such signals they should signal in turn to the other islands so that the signals from Rhodes should reach Wittingen's galley.

The degree of information that could be transmitted by smoke and fire was clearly limited. Cloud, fog or haze would have obscured visibility, and the system depended on a supply of firewood and the vigilance of the lookout guards.[18] An

---

[15]   The gulf intended may have been, probably, to the north or, possibly, to the south of the Tracheian Peninsula: F. Hild, 'Stadia und Tracheia in Karien', in K. Belke *et al.* (eds), *Byzantina Mediterranea: Festschrift für Johannes Koder zum 65. Geburtstag* (Vienna, 2007), pp. 231–43.

[16]   Malta, Cod. 355, fols 182–182v [180–180v].

[17]   On the remains there, [B]. Rottiers, *Description des Monumens de Rhodes* (Brussels, 1830), album LIV.

[18]   On various difficulties, Molin, pp. 50, 157–61, 251–2. Rottiers (p. 22) reported a tradition on Symi that the number of fires indicated the number of ships sighted. On the

understanding of the signal code was clearly essential. The written code, dated 27 May 1449, read:[19]

Queste sono le ordinatione deli segnali de focho e de fumo dela nostra isola de Rhodo e dele altre isole de nostra religione che se anno a fare e seruare de zorno e de notte.

Primeramente che tute garde de zorno, ouer merouigles, che uederano o descoprarano fusta o fust[e][20] de inimici tanto turchi como mori, o de qualseuoglia altra natione, tanto christiani como infideli, che se tegnano inimici dela religione, faceno tanti fumi sopra lo piu alto loco de lor guarda notificando per tali fumi como hauerano uiste fuste de inimici.

Item ciaschuna guardia sia tenuta de responder lo primer fumo descoperto e che auera uisto, e che la guarda che fara el fumo non cessa de far fumo fin che la altra guarda li respondera de fumo.

Item en caso che la ditta guarda che fara lo fumo ueda che l'altra guarda non li responde per fumo, la dita guarda che auera fato lo fumo sia tenuta de auisare lo turcopolo e lo viglocomi o quelli homini a chi tochara la garda la notte sequente.

Item che lo dito turcopulo o uiglocomi o garda dela notte sequente sia tenuto ha far la note sequente .ij. fochi, l'uno apresso del'altro, al modo de una porta fin che l'altra guarda li responde.

Item che la garda che uedera fare .ij. fochi, ala prima garda li responda con vno focho.

Item quando la prima guarda hauera uisto che l'altra li hauera responduto de vno foco alor li de far tanti fochi quante fuste de inimici la guarda hauera uisto.

Item la guarda dela notte sera tenuta similmente a far tanti fochi quante fuste la notte pora vedere e descoprire.

Item che l'una dele guarde secondo la bona costuma, uisto che hauera fuste de inimici ouer segnali de fochi fatti per altra guardia debia prestamente correre e andare al proximo casal e auisare lo turcopullo e li homini del ditto casal per saluatione de lor persone e beni.

Item ordenamo s'el se troua homo dela guarda del zorno o dela notte chi per negligentia non fazi o non responda cum fumi o fochi e per lengua secundo de sopra se contene, che aquello talle negligente sia per li officiali de mons[ignore] lo maestro, ouer castellani del casal e loco doue sera el dito negligente, li sia talliato la barba e li capilli, e habia .xxv. nel pilleri.

Scrito adi xxvij mazo mccccxluiiij.

---

remains of fire pits, Heslop, 'Southern Rhodes', p. 197.

[19]   Malta, Cod. 361, fols 362v–363 [374v–375]; the text is lightly edited.

[20]   Ms: *fusta*.

These are the regulations for fire and smoke signals for our island of Rhodes and the other islands of the Order that are to be made and observed by day and night.

Firstly, all day guards or *merovigles* who see or discover an enemy ship or ships, whether Turkish, Moorish[21] or of whatsoever other nation, whether Christian or infidel, that may be suspected of being enemies of the Order, should make as many smoke [signals] on the highest place of their guard, notifying through such smoke [signals] that they have seen so many enemy ships.

Item, each guard is bound to respond to the first smoke noticed and seen, and the guard that has made the smoke shall not cease to make smoke until the other guard has replied with smoke.

Item, in the case that the guard that makes the smoke sees that the other guard does not reply to it with smoke, the said guard that made the smoke is to advise the turcopole and the *viglocomi* or whatever men are bound to keep guard the following night.

Item, that the said turcopole or *viglocomi* or guard of the following night is bound on the next night to make two fires, one close to the other in such a way that one continues until the other guard replies to it.

Item, the guard that sees two fires to have been made shall reply to the first guard with one fire.

Item, when the first guard has seen that the other has replied to it with one fire, then [that guard] should make as many fires as the number of enemy ships that the guard has seen.

Item, the night guard is similarly bound to make as many fires at night as ships may have been noticed and seen.

Item, according to good custom, when enemy ships or the fire signals that have been made by another guard have been seen, one of the guards is to run at once to the next casale and inform the turcopole and the men of that casale for the salvation of their persons and goods.

Item, we command that if a man of the guard of the day or of the night has through negligence not made or replied with smoke, or fire or by word of mouth according to the above instructions, that negligent person is, whether by the Master's officials or by the castellan of the casale and place where the negligent person may be, to have his beard and hair cut off and have 25 in the pillory.[22]

Written on the 27 day of May 1449.

---

[21] Moorish: either Mamluk or Maghrebi.
[22] The '25' may have been days or sessions.

Chapter 11

Success and Failure in the Practice of Power by Pere Ramon Sacosta, Master of the Hospital (1461–67)

Pierre Bonneaud

In the fifteenth century two Catalans, Antoni de Fluvià (1421–37) and Pere Ramon Sacosta (1461–67), rose to become masters of the Hospital; they were the first Catalans to attain that office. In the case of Fluvià, there is scarce information about his career before he was elected to the highest rank in the Order. By contrast, we are able to trace in detail Sacosta's long career from his reception as a knight brother in about 1418 until his death in 1467 aged 64: his age is recorded on the tombstone which survives in the Vatican.[1] Sacosta was not born into an illustrious noble lineage or into a rich and influential family of Barcelona patricians. Several Sacostas are documented as unimportant knights or squires in the western part of Catalonia around the small town of Agramunt on the Urgel plateau, not very far from the Pyrenees and no great distance from the area where Antoni de Fluvià's origins, also in a humble lineage of knights, are to be found. In both cases the meagre incomes from their lordships made them dependent either on the offices they could obtain from the monarchy or on a religious career.

No licence from the master for the reception of the young Sacosta as a knight brother that actually gives his name is known, but he went to Rhodes on the same 'passage' as Galceran de Sentmenat, who was the beneficiary of just such a licence in 1418. Both men received the habit during the same solemn mass in the conventual church, probably sometime before the beginning of the Chapter General held at Rhodes in 1420.[2] Sacosta would have been 17 years old when he started his long career of forty-six years, only eight of which he spent in his Castellany of Amposta, the more important of the two Spanish priories in the Crown of Aragon.

We shall try to stress here his ambitions and character traits as shown in his quest for advancement, as well as in his strategy, conduct and political manoeuvres. His successes and failures depended greatly on how he conducted his relations

---

[1]   J.B. de Vaivre, 'Les tombeaux des grands maîtres des Hospitaliers de Saint-Jean de Jérusalem à Rhodes', *Monuments et mémoires publiés par l'Académie des inscriptions et Belles lettres*, 76 (1998), 64.

[2]   Malta, Cod. 342, fols 104r, 68v.

with the three sources of power that played a part in the career of any Catalan or Aragonese Hospitaller, that is to say with the complex system of government at Rhodes, the pope, and the king of Aragon.

The Three Sources of Power that might Lead to the Advancement of a Hospitaller Knight's Career

For a long time major decisions on brethren's careers had been taken at Rhodes by the master and his council with occasional approval from the Chapter General, although every *langue* had developed its own customary rules for the *cursus honorum* of its members. The influence of the convent, however, had been greatly diminished at the time of the papal schism, which explains several long periods in which the master was absent from the convent between 1379 and 1420. The Chapter General of 1420, presided over by Philibert de Naillac, and those of 1428 and 1433, under Antoni de Fluvià, had re-established clearly the prominence of Rhodes in the government of the Order. Access to commanderies and conventual offices were to be much more strictly subject to the rights of seniority acquired by a prolonged presence at the convent. Numerous brethren, among them many Catalans, became accustomed to meeting in the auberges of their *langues* to prove or question their rights of seniority. By the middle of the century, they were given the prerogative of declaring their *ancianitas* by mutual agreement, their decisions being subject, however, to the master's approval.[3]

At the same time, tensions mounted between the French *langues* of Auvergne, France and Provence, which had for long been paramount in the conduct of affairs and access to major offices, and the so-called 'minor' *langues* of England, Germany, Italy and Spain. Jean de Lastic and Jacques de Milly, the two masters who governed between 1437 and 1461, both of whom had been prior of Auvergne, suffered a weakening of authority as a result of acrimonious debates and criticisms. Sacosta, in his quest for advancement, could not of course ignore the authority of the master, but he needed to gain the backing of the brethren of his *langue*.

The pope, as the head of the Church, had played an important part in the government of the Hospital until the schism weakened the papacy's authority. After the Council of Constance, the master had secured from the new pope, Martin V, an assurance that he would refrain from granting commanderies and offices and from accepting appeals from Hospitallers dissatisfied with decisions taken in Rhodes. Martin's successor, Eugenius IV, was more inclined to intervene in the Order's affairs, especially in view of the protests against Lastic's financial measures and of the conflicts between the *langues*. His successors took different

---

[3]   P. Bonneaud, 'La règle de l'*ancianitas* dans l'ordre de l'Hôpital, le prieuré de Catalogne et la *Castellania de Amposta* aux XIVe et XVe siècles', in K. Borchardt, N. Jaspert and H. Nicholson (eds), *The Hospitallers, The Mediterranean and Europe, Festricht for A. Luttrell* (Aldershot, 2007), pp. 221–32.

stands, Nicolas V and Paul II being strongly in favour of imposing their choices on Rhodes, Calixtus III and Pius II claiming fully to respect the convent's autonomy.

The third major power involved was the king of Aragon, who in principle had no authority to intervene. But the kings had always made use of their influence with the pope, the master, the prior of Catalonia and the *castellán de Amposta* in favour of those families that gave them military or financial support. King Alfonso the Magnanimous, who reigned from 1416 to 1458, was desperately in need of knights and funds for his great aim of becoming king of Naples, an ambition that involved him in a long military conflict with the Angevin dynasty. He intrigued endlessly to obtain the best commanderies for the younger sons of families that gave him support and he even diverted to his full-time service in Italy about ten Catalan and Aragonese Hospitaller knights who used the incomes of their commanderies to finance their presence in the king's court and armies. Until he finally conquered Naples in 1442, Alfonso encountered the hostility of the popes, who favoured the Angevins, and he was often in conflict with the master who needed the brethren to be stationed in Rhodes rather than in Naples in order to repel attacks from the Mamluks and, after the fall of Constantinople in 1453, from the Ottomans.[4] The 1433 Chapter General declared that any brother who requested letters of recommendation in order to obtain a commandery or other office would lose his rights of seniority.[5] However, the king's requests and stratagems could not be ignored.

Sacosta's First Steps in the Order (1420–39)

A quick survey of Sacosta's career illustrates how he resorted alternately to the king of Aragon, the convent and the pope. In 1429, more or less ten years after Sacosta's reception in the Order, King Alfonso asked Master Antoni de Fluvià to grant Sacosta a commandery by magistral grace; probably he was not yet then entitled to receive his *cabimentum*, that is to say his first commandery due to seniority.[6] It is most likely that a member of his family, probably his brother and proxy Pons Sacosta, who became a royal officer a few years later, had requested and obtained such a recommendation. He obtained from Fluvià the commandery of Horta. In 1436 Sacosta further secured the office of bailiff of the island of Rhodes, a lucrative position that was conferred by the master and that introduced its holder into the circle of officials involved in the government of the Order. But when Fluvià died a year later, Sacosta feared that his successor, Jean de Lastic, would not confirm him in his new office, and he once again resorted to the backing of the monarchy. He

---

4   P. Bonneaud, *Le prieuré de Catalogne, le couvent de Rhodes et la couronne d'Aragon (1415–1447)* (Millau, 2004), pp. 260–65.

5   Archives départementales de la Haute-Garonne (Toulouse), H.13, Statut 55 du chapitre général de 1433, fol. 111v.

6   Archivo de la Corona de Aragón (Barcelona), Registro de la Cancillería real (ACA RC), 2578, fol. 75v.

had his brother Pons travel to France to visit the new master Lastic who had not yet embarked for Rhodes; Pons took with him a letter of recommendation, dated July 1438, from Queen Maria who was the king's lieutenant.

In 1436 Sacosta had secured his second commandery, Torrent de Cinca, as his *cabimentum*, thanks to his rights of seniority and to the support of the brethren of the *langue* of Spain. These brethren, who had a say in all matters concerning seniority, had to choose between two contenders and it was not a simple choice because both Galceran de Sentmenat, already commander of Valencia through the master's grace, and Sacosta had exactly the same rights since they had come to Rhodes on the same 'passage'. Sacosta successfully argued that he was better placed than Sentmenat because Torrent was geographically closer to his commandery of Horta than was Valencia: it would be easier for him to take effective care of both commanderies although neither of the two candidates was really considering leaving the convent. Sacosta was then immediately granted his new commandery by Fluvià.[7] However, for rather obscure reasons, shortly after the death of Fluvià in November 1437, he decided to resign Torrent.[8]

Sacosta, Drapier of the Convent (1439–44)

A majority of brethren also voted in Sacosta's favour when in 1439 he applied for the main office in the *langue* of Spain, that of Drapier, which was vacant after the death of the prior of Catalonia, Lluis de Gualbes, and the latter's replacement by the Drapier, Rafael Saplana.[9] The Drapier was the conventual bailiff of the *langue* of Spain and, as such, was a member of the master's council. He presided over the meetings of the brethren of the *langue* as he was the first in rank among them. The election of Sacosta as Drapier produced a state of turmoil with several Catalan and Aragonese brethren opposing him, mostly on the ground that they were more senior and more qualified. His main opponents were Felip d'Hortal of the priory of Catalonia, and Pedro Sarnes and Pedro de Linyan both of the Castellany. They appealed to the master and his council to annul the election, but were told that Sacosta, whose choice the master had ratified, complied with the statutory requirement of ten years of continued presence at the convent while *ancianitas* was not necessarily to be taken into consideration in the appointment of a conventual bailiff.[10] Sometime later, however, Lastic reversed his stand, and declared Felip d'Hortal Drapier.[11] It is probable that King Alfonso had used his influence with the master and with the Catalan and Aragonese brethren at the convent to give support to Hortal, in whose favour he had frequently intervened in the past.

---

7   Malta Cod. 352, fols 68v–69r.
8   Ibid., 353, fol. 68r–v.
9   Ibid., 354, fols 120v, 181v–182r.
10 Ibid., 354, fols 182v–184v.
11 Ibid., 355, fol. 97r.

Sacosta then decided to appeal to Pope Eugenius IV. This was a rather risky move as the pope had been dismissed by the Council of Basle in favour of the former duke of Savoy, Amadeus VIII, who took the name Felix V, and the king of Aragon, hostile to Eugenius, had ordered his subjects to observe a neutral stance and to refer to the crown any question pending with the papacy. It happened that Pedro Sarnes and Pedro de Linyan were among the ambassadors that the king had sent to the Council of Basle. Both were strong opponents of Sacosta, who must have gone in person to Rome where he gained Eugenius's total support. The struggle for the office of Drapier was affected directly by the conflict in the Church and by the difficult relationship between the pope and the king. At no time, however, did the master and the convent seem to have questioned Eugenius' legitimacy, and Sacosta was confident that only a decision of the pope in his favour would restore him to his office as Drapier.

At Sacosta's request, Hortal, Sarnes and Linyan were summoned to Rome. King Alfonso was infuriated by what he considered to be a breach of his declaration of neutrality in the papal schism and, as punishment, he requested Queen Maria to sequestrate Sacosta's commandery of Horta.[12] However, the situation changed radically once the king's troops had taken over Naples in 1442 and he had rapidly gained control of the rest of the realm. In 1443 the pope, faced with the unquestionable success of the king, recognized the legitimacy of Aragonese claims to the Neapolitan throne and Alfonso broke off all contacts with the Council of Basle and the schismatic Felix V. In view of the improved relations between the king and the pope, Hortal and Sarnes anticipated a papal sentence at the Roman *curia* in favour of Sacosta, made with Alfonso's agreement. They pressed Lastic to intervene in their favour, but the master considered that the decision re-establishing Sacosta as Drapier soon afterwards would be taken by the pope.[13] The king pardoned Sacosta, while Hortal, probably in compensation for having lost the office of Drapier, received the rich commandery of Cyprus in addition to his Catalan commandery of *Cases antigues de Lleida*.[14] As Cyprus was a capitular bailiwick, Hortal was entitled to remain a member of the master's council.

Sacosta, *castellán de Amposta* (1444–61)

Pope Eugenius again confirmed his support for Sacosta when a new opportunity for further advancement was offered after the *castellán de Amposta*, Joan de Vilagut, was drowned in a shipwreck off the coast of Malta in 1444.[15] It was customary in the *langue* of Spain that the offices of prior of Catalonia and *castellán de Amposta*

---

[12]   ACA RC 2652, fol. 2v; ACA RC 2615, fols 9v–10v.

[13]   Malta Cod. 355, fols 86r–87r.

[14]   ACA RC 2562, fol. 2v; ibid., 2615, fols 9v–10v; ibid,, 2528, fols 178v–179r; Malta Cod. 356, fols 86v, 211v–212v; ibid., 357, fol. 94r–v.

[15]   ACA RC 2653, fol. 61r.

should, when vacant, be offered to the Drapier, provided he belonged to the vacant priory. Sacosta, however, chose to run no risk, and he obtained a papal bull declaring him *castellán* rather than waiting for a provision from the master and the convent. The pope had stated in his bull that his choice had been strictly guided by the rules of the convent as well as by the advice of several Hospitallers present in the *curia* at that time. Such a papal bull could not be rejected by the master, and Lastic flatly refused to reissue a bull of provision that Sacosta further requested, because he considered the papal bull to be sufficient and, in any case, as over-ruling any other choice that might have been made in the convent.[16]

Sacosta's influence in Rome became obvious once more when in 1446 Eugenius IV, questioning Lastic's ability to face the financial crisis and the lack of support from the minor *langues*, decided to convene a Chapter General in Rome. In Lastic's absence, the chapter was presided over by a triumvirate consisting of the prior of France, the prior of England, and the *castellán* Sacosta. Sacosta easily secured from the assembly the provision of the Castellany for life and a redistribution of its commanderies. He seems to have pursued three main purposes: to penalize his former enemies, to satisfy the protégés of the King, and to obtain for himself outstanding career advantages. He was granted the right to hold five commanderies as his *camere* instead of the usual four and was promised the two important commanderies of Zaragoza and Miravet on the death of their commanders who were both very old. Sacosta also gave satisfaction to the brethren of the Castellany residing at the Convent by including in the chapter's decisions a new rule for promotions to commanderies that gave the Aragonese and the Catalan brethren the right to establish their seniority by agreements made among themselves.[17]

After the Chapter General at Rome, Sacosta, by then aged 42, had charge of one of the major Hospitaller priories, and had become one of the most influential officers in the collegiate government of the Order at Rhodes. Between 1446 and 1461 he spent eight years out of sixteen away from his Castellany, mostly in Rhodes. He had achieved his goals successfully and swiftly, thanks to his unwavering ambition and to his adroit manoeuvres. The full support of Eugenius IV had certainly been essential, but Sacosta had thus far managed in addition to gain the support of the brethren of his *langue*. As his bitter enemy Hortal declared, 'brother Pere Ramon, with his astute malignity and his public renown has gained a large devotion at the convent'.[18] But at the same time, Sacosta had bypassed Master Jean de Lastic's authority and by doing so had provoked his rancour. He had also been opposed to King Alfonso on several occasions and could no longer count on his goodwill.

Sacosta, after spending a couple of years in his Castellany, attended the 1449 Chapter General at Rhodes and did not return to Spain until the autumn of 1453. During this period he further increased his wealth by receiving Zaragoza in 1448, and then, in 1451, Miravet, both of which were included among his *camere*. For a

---

[16]   Malta Cod. 357, fol. 73r.
[17]   Bonneaud, 'La règle de l'*ancianitas*', pp. 228–9.
[18]   Malta Cod. 355, fol. 86v.

short time, Sacosta was receiving the rents of seven commanderies: Ascó, Caspe, Horta, Miravet, Ulldecona, Zaragoza and Aliaga, although according to the rules of the Order, he could hold at most five *camere*.

Major problems continued to be caused by King Alfonso's determination to provide his protégés with wealthy commanderies. Pope Eugenius IV had died in 1447 and his successor Nicolas V not only maintained excellent relations with Alfonso but was prepared to bypass the master's authority to grant commanderies at the request of the king. Sacosta suffered a defeat when in 1451 Juan Claver, a close familiar of Alfonso, who was already commander of San Stefano de Monopoli in Puglia and of Chalamera in the Castellany, claimed for himself the commandery of Ulldecona, one of the most prosperous *camere* of the *castellán*. He argued that Sacosta had accepted Horta as his fifth *camera* at the 1446 Chapter General and that he could not also hold Ulldecona as his fifth *camera* after having received Miravet as his fourth.[19] Sacosta had in fact agreed to renounce the less important Horta and had acknowledged the provision of that commandery made by Nicolas V to another protégé of the king, Ramon de Siscar. Claver, who acted as an emissary between the king and the convent, was influential with Lastic, who had recognized Claver's *ancianitas* although he had not been residing at the convent. The master had praised 'his endless industry and diligence in best serving the Order's interests at the king of Aragon's side and wherever else was necessary'.[20]

The brethren of the *castellania* in Rhodes requested that the question be submitted by the master to the 'complete council', which was the highest judicial body in the convent. The council was to decide which commandery, Horta or Ulldecona, should be considered as part of the castellan's *camere*. Sacosta was not satisfied with the sentence and decided to appeal to Pope Nicolas V, as he had done successfully with Eugenius IV when he had been deprived of his office of Drapier. He was then summoned in person to Rome by the pope, who refused his appeal and remitted the case to the master and the convent.[21] Ulldecona was granted to Claver as his *cabimentum* by the Chapter General of 1454, which Sacosta did not attend even though he had been summoned by Lastic together with the other priors in view of the critical situation of the Order that was bankrupt and threatened by the Ottomans following the fall of Constantinople.[22] His absence was possibly due to his knowledge that he could not avoid the loss of Ulldecona, a major setback especially since Horta was now in the hands of Ramon de Siscar. He was also in conflict with the treasury at Rhodes because he was refusing to pay the so-called ordinary *responsiones* for his *camere*, asserting that they had never been paid by his predecessors.[23]

---

[19]   Ibid., 363, fols 76v–77r.
[20]   Ibid., 361, fol. 102r.
[21]   Ibid., 361, fols 78r–v.
[22]   AHN OO.MM, Cod. 606, fols 3v–4v.
[23]   AHN OO.MM, Cod. 606m fols 5v–6r.

Sacosta, who had so far managed to obtain the backing of either the convent, the king or the pope, now faced direct opposition from all three. It was not until the death of Lastic shortly before the Chapter General and after having been summoned to Rhodes in 1455, that he gained the full support of the new master; Milly chose him as his lieutenant and allowed him not to pay the ordinary *responsiones*.[24] In 1458 King Alfonso of Aragon and Naples died. Sacosta was immediately sent back to Spain as Milly's lieutenant in the Hispanic priories, with the intention of recovering those Aragonese and Catalan commanderies that were in the hands of the king's familiars for the benefit of those brethren who had made their career on Rhodes. In particular Sacosta claimed that Horta was still his and not Siscar's.[25] But backed by Juan II, Alfonso's brother and successor, none of those who had been installed in these commanderies agreed to hand them over. By then Sacosta had regained the trust of most Hospitallers at the convent, which permitted his election as master on Milly's death in 1461.

Sacosta, Master of the Hospital (1461–67)

At the time Sacosta received news of his election as master, Catalonia had been plunged into a violent political crisis followed by a civil war that was to last ten years. Juan II found himself opposed by most of the Catalan urban patricians, church officials and knights assembled under the leadership of the Catalan *Generalitat*. When the king heard of Sacosta's election as master, he decided that the Castellany should go to one of his most trusted supporters, the Hospitaller Bernat Hug de Rocaberti, and he unsuccessfully ordered his officials to seize the two strategic Catalan fortresses of Miravet and Ascó on the Ebro river, both of them *camere* of the castellan. The king's attitude led Sacosta to give his support to the *Generalitat* and take the unprecedented decision to keep the Castellany for himself in spite of his new status as master. After holding a six-month-long assembly of Aragonese and Catalan brethren in Barcelona, Sacosta sailed east in March 1462 in the company of the prior of Catalonia and of most of the Catalan and Aragonese commanders. The commanderies were left in the hands of procurators and farmers at a moment when they were exposed to the uncertainties and turmoil of a bitter civil war.[26]

Sacosta was faced with great difficulties in the Levant. The Turks had increased their raids on the Hospitallers' islands. The Order, in need of more knights and weapons, could not manage its finances and had enormous debts. Furthermore, the strife between the *langues* in the convent had, under Milly, reached an unbearable point. Sacosta decided to stop at Rome on his way to Rhodes in order to obtain the pope's full support and allow him to impose his authority upon all levels within the

---

[24]   Malta Cod. 367, fol. 204r.

[25]   Ibid., 367, fol. 204v.

[26]   N. Coll Julia, *Juana Enriquez, Lugarteniente real en Cataluña, 1461–1468* (Madrid, 1953), pp. 148–67.

Hospital. This visit and his activities in the following two years were successful for the new master. The pope showed enthusiasm for Sacosta, whom he qualified in his memoirs as a man of brilliant intelligence and as a Spaniard who came at the right moment to govern the Hospital, while the previous French masters had only shown interest when the situation was peaceful and the finances prosperous.[27] In a series of bulls, Pius II gave his complete backing to Sacosta with regard to the financial crisis, and he declared he would in no way intervene by granting out commanderies and other offices.[28] He also approved Sacosta's retention of the Castellany.

Sacosta held a Chapter General at Rhodes in November 1462. Among an impressive number of decisions and statutes he reinforced the so called 'minor' *langues* by creating an eighth one, splitting the Spanish *langue* in two. He established strict control over those Hospitallers who owned ships in order to curb their piratical activities. The chapter also decided that a commission would work on 'correcting, shortening and putting in order the volumes and books of statutes which should be condensed into a single *compendium*', a task that was not carried out until twenty years later. Above all, the chapter ordered all commanderies to send to the treasury their full incomes for three years in order to pay the Order's debts.[29]

Pius II died in 1464 at Ancona while he was preparing to lead in person a crusade against the Ottomans with the support of Venice, which had declared war on the sultan. The crusade project, to which Sacosta would have contributed five galleys, was then abandoned.[30] The death of Pius, who was succeeded by the Venetian Paul II, marked the end of the success story of Sacosta as master. A sequence of major misjudgements and failures ensued. While it was left to Venice to lead the struggle against the common enemy, the master ordered the capture and the sequestration on Rhodes of two Venetian merchant galleys coming from Alexandria. Venice retaliated by launching a massive attack on Rhodes with its war fleet. Sacosta welcomed several Catalan pirates who sheltered their galleys in Rhodes harbour, probably to dissuade Turkish attacks.[31] At the same time, he pursued unsuccessful peace talks with the Ottoman sultan.[32] Obviously these policies would not meet with the approval of the pope or the Venetians.

In the convent, bitter criticism of Sacosta followed his demand for three years' income from the commanderies, a demand that was too heavy to be accepted easily and was altogether unrealistic. Furthermore, the master was accused of being financially grasping and of favouring his relatives. His major failure had in fact been to keep the Castellany for himself and have it administered by his brother, who resisted the king in the Castellany's two Catalan fortresses. It had also proved

---

[27]   Pius II, *I commentari*, ed. G. Bernetti (Siena, 1972–76) 5, lib. 12, pp. 118–24.

[28]   Malta Cod 1127, fols 64r–68v; ibid., 1129, fols 190r–193v.

[29]   Ibid., 282, fols 113v–114v, 117r–119r, 123v.

[30]   G. Bosio, *Dell'Istoria della Sacra Religione e Illustrissima Militia di San Giovanni Gierosolimitano*, 2nd edn (Rome, 1629), 2, pp. 287–90.

[31]   Malta Cod. 374, fols 228v–229r.

[32]   Bosio, pp. 292–7.

unwise to leave the commanderies without their commanders at a time of armed conflict in Catalonia. The king was winning the war, taking over the abandoned commanderies and requesting the pope to destroy Sacosta, whom he considered to be one of his major enemies.[33]

Paul II refused to support Sacosta and decided that a Chapter General should be held in Rome under his own control, thereby clearly indicating the master's mismanagement.[34] The chapter, held in December 1466 and January 1467 in Sacosta's presence, was in fact directed by five bishops from the *curia*. Three papal bulls were issued, which included the chapter's statutes, a reform of the expenses and administration of the convent, and a new five-year plan for the repayment of debts. The papacy asserted its full right to intervene in the Order's affairs, especially with regard to appointments to commanderies and other offices.[35] Such a situation marked a serious setback for the authority of the master who had to resign the Castellany. Shortly after the end of the chapter Sacosta died in Rome, on 21 February 1467.

The frustrating conclusion to Sacosta's short-lived exercise of power as master should not lead to an underestimation of his real achievements with the impressive corpus of reforms adopted at the 1462 Chapter General and his contribution to the defence of Rhodes, where he undertook the construction of the Tower of St Nicholas and possibly of the tower on the mole of the windmills, both of which commanded access to the harbour.

Altogether, Sacosta led a brilliant but conflicting career in which he displayed unquestionable talents as a skilful politician. However, he accumulated misjudgements, most of all by repeatedly appealing to the papacy without at the same time building up a strong and loyal support among a majority of the brethren at Rhodes. His supporters in the convent had been reduced to a small group of Catalan Hospitallers holding high office and Sacosta was probably considered to owe his success to his ambition and to his manoeuvres rather than to the conduct expected for a successful career or to any distinguished military activity. There is no doubt, either, that Sacosta, like many successful Hospitallers who came from impoverished knightly lineages, lost no opportunity to accumulate all sorts of incomes for himself and his kinsmen and in so doing created an image of greed.

---

[33]   P. Bonneaud, *Els Hospitalers catalans a la fi de l'Edat mitjana, L'Ordre de l'Hospital a Catalunya i a la Mediterrània, 1396–1472* (Lleida, 2008), pp. 376–7, 394, 396–402.

[34]   Malta Cod. 383, fols 28r–v.

[35]   Ibid., 283, fols 36v–47v.

Chapter 12

Battlefield Tourism: A Description of the 1480 Siege of Rhodes

Theresa M. Vann

Literate pilgrims who stopped at Rhodes en route for Jerusalem in the fourteenth and fifteenth centuries sometimes described the island and the city's main features in their accounts of the journey. Historians attempting to recreate the topography of the medieval city of Rhodes have found these descriptions inadequate for identifying buildings and fortifications. Michel Balard theorized that these visitors were so distracted by the presence of the Master and Convent of the Order of the Hospital that they overlooked the inanimate sites of Rhodes.[1] Rhodes, however, was not a destination for these travellers, just one more stop on the pilgrimage route. Pilgrims found the island's attractions scant in comparison with the glories of the Holy Land. Their interest in Rhodes evolved after the Order successfully defended the island against the Ottoman Turks in 1480. Afterwards, Guillaume Caoursin, the Hospitaller vice-chancellor, published the Order's official history of the siege, entitled *Descriptio Obsidionis Rhodie Urbis*. The Latin edition was printed in eight European cities between 1480 and 1483, and Hospitaller priories throughout Europe commissioned translations into the vernacular. At the same time, the familiar medieval manuscript genre of the pilgrim's journey to Jerusalem evolved into a complex work that catalogued attractions and novelties, similar to the modern guidebook.[2] One of the best examples of this new comprehensive pilgrimage account, Bernhard von Breydenbach's *Peregrinationes in Terram Sanctam* (1486), not only included an original engraving of the port of Rhodes but also reprinted Caoursin's *Descriptio*, providing visitors a comprehensive history of Rhodes along with a complete catalogue of its points of interest. A unique late fifteenth-century French manuscript account of the Siege of Rhodes (Paris, Bibliothèque Nationale, Dupuy 255) shows the influence of this new pilgrimage genre. The unknown author integrated the history of the siege of Rhodes with a description of the city to provide a guide to the sights of the battlefield.

---

[1]   A. Luttrell, *The Town of Rhodes 1306–1356* (Rhodes, 2003), p. 8; M. Balard, 'The Urban Landscape of Rhodes as Perceived by Fourteenth- and Fifteenth-Century Travellers', in B. Arbel (ed.), *Intercultural Contacts in the Medieval Mediterranean* (London, 1996), pp. 23–34.

[2]   F.T. Noonan, *The Road to Jerusalem: Pilgrimage and Travel in the Age of Discovery* (Philadelphia, 2007), pp. 8–9.

Rhodes was not a regular stop on the Jerusalem pilgrimage; it featured as a Christian port where ships could stop before journeying to their final destinations in Jaffa, Acre or Alexandria. Nor was it a destination for cultured travellers seeking Greek antiquities; for them, it only offered the former site of the lost Colossus of Rhodes.[3] Pilgrims stayed on Rhodes for periods of as short as two days or as long as a month, depending upon how long it took to arrange transportation, refurbish their vessels, or obtain a safe conduct.[4] Most visitors remained in the city or harbour area, although those with additional time explored other parts of the island.[5] While no one feature was mentioned in all the written accounts, for much of the fourteenth and fifteenth centuries there was a consensus that the walls were massive and the relics meagre.[6] The only relics of any note during this period were a thorn from the Crown of Thorns and a cross made from the basin that Christ used to wash the feet of his disciples, although particularly energetic pilgrims could travel to Philermo to view the icon of the Virgin painted by St Luke. This, combined with increasingly outlandish stories of the size of the Colossus, were all that pilgrims thought noteworthy about the city.[7]

Some pilgrims included the history and the exploits of the Order of the Hospital in their descriptions of Rhodes. Ludolph of Sudheim, who travelled in the East between 1336 and 1341, knew of the Hospitallers' recent history, which included the conquest of the island and their acquisition of Templar properties.[8] By the late fifteenth century, the tone of pilgrimage accounts changed to embrace the enthusiasm for the Order noticed by Balard.[9] Pilgrims recorded audiences with the master, hospitality in the auberges, impromptu tours of the city conducted by friendly knights, encounters with knights who spoke their native language, knights who enquired about mutual acquaintances, and knights who briefly travelled with them either to or from Rhodes.

Fifteenth-century visitors recorded their impressions of the hospital, the conventual church, and the magisterial palace. The pilgrims knew when Rhodes was preparing for war. Anselm Adorno, who visited the Holy Land in 1470–71, described the battle readiness of the city and its people during the year the Turks

---

[3]   This did not stop some from looking; nor did it stop Rhodians from pointing out where it had been. See Luttrell, *Town of Rhodes*, pp. 219–21, 281; E. Borsook, 'The Travels of Bernardo Michelozzi and Bonsinore Bonsignori in the Levant (1497–98)', *Journal of the Warburg and Courtauld Institutes* 36 (1973), 164, 170; Bernhard de Breydenbach, *Sanctarum Peregrinationum in Montem Syon* (Mainz, 1486), fol. 17v.

[4]   Santo Brasca, *Itinerario alla santissima città di Gerusalemme* (Milan, 1481), p. 22; Borsook, p. 164.

[5]   Ibid., p. 170.

[6]   Balard, pp. 25–7.

[7]   Nicolas de Martoni ('Relation du pèlerinage a Jérusalem de Nicolas de Martoni, Notaire Italien (1394–1395)', ed. L. le Grand, *ROL*, 3 (1895), 585–6) thought the Colossus spanned the large commercial harbour of Rhodes, as opposed to the smaller Mandraki.

[8]   Luttrell, *Town of Rhodes*, pp. 216–19.

[9]   Balard, pp. 24–5, 34.

captured the island of Negroponte.[10] Fr Felix Faber's ship was the first Christian ship to enter the harbour immediately after the siege of 1480, and there were some tense moments when the Rhodians mistook his ship for the Turks. Fr Felix found the shore covered with the bodies of dead Turks, the walls and towers ruinous from the shelling.[11]

The unique observations recorded by individual pilgrims circulated in manuscript format, but this literary genre was superseded in the fifteenth century when the invention of printing facilitated the circulation of popular itineraries that, like modern guidebooks, standardized the visitor's experience. The most significant work of this type was Bernhard von Breydenbach's *Peregrinationes in Terram Sanctam*, an illustrated guidebook based on the author's 1483 journey to the Holy Land, which was translated and reprinted many times.[12] Breydenbach's work listed the places he visited along the pilgrimage route, a brief history of each location, and sites of interest to pilgrims. The book also contained works by Jacques de Vitry, Vincent of Beauvais, Petrus Alphonsus, Bartholomew de Lucca, Patriarch Isidorus of Constantinople, Balthasar Perusino and Guillaume Caoursin describing the Holy Land, the Muslims, the Jews and non-Latin Christians, and descriptions of the sieges of Constantinople, Negroponte and Rhodes.[13] The woodcuts are attributed to Erhard Reuwich, a painter from Utrecht, who travelled with Breydenbach and drew the cities from life.[14]

Breydenbach's words and images informed his readers what they should know about Rhodes. After describing his party's arrival, he catalogued the relics in the castle and the conventual church. The Sultan Bayezid had sent the Order the arm of John the Baptist in 1484, and Breydenbach duly listed it along with all the other relics available for view, to wit: one of the 30 pieces of silver that Judas received for betraying Christ; the heads of Sts Philomena, Eufemia and Polycarp, plus a head from one of Ursula's 10,000 virgins; the hands of Sts Claire and Anne, and, in addition to the arm of John the Baptist, portions of the arms of Sts Blaise, Stephen, George, Thomas, Katherine and Leodegard.[15]

The reader of the *Peregrinationes in Terram Sanctam* could identify Rhodes from Reuwich's woodcut, an oblique view of the city as seen from the ship. Breydenbach's written description of the city did not say more than could be discerned in the picture. He described Rhodes as a fortified city with towers and walls, counted thirteen windmills on the mole of the windmills, and concluded

---

[10]   *Itinéraire d'Anselme Adorno en terre sainte (1470–1471)*, eds and trans. J. Heers and G. de Groer (Paris, 1978), pp. 362–5.

[11]   Felix Faber, *The Wanderings of Felix Fabri*, trans. A. Stewart, PPTS (London, 1896), 1, p. 31.

[12]   H.W. Davies, *Bernhard von Breydenbach and his Journey to the Holy Land* (London, 1911), pp. vi–vii; Noonan, pp. 35–44.

[13]   Davies, pp. ix–xi.

[14]   Ibid., pp. iii, xxi.

[15]   Bernhard von Breydenbach, fol. 17r.

with a brief account of the history of Rhodes between the Trojan War and the 1480 siege.[16] For more information on the Turkish expansion in the Levant, his readers could turn to the back of the book and read the unattributed accounts of Isidorus, patriarch of Constantinople, on the fall of Constantinople in 1453, Balthasar Perusino on the capture of Negroponte in 1470, and Guillaume Caoursin's description of the siege of Rhodes in 1480.[17] These appendices became part of the text of Breydenbach's book, and were included in subsequent reprints and translations.

The successful publicity of the Order's victory, combined with the publication of printed guides to the Holy Land, changed the pilgrims' expectations of Rhodes. Before 1480, hospitality consisted of a good meal and a viewing of the relics. After 1480, pilgrims expected to see the places in Rhodes where the events of the siege took place. Rhodes remained a prominent Christian outpost against the Turks through the 1480s, even after the death of Sultan Mehmed II in 1481. Mehmed's two sons, Bayezid and Djem, fought over the accession. His younger son, Djem, arrived in Rhodes in 1482 after losing to his brother, the sultan Bayezid II. The Hospitallers negotiated a treaty with Bayezid to keep Djem in custody.[18] Until his death in 1495, Djem remained a high-profile hostage, and the Hospitallers enjoyed a truce with the Ottomans. Immediately after Djem's death, the Order prepared for the resumption of hostilities.

In 1496, Caoursin published a new edition of his *Descriptio* as part of a volume entitled *Rhodiorum historiae*.[19] Caoursin himself edited this new, illustrated edition, which collected all of his historical works into one volume. Starting with his *Descriptio* of the 1480 Ottoman siege, the volume included the account of the 1481 earthquake on Rhodes, the death of Mehmed II, and finished with the story of Djem's arrival in Rhodes. The timing and format of the publication suggests a deliberate effort on the part of the Order to remind Europeans of their immediate history with Mehmed and his sons. Breydenbach and his numerous appendices may have inspired the comprehensive nature of the collection. The volume survives in numerous copies, an indication of its popularity.[20]

Caoursin's *Descriptio* spawned translations and adaptations, most notably by Adhemar du Puy and John Kay.[21] But Paris BN Dupuy 255, entitled *Histoire journalière de ce qui se passa, soubs la conduite de Fr. Pierre d'Aubusson, grand-maistre de Roddes, au siège de la ville de Roddes, faict par Mahomet II, empereur*

---

[16]   Ibid., fols 17r–18r.

[17]   Davies, pp. ix–xi.

[18]   N. Vatin, *L'Ordre de Saint-Jean-de-Jérusalem, l'Empire ottoman et la Méditerranée orientale entre les deux sièges de Rhodes 1480–1522* (Louvain, 1994), pp. 161–87.

[19]   Guilelmus Caorsin, *Rhodiorum historia (1480–89)* (Ulm, 1496).

[20]   The British Library Incunabula Short Title Catalogue (ISTC no. ic00113000) lists seventy-six surviving copies; more may be held in private collections.

[21]   Adhemar du Puy, *Le siege de Rhodes* (Lyons, 1480; Audenarde, 1482); John Kay (trans.), *The Siege of Rhodes* (Westminster/London, 1482–83).

des Turcs, en l'année 1480, which appears at first glance to be another adaptation of Caoursin, is in fact dissimilar in style from any other account of the siege, and its first seven folios are similar to a pilgrim's description of the city. The third-person narrative conducts the reader on a walking tour of the city of Rhodes. The tour begins at the harbour, where the anonymous narrator names all the towers, their purpose, their size and distance from each other. But the narrator is clearly standing on the shore looking out into the harbour, and not a pilgrim gazing at the city from aboard a ship. For example, Anselm Adorno wrote that the mole of the windmills was on the left, but the *Histoire journalière* places it on the right.[22] The manuscript also identifies the location of the Turkish artillery during the siege and which parts of the harbour defences were destroyed. The narrator then points out parts of the landscape, mentioning Mount Philermo, with its icon. Here the narration switches between third-person and first-person plural to describe how the Rhodians appealed to the Lady of Philermo for succour during the siege.[23]

The narrator provides the reader with basic information about the composition of the Order. He lists the *langues* and their responsibilities, with particular emphasis on the Hospitaller, the head of the *langue* of France. He then locates each auberge in the city, the master's palace, the conventual church (omitting its relics), and each *langue*'s post on the walls of Rhodes, explaining where they are in relation to each other and how far apart they are. The narrator begins a day-by-day recitation of the events of the siege on folio 12, which concludes on the second to last folio with the departure of the Turkish fleet. The same hand squeezes in on the final folio two events that happened after the conclusion of the siege: Djem's arrival in Rhodes in 1482 and the earthquake of 1481.

The damaged colophon dates the manuscript to 1495, but the name of the scribe and the location of the scriptorium are missing.[24] The name 'Francoys Guysmier' written by a later hand on folios 1r, 54r and 70v identifies an owner, not the author or the scribe. The text, written in an ornamented semi-gothic script, leaves numerous spaces for illustrations. One illustration is filled in, but not, apparently, by a professional artist. It will require more codicological study

---

[22] *Itinéraire d'Anselme Adorno*, p. 362: 'In sinistra vero parte in oppositum hujus, alia turris etiam multum fortis tutelisque defensoriis fulcita, prope quam supereodem aggere sexdecim, si bene numeraverim, molendina ex petris inferius cum sex alis sive veils constructa sunt.' Paris BN Dupuy 255, fol. 5r: 'Ala dextre' a ung | molle dedens lamer fait a chaulx et a sabloy en fasson | duy mus qui contient en viron trois cens pas de | longuens et trente' pas de largeur. Sur le quel | molle a xiii mollins a vent ariengiez par quoy est | appelle le molle des mollins.'

[23] Ibid., fol. 6v: 'Autemps de laguerre loy aporta ledit ymaige en grant solem|pnite en procession en la cite et lappellon nostre Dame de Phillerme Durant nostre tribulacion elle a este sonnentes foiz piteusemene.'

[24] Ibid., fol. 127v (colophon), dated 13 September 1495. The manuscript consists of 123 folios; the first part is missing (the enumeration begins with fol. 5 and ends with fol. 127).

to establish the provenance of the manuscript.[25] The writer could have worked from a French translation of Breydenbach, which included Caoursin's *Descriptio*, to create an imaginative description of the island and the siege.[26] But the 1495 colophon mentions two events that Caoursin only recorded in the illustrated history of Rhodes, published in 1496. This, combined with the occasional use of the second person plural in the manuscript, suggests that Dupuy 255 may have been composed in Rhodes; the use of French and the emphasis on the *Langue* of France suggest some connection with that priory. The didactic nature of the narrative, which describes the Order's home and explains their international organization and cooperation, their hospitaller mission and their actions during the siege of 1480, suggests an external audience. There is no proof that it ever served such a purpose, but one could imagine the text serving as a script for a guided tour of the city. All speculation aside, the text is evidence that someone was fascinated by the events of 1480 and the places where they happened; and that, after the siege, visitors came to Rhodes not only to stop on the way to Jerusalem but to see where one of the great victories of the Order of the Hospital had taken place.

---

[25]   Laurent Vissiere is at work on the text of Paris Dupuy 255 and generously shared some of his findings with me. All speculation on the possible purpose of the manuscript is entirely my own.

[26]   Bernhard von Breydenbach, *Des sainctes peregrinations de iherusalem*, trans. Nicolas le Huen (Lyons, 1488); idem, *Le saint voyage et pelerinage*, trans. Jean de Hersin (Lyons, 1489), used the original woodcuts and included the appendices.

Chapter 13

Woven Tapestries: Manifestations of Grandeur, Politics and Power as well as Pictorial Sources for Hospitaller History: A Re-identification

Robert L. Dauber

From the Middle Ages onwards tapestries (*tapisseries*) were not simply *objets d'art* but also very expensive statements of grandeur, politics and power for both secular and ecclesiastical rulers.[1] Historical tapestries were large and elegant woven pictorial wall coverings of great splendour and value, enormously appreciated by observers. They were used to adorn palaces and churches in order to impress contemporary and future statesmen and peoples. Series of tapestries would often document important battles and historical events with the specific purpose of extolling and publicizing the fame of the princes and dynasties concerned. Tapestries were not only commissioned and displayed by secular rulers, but also, among others, by the heads of the Order of the Hospitallers, the grand masters of Rhodes and Malta. We know for instance that in 1493 Grand Master Fr Pierre d'Aubusson had a set of tapestries with his magisterial coat of arms manufactured in Flanders. In about 1600 this cultural and historical artefact could still be seen in Valletta by the famous historian of the Order, Fr Giacomo Bosio, as he himself noted in his celebrated *Istoria*.[2]

Sometime after 1510 his successor, the Grand Master Fr Aymeric d'Amboise, ordered and then exhibited in Rhodes a series of most beautiful tapestries, which similarly showed his coat of arms and which commemorated his and the Order's naval exploits in general and, more specifically, the strategic victory at Laiazzo (or Ayas, now Yumurtalik) in the Gulf of Alexandretta (Iskenderun). Bosio reports that he himself had also seen this series in the grand master's palace in Malta.[3] One piece that survives from Aymeric's series will be considered below.

---

[1] *Der Kriegszug Kaiser Karls V. gegen Tunis. Kartons und Tapisserien*, ed. W. Seipel (Vienna and Milan, 2000), p. 56 (catalogue of the international exhibition on Charles V shown in Ghent, Bonn, Vienna and Toledo).

[2] G. Bosio, *Dell'Istoria della Sacra Religione e Illustrissima Militia di S. Giovanni Gierosolomitana* (Venice, 1695), 2, p. 419.

[3] Ibid., 2, p. 495.

After 1697 the grand master in Malta, Fr Raymondo Perellos y Rocaful, ordered a series of no less than twenty-nine splendid tapestries from the tapestry manufacturer Jodocus de Vos in Brussels. This series is nowadays exhibited in the palace and in the co-cathedral of St John in Valletta.

It is true to say that, with the passage of time, the historical content of the tapestries and other depictions, or important details of it, have in some cases been forgotten and lost beyond recall. It is also true that tapestry specialists and art historians, who quite often are members of staff in the museums where they are kept, specialize mainly in the study of how tapestries were made, in their physical attributes, in the workshops that produced them, and in their subsequent histories, but are seldom concerned with the detailed historical background of the themes depicted. On the other hand, general historians and specialists in the history of the military orders rarely have an intimate connection with, a knowledge of, or access to, the relevant tapestries. So there would seem to be a gap between academic disciplines and a surprisingly large scope for research into old tapestries and in other types of pictorial representation and for the re-identification and rediscovery of their historical contents. What follows is based on my own specialized and detailed research into the naval history of the Hospitallers, and is concerned with just one example of a rediscovery and re-identification of the forgotten or, rather, misinterpreted subject-matter of a tapestry.

*   *   *

In 1970 the Council of Europe sponsored an exhibition in Valletta with the title, 'The Order of St John in Malta'. The catalogue entry for exhibit no. 283 describes a beautiful tapestry of silk and wool measuring 410cm by 810cm that had been lent by the Museo de Arte de Barcelona and which is there entitled 'The Virgin Mary appears during the Siege of Rhodes 1480'.[4] Today this tapestry is on display with a similar caption – 'El Sitio de Rodas 1480' – in the Museo Textil i d'Idumentaria in Barcelona (Fig. 13.1).[5] This tapestry had been bought in 1589 by the Taula de Canvi (a bank established as far back as 1401) in Barcelona for 140 *libras*, and was given by its legal successors to the Museo de Arte in Barcelona in 1866. Records show that, besides the 1970 Valletta event, it has frequently been included in public exhibitions both in Barcelona itself (1877, 1906, 1930, 1947, 1971, 1981 and 1989) and elsewhere: Ghent (1955), Toledo (1958), Brussels (1985) and Valencia (1997). In all these exhibitions, the captions and the published

---

[4]   *The Order of St. John in Malta: Catalogue of the Thirteenth Council of Europe Art Exhibition, 1970* (Valletta, 1970), p. 278, exhibit 283.

[5]   Museo Textil i d'Idumentaria, Barcelona, Inventory number 37.782. The author wishes to express his thanks to the museum staff and the curator, Rosa M. Martin i Ros, for their unstinting assistance, for supplying numerous photocopies from past exhibition catalogues and for granting permission to publish a photograph of this tapestry.

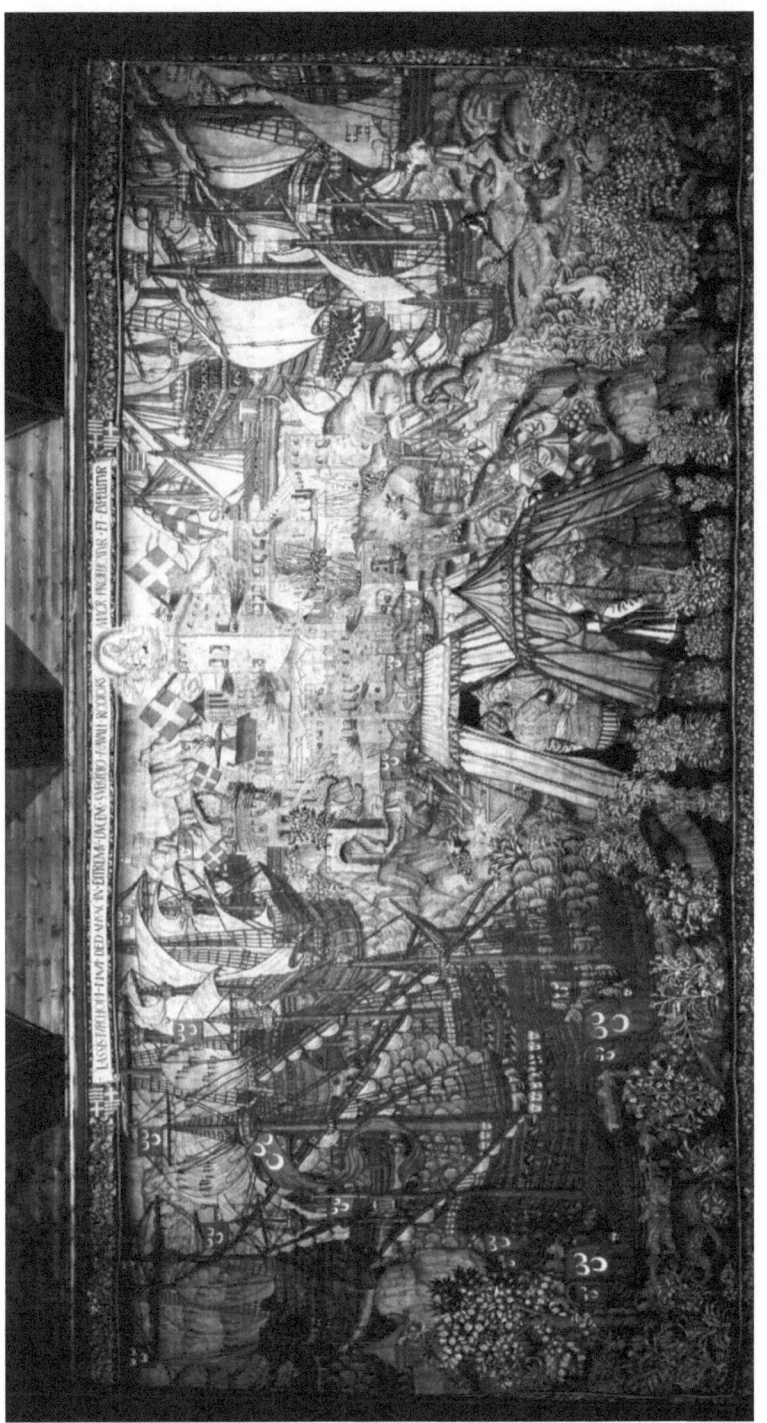

Figure 13.1   Grand Master d'Amboise's squadron and the relief of Rhodes (© Museo Textil i d'Idumentaria, Barcelona)

commentaries employ the title 'El Sitio de Rodas 1480' – 'The Siege of Rhodes 1480'.[6]

The catalogue of the 1970 Malta exhibition indicates that this tapestry presents at the top the coat of arms of Grand Master d'Amboise and an inscription in Latin that, when expanded, reads: 'CLASSIS TURCHORUM NAVI DEDANS AC IN EXTREMUM DUCENS SUBSIDIO NAVALI RODIORUM MAGISTRI (MAǦR) PROFLICATUR ET EXPELLITUR', a possible translation being: 'The Turkish fleet, yielding to a *navis* and coming to the limit, is dispersed and driven off with the naval assistance of the master (or MAǦR – by Mary's grace) of the Rhodians'.

The tapestry shows a sort of peninsula with a sea fortress adorned with flags of the Order, that is evidently Rhodes (or possibly, in the author's view, Kos). On the left side of the fortress there appears a large fleet with Turkish flags, and on the right there are galleys and small and great sailing ships flying the flag of the Order of Rhodes together with that of Grand Master d'Amboise and of two conventual bailiffs.

The Malta exhibition catalogue describes the scene as the Siege of 1480 and gives the tapestry the title: 'The Virgin Mary appears during the Siege of Rhodes 1480'. It also states with reference to the inscription and d'Amboise's coat of arms: 'The tapestry changes the story by suggesting, that the siege was raised by the arrival of a relieving flotilla under Amboise. It is true that Rhodes was succoured in 1480 when two ships of Naples broke through the blockade, and Amboise, then a relatively junior member of the Order, may have been with them'; and further down: 'None of the military and naval actions of Amboise's magistracy fits the events depicted in the tapestry'. The statements of the 1970 Malta catalogue broadly speaking repeat the gist of other exhibition catalogues, captions and commentaries that stretch back over a century to 1877. It should be noted that none of these catalogue entries or other publications has been authored by a historian who could claim to be a specialist in the history of the Hospitallers. Calling the d'Amboise tapestry 'El Sitio de Rodas 1480' is simply the commonly received view.

On the basis of my own detailed researches into Hospitaller naval history, which in recent years has focused especially on the period on Rhodes,[7] I am now for the first time in a position to say that the longstanding assumptions and repeatedly published statements to the effect that the tapestry represents an episode during the 1480 siege of Rhodes, but that it alters the story by suggesting that the siege was raised by a flotilla under d'Amboise and that none of the military and naval actions of d'Amboise's magistracy fits the events depicted in the tapestry, are all totally mistaken.

---

[6]   Information courtesy of the staff at the Museo Textil i d'Idumentaria, Barcelona.

[7]   The author is currently preparing a three-volume monograph, *Classis et Castra: Marine und Seefestungen der Johanniter von Rhodos*, on the structure and chronology of the Hospitaller navy during their rule on Rhodes. Volume 3 (the first to appear) was published in Gnas (Austria) in 2010: *Seeoperationen und Seefestungen (1309–1523)*.

A detailed analysis of the pictorial content of the tapestry, the biography of Grand Master d'Amboise, and of the military and naval situation generally during his magistracy, has convinced me that the event depicted represents an episode described by Bosio: the initial, grandiose arrival of the strong naval squadron commanded by the newly elected grand master Fr Aymeric d'Amboise from France at Rhodes (or perhaps in Kos) at around 9 o'clock in the morning on 1 September 1504. We know that the naval squadron was composed of two great ships (*gran navi*) belonging to the Order, one chartered carrack from Genoa and two *barcie* (heavily armed sailing ships), a total of five sailing ships plus two accompanying French galleys and finally three galleys of the Order, which had escorted the incoming squadron from the island of Kos as far as Rhodes itself.[8]

Looking at the tapestry, we can clearly see repeated examples of d'Amboise's flag, thus showing that he was in command of the squadron, and at least five sailing ships and some galleys decked out with flags of the Order and with those of two of its bailiffs. One such flag flown on a sailing ship represents that of the then *magister domus* (the master of the grand master's household), Fr Jean de Villiers de l'Isle-Adam, who was later to be the last grand master of Rhodes and first of Malta.[9] (His ship and flag are to be found on the right side of the tapestry towards the bottom.)

The mass of Turkish ships on the left-hand side of the tapestry would seem to represent the attacks and threats of the Ottoman fleet, which had persisted since the end of the great Christian naval alliance and the death of the grand master Fr Pierre d'Aubusson, both of which occurred in 1503. Fr Aymeric d'Amboise had been elected the new grand master at a time when he was in France. That same year, 1503, the Turks had attacked Rhodes with 16 *fuste* (biremes, oared fighting ships rowed in the *à zenzile* style), had sacked and burnt a number of villages and had taken 160 inhabitants as slaves. A subsequent engagement with a pursuing Rhodian naval squadron ended in a Hospitaller victory.[10] In 1504 friction and provocation involving Rhodes and the Turks had continued. The Turks made threats, and the Hospitallers, fearful of further attacks, prepared for war, and turned to Europe, where the new grand master had remained, for assistance.[11]

This, then, was the precarious situation in Rhodes into which the grand master arrived with a strong naval force in September 1504. Despite the Turks' aggressive policy and the attacks on Rhodian territory, his personal presence and the military measures he took in conjunction with the council of the Order led to a stabilization of the military situation. Moreover, as Bosio was to emphasize, in the period of

---

[8]   Bosio, 2, pp. 419, 479.

[9]   See J. Sarnowsky, *Macht und Herrschaft im Johanniterorden des 15. Jahrhunderts. Verfassung und Verwaltung der Johanniter auf Rhodos (1431–1522)* (Münster, 2001), pp. 661–2.

[10]   Bosio, 2, pp. 473–4.

[11]   Ibid., 2, p. 475.

Amboise's magistracy, the Hospitallers' fleet secured more victories at sea than at any other time in the history of the Order and its navy.[12]

As mentioned above, the tapestry discussed here is in all probability one of the series ordered by Grand Master d'Amboise after his decisive victory over an Egyptian fleet at Laiazzo in 1510. Contrary to the statements to be found in the Malta exhibition catalogue and the other exhibition publications, it would seem beyond any reasonable doubt that this tapestry does not represent an episode involving the grand master d'Aubusson during the Siege of Rhodes 1480, but that the episode the tapestry illustrates relates to the dangerous situation facing Rhodes and its relief thanks to the arrival of a powerful naval squadron under Grand Master d'Amboise in September 1504: with his arrival and the military measures he and his government instigated, the naval pressure exerted by the Turks was contained, thereby fully justifying the inscription on the tapestry.

This research provides an example of how an analysis of a picture of a naval episode by a specialist in maritime history can correct misinterpretations that go back at least 150 years. I can think of at least four other illustrations that, as a specialist in naval history and Hospitaller biography, I can re-identify. The point is that representations of naval matters relating to the military orders should be analysed by historians specializing in their navies; flags should be analysed by vexillologists, coats of arms by heraldry specialists, and so on – and not by general historians or even art historians. Closer interdisciplinary cooperation in this field of research is needed.

One final question: is it possible that any of the other pieces from the set of naval tapestries that, according to Bosio, d'Amboise commissioned after 1510 survived, and, if so, where are they and have they been identified as such?

---

[12]   Ibid., 2, p. 490.

Chapter 14

Politics and Power in Grand Master Verdalle's *Statuta Hospitalis Hierusalem* (1588)

Emanuel Buttigieg

By 1581, the Order of St John had been on Malta for just over fifty years. It had survived the loss of Rhodes in 1523, seven years of homelessness and the Great Siege of 1565, and had founded the new city of Valletta. Yet for all its residual strength, the Order was nearly torn apart by an internal power struggle during the summer of 1581. It took the determined intervention of the papacy to avert disaster, and the man assigned the task of picking the pieces was Fr Hughes Loubens de Verdalle (1582–95). This paper focuses on the policies and authority of this grand master, whose reign of thirteen years was marked by an increasingly absolutist drive to assert the powers of his office. The new grand master embarked on an ambitious architectural and iconographic programme, which included the production of a lavishly illustrated new edition of the statutes of the Order, the *Statuta Hospitalis Hierusalem* (in short *Statuta*) of 1588. This paper adopts a 'visual culture' approach in which the documentary and non-written sources are considered alongside each other with the aim of forming a more holistic understanding of Verdalle's politics and power and the role of the *Statuta* therein.

Source Analysis and Methodology

The *Statuta* was produced in Rome and incorporates engraved book illustrations by the artists Philippe Thomassin (1562–1622) and Giuseppe Cesari, *Il Cavalier d'Arpino* (1568–1640).[1] Some of these images are familiar to Hospitaller scholars; three in particular have often been reproduced in books. These are the images showing the reception of nuns into the Order, depicting Hospitallers processing

---

[1]   L. Ronzon, 'Appunti per una storia della grafica dell'Ordine Gerosolimitano', in L. Corti, F. Amendolagine and M. Doglioni (eds), *Lungo il tragitto crociato della vita* (Venice, 2000), pp. 263–4; G. Bonello, *Histories of Malta: Mysteries and Myths* (Malta, 2007), pp. 75–84. On sixteenth-century engravings, K.L. Bowen and D. Imhof, *Christopher Plantin and Engraved Book Illustrations in Sixteenth-century Europe* (Cambridge, 2008).

towards the Conventual Church, and illustrating the Holy Infirmary.[2] It is easy to see why these images are so popular: their beauty and craftsmanship are truly remarkable.[3] Nonetheless, images from the past have more to tell us when considered as sources in their own right, and not just as illustrative material; there is a need for the visual component of the *Statuta* to be ploughed back into the mainstream of Hospitaller historiography. In the main, historians have concentrated on written documents, and source criticism is an integral part of historians' training.[4] On the other hand, the use of images and objects as primary matter for historical consideration has only slowly come to be adopted as part of a historian's intellectual apparatus, partly because of a perception that non-written material is the province of art historians and archaeologists.[5]

The challenge facing historians is how to avoid misinterpreting the significance of an object by reading things that were not originally there while, at the same time, missing meanings that would have been easily recognizable by contemporaries. In this vein, there are tools that historians can use to assist their understanding of non-written sources. For instance, the series called *Dizionari dell'Arte* (Mondadori Electa) is an example of a highly useful aid. Each volume in this collection is dedicated to a specific theme, such as the meanings of gestures or the symbols of power.[6] These are of great utility in aiding understanding of significant subtleties. Another example of a tool that can assist historians in understanding non-written sources is an online project developed by the Centre for History and New Media, George Mason University, and the Department of History, University of California, for the *American Historical Review*.[7] This website is dedicated to images of the French Revolution; its most original feature is what is called an 'Image Tool'. This Image Tool allows the user to choose images about the French Revolution from a database and to 'play' about with them: it is possible to have different images next to each other, to juxtapose images, to fade them into each other. There are essays by experts in the field and a space for discussion. Questions discussed vary

---

[2]   *Statuta Hospitalis Hierusalem* (Rome, 1588), pp. 12, 18, 30. For reproductions, R. McHugh, *The Knights of Malta: 900 Years of Care* (Dublin, 1996), p. 140; G. Bonello, *Histories of Malta: Figments and Fragments* (Malta, 2001), p. 80; J. Riley-Smith, *Hospitallers* (London, 1999), p. 28.

[3]   On the influence of the *Statuta* on Caravaggio, D.M. Stone, 'The Context of Caravaggio's "Beheading of St John" in Malta', *The Burlington Magazine*, 139 (1997), 165–6.

[4]   P. Burke, *Eyewitnessing: The Use of Images as Historical Evidence* (London, 2001), pp. 14–15.

[5]   I. Gaskell, 'Visual History', in P. Burke (ed.), *New Perspectives on Historical Writing* (Cambridge, 2001), p. 211; I. Hodder and S. Hutson, *Reading the Past: Current Approaches to Interpretation in Archaeology* (Cambridge, 2003), pp. 128–42.

[6]   P. Rapelli, *Simboli del potere e grandi dinastie* (Milano, 2004); R. Giorgi, *Simboli, protagonisti e storia della Chiesa* (Milan, 2004).

[7]   'Imaging the French Revolution', at http://chnm.gmu.edu/revolution/imaging/home.html (accessed 14 November 2011).

from the relationship between images and texts, to the power of images in the historical imagination. Such a website points out ways in which images can be used to inform historical analysis and is indicative of how the historical discipline is evolving, both in terms of teaching as well as research. Together, these two examples alert historians to the need to be aware of the potential of images as sources for social commentary.

According to Peter Burke, non-written sources 'are neither a reflection of social reality nor a system of signs without relation to social reality, but occupy a variety of positions in between these extremes'.[8] It was the early cultural historian Aby Warburg who initiated a reconsideration of the relationship between documentary and non-documentary forms of evidence. Warburg's work and library highlight his belief that the past speaks to us through both documents and images.[9] A more recent figure in the reappraisal of non-written sources for historians was Svetlana Alpers. She has argued that, as images were an integral part of early modern society, taking the Dutch as her particular example, the neglect of images as sources can only produce a partial story.[10]

The study of non-written sources has helped to move artefacts belonging to the masses from the fringes of folklore studies to their place within a revitalized examination of popular culture.[11] At the same time, there has been a re-examination of the material culture of the more affluent sectors of society. Hence this paper is influenced by two key studies: Quentin Skinner's *Hobbes and Republican Liberty* (2008) and Peter Burke's *The Fabrication of Louis XIV* (1992). Skinner's work outlines the depth of meaning to be found in the frontispiece of Thomas Hobbes's *Leviathan* (1651). Hobbes had a lifelong concern with the visual representation of his philosophical ideas and the emblematic frontispiece attempted to summarize the arguments of the ensuing text. Hobbes' use of images was derived from his early education centred on the Renaissance syllabus of the *studia humanitatis*.[12] The *Statuta* of Verdalle was a product of these same humanist notions that believed in the effectiveness of the use of images to help people understand ideas. On the other hand, Burke's study of Louis XIV through contemporary representations shows how the image of the sun king was built up from depictions of him in stone, bronze, paint, poems, and other media.[13] It is therefore a case study of the relations

---

[8]   Burke, *Eyewitnessing*, p. 183. See also B. Wilson, *The World in Venice: Print, the City and Early Modern Identity* (Toronto and London, 2005), pp. 20–22.

[9]   A. Warburg, *The Renewal of Pagan Antiquity: Contributions to the Cultural History of the European Renaissance* (Los Angeles, 1999), p. 56.

[10]   S. Alpers, *The Art of Describing: Dutch Art in the Seventeenth Century* (London, 1983), p. xxv.

[11]   Gaskell, p. 190.

[12]   Q. Skinner, *Hobbes and Republican Liberty* (Cambridge, 2008), pp. 190–98; idem, 'Seeing is Believing: The Words and Imagery of Thomas Hobbes', *BBC History Magazine*, 9 (2008), 53–7.

[13]   P. Burke, *The Fabrication of Louis XIV* (New Haven, CT, 1992), p. 1.

between art and power.[14] In a similar vein, the *Statuta* of Verdalle formed part of a range of means – buildings, palaces, rituals – that helped to build up the image of Verdalle as cardinal-grand master.

This paper adopts a 'visual culture' approach to understand the role of images and objects in the practices of the Hospitallers, including politics and power. To reconstruct the specific historical context of these visual conventions it is necessary to be aware of shifting aesthetic predispositions among producers, subjects, patrons and viewers.[15] There are also limits to the weight of connoisseurship: theories about authorship of an object and its uses tend to be subject to periodic revisions or to intellectual fossilization.[16] The medium for describing pictures and objects is language; speech acts engage with the visual to elicit an understanding of the object and of its effects on people.[17] These considerations act as bearings on the path to a better understanding of the context in which an object was produced and perceived.[18]

The Events of 1581 and Verdalle

Having outlined the methodological approach of this paper, it is now necessary to set out the peculiar political situation in which Verdalle came to power. In 1581 the disciplinarian grand master Fr Jean Levesque de la Cassiere (1572–81) ordered the expulsion of all prostitutes from Valletta; the immediate reaction of many Hospitallers was to revolt, depose the grand master and appoint Fr Marthurin d'Aux de Lescaut dit Romegas as lieutenant in his place.[19] Underneath an apparently trivial motive for a political coup lay deeper considerations, in particular the aspirations of various senior Hospitallers to attain the grand magistracy. A grand master who lived too long constituted a problem for those persons and factions set on controlling the Order and promoting their own interests and ideas. Similar feelings surfaced during an overly long pontificate, such as that of Pope Urban VIII (1623–44).[20] Therefore the 'politics of age and ageing' constituted a prime mover in the deposition of la

---

[14]   Ibid., p. 13.

[15]   J. Spicer, 'The Renaissance Elbow', in J. Bremmer and H. Roodenburg (eds), *A Cultural History of Gesture: From Antiquity to the Present Day* (Cambridge, 1991), pp. 118–20.

[16]   Gaskell, p. 198.

[17]   On the complexity of describing images, M. Baxandall, *Patterns of Intention: On the Historical Explanation of Pictures* (New Haven, CT, and London, 1985), pp. 1–11.

[18]   S. Kemal and I. Gaskell, 'Art History and Language: Some Issues', in S. Kemal and I. Gaskell (eds), *The Language of Art History* (Cambridge, 1993), p. 1.

[19]   B. dal Pozzo, *Historia della sacra religione* (Verona, 1703–1715), pp. 179–80; A.P. Vella, *The Tribunal of the Inquisition in Malta* (Malta, 1973), pp. 23–5; M. Galea, *Grand Master Jean Levesque de La Cassiere, 1572–1581* (Malta, 1994), pp. 73–95; C. Testa, *Romegas* (Malta, 2002), pp. 177–222.

[20]   P. Rietbergen, *Power and Religion in Baroque Rome* (Leiden, 2006), p. 10.

Cassiere.[21] Furthermore, the actions of the conspirators were carried out within the Council of the Order, the corporate body alongside which the grand master was meant to rule. The development of conciliar ideas of government within the Order reflected wider European trends where various bodies and corporations tried to resist monarchs, including the papacy, that were asserting an absolute style of politics against other forms of power. For instance, the College of Cardinals, traditionally a senate composed of vocal political counsellors, was steadily sidelined by the increasingly monarchical papacies of the Renaissance and Counter-Reformation.[22] The combination of all these factors meant that the situation in Malta rapidly deteriorated; however, strong papal intervention and the sudden demise of both la Cassiere and Romegas brought the situation under control.

Pope Gregory XIII (1572–85) put forward the names of three French Hospitallers from which the Order – supervised by Cardinal Gaspare Visconti – was to elect the new grand master. The choice fell on Verdalle.[23] It is generally acknowledged that he immediately and fiercely implemented disciplinary measures and sought to assert his authority.[24] Verdalle insisted that the Hospitallers had to be more restrained in their behaviour and that obedience to him was sacrosanct; to achieve these aims he projected his politics and power in various and imaginative ways. On the one hand, he showed his religious devotion and zeal by founding and patronizing a convent of cloistered Hospitaller nuns and by bringing to Malta two prominent Counter-Reformation religious orders, the Capuchins and the Jesuits, as well as collecting relics, supporting religious art and encouraging devotion to saints such as the Hospitaller saints Ubaldesca and Hugo.

On the other hand, to project his sovereign authority, he minted coins and medals; one particular medal carried the motto, 'He drives away the gathered clouds and brings back the sun'.[25] Like his predecessors, he organized a grand entry into Mdina, the old city of Malta, for which he was attired in expensive black velvet. Furthermore he carried out extensive work on the magisterial palace in Valletta and built the country palace that still carries his name. Inside the main hall of Verdala Palace, Verdalle commissioned the artist Filippo Paladini to create a series of frescoes depicting the main episodes of his life.[26] Such images belong to a genre known as bio-depictions and represent a sixteenth-century artistic fashion

---

[21]   Pozzo, *Historia*, p. 180.

[22]   J.M. DeSilva, 'Senators or Courtiers: Negotiating Models for the Colleges of Cardinals under Julius II and Leo X', *Renaissance Studies*, 22 (2008), 154–73.

[23]   Galea, *Grand Master La Cassiere*, p. 79.

[24]   M. Galea, *Grand Master Hughes Loubenx de Verdalle, 1582–85* (Malta, 2000), pp. 38–9; A. Blondy, *Un prince de la Renaissance à l'aube de la Contre-Réforme. Hugues de Loubens de Verdalle (1531–1582–1595): Cardinal et Grand Maître de l'Ordre de Malte* (Saint-Denis, 2005), p. 63.

[25]   Galea, *Grand Master Verdalle*, p. 20: 'Collectasque fugat nubes solemque reducit'.

[26]   On Paladini, K. Sciberras and D.M. Stone, 'Saints and Heroes: Frescos by Filippo Paladini and Leonello Spada', in A. Ganado (ed.), *Palace of the Grand Masters in Valletta* (Malta, 2001), pp. 139–43.

that combined autobiography, adventure, allegory and costume.[27] Crucially, in 1589, he commissioned Fr Giacomo Bosio to write a history of the Order from its foundation to 1571 (when the Order moved to Valletta).[28] In this way he extended his patronage on to an enterprise that would serve to preserve the Order's collective memory.[29] Hence Verdalle's stamp on the political, historical and sacral landscape of the Order was indelible as he sought to bolster his authority and the obedience that was due to him.[30]

In November 1582, Verdalle read out to the council a papal bull in which the pope made the Order acknowledge its mistake in deposing la Cassiere: the aim here was not only to rehabilitate the late grand master but also to bolster the position of the new one.[31] He sought to impress upon the members of the council that their role had to be one of subservience to him, rather than criticism. When he summoned the Chapter General in 1583, he used it to pass measures that augmented further his authority. For instance, the Chapter General agreed to increase both the income and powers of the grand master over the everyday running of the Order.[32] Then, in December 1587, Pope Sixtus V (1585–90) elevated Verdalle to the rank of a cardinal. His trip to Rome was marked by much ritual and celebration, all of which were meant to emphasize Verdalle's importance and authority. In this way, his status within the Order dramatically increased and his position vis-à-vis the bishop and the inquisitor in Malta was also amplified. In a world and institution so thoroughly obsessed by rank and hierarchy, such titles mattered a great deal. To commemorate the occasion of his promotion, Verdalle erected a column in the square of the magisterial palace, which was topped by a wolf (a motif from his coat of arms) and which faced towards the bishop's residence.[33]

As cardinal-grand master, Verdalle summoned another Chapter General in 1588 to deal with various financial and administrative issues. The proceedings of this Chapter General were transformed into the lavishly illustrated *Statuta*. This consists of a relatively slim and portable volume, copies of which are available at a number of locations, including the National Library of Malta, the *Biblioteca Casanatense* in Rome, Cambridge University Library and Harvard University Library. There are slight variations in these copies, possibly due to the necessity of integrating the illustrations in the text.[34] The volume commences with an

---

[27]   V. Groebner, 'Inside Out: Clothes, Dissimulation, and the Arts of Accounting in the Autobiography of Matthäus Schwarz, 1496–1574', *Representations*, 66 (1999), 108.

[28]   G. Bosio, *Dell'istoria della sacra religione* (Rome, 1594).

[29]   For bibliographic details, see R. Mortimer, *Italian 16th-century books* (Cambridge, MA, 1974), 1, pp. 117–20.

[30]   Descriptions of the various projects carried out by Verdalle can be found in Galea, *Grand Master Verdalle* and Blondy, *Un prince*. See also M. Buhagiar, *Essays on the Knights and Art and Architecture in Malta 1500–1798* (Malta, 2009), pp. 9–12.

[31]   A.S.V., Segr. Stato, Malta, Ms.1, fol. 126r, 12 November 1582.

[32]   Blondy, *Un prince*, p. 70.

[33]   On Verdalle's column, Galea, *Grand Master Verdalle*, p. 49.

[34]   Mortimer, 2, pp. 395–8.

elaborate allegorical frontispiece, where the title '*Statuta Hospitalis Hierusalem*' hovers above the figures of Faith, Charity and Hope (Fig. 14.1). Above the title is the figure of St John the Baptist with hospital beds, thereby indicating the medical origins and dimension of the Order. In the left-hand corner is an imagined miniature reproduction of the first statutes of the Order compiled under its second Master, Raymond du Puy, while in the right-hand corner is a stylized cityscape, most probably Jerusalem. These two components indicate that Verdalle's *Statuta* was derived directly from the original compiled in Jerusalem under du Puy: hence its authority was based upon a solid continuous tradition.

The frontispiece is followed by an image of Verdalle receiving a Hospitaller cap and a sword from Pope Sixtus V, and another image showing Verdalle flanked by the figures of Faith and St George slaying the dragon. The fourth image shows Sixtus V placing the cardinal's hat on Verdalle's head; the depiction is topped by the phrase, 'He is dressed in splendour'.[35] This is followed by an image bursting with angels and cherubs playing musical instruments and solemnly proclaiming 'Images of the Masters of the Hospital'.[36] There are then medallion portraits of all the grand masters down to Verdalle himself, who is crowned with a baronial coronet and the cardinal's hat. Through this sequential visual representation of all the masters of the Order, Verdalle was again emphasizing how he was part of a venerable tradition stretching all the way back to the founder of the Order, the Blessed Gerard (d. 1120). The legacy of these masters of the Order, real or fabricated, was to constitute a fundamental element in the veneration of the past that became prevalent within the Order after about 1581.[37] Verdalle presented himself as the culmination of this series of illustrious leaders, so that any challenge to his authority would be a challenge to an esteemed and legitimate line of succession. The medallion portraits of the masters are followed by the various chapters making up the *Statuta*, each of which is preceded by an illustrative plate that reflects the essence of that particular chapter.

The *Statuta* – in both its written and visual form – was meant to put across the message that obeying established authority, that is, the grand master, was a sacred duty. The *Statuta* constituted an exercise in 'translation', that is, in rendering ideas in a clearer and more attractive format.[38] The images were pictorial renderings of the ideas and information contained in the written chapters and they illustrated the workings of noble, religious and masculine dictates within the Order. Each of the chapter frontispieces is surrounded by a border design, which includes a Latin motto. Qualitatively, there is a marked difference between the frontispiece and its border: the latter is in fact generally omitted when reproduced in modern-

---

[35]   *Statuta Hospitalis Hierusalem*, p. 7, '*DECOREM INDVTVS EST*'.

[36]   Ibid., p. 6, '*EFFIGIES MAGISTROR HOSPITALIS*'.

[37]   See for instance G. Bosio, *Le imagini de' beati, e santi della Sacra Religione di S. Gio. Gierosolomitano* (Rome and Palermo, 1633); and Malta, Cod. 1697, 1670, fol. 4v.

[38]   On the dynamics of early modern translation see P. Burke and R. Po-Chia Hsia (eds), *Cultural Translation in Early Modern Europe* (Cambridge, 2007).

Figure 14.1   Frontispiece of the *Statuta Hospital Hierusalem* (1588) (© National
                Library of Malta)

day publications. Nonetheless, it is useful to consider the border designs, for they
are replete with symbolism associated with the Hospitallers: armour, weapons,
trophies of war, Muslim slaves, angels blowing triumphal trumpets, hospital
scenes and images from the life of St John the Baptist. In each of these borders,
and occasionally in the images themselves, one can also see the coat-of-arms of
Verdalle. In at least two instances, the grand master in the image is even given
the facial features of Verdalle himself: p. 96, 'On the Master', and p. 144, 'On the
commanderies and on the administration'. Thus, Verdalle's stamp is deeply set in
this book and the *Statuta* of 1588 is unmistakably *the* statutes book *of* Verdalle. His

Figure 14.2   On the common treasury, *Statuta Hospitalis Hierusalem* (© National
Library of Malta)

persona – including his body – became one with the written word upon which the
entire functioning and way of life of the Order was founded. Therefore, Verdalle's
image and coat-of-arms were reproduced a number of times to emphasize how
all authority rested and emanated from his office and in his own physical person.

This paper will now focus on one particular frontispiece – dedicated to the
Common Treasury (p. 40) – as a case study (Fig. 14.2). This image is not among
those commonly reproduced in modern publications; nonetheless, it is probably
the one image that best captures the essence of the Order's functions. It is an
image bustling with activity and made up of a number of constituent parts. On the

left-hand side is a representation of the entrance to the Holy Infirmary. Under the inscription 'peace be with visitors', one can see a young unbearded Hospitaller, most likely a novice, distributing alms to the destitute and the infirm. On the right-hand side, under a banner with the words 'warfare on the enemy', there are armoured soldiers; in the background can also be seen a number of galleys. In the centre, three bearded men are counting money. The figure in the middle wears a beret reminiscent of the one customarily worn by grand masters as the symbol of their office. Hanging above the three treasurers, on the façade of a building, is the coat-of-arms of Verdalle. Once again, the reader–viewer is reminded that this is Verdalle's world.

On one level, this image captures the dual role of the Order as an institution that nursed and fought. The 'motto' of the Order was 'Serve the Poor and Defend the Faith', outlined in the first rule of the Order that was drawn up by du Puy in the first half of the twelfth century and upheld to this day.[39] The placing of the treasurers at the heart of the image signified that a sound financial situation was crucial to the continued functioning of the Order. In fact, during the two Chapter Generals of 1583 and 1588 that he had convened, Verdalle paid considerable attention to measures to strengthen the Order's Common Treasury.[40]

On another level, the image is also an allegory of the ages of man. The young unbearded novice fulfils the role of the meek servant of the poor. The bearded middle-aged leader with the high-plumed helmet and outstretched arm stood for the indomitable military warrior. Furthermore, his virility was unashamedly proclaimed through a codpiece. Across early modern Europe, particularly in the sixteenth century, men often wore outlandish codpieces to the consternation of moralists who considered them a form of nudity. It was both a sexual object as well as a joke, pushing the boundaries of masculinity and morality.[41] The insertion of the codpiece in this image was unequivocally meant to convey a message of manly potency and aggression. This was further emphasized through the tight-fitting leggings and broad shoulders of the soldiers; well-formed legs were characteristic of depictions of French kings as an expression of their monarchical potency.[42] Finally, the old treasurers in the middle of the image, with their particularly long beards, represented the wisdom and authority vested in old men. Such long beards were thought to be associated with the highly authoritative Old

---

[39] 'Obsequium Pauperum et Tuitio Fidei'; G. Scarabelli, 'Il ruolo della marina giovannita nel Mediterraneo dal medioevo all'epoca moderna', in *I cavalieri di San Giovanni e il Mediterraneo* (Taranto, 1996), pp. 36–7.

[40] Blondy, *Un prince*, pp. 67–70, 137–8.

[41] On the codpiece, L. Roper, *Oedipus and the Devil: Witchcraft, Sexuality and Religion in Early Modern Europe* (London and New York, 1994), pp. 117–18.

[42] K.B. Crawford, 'The Politics of Promiscuity: Masculinity and Heroic Representation at the Court of Henry IV', *French Historical Studies*, 26 (2003), 227. See also M. Hayward (ed.), *Dress at the Court of King Henry VIII* (Maney, 2007), p. 95.

Testament prophets.[43] Facial hair – the beard in particular – was a critical part of the Hospitallers' identity, hence its prevalence in this image and its centrality in this analysis. Beards served to emphasize differences between adult males and boys, and they conferred a sense of authority, manly sobriety and forceful military presence.[44] As in the rest of the *Statuta*, the message here emphasized the importance of hierarchical obedience to established authority. The politics of age had been a major factor contributing to the downfall of Grand Master la Cassiere. The point of this image, then, was that through obedience, respect and mutuality, harmony could be preserved within the Order so that it could continue to fulfil its mission of defending the faith and serving the poor. Verdalle believed that deference by the young towards the old was the key to reinforce his authority; and the older he became, the more sensitive he was of the precariousness of the politics of age in a multi-ethnic institution.

Conclusion

In general, up to 1592, if there were complaints against Verdalle, the papacy listened and advised Verdalle accordingly, but in the end it was mostly concerned about bolstering his authority and maintaining discipline among the Hospitallers. The advent of Pope Clement VIII (1592–1605) saw a change in policy in that criticisms of Verdalle were taken more seriously.[45] Verdalle could not prevent Hospitallers from alerting the papacy about misgivings with regards to his government and many complaints are still preserved in the Vatican Secret Archive. Various Hospitallers openly declared that under Verdalle the Order, instead of being a place characterized by fraternity and charity, felt more like hell.[46] As he became older and increasingly ill, Verdalle faced a growing number of such complaints.

The *Statuta* of 1588 represented the climax of Verdalle's efforts to fortify magisterial authority and propel the Order towards absolutism. In 1594, the first part of Bosio's *Dell'Istoria della Sacra Religione* was published in Rome. Its elaborate frontispiece represents a continuation of the visual lessons gained in the *Statuta*. It shows the figure of St John the Baptist, the papal tiara and crosier, thereby emphasizing the religious dimension of the Order. There is also an armoured Hospitaller soldier, a helmet and a sword, thereby emphasizing the

---

[43]   Such a link was particularly popular among Protestants, but was found across Europe: U. Rublack, *Reformation Europe* (Cambridge, 2005), p. 181; D. MacCulloch, *Reformation: Europe's House Divided, 1490–1700* (London, 2004), pp. 254, 260.

[44]   On the cultural significance of beards, W. Fisher, 'The Renaissance Beard: Masculinity in Early Modern England', *Renaissance Quarterly*, 54 (2001), 155–87; M.J. Zucker, 'Raphael and the Beard of Pope Julius II', *The Art Bulletin*, 59 (1977), 532.

[45]   Blondy, *Un prince*, pp. 184–8.

[46]   A.S.V., Segr. Stato, Malta, Ms.6, fols 13r–25v, 29 May–21 July 1594 and fol. 18r, 7 July 1594.

military dimension. Verdalle's coat-of-arms, crowned by both a crown and a cardinal's hat is at the heart of the frontispiece. Vicious-looking wolves – derived from his family emblem – pounce upon the lunar crescents of Islam and upon two hapless Turks. The image is one of power and glory.

In effect, by 1594, Verdalle was more comparable to the Turk being attacked by the wolf, than the other way round. Exhausted and worn down by illness and by administrative pressures, he died in 1595. In the short term, Verdalle's efforts to strengthen the authority of his office against that of the council and other dissenting voices may appear to have been ineffective. Nonetheless, his labours, including the production of the *Statuta*, were the critical foundations upon which later grand masters increased their powers. This is not to imply a notion of a teleological account; rather, focused case studies such as this one help to anchor long-term trends in empirical analyses. One gauge of the increased authority of grand masters is the frequency with which the Chapter General met before and after Verdalle: twelve meetings in sixty-five years (1530–95), as against five meetings in 202 years (1596–1798). The trend towards absolutism culminated in the figure of Grand Master Manuel Pinto de Fonseca (1741–73) in the eighteenth century, whose portrait as the supreme sovereign of the Order modelled on that of Louis XIV (1638–1715) remains iconic.[47] As in the case of written sources, images are not neutral records of the past. They are alive and full of meanings about Hospitaller lives. Excluding this visual and non-written element can only produce a partial account of a rich and varied tapestry. Moreover, in historiographical terms, the *Statuta* alert historians to the need to understand the visual practices of the Hospitallers because images are not just reflective of 'reality'; they are productive in themselves. The development of Hospitaller imagery from the *Statuta* to the Pinto portrait forms a central and indispensable part of the story of the politics and power of this institution.

---

[47]   For a reproduction of this painting see the website of St John's Co-Cathedral, Valletta, at http://www.stjohnscocathedral.com/index.php?id=36, section 23 Mgr Coleiro Hall (accessed 23 November 2009).

Chapter 15

Towards the End of the Order of the Hospital: Reflections on the Views of Two Venetian Brethren, Antonio Miari and Ottavio Benvenuti

Victor Mallia-Milanes

Introduction

The writings of perceptive contemporary observers are relevant to the reconstruction of the past as they provide source material and, more often than not, determine the pattern that later researchers adopt.[1] Such literature has exercised an enormous influence and continues to echo through the writings of present-day historians. This paper focuses on two such eyewitnesses who lived through the trauma of the agonizing last years of the Order of St John on Malta and whose correspondence constitutes a firsthand account of contemporary developments as they evolved or as they were seen to evolve. Both observers were high-ranking Venetian members of the Order. One is Antonio Miari, from Belluno, de Rohan's secretary for Italian affairs and, since 1 February 1793, Resident Minister for the Venetian Republic at the magistral court in Valletta. His attainments are not sufficiently or adequately documented. He has been called an accomplished scholar, but what his true intellectual pursuits were remains obscure. It has been claimed that he was a 'distinguished diplomat',[2] but all that can be established is that, after the fall of Hospitaller Malta to the French, he acted as the Order's envoy at Vienna in 1815 and at Aix-la-Chapelle in 1818, where his performance does not appear to have been very impressive: indeed, his achievements there have been described

---

[1] The detailed *relationi* of Venice's ambassadors and those of her *Terraferma* rectors, drawn up at the end of their mission, are classic examples. On the former, see, for example, E. Albèri, *Le relazioni degli ambasciatori veneti al Senato durante il secolo XVI.*, 3rd ser. (Florence, 1844); on the latter, V. Mallia-Milanes, 'Venice's Terraferma Rectors and the Republic's Trade in Salt', in P. Xuereb (ed.), *Karissime Gotifride: Historical Essays Presented to Professor Godfrey Wettinger on his Seventieth Birthday* (Malta, 1999), pp. 81–8.

[2] H.P. Scicluna, 'Notes on the Admiralty House, Valletta', *Archivum Melitense*, 9/2 (1933), 71.

as almost negligible.[3] His ministerial mission on Malta was to look after Venice's interests in the central Mediterranean in general and in the Hospitaller principality in particular. After the *rapprochement* of the 1760s,[4] political relations between the two states grew steadily more cordial.[5] Venice's major concerns included her declining commerce, the security of navigation, and the safety of her merchants and sailors. Through his regular letters from Malta, Miari would keep the doge and the esteemed *Cinque Savii alla Mercanzia* well informed about the prevailing conditions in the area, about developments in current issues and if and when they were resolved, and about problems that might yet emerge.[6] This task he appears to have accomplished admirably, and it is probably because of his correspondence that we are more familiar with this aspect of his career than with any other. Above all, he would write with fervour on the harmful impact that the Revolution in France was having on his Order, a passionate manifestation of where his true loyalties lay. Within the limitations of our factual knowledge of the man, the surviving collection of his original letters to Venice provides valuable insights into his thought and personality.

The second eyewitness is Ottavio Benvenuti, from the Lombard city of Crema. His career within the Hospital is also largely unknown, except that in 1795 the Venerable Council elected him minister plenipotentiary and resident receiver in Venice.[7] His letters, dispatched from the Adriatic Republic, were addressed to the Lords of the Common Treasury on Malta and cover the long torturous days from 6 May 1797 to 14 April 1798.[8] Benvenuti's correspondence offers a complementary,

---

[3]   H.J.A. Sire, *The Knights of Malta* (London, 1994), pp. 247–9.

[4]   See V. Mallia-Milanes, *Venice and Hospitaller Malta 1530–1798: Aspects of a Relationship* (Malta, 1992), pp. 221–69. For the long years before the 1760s, V. Mallia-Milanes, 'Corsairs Parading Crosses: The Hospitallers and Venice, 1530–1798', in *MO* 1, 103–12.

[5]   For an overview of the relationship, V. Mallia-Milanes, *In the Service of the Venetian Republic: Massimiliano Buzzaccarini Gonzaga's Letters from Malta to Venice's Magistracy of Trade 1754–1776* (Malta, 2008), pp. 1–100.

[6]   ASVen, Cinque Savii alla Mercanzia, prima serie (hereafter CSM, p.s.), Diversorum, busta 403, filza 76, Malta, 21 March 1793; ASVen, Senato, Secreta, filza 9, Dispacci Malta, *Lettere dell'agente veneto in Malta da 27 marzo 1793 sino 30 marzo 1797 da Venerando Antonio Miari*.

[7]   As receiver, he succeeded Antonio Colleoni, assuming office on 21 December 1795. He was apparently the only Hospitaller receiver in Venice to have been simultaneously appointed also minister plenipotentiary. On the role of the Order's receiver in Venice, V. Mallia-Milanes, 'The Hospitaller Receiver in Venice: A Late Seventeenth-century Document', *Studi Veneziani*, n.s., 44 (2002), 309–26. Also P. Scarpa, 'Ricevitori e Rappresentanti dell'Ordine di Malta a Venezia in Epoca Moderna nelle *Esposizioni* del Collegio', *Archivio Veneto*, 5th ser., 165 (2006), 191–210.

[8]   The letters are in Malta, Cod. 1632, *Fasci di lettere riguardanti le finanze dell'Ordine Gerosolimitano spedite ai Procuratori del Tesoro da Bologna, Rimini, Venezia (1797–98)*. On Benvenuti and the fall of Venice, V. Mallia-Milanes, '"Guardando la loro uscita dalla

though different, perspective to Miari's. Not only did he witness the violent overthrow of the Venetian Republic in May 1797, he experienced at first hand the disastrous impact of Napoleon's systematic exploitation of Hospitaller estates in the Veneto, including the Grand Priory itself. While the grand master and his Sacred Council were engrossed in discussions on the potential threat to the security of their convent on their Mediterranean principality and how best to deal with it, Benvenuti's series of letters presents a graphic depiction of the slow disintegration of innumerable estates in Venice's mainland territory that had for ages belonged to the Hospital and had helped pay for its activities. Now confiscated, these lands, like all other lands in *Ancien Régime* Europe, would now directly finance France's enormous military effort and dramatic expansionism.

The striking picture that so vividly emerges from both sets of correspondence – Miari's and Benvenuti's – is one of explicit realism, the sorry state of the Hospitaller institution in the mid- and late-1790s, a living and visual portrayal of the collapse of what had been so piously assembled, so diligently consolidated and sustained, and so fiercely defended over eight centuries in the service of Man.

Antonio Miari

At the beginning of 1796, Antonio Miari confessed to the Serenissima that the Order was endeavouring to curtail its expenses, as far as it could, so as to balance them against its revenues that had shrunk considerably since the outbreak of the French Revolution. That would be extremely difficult to realize without resorting to the extortion of further funds from its estates in Europe.[9] Three years earlier he had already pointed out that the Order's common treasury in Valletta experienced 'a serious shortage of capital', which not even a loan of 400,000 scudi would be able to mitigate.[10] Nor would the situation be relieved if all the brethren made, as they had been asked to make, generous contributions to their Order, each according to his own ability. There would still be the need to raise the value of the responsions, the fixed portion of the annual revenue from the Hospitaller estates that belonged to the treasury.[11] He admitted that, although this was the surest, perhaps simplest, method of achieving the desired effect, it was, nevertheless, insensitive to the

---

storia": Venezia e l'Ordine Ospedaliero di San Giovanni alla fine del Settecento', *Studi Veneziani*, n.s., 43 (2002), 389–98.

    [9]   ASVen, p.s., Diversorum, busta 403, filza 76, 25 February 1796. Antonio Miari is referred to as 'Persona della Republica in Malta'. The only 'calcolabile accrescimento', he wrote, was precisely 'quello di aumentar l'imposizioni e li diritti a favour della Religione sulli stessi suoi beni e proprietà'.

    [10]   On the loan, Malta Cod. 274, fol. 223v, 5 July 1793. Also F. Panzavecchia, *L'ultimo periodo della storia di Malta sotto il governo dell'Ordine Gerosolimitano* (Malta, 1835), pp. 321–2.

    [11]   ASVen, p.s., Diversorum, busta 403, filza 76, 11 April 1793.

current needs of all members of the Order, who themselves were experiencing a host of other locally imposed financial burdens.[12] The Order would delay resorting to these expedients for as long as possible.[13] Miari feared that measures as extreme as these would announce with signal clarity 'the final stages of our existence'. The financial deficit created by the Revolution in France was itself exceedingly difficult to contain. The situation grew worse with the certain prospect of war: the island had to be placed in a state of defence, with the Hospitallers having to build and man what he called 'new forts that would guarantee security against any hostile attack'.[14] France in fact declared war on Austria on 20 April 1792. Worse was yet to come. Between 19 and 22 September that year, the *loi spoliateur* nationalized all Hospitaller estates on France, the French monarchy, the Order's greatest patron and protector, was abolished, and France was declared a Republic. Within a month of these fateful events, on 22 October, the National Convention decreed the end of the Order of the Hospital in France.[15]

There is no doubt that Miari's views reflected those of Grand Master Emanuel de Rohan, his Venerable Council, and all those who were devoted to the Order. His was a very good pen – fluent, profound and penetrating. Those who remained loyal to the Order, who still felt genuinely attached to it, sought to find long- or short-term measures that would effectively provide a way out of the crisis.[16] One of the more promising proposals involved the revival of the Order's corsairing activities in the Levant, a venture that would guarantee no mean return.[17] It may perhaps not be historically accurate to claim that by then privateering activity in the Mediterranean had become almost 'negligible'; there was an evident resurgence vigorously instigated by the North African regencies.[18] Miari was

---

[12]   'già aggravate ... per li pesi locali e per l'imposizioni de respetti sovrani'. Ibid., 25 February 1796.

[13]   Ibid.

[14]   Ibid., 11 April 1793. In fact, a new redoubt, known as Fort Tigné, was constructed on Dragut point in the Grand Harbour a year before, in 1792, consisting 'of four wings at right angles to each other and a round tower towards Valletta, surrounded by a ditch and well mined throughout'. Alison Hoppen, *The Fortification of Malta by the Order of St John 1530–1798* (Edinburgh, 1979), pp. 97–8.

[15]   For a general history of the Order during these turbulent years, F.W. Ryan, *The House of the Temple: A Study of Malta and its Knights in the French Revolution* (London, 1930); M. Miège, *Histoire de Malte* (Paris, 1840), 2, pp. 267–480 and 3, pp. 1–13; R. Cavaliero, *The Last of the Crusaders: The Knights of St John and Malta in the Eighteenth Century* (London, 1960).

[16]   'Frattanto tutto quello ch'è qui attaccato a questa Religione di spirito zelante, attivo, ed inventore non in altro s'occupa puramente che in suggerire dei mezzi di risorsa per la Religione più o meno efficaci durevoli ed anche quasi momentanei poiché essa ò in caso per l'attuali sue ristrettezze ò veramente gravissime di profittare di tutto.' ASVen, CSM, p.s., Diversorum, busta 403, filza 76, 25 February 1796.

[17]   'far rinovar a bastimenti della Religione il *corso* in Levante'. Ibid.

[18]   See, for example, Sire, p. 98.

knowledgeable about the recent history of Malta's privateering and its sad impact on the Order's relations with the Venetian Republic.[19] It was almost certain, he explained, that these adventures in the eastern Mediterranean were about to be revived.[20] The subject was brought up again in another letter. The tone this time was more reassuring. Licenses issued by the Order's chancellery in Valletta to cover Maltese corsairs included clear instructions to keep away from Venetian waters. Severe penalties would be imposed on transgressors.[21] The grand master was determined not to allow the good relations his Order and principality enjoyed with the Adriatic Republic to be weakened or destabilized. New measures were issued to regulate the activity in the Levant, including 'the most rigorous and positive' penalties for any Maltese corsair captain who dared, for whatever motive, sail into Venetian waters and anchor in any of the Republic's ports. If someone, out of sheer provocation, approached forbidden waters 'with no positive or visible need', Venice would, in line with the grand master's wishes, be fully reimbursed for any damage the Republic or its subjects might suffer. If, on the other hand, any Maltese corsair was driven there against his will, Venice's public representatives were directed to limit themselves to supplying the corsair solely

---

[19]   'veramente era in altri tempi il *corso* per quest'Isola una vera sorgente di dolizie.' ASVen, CSM, p.s., Diversorum, busta 403, filza 76, 25 February 1796.

[20]   Miari was here seeking advice from Venice, his accrediting state, how to respond if he were to be approached by the grand master on the matter, as he knew for certain that he would. 'Supplico umilmente la sua benignità e saviezza', he petitions the Doge, 'prescrivermi la condotta che dovrò osservare per essere più sicuro di non far cosa ch'esser possa di suo spiacimento'. Ibid. From the Venetian point of view, what was at stake was whether the new magistral licenses would allow the Hospitallers to enter Venetian waters and would the reintroduction of the Maltese *corso* in the Levant sour the relationship between the two states as in the past? On 26 March 1796, Miari's letter was sent to the Venetian magistracy of trade, the *Cinque Savii alla Mercanzia.*

[21]   'si sono renovate a' medesimi [corsari maltesi] le proibizioni più positive colla minazione dei più severi gastichi a chi di essi ardisse di trasgredire gli ordini ora ripetuti.' Undated letter, apparently in April 1796, in ibid. In this second letter, Miari refers to the first as having been written 'quasi due mesi sono'. Notwithstanding the rigour of these precautions, incidents defying such prohibitions occurred. One involved a Maltese *xebec* that preyed upon a *martigo* from Cephalonia in the waters of Prodano. 'Il Capitano del sciabecco maltese trovasi gravemente malato in questo Spedale, sebbene in luogo di carcere, e il di lui Tenente, ch'è stato il più colpevole di tutti non è pervenuto nelle mani della giustizia, che avanti ieri, essendosi egli prima rifuggiato in Chiesa.' De Rohan, explained Miari, was determined 'nella sincera intenzione di dare sodisfazione alla Serenissima Republica proporzionalmente all'offesa fatta e al suo decoro e dignità'. Ibid. On another occasion, the same Maltese corsair, unidentified in the document, had also been reported to have attacked Gerosino Metaxà, a Venetian subject from Cephalonia – 'aggredito e spogliato al Prodano'. Details about this incident had reached Miari through a letter, dated 11 April 1796, sent by Venice's Provveditor General da Mar. Miari demanded the reimbursement of 'seicento Tallari'. Ibid.

with the barest essentials.[22] Miari appears to have had first-hand knowledge of the social context and conditions in which Maltese corsairs lived. He was aware how extremely difficult it would prove to try to extract any reimbursements from them. The rigidity and rigour with which de Rohan assumed this stance was a clear indication that, ever since its revival under his magistracy, privateering activity in that area appears to have spread considerably. Constant vigilance or surveillance would not have been necessary for isolated cases. Numbers must have been a determining factor. In fact, in 1796 alone, the Maltese *corso* yielded some 117,000 scudi, while the average annual income from the same source for the decade 1787-97 had been slightly over 65,500 scudi.[23]

By April 1796 the pitiable situation within the Order remained unchanged, marked solely, observed Miari, by a vicious discord between the French and the other nations on Hospitaller Malta – all constituting 'this moral institution'.[24] There were three major categories of French knights – the republicans; the moderates; and the counter-revolutionary *émigrés*, some of whom sought refuge on Malta. [25] Having lost practically everything, the three French *langues* – Provence, Auvergne and France – were, so we are told, deeply sunk in a pervasive spirit of despair, resigned to the oncoming tide of total destruction. Three years earlier Miari had already called them 'exceedingly uncompromising', employing every effort to coerce the Order into declaring war on the new Republic.[26] De Rohan, though French, and all the others, who were not and had some influence in the government of the Order, were determined against that stand. 'And I dare say', wrote Miari, 'that we will never ever accede to such an acute measure, even if we were all indiscriminately to suffer a sudden attack of lunacy. Each one of us is prepared to defend himself to the very end' if any nation, whichever that might be, decided to invade the island. Some, too, feared the English as potential aggressors.[27] There were fairly widespread rumours, he claimed, that the Order had earlier been secretly seeking protection from Great Britain, partly in the form of an award of a lifelong pension for the French Hospitallers who had found themselves impoverished overnight by the Revolution.[28] In return, the Order was prepared to cede its strong strategic fortress of Malta with its superb harbours in the central Mediterranean to Britain,

---

[22]   'soltanto il pure indispensabile per il sostentamento'. Ibid.

[23]   A. Luttrell, 'Eighteenth-Century Malta: Prosperity and Problems', *Hyphen* [Malta], 3/2 (1982), 45.

[24]   ASVen, CSM, p.s., Diversorum, busta 403, filza 76, 25 February 1796.

[25]   Ryan, p. 170. The moderates were the 'Feuillants', those 'ready to stand by the Constitution, the King, the nation and the law'.

[26]   ASVen, CSM, p.s., Diversorum, busta 403, filza 76, 11 April 1793.

[27]   'poichè ora sembra che si voglia vivere in sospetto anche degli inglesi.' Ibid.

[28]   'È possibile che costà giunga una voce, che ha corso in qualche altra parte, cioè che l'Ordine Gerosolimitano fosse sul punto di concludere un trattato colla Gran Bretagna di cession ad essa di quest'isole, sotto certe condizioni non ben enunciate se non in quanto alle Pensioni vitalizie che dovrebbonsi accordare ai Cavalieri Francei, rimasti nell'indegenza per la Rivoluzione succeduta nelloro Regno.' ASVen, CSM, p.s., busta 601, fascicolo

then at war with France since 1 February 1793.[29] According to Miari, these were foul and filthy rumours, that he had no hesitation in denying categorically.[30] Were they perhaps maliciously fabricated by the French Hospitallers themselves, who, in his view, were unable to entertain anything positive or constructive? They tended to obstruct any proposed project believed to be potentially useful for the economy by rejecting them outright.[31] But what exactly were these suggested reforms?[32] Miari does not specify; he only points vaguely to their economic nature. That was the Order's overwhelming preoccupation after 1792. The proposed increase in responsions had been one such radical 'reform'. Another pertained to the navy, which had already been cut back in the 1780s. Now, in the 1790s, it was the turn of the galley crews to be drastically reduced.[33] This change does not find an echo in any of Miari's letters. Then there were the grand master's forceful attempts at reaching agreements with Great Britain, Catherine II's Russia, and the United States of America to counter the impact of the loss of the three French *langues*.[34] None of the intended outcomes was realized. Reason, the tone of his remarks seems to have suggested, no longer inspired the current climate of opinion. Should the Order depart from its traditional practices and beliefs? Should it now abandon the sacred principle of neutrality it had always professed and indiscriminately practised towards all Christian States?

On Malta, talk of war was widespread, in Miari's words, 'now perhaps more than ever'.[35] In reality, according to a confession he made in one of his letters, not

---

Lettere Miari, 24 May 1793. Also Miari's letter to the Doge, ASVen, Senato, Secreta, filza 9, Dispacci Malta, 27 June 1793.

[29]   See R. Vella Bonavita, 'Britain and Malta 1787–1798', *Hyphen* [Malta], 1/1 (1977), 3–4.

[30]   'Io non ho però difficoltà', he told the Cinque Savii, 'di assicurare Vostre Eccellenze nella maniera piú positive, che una tale imputazione è della piú grande falsità; e che la Religione non ha mai sognato una cession di questa natura, che certissimamente non potendosifare con grande segretezza, sarebbe impossibile che neppur si tentasse dai meno considerate, giaché risapendole le persone adette al servigio dell'altre Potenze, costituirebbero l'Ordine nella piú precesa necessità di preventivamente ottenere il rispettivo consenso, che non credo sarebbe accordato da alcuna Potenza Protettrice conosciuta dell'Ordine'. ASVen, CSM, p.s., busta 601, fascicolo Lettere Miari, 24 May 1793.

[31]   'Ogni utile progetto di riforma e d'economia è rigettato dal partito che per se non ha più che l'annihilamento e la disperazione.' ASVen, CSM, p.s., Diversorum, busta 403, filza 76, 25 February 1796.

[32]   I owe this point to Dr Theresa Vann, who asked the question at the end of my lecture at Cardiff.

[33]   U. Mori Ubaldini, *La marina del Sovrano Militare Ordine di San Giovanni di Gerusalemme di Rodi e di Malta* (Rome, 1971), pp. 508, 512.

[34]   For Britain, see nn. 28 and 29; for Russia and America, see nn. 54–7 and 58 respectively.

[35]   'Qui si continua a parlare di dichiarazione di Guerra, ed ora anzi più che mai.' ASVen, CSM, p.s., Diversorum, busta 403, filza 76, 25 February 1796, 25 April 1793.

only were there no sufficient funds in the common treasury to finance a long and hugely expensive war, there were hardly any to sustain the Order's own existence. The direction the Revolution was taking overwhelmed the Order, leaving what Miari called an intoxicating effect over all its brethren. 'It is blinding' the entire Hospitaller community, he wrote, making it waste the little precious time it has left, instead of seeking solutions, instead of undertaking any necessary reforms that would yield positive results for the economy. Rather, this precious little time was callously dissipated in enhancing discord and dissension, in the joy of annihilating 'whatever remnants of tranquillity our [external] enemies have left us', indeed 'in upsetting the very confidence of being able to hope that someone might come to our rescue'.[36] On 10 October 1793, shortly after Naples had declared war on France, de Rohan, in a formal manifesto,[37] declined to acknowledge the French Republic.[38]

These were disquieting issues. There were others of a different nature and just as unsettling. 'My only hope', he confessed, 'lies almost in [the onset of] a [new] crisis'. This is 'always dangerous', he admitted, 'and often fatal'. As he pointed out, it was only recently that the Order had just survived one such crisis at Rome: 'The pope has been so furious with us', he explained, 'or rather with our head [the grand master], that he seriously threatened to extend to us the same treatment his immediate predecessor had dealt the Jesuits'.[39] Miari's comparison was apposite. Pope Pius VI (1775–99) was indisputably contemplating the dissolution of the Order of the Hospital,[40] in the same manner that Clement XIV (1769–74) and, centuries earlier, Clement V (1305–1314) had, for different reasons, suppressed the Society of Jesus and the Templars respectively.

It would appear difficult, indeed impossible, wrote Miari, to understand how a man as enlightened as Grand Master de Rohan should have exposed his Order to such potential danger simply on account of an ecclesiastical jurisdiction that,

---

[36]   Ibid.
[37]   See A.V. Laferla, *The Story of Man in Malta*, 4th edn (Malta, 1972), p. 131. The manifesto is reproduced, in translation, in Ryan, p. 220, and in Panzavecchia, pp. 323–4.
[38]   On 25 April 1793, Miari wrote thus: 'Coll'ultimo ordinario è giunta a questo Gran Maestro una lettera di Monsieur Fratello dell'infelice Luigi XVI, con cui gli partecipa l'assunta Reggenza del Regno di Francia in occasione della minoretà di Luigi XVII ed invitandolo a dare delle publiche e solecite prove dell'attaccamento dell'Ordine alla Corona di Francia ed ai suoi Rè. Questa notificazione, che si è saputa dal Publico, ha cagionato una sensazione e consolazione grandissima nell'animo della maggior parte, ma che pur non riflette al vero stato ed alle circostanze dell'Ordine. Il Governo però temendo di compromettersi, ha differito la sua risposta sin tanto che non siano meglio conosciuti li sistemi dell'altre Potenze; massimamente delle nostre Protettrici, allequali, permettendolo la nostra situazione per ogni notevole motivo dobbiamo cercar di uniformarsi; potendo una troppo sollecita risoluzione portare tra tanti altri inconvenient il sagrifizio di molti nostri individui che tuttavia si ritrovano in Francia.' ASVen, CSM, p.s., Diversorum, busta 403, filza 76, 25 April 1793.
[39]   ASVen, CSM, p.s., Diversorum, busta 403, filza 76, 25 February 1796.
[40]   'si trattava realmente della nostra non-esistenza.' Ibid.

contrary to the holy wishes of the papacy, he was so vigorously determined to reform. In the end it was precisely his enlightened vision and experience that gave him sense to realize the magnitude of the pope's threat. 'We are too small', remarked Miari, to resist adamantly papal admonitions, to persist in our direct confrontation with so powerful and respectable a sovereign',[41] who, though 'commanding very little authority', as Georges Lefebvre claimed,[42] was nevertheless the Order's ultimate head: 'il nostro Supremo Superiore'.[43]

Was Miari accusing de Rohan of acting out of character or of being passionately driven by an over-ambitious objective? The answer to both questions is, in my view, in the negative. There can be no doubt that de Rohan was a man of reform, of progressive ideas and ideals, as indicated in the first step he took on being elected grand master. In perfect harmony with his predecessor's desire, he summoned a chapter general, the first for 144 years. This was a daunting task, a challenge with potentially devastating consequences for the magistracy. Any attempt at a massive legislative overhaul of the Hospitaller institution, any serious endeavour to revisit the security of the fortress-principality, and any effort to address the several unforeseen demands made by the substantive grievances that had been allowed to accumulate over the years, from individual brethren, from priories, from *langues*, as any chapter general would have envisaged, needed courage and stamina, an iron will and unfaltering determination. In a sense, it also manifested a genuine expression of love for the institution – intrepidly acknowledging the Order's need to return to its magnificent origins, to its sacred mission and charitable practices, to revisit its sacred commitments to hospitality in the midst of such social, political and intellectual upheaval. Grand Master Ximenes de Texada (1773–75), de Rohan's immediate predecessor, had felt the urgency of such reform and resolved to summon a chapter general, but did not live to realize it.[44]

Partly under the overbearing influence of his enlightened advisors,[45] de Rohan endeavoured to rationalize the legal and judicial framework within which the state in all its traditional institutions functioned. The jurisdiction of secular tribunals deserved a large measure of autonomy that would restrict the competence of ecclesiastical courts and curtail the powers of the Inquisition. The ultimate objective appears to have been to loosen the traditionally close ties, old and anachronistic by late eighteenth-century European norms, of the local church to Rome. Even if the

---

[41]   Ibid.

[42]   G. Lefebvre, *The French Revolution, from its Origins to 1793*, trans. E.M. Evanson (London, 1971), p. 167.

[43]   ASVen, p.s., Diversorum, busta 403, filza 76, 25 February 1796.

[44]   Mallia-Milanes, *In the Service of the Venetian Republic*, p. 78 and corresponding notes.

[45]   Nicolò Muscat was one such. See, for example, F. Ciappara, 'Gio. Nicolò Muscat: Church–State Relations in Hospitaller Malta during the Enlightenment, 1786–1798', in V. Mallia-Milanes (ed.), *Hospitaller Malta 1530–1798: Studies on Early Modern Malta and the Order of St John of Jerusalem* (Malta, 1993), pp. 605–58.

papacy was constitutionally the ultimate head of the Order of the Hospital, why should Rome, too, be allowed to dictate the domestic and foreign affairs of the island-principality? Similar issues had emerged during Manoel Pinto's magistracy, in perfect harmony with Bernardo Tanucci, the kingdom of Naples' prime minister. In secular matters, why should Hospitaller Malta not 'resist the will of the papacy' and allow itself 'to drift away from Rome'? Pinto had fully supported Neapolitan political endeavours to have Rome's presence on the island reduced to the barest minimum.[46] They also found favour with de Rohan. In 1793, the Hospitaller Lorenzo Grimaldi, Naples' resident minister on Malta, submitted to de Rohan a plan, consisting of fourteen articles, designed to reform the ecclesiastical jurisdiction on the island. In the view of Bishop Vincenzo Labini, to quote one historian, its purpose was 'to destroy ecclesiastical immunity entirely'. According to the same historian, the grand master was determined to implement it.[47]

More radical reforms of this nature had been successfully introduced in Catholic kingdoms like Austria and Spain, Portugal and Naples, not to mention enlightened, revolutionary France. In the mid-1790s, the Hospitaller principality felt small and stood alone. The intricately pronounced role the Church had been playing on the island ever since early medieval times and its vigorous resistance to such reform measures, which it viewed with suspicion, not only generated deep animosity. It constituted a bitter, implacable obstacle and a fertile source of serious grievances and litigation.[48] Miari was not questioning the soundness, legitimacy or wisdom of de Rohan's programme for local church reform. Beneath his seemingly genuine and considerate apprehension lay a very subtle irony that only a perceptive Venetian observer of manners knew how to conceal in eloquently restrained and courteous language. If there was a flaw in de Rohan's intriguing stand, it could not be attributed either to the Order's frailty at this point in time, or to the diminutive territorial extent of the Hospitaller island. Rather, Miari's fears and criticism rested on its timing. From his perspective, de Rohan's stand was not entirely absurd. The only problem was that, within the framework of the turbulent international situation of the mid-1790s, his grand master no longer enjoyed the power and the means to sustain the stance he had assumed. With hindsight, historians of the Order have viewed the institution and its marvellous powers of resilience as an astonishing phenomenon. Contemporary eyewitnesses, like Miari, did equally marvel at it. He was sufficiently realistic, however, to comprehend the forces behind the fast-evolving political drama in Europe. Conditions, he admitted, were not the same as they had been in the past. They had changed radically to the Order's discomfiture. His institution could no longer boast the powerful patronage it had enjoyed before. Solid, deep-seated moral support and formidable political

---

[46]   For a fuller treatment of the subject, F. Ciappara, 'Malta, Napoli e la Santa Sede nella seconda metà del '700', *Mediterranea: Ricerche Storiche*, 5 (2008), 173–88.

[47]   Ciappara, 'Malta', p. 187.

[48]   On the issue of de Rohan's reforms, F. Ciappara, *The Roman Inquisition in Enlightened Malta* (Malta, 2000), *passim*.

protection, both vital components of its long and chequered history, no longer lay comfortably at its disposal.[49] 'It is not in our power', he pointed out, either 'to change current circumstances', or even less to grind the revolutionary drive effectively to a halt. The grand master, with an exquisite sense of balance, had no choice but to give in. Resistance to Rome disintegrated, with his leading advisor, Nicolò Muscat, stubborn, enlightened, anticlerical, apparently the true author of the plan, dismissed on instructions from the Holy See.[50] If de Rohan's allegedly anti-ecclesiastical measures had not been intended to undermine the 'positive and real rights of the Church' on Hospitaller Malta, commented Miari, they certainly were meant to restrict drastically both its privileges and the traditional practices it had always laid claim to. His ultimate objective was not, it would appear, to curtail the local church's prerogatives, but to defend the autonomy of the secular tribunals against the encroachment of the Roman curia. Not much to the satisfaction of the kingdom of Naples, Rome had had its way. Its threat to the Order's existence appeared to have been dispelled. For Miari, the crisis was over. In this sense, peace and tranquillity returned.[51] The crisis, as Miari had hoped it would, had indeed solved the Order's immediate problem of its continued existence. A whole wide gulf still yawned, however, between the present and the past. Would the passage of de Rohan's revolutionary measures through his Venerable Council have offered the institution a ray of hope, implicitly invoking the future rather than the past?

Miari grasped every opportunity to reflect on the situation then prevailing within his Order, even though the issue might have been of no particular concern of the Venetian authorities. In 1793, for example, Bailiff Giovanni Battista Tommasi, a Hospitaller grand cross from Arezzo in the upper Arno region and captain general of the ship of the line squadron in 1784–85, was appointed resident minister for the Grand Duke of Tuscany at the grand master's court in Valletta. On several occasions in the past, pointed out Miari,[52] he was known to have stood up for Venice's interests and those of her subjects on the island. From what he wrote in one of his letters to the Adriatic city, Miari considered Tommasi an admirable Hospitaller, one of the few currently within the Order whom he respected as an exceedingly worthy person – upright, unwavering, and above all, he said, so graciously devoted, so firmly committed, to his Order. Miari believed that, in 'these critical times', Tommasi would have achieved much for the Order had he enjoyed wider internal support and had his innate characteristic determination and deep sense of purpose been more pervasive within the institution. Miari regretted that these rare qualities were not to be found so profusely, so lavishly within the

---

[49]   'le circostanze non sono più quelle de' tempi passati, quando avevamo degli appoggi fortissimi quasi a nostra disposizione.' ASVen, CSM, p.s., Diversorum, busta 403, filza 76, 25 February 1796.

[50]   Ciappara, 'Church–State Relations', *passim*.

[51]   ASVen, CSM, p.s., Diversorum, busta 403, filza 76, 25 February 1796.

[52]   Ibid.

convent. Tommasi's rivals portrayed him as of meagre intellectual abilities.[53] Miari appears to have been a gifted observer of manners. His brief pen-portrait of the Tuscan minister is interesting in view of the turbulent developments within the Order after the loss of Malta. Pope Pius VII appointed Tommasi grand master on 9 February 1803 at the age of 72. He died two years later.

The Second Partition of Poland (1793) had two immediate consequences for the Hospital. First, what had practically made up the Grand Priory of Poland within Volhynia went to Russia as part of Catherine II's share of the spoil. Secondly, the responsions that the Priory owed the common treasury were suspended. It was at this point that de Rohan was trying to promote what Miari defined as 'a novel spirit of cordiality' with Orthodox Russia, one designed to create 'new Muscovite *langues*' as a substitute for those of Provence, Auvergne and France. Miari's letters on this issue show him as a sharp critic of this potential alliance 'with Moscow'. In fact, his only consolation in 1793 was that, at several levels, the project would prove extremely difficult to materialize. In his view, it infringed upon the statutes of the Hospital, the Order's state council 'was far from unanimous' over the question, and, thirdly, Rome would definitely create 'insuperable' obstacles. In the end, by the time of Catherine II's death on 16 November 1796, Giulio Litta's mission to St Petersburg had failed to produce the desired results.[54] The czarina's indifferent attitude towards the Hospital in the 1790s may have probably owed its origin to the Order's negative response, on grounds of neutrality,[55] to Catherine's earlier requests to Grand Master Pinto to have her mighty Muscovite squadron anchor in Malta's harbour, to be allowed to use the port as a base of operations 'against the common enemy' during the Russo-Turkish war of the early 1770s, and to have the Hospitaller squadrons participate in the war against Turkey.[56] It was Russia's objective to seize the Dardanelles and guarantee secure access to the Mediterranean.[57]

To de Rohan's two abortive attempts at mitigating the impact on his Order of the loss of the three French *langues* – the offer to Great Britain of unrestricted access to limited naval support and other related facilities available at Malta's harbour (like troops, seamen, munitions, stores, and so on), and the request to Catherine II to have new Hospitaller *langues* established in Moscow – a third should be added. A few contemporary documents refer to de Rohan's endeavour to

---

[53]   Ibid.

[54]   See AS Ven, Senato, Secreta, filza 9, Dispacci Malta, 27 June, 8 August, 14 November 1793, 2 and 28 January, 11 June 1796.

[55]   For the Hospitaller Commission's report on Catherine II's request, Malta, Cod. 272, fols 185–6, 31 January 1770.

[56]   For Catherine II's letter, ibid., 272, fol. 184, 18 July 1769; Panzavecchia, pp. 17–18; A. Mifsud, *Knights Hospitallers of the Venerable Tongue of England in Malta* (Malta, 1916), p. 282. For the Russian Admiral Spiritoff's letter, Malta, Cod. 272, fol. 184, 26 December 1769; Panzavecchia, p. 18; Mifsud, p. 282.

[57]   On Hospitaller Malta and the Russo-Turkish war, Mallia-Milanes, *In the Service of the Venetian Republic*, pp. 79–85.

reach an agreement with the United States of America in 1794. In return for landed estates there for the mass of Maltese unemployed 'to clear, cultivate, and settle thereon', Hospitaller Malta would make a similar offer as that extended to Great Britain. Two years later, either in response to a real commercial need, or simply as a diplomatic gesture, the grand master appointed one William England American consul on the island.[58] We have no further archival evidence for how things developed. What is odd, however, is that Miari's correspondence, so eloquent on other issues, remains entirely silent on this particular question.

Ottavio Benvenuti

Ottavio Benvenuti's first letter, dated 6 May 1797,[59] defines his mission to Venice and sets the running theme and general tone of his subsequent correspondence. With the French invasion of the Republic having rendered the situation particularly tense, it was the purpose of his mission to keep the Lords of the Treasury in Valletta knowledgeable about developments. The Order owned vast estates within Venice's *Stato da Terra* and therefore had vested interests there. The current circumstances in the Republic now dictated, as they had done on Hospitaller Malta, a resort to extreme measures. Public funds were exhausted. All interest on deposits had had to be withheld, creating chaos and undermining public trust. Commerce ground to a halt. The impact on Hospitaller interests on the Venetian mainland territories was predictable. The collection of outstanding debts depended on the outcome of the pending peace negotiations. The day Benvenuti was writing his first letter in this collection, 6 May, Venetian deputies were expected to meet General Bonaparte in the hope of reaching an amicable solution.[60] A gripping fear of the unknown and the ferment of public opinion caused a general exodus of patricians from the city, including ambassadors and other diplomatic representatives. His 'intimate and inexpressible attachment' to the Order and his belief that his presence in the city was vital for it, declared Benvenuti, left him calm, almost impervious to passion or emotion. Such qualities, he confessed, made him more capable of speculating on the potential advantages his Order could gain from the widespread chaos than of assessing the scope of his personal danger. He refused to leave the city against the advice of many of his colleagues.[61] The prevailing atmosphere of 'uncertainty, bewilderment, and ill-defined apprehension'[62] rendered it difficult to have daily events reported with any modicum of accuracy or precision. Indeed, no sooner had he announced his determination to remain in Venice, than the threat of an imminent popular insurrection constrained him to flee the city, to follow swiftly

---

58   See Mallia-Milanes, *Venice and Hospitaller Malta*, p. 290.
59   Malta, Cod. 1632, *filza* 'Venezia', 6 May 1797.
60   Ibid.
61   Ibid.
62   J.J. Norwich, *Venice: The Greatness and the Fall* (London, 1981), p. 371.

on the heels of the Nuncio, Mgr Scotti. His fears were well founded. On 12 May a fierce popular insurrection broke out. The *Maggior Consiglio* had abdicated. The sovereign authority in what was now termed a provisional democratic municipality lay with the newly formed Committee of Public Safety.[63]

The moment the uprising was suppressed, Benvenuti returned to the city to discover a completely different Venice from the one he had known two or three days earlier. The fast, dramatic and sensational pace of events was changing the physiognomy of the city with incredible speed into a totally unfamiliar shape – a city 'shivering', he said, and 'deformed', 'as befalls any metropolis in a state of revolution'. On the first hint of insurrection, he reported, the populace, a wild and terrible force, accustomed as it had been for long centuries to servile obedience and to considering itself divested of any rights, went on the rampage, ravaging everything that it could lay hands on. Such radical upheaval was greeted in Venice by a mixture of youthful enthusiasm on the one hand, and the excitement and agitation of divergent opinions and contrasting factions on the other. Benvenuti, addressing the cream of European nobility stationed on distant Malta, underscored what he described as the shrieking grievances of the 'miserable aristocracy', who claimed to have been deceived by illusions of security and seduced by false promises of solid financial support.[64] Benvenuti was sending the convent on Malta a clear message. The fears that Miari entertained on the Mediterranean island were sustained by Benvenuti's accounts from Venice. Was he unwittingly anticipating a similar fate to Hospitaller lands elsewhere in Europe? After all, what was happening in Venice was a perfect replica of what had already occurred throughout France.

To the popular mind, the Republic's failure to offer any modicum of resistance, symbolized in the Great Council's abdication, not much unlike the Order's unavoidable surrender a year or so later, was a spontaneous expression of shame and exasperation. These were moments when no one felt safe or secure, when official ministerial residences were potential prey to popular assault and plunder. The Hospital's Grand Priory, explained Benvenuti, was a more likely target than any other because of its reputation for excessive wealth and privilege.[65] It was precisely the fate of the Grand Priory that Benvenuti was mostly concerned about. He claimed to have been doing his utmost to spare the priory 'the fatal consequences impending over us'. In several mainland localities, particularly at Treviso, the French commissioners had demanded a detailed list of all Hospitaller estates. This was forwarded immediately on request, accompanied by documentary evidence of their respective rents and revenues. Their tenants or leaseholders were next ordered to pay what he called *un' intiera annata brutta*, holding the Hospitaller Commanders responsible for the usual payment of the annual dues. Thirdly, they demanded all the silver to be found in Hospitaller churches. This, too, was handed

---

[63]   NLM, arch.1632, *filza* 'Venezia', 20 May 1797.
[64]   Ibid., arch.1632, *filza* 'Venezia', 27 May 1797.
[65]   Ibid.

over forthwith, fearing, as they threatened they would, the confiscation of the entire property if instructions were not promptly executed.[66]

Notwithstanding the threat of heavy impositions (*gravissime contribuzioni*), which would have rendered the Hospitallers in the Veneto unable to satisfy their statutory responsions, by 8 July 1797 none had in fact been extracted. In the prevailing state of moral and physical violence and insecurity, when things changed overnight from bad to worse, the slightest delay in the execution of such orders appeared to offer a pale glitter of hope.[67] But in 'the most calamitous circumstances' such as these, frail hopes, too, perhaps more than anything else, proved illusory and short lived. On 15 July Benvenuti could accurately predict the certain confiscation of the Order's estates in the Veneto: for if the French were adopting the same policy everywhere, why would they exempt Hospitaller property there?[68] Indeed, on 2 September he wrote of the appropriation of Hospitaller lands in the Trevigiano on specific directives from Napoleon. Benvenuti instructed the Hospitaller Bertolini, residing in Udine, to submit a remonstrance to the French General in the vain hope of having this measure somehow modified. He knew only too well that he was clutching at the flimsiest of straws. 'I would have proceeded to Udine myself', he confessed, 'had it not been too expensive and had there been the slightest hope of achieving anything positive'.[69] He was right. Within less than a week, not only had Napoleon ordered the confiscation of Hospitaller property lying within the province of Udine; its Municipality issued without delay the necessary instructions to have that property sold.[70] Also suppressed were the *juspatronats* that the Cornaro and the Lippomano families had been enjoying since 1588 and 1598 respectively.[71]

What had for long been painfully anticipated carried the same shocking effect on Benvenuti as the unknown or the unexpected. The city of Venice was about to experience what its mainland territories had already suffered. 'I thought promptly', he wrote, 'of what means I could take to redress in part the fast approaching scourge that would totally destroy the Order'.[72] He set up a three-man commission to plead the Order's cause personally with Napoleon. 'Who knows', he wrote on 9 September, 'what outcome this would have'.[73] Meanwhile, he had all the papers documenting the Order's ownership rights over its confiscated lands in the former State of Venice packed in a case and stored in a safe place. Once tranquillity and normality returned, they would be found necessary supporting evidence for any eventual claim the Order might wish to make.

---

[66]   NLM, arch.1632, *filza* 'Venezia', 1 July 1797.
[67]   Ibid., arch.1632, *filza* 'Venezia', 8 July 1797.
[68]   Ibid., arch.1632, *filza* 'Venezia', 15, 22 July 1797.
[69]   Ibid., arch.1632, *filza* 'Venezia', 2 September 1979.
[70]   Ibid., arch.1632, *filza* 'Venezia', 9 September 1797.
[71]   Ibid. See also Mallia-Milanes, *Venice and Hospitaller Malta*, pp. 186, 190.
[72]   Ibid.
[73]   Ibid.

The three-man deputation consisted of the Fr Fulvio Alfonso Rangoni, the grand prior's lieutenant; Fr Antonio Rota Merendi, Benvenuti's secretary; and Commendatore Bertolini of Udine. Its mission was to suspend or delay 'the impending curse'. The first two left Venice and proceeded to Udine in the hope of finding in Bertolini, familiar as he was with recent developments in that city, a helpful guide and a much needed psychological support. According to Benvenuti, Bertolini was caught in a dilemma: should he act against the interests of his own native land and draw upon himself the certain consequences of the Municipality's resentment? Or should he fail to honour the solemn vows he took when he had originally joined the Hospital? He decided to help his brethren cautiously and covertly. Rangoni and Merendis therefore proceeded alone to Passariano (the town where the huge luxurious villa, owned by Lodovico Manin, the last Venetian doge, was located) to meet Napoleon on the day fixed for his congress at Udine. Napoleon's manners and behaviour on that occasion, we are told, contrasted sharply with those he had displayed to the courteous Venetian patrician Pietro Pésaro earlier on. Without betraying any sense of indignation or disdain, he received them well and promised he would consider their case; indeed, he showed them some form of generous disposition by extending to both an invitation to the congress dinner. They were then dismissed amicably and politely. For both of them, this was refreshingly promising.

It was an exercise in delusion. They returned to Passariano several times for a definite reply. On these occasions they were treated with indifference. In the end, reported Benvenuti, Napoleon explained that in principle he could not, and would not, act differently from the way he had acted through the greater part of Italy. Hospitaller property was national property. The nations had full right to avail themselves of it according to their needs, such as those of maintaining their troops.[74]

On 14 October 1797, three days before the treaty of Campoformio was concluded, Benvenuti claimed with some confidence that Venice appeared serene and peaceful. No Hospitaller estates in Venice had yet been confiscated, and he attributed this to what he termed the friendly disposition that the provisional municipality of Venice had all along been showing towards the Order of St John. Indeed, he claimed, it endeavoured 'with all its might to save us and protect us'.[75] Were things changing for the better? 'Peace has returned, with universal joy, at long last', he wrote. He felt tempted to claim that 'our ills (misfortunes) have, at least in part, come to an end', and that it was possible to identify which parts of his institution would survive, which hopes to nourish for an eventual compensation.[76] The Grand Priory had been transferred to Austrian sovereignty. In his letter of 23 December 1797, he made it clear to the Lords of the Treasury in Valletta that, if he had failed in everything, he had at least succeeded in saving the Grand Priory

---

[74]   For the whole episode, NLM, arch.1632, *filza* 'Venezia', 23 September 1797.
[75]   Ibid., arch.1632, *filza* 'Venezia', 21 October 1797.
[76]   Ibid.

of Venice from complete destruction.[77] There was no reason, therefore, why the Priory should not participate, along with the other foreign missions, in the city's celebrations in honour of His Imperial Majesty. The façade of the palace where he resided in Calle Malta on the Calle dei Furlani, and which still belongs to the Order today, would be illuminated as a formal sign of gratitude.[78] The joyful atmosphere was short lived. With Austria's humiliating peace of Pressburg of 1805 after the battle of Austerlitz, Venice and its mainland territories were ceded to the 'Kingdom of Italy', in what has been called 'a second conquest by the French'.[79] Four months later, on 30 April 1806, the Grand Priory was suppressed on Napoleon's instructions, its property confiscated by the State.[80]

Conclusion

The two collections of letters upon which the present paper has been mostly based, like similar ones, published or unpublished,[81] have much to offer the historian of the Hospital – not only for the facts they recorded and the views their authors entertained, often with an unavoidable modicum of bias, but rather as a valuable source of contemporary personal reflection on the spirit of the times. It has almost become established practice for traditional historians of the Order to view the phenomenon of the Hospital's decline and fall as the logical outcome of a long-drawn-out process of anachronism and internal decadence. By the eighteenth century, claimed the influential Elisabeth Schermerhorn in 1929, the magistracy no longer enjoyed divine creative inspiration; 'and its faith burned low'. Her description of 'the brilliancy of Valletta's court' as 'a gilded shell, ready to collapse under the first determined fingers that grasped it' is a classic example

---

[77]   Ibid., arch.1632, *filza* 'Venezia', 23 December 1797.

[78]   'Trovandosi ora questa città in un destino felice, vuole dar saggio della sua esultanza, e riconoscenza, facendo dimani a sera illuminazioni e feste. Ad esempio di alter Corti ministerial, e della città tutta, faro ancor io la illuminazione della ministerial Residenza. Se non fosse qui impulso, benché bastante, dovrei in vista particolarmente degli ottimi sentimenti manifestatimi dai Commandanti Austriaci per la Sacra Religione di dare contrasegni di giubilo e gratitudine. Questi giusti riguardi saranno dalla loro considerazione, che bens a rispettare I rapport vantaggiosi pella nostra Comun madre, approvati pienamente.' Ibid., arch.1632, *filza* 'Venezia', 20 January 1798.

[79]   Sire, p. 173.

[80]   See M. Celio Passi, *Il Gran Priorato di Lombardia e Venezia* (Venice, 1983).

[81]   The present author's recent edition of Massimiliano Buzzaccarini Gonzaga's correspondence is one example. See also *Des Nouvelles de Malte: Correspondance de M. L'Abbé Boyer (1738–1777)*, ed. A. Blondy (Brussels, 2004). The National Library of Malta, in Valletta, holds a superb collection of original letters written by the Order's resident ambassador to the Holy See to the grand master from 1596 (AOM cod. 1249) to 1790 (AOM cod. 1373).

of wild Romantic fantasy let loose.[82] She cites lines from the equally influential Patrick Brydone's account[83] of his visit to Malta in support of the general claim of immorality among the Order's younger generation. She then goes a step further. The Order 'could not weather the shock of the French Revolution ... because its Treasury was bankrupt'.[84] The reverse is correct. The Hospital's treasury went bankrupt as a direct outcome of the Revolution and its wars. There were earlier spells, of course, when the Order's common treasury was marked by a consistently downward trend, but this happened with the finances of almost any state. By the time of the fall of the Bastille, however, the state of the Order's finances had been restored. The Hospital's vast estates in Europe, we are told,[85] 'presented an impressive spectacle, rich with the accumulations of centuries'. By the mid-eighteenth century, its brilliant naval task of restraining the Barbary corsairs to near-negligible proportion, with the exception perhaps of Algiers, was practically complete, thereby reducing its own performance to seeming inactivity.[86] By then, too, there was political peace on the Hospitaller principality, in the convent and outside it, with a steadily increasing population enjoying a standard of living that compared fairly favourably with neighbouring Mediterranean centres; there was economic prosperity, there was a general state of 'exceptional tranquillity', with no evidence that the Order was keeping 'afloat with difficulty'.[87] What the futile and isolated uprising of a small sector of the local clergy had in fact achieved, for example, in September 1775 was to disturb briefly the domestic peace and quiet and somehow shake the government's full confidence in the people's loyalty. It failed to inspire any popular response.[88] Before the enlightened revolutionary doctrine on the absurdity of privilege was preached in France and permanently enshrined along with the other principles of 1789, no traits of dissolution could be observed in the structural unity of the Hospitaller institution; nor were forces of early nationalism evident to have been gnawing at the centuries-old administrative divisions of the Hospital, except perhaps for a few occasional minor squabbles in the streets of Birgu and Senglea with no serious consequence. Nor has any convincing evidence been produced to show that the Hospitaller community was at any time before 1789, to use Eileen Power's phrase, 'stricken by disease',[89] blatantly breaking their three monastic vows, neglecting their statutory charity

---

[82]   E. Schermerhorn, *Malta of the Knights* (Surrey, 1929), p. 277.

[83]   P. Brydone, *A Tour through Sicily and Malta in a Series of Letters to William Beckford Esq.* (London, 1773).

[84]   Schermerhorn, p. 277.

[85]   Sire, p. 111.

[86]   Ibid., pp. 97–8.

[87]   A. Bartolo, 'History of the Maltese Islands', in *Malta and Gibraltar Illustrated*, ed. A. Macmillan (London, 1915; facsimile edn, Malta, 1985), p. 116. F. Ryan's chapter on the Maltese Renaissance in his *The House of the Temple*. Also Luttrell, pp. 37–51.

[88]   See Mallia-Milanes, *In the Service of the Venetian Republic*, pp. 43, 78, 598–601 and notes.

[89]   Term quoted from E. Power, *Medieval People* (New York, 1992; 1st edn, 1924), p. 3.

and hospitality, and defiantly abandoning their holy commitments to Catholic Europe in their various manifestations. The brethren never seriously threatened or questioned the institution of the magistracy, and recruitment of members into the Order never betrayed insurmountable difficulties of consistently dwindling numbers. It was only from 1792 that real, intractable problems began to emerge with the confiscation of all Hospitaller estates in France, which had made up for over half the Order's revenue from its European sources.

The end of the Order of the Hospital on Malta in June 1798 was inevitable, but in no way was it, in the long term, predictable. It was dramatic and sudden, violent and almost instantaneous. And if by then the unceremonious dislodgement of the Order from its secure Mediterranean island-fortress might not have appeared very astonishing, it was because the upheaval in France had broken out a whole decade earlier and had already spread through most of Europe. Although in the long-term perspective of historical development a decade is hardly significant, within the framework of daily human relations, ten whole years of radical devastation, terror and bloodshed are far too long. By 1798, the conscious rejection of the past had almost become the norm. Rather than being the obvious result of slow and steady disintegration, what was in fact remarkable in 1798 was the speed with which the institution disappeared from behind the admirably massive stone walls of Valletta. By the middle of the eighteenth century, the military–religious Order was still a healthy institution and able to convince its powerful patrons that the threat confronting Christian Europe was real and that its role in trying to contain it was essential. To prolong its own survival, it successfully endeavoured to keep the crusade alive by promoting and sustaining the fearful image of the common enemy.

In the massive correspondence between Venetian resident ministers on Hospitaller Malta and the Adriatic Republic and between the Order's resident receivers in Venice and the Lords of the Common Treasury, it was only in the 1790s that talk was made of the serious dangers challenging the Order's existence. It was only the direct impact of the Revolution. There was no serious talk of it before. There was no single reference to any conceptual symptoms of decline in Massimiliano Buzzaccarini Gonzaga's long and detailed letters from Malta between 1754 and 1776. On the contrary, in one of his later letters, he could still perceive the Hospitaller principality in the central Mediterranean as fulfilling its professed and accomplished commitments – to physically and spiritually rehabilitate the sick, the poor and the needy through charity and hospitality at what had become one of the leading hospitals in Europe, and to extend their naval and military establishments and their medical knowledge and expertise in defence of Christian Europe as much against Islam as against the plague and other natural catastrophes. The Order's immediate response to the earthquake that devastated Sicily and Calabria in 1783 is a classic example.[90] Nowhere is there in these letters, which were at times fairly critical of the Order and the magistracy, the slightest or vaguest suggestion that the Hospitaller institution was approaching its end. There

---

[90] Bartolo, p. 117.

is no such intimation either in Alviero Zacco's equally thorough and exhaustive correspondence written from Malta when Buzzaccarini Gonzaga was on extended leave of absence, visiting his home town, and after the latter had passed away in 1776. It is quite revealing that it was not the myth of the aging process that struck almost fatally at the Order in the 1790s. Nor ironically was the blow delivered by the Ottoman Empire, its traditional enemy, which, under the progressive Selim III, had unsuccessfully tried to reach some form of a peace settlement and trade agreement with the Order in 1796.[91] As the two Venetian brethren so eloquently acknowledged in their correspondence, the swipe came from revolutionary France. The enlightened doctrine of 1789, encompassing the powerful concept of equality, challenged the old principle of privilege and destabilized the entire social structure. In so doing, it sounded the death-knell of the *Ancien Régime*, and announced the collapse of the Hospital of which it had formed so intimate a part.

If by the time the Order resolved to sojourn temporarily on Cyprus five centuries before the fall of the Bastille and the collapse of the system it symbolized the defence of the Holy Land had for long been one of its main reasons for existence, then its loss and the loss of any hope of retaking it had indeed, as Jonathan Riley-Smith argued years ago, turned the Order into an anachronism.[92] The disastrous surrender of Jerusalem to Saladin in 1187 and of Acre to the Mamluks in 1291 had been grievous, humiliating and embarrassing to the crusaders and the military orders, frustrating to their patrons in the Latin West, themselves embroiled in their own territorial wars and politics. But the story, as we all know, did not end there. The Order of St John survived. Hospitaller Rhodes would soon evolve into a formidable fortress-state, a significantly strategic base of operations against an expanding Ottoman Empire, converting the whole of Christendom into a new and more extensive 'holy land' whose defence against the Saracens' successors would again render the Order and the role it now assumed not only as politically relevant but as indispensible to Europe as it had hitherto been. Its spirit of resilience, its ability to recover readily from any temporary setback, or relentlessly resist being affected by it, had once more succeeded in transforming what its detractors denounced as weaknesses into strengths. Historically, the same argument would be as valid for developments after 1522, 1798, and the chaos that distinguished the early half of the nineteenth century leading to the constitutional restructure of the institution in a brave response to novel demands from a new world. The Hospital's chequered past was consistently marked by alternating stages of struggles against unexpected obstacles emerging at remote intervals and successful attempts at

---

[91]   ASVen, Senato, Secreta, filza 9, Dispacci Malta, 22 September 1796. Also Malta, Cod. 275, fol. 24, *Lettera del Principe della Pace che partecipa al Gran Maestro le premure fatte al Re dalla porta per stabilire una tregua tra la medesima e l'Ordine*, 31 July 1796. Godoy's letter is reproduced in Panzavecchia, pp. 340–42; Mallia-Milanes, *Venice and Hospitaller Malta*, pp. 291–4.

[92]   This has been discussed in J. Riley-Smith, *The Order of St John in Jerusalem and Cyprus, c.1050–1310* (London, 1967), pp. 475–6.

prevailing over them. The French Revolution was one such. At no time was there a complete rupture in its over 900 years of history. The Order, like any living organism, knew how to adapt to radically changing conditions.[93] Its innate powers to exploit adverse circumstances stubbornly to its own advantage helped it to prolong its existence indefinitely. Survival was a historically permanent triumph, not a shameful symptom of failing strength.

[93]   See J. Riley-Smith, 'Towards a History of Military–Religious Orders', in K. Borchardt, N. Jaspert and H.J. Nicholson (eds), *The Hospitallers, the Mediterranean and Europe: Festschrift for Anthony Luttrell* (Aldershot, 2007), p. 284.

PART 3
The British Isles

Chapter 16

The Military Orders in Wales and the Welsh March in the Middle Ages[1]

Helen J. Nicholson

In the later medieval centuries the Hospitallers' estates in Wales were among the most extensive of any religious corporation there. In 1535, just before the dissolution of the monasteries, the commandery at Slebech was the third richest monastic house in Wales, after the Cistercian abbeys at Tintern and Valle Crucis. The next richest house after Slebech was another Cistercian house, Margam Abbey, followed by the Benedictine priory at Abergavenny.[2] Slebech was also wealthy by comparison with other Hospitaller houses in England and Wales. In 1338 it received the largest income of any Hospitaller house in England and Wales, apart from the main house at Clerkenwell just outside London,[3] while in 1535 it had the fourth highest net value of the Hospitallers' twenty-two houses in England and Wales, after Clerkenwell, Buckland and Ribston.[4] With such comparative wealth, we might expect the Hospitallers to have held great authority and power in Wales, and their Welsh property to have been very significant within the Order.

In contrast, the Templars held very little property in Wales. In 1308, when the Templars in the British Isles were arrested on the order of King Edward II of

---

[1]   I am very grateful to Philip Handyside, Kathryn Hurlock and Paul Sambrook for their assistance with certain points in this paper.

[2]   The annual net income of Slebech was £184, after Tintern's £192 and Valle Crucis's £188; the annual net income of Margam was £181 per annum, while Abergavenny's was £129. D. Knowles and R.N. Hadcock, *Medieval Religious Houses: England and Wales*, 2nd edn (London, 1971), pp. 52, 114, 301; cf. R.K. Turvey, 'Priest and Patron: A Study of a Gentry Family's Patronage of the Church in South-West Wales in the Later Middle Ages', *Journal of Welsh Ecclesiastical History*, 8 (1991), 7–19, here p. 9; G. Williams, *The Welsh Church from Conquest to Reformation* (Cardiff, 1976), p. 563.

[3]   Figures in *The Knights Hospitallers in England, Being the Report of Prior Philip de Thame to the Grand Master Elyan de Villanova for AD 1338*, ed. L.B. Larking, intro. J.M. Kemble (Camden Society, 1st ser., 65) (1857), summarized in S. Phillips, *The Prior of the Knights Hospitaller in Late Medieval England* (Woodbridge, 2009), p. 21.

[4]   G. O'Malley, *The Knights Hospitaller of the English Langue, 1460–1565* (Oxford, 2005), pp. 362–3. By 1540, when the Order was dissolved, Slebech had slipped to ninth place out of a total of 21.

England, no Templars were arrested in Wales.[5] It is doubtful whether there had ever been any Templars in Wales itself for any length of time. However, the Templars did hold more extensive estates in the Welsh March, where seven Templars had been living prior to the arrests.[6] When in the mid-fourteenth century the officials of the borough of Montgomery compiled a collection of the statutes of England, starting with King Edward I's reissue of the Magna Carta, they included a copy of King Edward II's statute of 1324 stating that Parliament had agreed that all the properties formerly belonging to the Order of the Temple should be assigned to its sister Order, the Hospital of St John of Jerusalem (in accordance with the papal bull of 1312).[7] Clearly the Templars' estates in the Montgomery region were sufficiently extensive for the question of who owned them and who had taken over the Templars' rights and duties after their dissolution in 1312 to be important to the people there.

The historiography of the military orders in Wales and the Welsh March is slender. The work of William Rees, professor of History at Cardiff University, entitled *A History of the Order of St John of Jerusalem in Wales and on the Welsh Border, including an account of the Templars* (Cardiff, 1947), has not been superceded. However, Rees did little to situate the orders in Wales within their operations within the rest of Britain, let alone within their respective order as a whole, and neither did he compare them to the other religious orders within Wales. There have been some studies of individual houses in Wales and the March, especially Slebech.[8] Most recently, Greg O'Malley's study of the Hospitallers'

---

[5]   *Fœdera, conventiones, litteræ, et cujuscunque generis acta publica, inter reges Angliæ et alios quosvis imperatores, reges, pontifices, principes, vel communitates: ab ingressu Gulielmi I. in Angliam, A.D. 1066. ad nostre usque tempora habita aut tractata*, eds T. Rymer and R. Sanderson, new edn eds A. Clark, F. Holbrooke and J. Caley, 2.1 (London, 1818), pp. 23–4; H.J. Nicholson, *The Knights Templar on Trial: The Trial of the Templars in the British Isles, 1308–1311* (Stroud, 2009), pp. 205–17.

[6]   Two were arrested at Lydley: Henry de Halton, lieutenant commander, and Stephen de Stapelbrugge. Eileen Gooder, *Temple Balsall* (Chichester, 1995), p. 149. However, according to Oxford, Bodleian Library, MS Bodley 454, fol. 115v, there were normally three brothers there: these were probably Henry de Halton, Stephen of Stapelbrugge (commander, see MS Bodley 454, fol. 96v) and Michael de Karvile (MS Bodley 454, fol. 144v). Two were arrested at Garway – MS Bodley 454, fol. 58v: Philip de Meux, knight, and William de Pokelington; and two at Upleadon – MS Bodley 454, fol. 58v: Thomas of Toulouse, the commander, a knight-brother, and Thomas the Chamberlain.

[7]   'Statuta Angliae: leges et consuetudines de Montgomery', Cardiff City Library MS L.385, fols 293v–299v; for description see N.R. Ker and A.J. Piper, *Medieval Manuscripts in British Libraries* (Oxford, 1969–92), 2, pp. 350–53. See also the enrolled copy of this statute, translated in *Calendar of the Close Rolls preserved in the Public Record Office, Edward II, 1323–1327* (London, 1898), p. 91.

[8]   J. Evans, 'Yspytty Ifan, or the Hospitallers in Wales', *Archaeologia Cambrensis*, 3rd ser. 6 (1860), 105–24; J. Rogers Rees, 'Slebech Commandery and the Knights of St John', *Archaeologia Cambrensis*, 14 (1897), 85–107, 197–228, 261–84; 15 (1898),

English *langue* included an assessment of the Welsh commanderies within the context of the whole *langue*.[9]

The sources of information for the military orders in Wales are a number of bishops' confirmations of earlier donations, notably Bishop Anselm of St Davids in 1230;[10] a few surviving charters of donation, some of which are in the Glamorgan Archives,[11] part of the cartulary of Slebech commandery that is preserved in the National Library of Wales,[12] and a report that Philip de Thame, then prior of the Hospital in England and Wales, sent to the Order's headquarters on Rhodes in

---

33–53; 16 (1899), 220–34, 283–98; E. Hermitage Day, 'The Preceptory of the Knights Hospitallers at Dinmore, Co. Hereford', *Transactions of the Woolhope Naturalists' Field Club* (no volume number) (1927), 45–76; J. Fleming-Yates, 'The Knights Templar and Hospitallers in the Manor of Garway, Herefordshire', ibid., pp. 86–101; W. Rees, 'The Templar Manor of Llanmadoc', *Bulletin of the Board of Celtic Studies*, 13/3 (1949), 144–5; E.J.J. Davies, 'The Church of St John the Baptist, Ysbyty Ifan', *Caernarvonshire Historical Society Transactions*, 56 (1995), 37–46; J.M. Parry, *The Commandery of Slebech in Wales of the Order of the Hospital of St John of Jerusalem* (no place, 1996); H.J. Nicholson, 'Margaret de Lacy and the Hospital of St John at Aconbury, Herefordshire', *Journal of Ecclesiastical History*, 50 (1999), 629–51, repr. A. Luttrell and H.J. Nicholson (eds), *Hospitaller Women in the Middle Ages* (Aldershot, 2006), pp. 153–77; R. Thomas, 'Slebech: A Study of a Commandery of the Order of St John in Wales', unpublished MA thesis, Cardiff University, 2002; H. Nicholson, 'The Sisters' House at Minwear, Pembrokeshire: Analysis of the Documentary and Archaeological Evidence', *Archaeologia Cambrensis*, 151 (2002), 109–38. Also, see now Philip Handyside, 'The Hospital of St John in Wales: Donors and Properties', unpublished MA thesis, Cardiff University, 2009.

9   O'Malley, *passim*. There is also some discussion of the Templars in Wales and the March in E. Lord, *The Knights Templar in Britain* (Harlow, 2002), pp. 110–17, and of the Hospitallers in Wales by H. Nicholson, 'The Knights Hospitaller on the Frontiers of the British Isles', in J. Sarnowsky (ed.), *Mendicants, Military Orders and Regionalism in Medieval Europe* (Aldershot, 1999), pp. 47–57.

10   Cardiff City Library MS 4.83 (was Middle Hill MS 19880); also copied by Richard Fenton and printed by Rogers Rees (1897), pp. 99–102; *St Davids Episcopal Acta, 1085–1280*, ed. Julia Barrow (Cardiff, 1998), no. 108, pp. 123–8, and see index, p. 193.

11   I am grateful to Philip Handyside for the following references: Cardiff, Glamorgan Archives, CL/DEEDS I/3656 (formerly Pembroke Deed 1053 in Cardiff Library): Reimundus fitz Martin gives to the Hospitallers of Slebech land around Beneg[er]duna and Minwear (no date); CL/DEEDS I/3658 (formerly Deed 1055), Walter Marshal, earl of Pembroke (d. 1245), grants liberties to the Hospital of St John the Baptist; CL/DEEDS I/3667 (formerly Deed 1063): John Bonesant, son of Philip from Patrick's Mount, gives to the brothers of the Hospital of St John of Slebech all the interest that accrued to him from the death of Joan, daughter of his brother Philip, in a moiety of land in Patrickshill in the tenement of Marteltwy: 1 May 1273.

12   Aberystwyth, National Library of Wales, Slebech: Estate and family records of Barlow, Phillips and de Rutzen of Slebech, Co. Pembroke, nos. 233, 283, 869, 3139, 3144, 3146, 11438, 11477.

1338.[13] There is also a history of the Hospitallers written in 1434 by the Hospitaller John Stillingflete that lists a number of donations to both orders in Wales and the Welsh March.[14] In addition, there are also the inventories that the royal officials took of the Templars' lands in 1308 when the Templars were arrested and their subsequent accounts for these lands – but very little actually relating to Wales[15] –, a rental of the Hospitallers' commandery of Dinmore and Garway from 1505,[16] various references in the English government records, and the record made in 1535 just before the dissolution of the monasteries in England and Wales, the *Valor Ecclesiasticus*.[17]

This paper will assess what property the Hospitallers and Templars held in Wales and the Welsh March, review motivations for donation, and consider how these orders made use of their properties here.

The Military Orders' Estates in Wales and the Welsh March

The Hospitallers and Templars had received their first donations of land in Wales and the Welsh March before 1150. The precise date of the Hospitallers' first acquisition is unclear. Sometime between 1176 and 1198, Bishop Peter of St Davids confirmed all the gifts to the Hospitallers within his diocese, and mentioned that his three predecessors, Wilfrid, Bernard and David, had allowed the Hospitallers to remove any chaplain or clerk from their churches. This could indicate that the Hospitallers had received responsibility for churches in south-west Wales by 1115, when Bishop Wilfrid died.[18] This was only two years after Pope Paschal II had acknowledged the Hospital of St John as a religious order; the Hospitallers did not

---

[13]   *Knights Hospitallers in England*, pp. 30–33 (Dinmore), 34–7 (Slebech), 38–40 (Halston), 195–200 (Upleadon, Garway and Stanton Long), 213 (Lydley and Pencarn).

[14]   John Stillingflete, 'Liber Johannis Stillingflete de nominibus fundatorum Hosp. S. Johannis Jerusalem in Anglia', in *Monasticon Anglicanum*, eds W. Dugdale *et al.* (London, 1846), 6.2, pp. 831–9. Dinmore appears on pp. 836 and 839, and Slebech pp. 826–7.

[15]   Kew, The National Archives: Public Record Office (hereafter cited as TNA: PRO), E358/18, rots 2, 44; E358/19, rots 25, 47d, 50–51 (Garway and Upleadon); E358/18, rots 4, 54; E358/19, rot. 36; E358/20, rots 5–6 (Lydley, Cardington and Stanton Long); E358/20 rot. 10 (Llanmadoc); SC 6/1202/3 (Llanmadoc).

[16]   Hereford Record Office, A63/III/23/1: Rental of Dinmore and Garway, 20 Henry VII.

[17]   Slebech: *Valor Ecclesiasticus temp. Henr. VIII auctoritate Regis institutis*, ed. J. Caley (London, 1810–34), 4, pp. 388–9; Halston is at 4, pp. 455–6. O'Malley (p. 71) notes that the entry for Dinmore is missing.

[18]   Rogers Rees (1897), pp. 106–7 n. 9; Rees, *History*, p. 25, both citing Cardiff Library MS 4.83 (was Middle Hill MS 19880); *St Davids Episcopal Acta*, nos. 21, 37, 45, 46, pp. 49–50, 61–2.

begin to receive gifts of land in England until 1128.[19] Yet it is not impossible that the Hospitallers had received some gift in Wales by 1115, for, as Anthony Luttrell has shown, the Order had received donations in southern France very soon after the First Crusade.[20] However, without other evidence, it is most likely that there is an error in Bishop Peter's confirmation, and that the Hospitallers obtained their first properties in Wales in the 1130s and 1140s.

Their earliest acquisitions were in south-west Wales, and were small manors and parcels of land and churches, given by local small landowners with Norman names, such as Philip de Kemeys and Richard son of Tancred.[21] The Hospitallers already had a fine scattering of possessions in Morgannwg and Pembrokeshire in south Wales by the time they received the lands that formed the bases of their two centres in the Welsh March (Fig. 16.1): at Dinmore in Herefordshire, in the 1180s, and at Halston in Shropshire, perhaps before 1187.[22] The Templars arrived later than the Hospitallers: in Shropshire at Lydley and Cardington, immediately to the south of Lydley, in the late 1150s; in Llanmadoc in the Gower in 1156; and at Garway in Herefordshire in the 1180s.[23]

The military orders' lands were clustered in the areas of Wales that were settled by the Normans, English or Flemish: that is, south Wales and the Welsh March. The Hospitallers also received some properties in the parts of central Wales that were temporarily under Norman domination, and a little in north-west Wales, which remained under Welsh lordship. However, while the Hospitallers were building up the estates of Slebech in south-west Wales to become one of the wealthiest religious houses in Wales, the Templars received very little land within Wales. Even when they were given land here they did not develop it.

For example, the Templars had 40 acres at Caerwigau, part of the fee of Bonvilston, now in the Vale of Glamorgan. This is in the River Ely valley, on a local communications route and a short distance from the old Roman road, which runs to the south. The Templars rented this land out to the Cistercian monks at Margam, and by the end of the twelfth century had transferred the land to the Abbey.[24] They received some land at Pencarn, now near Newport, on the west bank

---

[19]   Pope Paschal II, *Pia postulatio voluntatis* (15 Feb. 1113), *CH*, 1, pp. 29–30, no. 30; M. Gervers, 'Donations to the Hospitallers in England in the Wake of the Second Crusade', in (ed.), *The Second Crusade and the Cistercians* (New York, 1991), pp. 155–61.

[20]   A. Luttrell, 'The Earliest Hospitallers', in B.Z. Kedar, J. Riley-Smith and R. Hiestand (eds), *Montjoie: Studies in Crusade History in Honour of Hans Eberhard Mayer* (Aldershot, 1997), pp. 37–54, here p. 49.

[21]   Rees, *History*, pp. 105–6.

[22]   Rees, *History*, pp. 120, 127; Knowles and Hadcock, pp. 303–4.

[23]   Rees, *History*, pp. 124, 126–7; Knowles and Hadcock, p. 294.

[24]   See the confirmation charter by Countess Isabel of Gloucester, 1189–99, *Earldom of Gloucester Charters: The Charters and Scribes of the Earls and Countesses of Gloucester to A.D.1217*, ed. R.B. Patterson (Oxford, 1973), no. 137. In 1203 Pope Innocent III confirmed to Margam all its possessions, including the 40 acres that the abbey had received from the Templars. Rees, *History*, p. 53.

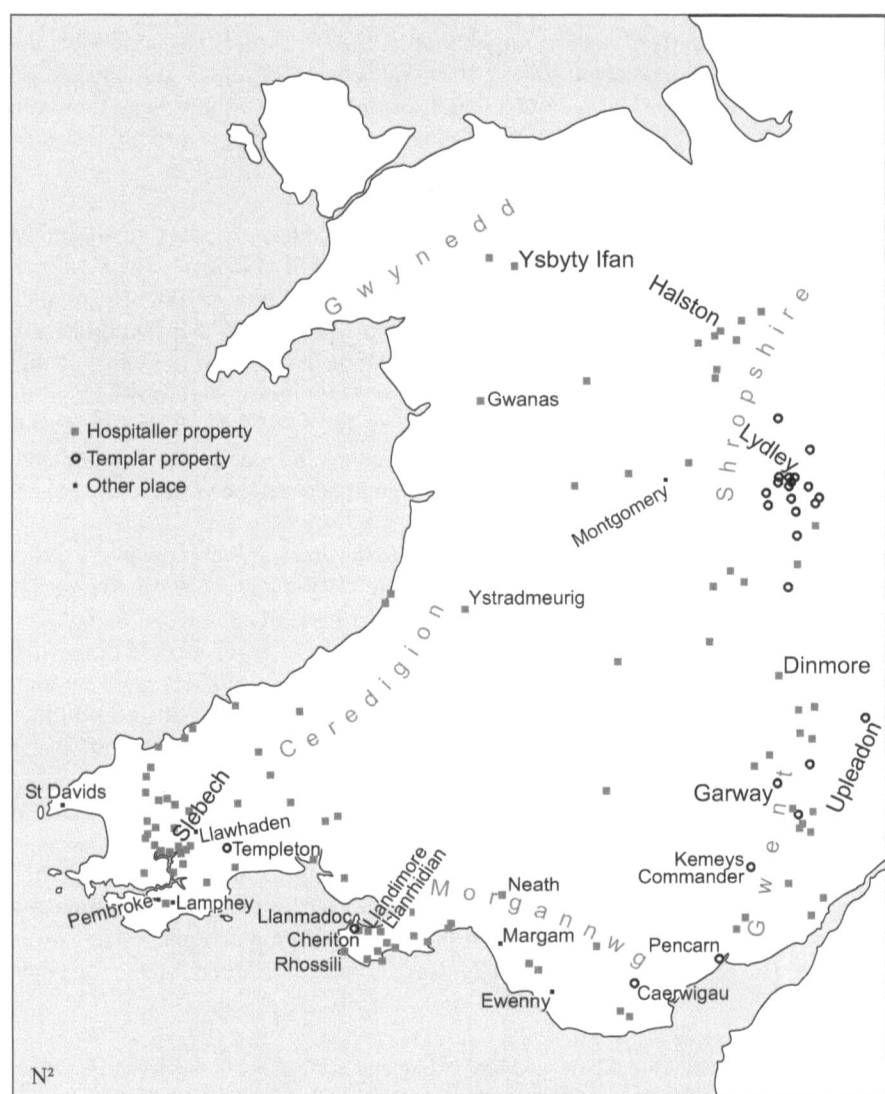

Figure 16.1   The military orders' properties in Wales, indicating the places
mentioned in the text. Based on the maps by William Rees, *A
History of the Order of St John in Wales and on the Welsh Border,
including an account of the Templars* (Cardiff, 1947) (graphics: N²
Productions)

of the River Ebbw where the Roman road forded the river. Yet there is no mention of this in any surviving Templar records, and the first evidence that the manor had been held by the Templars appears in 1338 when the Hospitallers complained that it was being held by the earl of Gloucester.[25] The Templars did maintain the church they had received at Kemeys Commander, at a crossing point on the River Usk in Gwent, although they did not develop any significant settlement there.[26] They also received land at Templeton, a new town founded in Pembrokeshire in the twelfth century, on the land of the lords of Narberth. In 1283 Templeton was called *villa Templariorum*, and was described as a possession of the late Roger Mortimer, lord of Narberth. The name implies that it had or did belong to the Templars, but either it was no longer a Templar property by 1283 or they held it by rent, not in full right. In any case, there is no record that it was administered by royal officials after the arrest of the Templars in 1308, and it did not pass to the Hospitallers with the Templars' other estates.[27] Finally, the Templars held the church of Llanmadoc in the Gower and maintained a small house there, which was apparently let out to tenants at the time of the Templars' arrest in January 1308. In addition to rent from tenants, they had a water mill, some arable land and some pasture, and received some court fees. At Michaelmas 1308, the sheriff reported receipts since January of thirteen pounds, seven shillings and five pence.[28] In 1338 (under the Hospitallers) its annual gross income, including income from the mill and court fees, was ten pounds, sixteen shillings and fourpence;[29] in 1505 the Hospitallers' income from rents, the mill and the church amounted to five pounds, thirteen shillings and three pence.[30]

The Templars' estates in England were generally in low-lying or gently rolling farming country, where they grew large quantities of grain and kept sheep and cattle.[31] Perhaps the Welsh terrain was not suitable for their farming needs; or perhaps the Welsh and the Anglo-Norman or Cymro-Norman families of Wales were not interested in giving to them. I will return to this point later.

By the end of the thirteenth century, the Hospitallers had commanderies at Slebech in south-west Wales, 'North Wales' (based at Ysbyty Ifan in the Welsh kingdom of Gwynedd) and Halston in Shropshire – by 1338 these two were administered as one commandery –, and Dinmore in the Welsh March on the

---

[25]   *Knights Hospitallers in England*, p. 213; Rees, *History*, p. 53.

[26]   Ibid., p. 52 for the crossing point. Note the pattern of public rights of way on the Ordnance Survey map.

[27]   *Calendar of the Close Rolls preserved in the Public Record Office: Edward I, AD 1279–1288* (London, 1902), p. 200; Rees, *History*, p. 55; R.F. Walker, *Pembrokeshire County History*, 2: *Medieval Pembrokeshire* (Haverfordwest, 2002), p. 143.

[28]   TNA: PRO E358/20, rot. 10 and SC 6/1202/3; ed. and trans. Rees, 'Templar Manor of Llanmadoc'.

[29]   *Knights Hospitallers in England*, p. 197. Note that the church itself was not mentioned in 1308 or 1338.

[30]   Hereford Record Office, A63/III/23/1, fols 33r–34r.

[31]   Nicholson, *Knights Templar*, pp. 218–49; TNA: PRO E358/18–20.

Gwent/Herefordshire border. The Templars had three commanderies in the Welsh March: at Lydley in Shropshire, Upleadon in east Herefordshire, and Garway on the Gwent/Herefordshire border, from which they administered their property in south Wales.

In his bull *Ad providam* of 2 May 1312 Pope Clement V declared that all the Templars' properties, except for those in the Iberian Peninsula and Mallorca, should pass to the Hospitallers.[32] In England and Wales, this bull was not put into statute until 1324, when Parliament agreed that all the properties formerly belonging to the Templars should indeed be assigned to the Hospitallers.[33] Even following this statute the Hospitallers were not able to obtain all the former Templar estates: those at Lydley, for example, were taken back into the hands of the FitzAlan family, which originally gave Lydley to the Templars.[34] The Hospitallers administered the remaining properties and churches that had belonged to this commandery from Upleadon.

Initially the Hospitallers continued to maintain the former Templar commanderies at their original status, listing them separately in their records.[35] Commanders of Garway were appointed until 1337;[36] and Upleadon was a commandery until at least 1381.[37] During the course of the fourteenth century the Hospitallers rationalized their administration in Wales and the March. Garway was amalgamated with Dinmore – for obvious reasons, as the two were very close together – and by the end of the fourteenth century Upleadon was also amalgamated with Dinmore.[38]

---

[32]   M. Barber, *The Trial of the Templars*, 2nd edn (Cambridge, 2006), p. 271.

[33]   Cardiff City Library MS L.385, fols 293v–299v.

[34]   *Victoria County History: Shropshire*, ed. C.R.J. Currie (Oxford, 1998), 10, pp. 27, 89.

[35]   See, for example, the list of 1319–22, published by J. Miret i Sans, *Les cases de Templers i Hospitalers a Catalunya: aplec de noves i documents històrics*, new edn with introduction by J.M. Sans i Travé (Lleida, 2006), pp. 400–401, in which Denemoz (Dinmore), North Uuallia (North Wales) and Slebethe are listed with other Hospitaller properties and 'Garellzy' (mistranscription of Garway) is listed at the end with the former Templar properties); and AHN OO.MM Sección de Codices 602B (= Registro Amposta, iv), fols 10v–12, a document of 1357 in which Dynemor (Dinmore), Norwalles (North Wales) and Slebech are listed with the Hospitaller properties and Garway at the end, with the former Templar properties (I am very grateful to Dr Antony Luttrell for drawing my attention to the former and supplying me with a transcription of the latter).

[36]   On 16 October 1337 Robert Cort was commander of Dinmore and Garway (at the General Chapter of that year: NLM, Archives of Malta (henceforth AOM), 250 fol. 42r).

[37]   18 Nov. 1381 Robert Hales died still holding, among others, the prioral commandery of Upleadon: the grand master gave this and his other bailies and commanderies to John Radington: NLM, AOM 321, fol. 145r–v. On 16 July 1399 William Hullus was commander of Temple Combe and 'Hopdelm' (possibly a variant spelling of Upleadon): NLM, AOM 330, fol. 71r (new foliation).

[38]   O'Malley, p. 62, n. 17; Phillips, p. 134.

Donors

Some donations to the Hospitallers and Templars in Wales and the Welsh March came from the English monarch. In the 1180s King Henry II of England gave the Templars land at Garway and allowed them clear forest here and at Botewood in Shropshire; he also gave the Hospitallers some land at Dinmore. His son Richard I gave the Hospitallers a hospital at Hereford – and, according to the Hospitaller historian John Stillingflete, writing in the 1420s, he was responsible for endowing the Hospitallers with the bulk of their land at Dinmore.[39]

William Rees's study of donors to the Hospitallers and the Templars in Wales, revealed that the majority of donations were from donors with Norman, English or Flemish names.[40] Motivations for donation are notoriously difficult to gauge, but it is clear that, because the military orders were religious orders, donors gave them gifts for the same pious reason that all religious orders received gifts. The military orders were also relatively cheap to endow – they would accept small parcels of land, small money gifts, even horses, armour and clothing – and this would have made them attractive to the relatively poor landowners of Wales. The military religious orders' particular appeal lay in their involvement in helping pilgrims to the Holy Land and in crusading campaigns. Those who had been on pilgrimage to the East or on crusade would give them gifts in thanks for their help; those who could not go – because they were too poor, or could not leave home for security reasons – could nevertheless assist the cause by making a donation.[41] While the Holy Land was seldom specifically mentioned in donation charters as a motivation behind a gift, donors may have considered it too obvious to mention. The military orders' role in protecting pilgrims to the Holy Land and helping crusaders was the obvious motivation for the donations by William Marshal (who had been to the East in the mid-1180s) and his family, and those by an earl and countess of Warwick.[42]

---

[39]   Rees, *History*, pp. 44, 47, 51, 120–21, 126; John Stillingflete, pp. 836, 839.

[40]   Rees, *History*, pp. 104–28 (list of donors).

[41]   For donations and motivation see, for example, M. Barber, *The New Knighthood: A History of the Order of the Temple* (Cambridge, 1994), pp. 13–14, 19, 24–7.

[42]   D. Crouch, 'Marshal, William (I), Fourth Earl of Pembroke (*c.*1146–1219)', in *Oxford Dictionary of National Biography*, eds H.C.G. Matthew and B. Harrison (Oxford, 2004) (hereafter cited as *ODNB*), 36, pp. 815–22. Donations by the Marshal family: Knowles and Hadcock, p. 193; Rees, *History*, pp. 52–3, 55, 105, 107, 117; *Records of the Templars in England in the Twelfth Century: The Inquest of 1185 with Illustrative Charters and Documents*, ed. B.A. Lees (London, 1935), p. 142. A privilege by Aymer de Valence, earl of Pembroke and great-grandson of William Marshal (d. 1324) to Slebech in 1323 survives in National Library of Wales, Slebech MS 11438. See Richard Fenton, *Historical Tour through Pembrokeshire*, 2nd edn (Brecknock, 1903), p. 326. Earls of Warwick and crusades, see E. Mason, 'Fact and Fiction in the English Crusading Tradition: The Earls of Warwick in the Twelfth Century', *Journal of Medieval History*, 14 (1988), 81–95.

However, for some noble donors there was no obvious crusading motivation. For example, the de Clares – an important Anglo-Norman noble family with several branches, holding the earldoms of Hertford, Pembroke and later Gloucester – were not a great crusading family, but had long been donors to religious institutions and had taken up the new religious movements of the twelfth century, the Augustinian canons, the Cistercians and the military orders; in the thirteenth century they took up the new orders of friars. Roger de Clare, earl of Hertford, his parents, his cousin Richard Strongbow, earl of Pembroke, and Roger's wife Matilda, daughter of James de St Hilaire and a major landowner in England and Brittany, gave property to the Hospitallers in eastern England, Ireland, Ceredigion and elsewhere.[43] The Clares' endowment of the Hospitallers may have been initially inspired by their work in the Holy Land, but there is no specific evidence for this.

Ceredigion had been captured by Roger de Clare's grandfather Gilbert de Clare in the early twelfth century. In 1136 the Welsh recaptured it, but Roger recovered it in 1158. He then gave some land there to the Hospitallers, perhaps as a thanks-offering to God. There was also the possibility that, even if the Welsh should attempt to recover the land, the Hospitallers would continue to support the earl's interests there. In the event, Prince Rhys ap Gruffudd of Deheubarth, Prince of South Wales, recovered Ceredigion in 1164 and confirmed the Hospitallers' holdings, thereby strengthening his own hold on the region.[44]

The majority of donations came from local lesser nobility, notably in the Gower of the de Turbervilles, who were also donors to the Cistercians at Neath.[45] The Turbervilles gave the Hospitallers the churches of Llanrhidian, Landimore and Cheriton, and Rhossili.[46] Again, here there was no obvious active crusading connection. Even though many people from the Marches and Wales took the cross during Archbishop Baldwin's preaching tour of 1188, so far as is known no one

---

Donations, see Rees, *History*, pp. 115, 127. My thanks to Philip Handyside for discussions on this point

[43]   J.C. Ward, 'Fashions in Monastic Endowment: The Foundations of the Clare Family, 1066–1314', *Journal of Ecclesiastical History*, 32 (1981), 427–51, here pp. 443–5; family trees in M. Altschul, *A Baronial Family in Medieval England: The Clares, 1217–1314* (Baltimore, MD, 1965), following p. 332.

[44]   *The Acts of Welsh Rulers, 1120–1283*, eds H. Pryce and C. Insley (Cardiff, 2005), pp. 166–7; R. Mortimer, 'Clare, Roger de, Second Earl of Hertford (*d.* 1173)', *ODNB*, 11, pp. 767–8, citing J.E. Lloyd, *A History of Wales from the Earliest Times to the Edwardian Conquest*, 3rd edn (1939; repr., 1988).

[45]   Glamorgan Archives, DXGC 115/1: Gilbert de Turberville to Neath Abbey; DXGC 115/2: a second grant to Neath Abbey. I am very grateful to Philip Handyside for these references and for allowing me to read his copies of these charters. See also D. Crouch, 'Turberville Family (*per. c.*1125–*c.*1370)', *ODNB*, 55, pp. 570–71.

[46]   Rees, *History*, p. 116.

from this region actually travelled to the Holy Land at that time.[47] Perhaps they gave to the Hospitallers instead of joining the crusade.

Welsh Donations

The Welsh were great pilgrims to the East, but generally did not join crusades: in her study of Wales and the crusades, Kathryn Hurlock has identified only a handful of Welsh people who actually joined crusades to the Holy Land.[48] William Rees identified thirteen donors to the Hospitallers in Wales and the March who had Welsh names, ten of whom were in the commandery of Slebech, one in Shropshire, and two in Gwynedd.[49] Most of these donors were men unknown elsewhere. The only certain princely donor was Rhys of Deheubarth, who supported the Hospitallers of Slebech.[50] There was a possible strategic advantage for him in fostering the Hospitallers' presence in this disputed region, but it is also possible that he specifically wished to support the Hospitallers' work helping pilgrims to the Holy Land.[51]

It is further possible that Prince Llywelyn ab Iorwerth 'the Great' of Gwynedd was a patron of the Hospitallers, as a charter survives of around 1225 in his name to 'the house of the Hospital of Jerusalem and the brothers serving God and St John there'. However, the validity of the charter is not certain: Huw Pryce, in his study of the acts of Welsh rulers 1120–1283, argued that the donation might be genuine but the style and dating clause are suspect.[52]

---

[47]   P.W. Edbury, 'Preaching the Crusade in Wales', in A. Haverkamp and H. Vollrath (eds), *England and Germany in the High Middle Ages: In Honour of Karl J. Leyser* (Oxford, 1996), pp. 221–33.

[48]   K. Hurlock, *Wales and the Crusades, c. 1095–1291* (Cardiff, 2011), pp. 214–32. I am very grateful to Dr Hurlock for her help with this point.

[49]   Anarawd ap Gruffydd (Rees, *History*, p. 107: Benegerdune, nr Minwear); Maelgwn ap Maelgwn (ibid., p. 117: Merthyr Cynlas on the Eastern Cleddau, downstream from Minwear); Owain ap Gruffudd (ibid., pp. 113, 118: land in Ceredigion at Moyl'on and at Riostoye – both unidentified); Cadwgan ap Gruffudd (ibid., p. 114: the land of Betmenon – unidentified). In Rhadnorshire: Meurig ab Adam (ibid., p. 115: the church of Llanfihangel nant Melan); at Swansea, Einon and his brother Goronwy, sons of Llywarch, gave the burgage of William fitz Palmer (ibid., p. 115); John Blaencagnel (ibid., p. 119: church of Penmaen in the Gower, and land); Gruffydd Goch or Madoc of Sutton (ibid., p. 128: the church of Kinnerley in Shropshire); Prince Llywelyn ap Iorwerth of Gwynedd and Ifan ap Rhys (ibid., p. 128); for donations by Prince Rhys of Deheubarth see the following note.

[50]   *The Acts of Welsh Rulers*, pp. 166–7; Rees, *History*, pp. 28, 112, 113; H. Pryce, 'Rhys ap Gruffudd (1131/2–1197)', *ODNB*, 46, pp. 614–16.

[51]   Gerald of Wales, 'Itinerarium Cambriae', ed. J.F. Dimock (London, 1868), in *Giraldi Cambrensis Opera*, eds J. Brewer *et al.*, RS 21 (London, 1861–91), 6, p. 15; Rees, *History*, pp. 28, 112, 113.

[52]   Ibid., p. 63; *The Acts of Welsh Rulers*, pp. 419–21; A.D. Carr, 'Llywelyn ab Iorwerth (*c.*1173–1240)', *ODNB*, 34, pp. 180–85.

The Roles of the Military Orders in Wales and the Welsh March

Military

Because they were military–religious orders, some writers have assumed that the Hospitallers and Templars must have played a role in the wars of Wales during the Middle Ages. In fact the evidence for the orders' involvement in military activity is sparse, and limited to the wars of King Edward I in Wales. On 24 May 1282, Brother Richard Poitevin, then lieutenant-commander of the Temple in England, was granted by King Edward I 'protection with clause *nolumus*, with reference to the king's army in Wales', which suggests that he was involved in military action for the king in the war against Llywelyn ap Gruffudd.[53] In December 1294 and January 1295 Brother Odo de Nevet or Ednyfed, commander of the Hospitallers' commandery of Halston, and Madoc ap Dafydd of Hendwr were reimbursed a total of £500 from the King's Wardrobe for paying a force of infantry stationed at Penllyn in Merioneth. It appears that the Hospitaller commander had been involved in putting down the Welsh rebellion in that year.[54]

The Hospitallers and Templars did not own any castles in Wales or the Welsh March, and none of their properties in the region shows any indication of having been fortified. Considering that other religious did fortify their dwellings in Wales – such as the Benedictine priory at Ewenny, in the Vale of Glamorgan, and the bishops' palaces at Lamphey and at Llawhaden in Pembrokeshire – this could be regarded as surprising. The tower of the Templars' church at Garway was originally built freestanding, separate from the church, and it is sometimes suggested that this was so that it could be used as a defensive building; but it was not unusual in that region for church towers to be freestanding. Other church towers built separately from their churches include those at Ledbury, Bosbury and Ewyas Harold (now joined to the main church) in Herefordshire, Westbury on Severn and Berkeley in Gloucestershire, and – on the other side of England – East Dereham in Norfolk.[55] While it is not clear why this was done, clearly it was not for some military purpose specific to the Templars.

---

[53]   *Calendar of the Patent Rolls preserved in the Public Record Office: Edward I, AD 1281–1292* (London, 1893), p. 24; apparently acting as lieutenant for grand commander Robert of Turville, who had gone to Scotland on 16 May (p. 20). However, he is not mentioned by J.E. Morris, *The Welsh Wars of Edward I* (Oxford, 1901), pp. 154–85.

[54]   *The Book of Prests of the King's Wardrobe for 1294–5, Presented to John Goronwy Edwards*, ed. E.B. Fryde (Oxford, 1962), pp. xx, 58, 59, 61, 186 n. 3. I am very grateful to Adam Chapman for these references. For Odo as commander of Halston, see *Calendar of Chancery Warrants Preserved in the Public Record Office, 1, 1244–1326, Prepared under the Superintendence of the Deputy Keeper of the Records*, ed. R.C. Fowler (London, 1927), p. 45. For the uprising of 1294 see Morris, pp. 240–66.

[55]   The renowned architectural historian Nicholas Pevsner wrote of Ledbury that the reason for the tower being built separately 'remains obscure' (N. Pevsner, *The Buildings of England: Herefordshire* (Harmondsworth, 1963), p. 125). His colleague David Verey

In 1338 the Hospitallers of Halston recorded that they were incurring a cost of 100 shillings a year (five pounds) giving 'gifts to various lords and their seneschals and officials for having and maintaining the Hospital's liberties and having and expediting their aid and favour and friendship'. Again, at Slebech the Hospitallers paid four pounds a year to 'two magnates of Wales, for maintaining and protecting the *bailie* from the bandits and malefactors in Welsh parts, who are fierce there: viz., forty shillings to Richard Penres and forty shillings to Stephen Perot'.[56] It is clear that the Hospitallers in Wales were not a regular military force, as in 1338 they were unable to defend themselves with weapons and had to bribe powerful officials not to attack them or to protect them. That said, Roger Turvey has pointed out that forty shillings a year would have been a welcome income to the Perot family income, as would the prestige of being appointed protector of this wealthy commandery of an influential supra-national religious order.[57]

Lordship

The Templars were not supported by the princes of Wales, but did receive generous donations from the kings of England. Their major holdings in the Welsh March were around Garway, most probably given to them by King Henry II.[58] Garway and its dependent church at St Wolstan are on the ridge that runs north-west along the east bank of the River Monnow, forming a physical barrier along the border between Wales and England. The royal castles of Skenfrith, Grosmont and White Castle are nearby. It is possible that King Henry II established the Templars here as part of his strategy of controlling this area. The Templars were close to the king – Templars were regularly at his court and acted as his advisers, as messengers, and provided financial services.[59] Henry could have located them here as men he knew he could trust, to represent royal interests in this frontier area. His major concern

---

considered that the separate towers at Berkeley and Westbury were for defensive purposes: at Berkeley '[t[he tower was placed on the N side of the churchyard to minimize danger if it should be captured' while at Westbury the tower 'was built c. 1270 as a garrison or watch–tower' (D. Verey, *The Buildings of England: Gloucestershire*, 2: *The Vale and the Forest of Dean*, ed. N. Pevsner (Harmondsworth, 1970), pp. 98, 399). None of these churches had any connection to a military religious order.

[56]   'Et in donis datis ibidem diversis dominis, et eorum senescallis, et eorum secretaries, pro libertate hospitalis habenda, et manutenenda, et eorum auxiliis, et favour, et amicitia habenda et expedienda'. *Knights Hospitallers in England*, pp. 39–40; 'Et soluto ij magnatibus Wallie, ad maintenendam et protegendam bajuliam, pro insidiatoribus et malefactoribus in partibus Wallie, qui sunt ibidem feroces; videlicet, Ricardo Penres xl s. et Stephano Perot xl *s.*'. Ibid., p. 36.

[57]   Turvey, p. 9.

[58]   Rees, *History*, p. 52; *Knights Hospitallers in England*, p. 197.

[59]   See, for example, M.L. Bulst-Thiele, 'Templer in königlichen und päpstlichen Diensten', in P. Classen and P. Scheibert (eds), *Festschrift Percy Ernst Schramm* (Wiesbaden, 1964), 1, pp. 289–308.

in this region would not have been the Welsh princes, whose powerbases were more to the west and north, but rather the Anglo/Cymro Norman marcher lords such as the de Clares, the de Lacys and the de Braoses.

This close relationship between the kings of England and the Templars – which until the 1250s was much closer than their relationship with the Hospitallers[60] – may cast light on why the Templars received so few gifts within Wales itself, and none in Welsh Wales, while the Hospitallers were more generously endowed. Because the Templars were servants of the kings of England, a gift to the Templars was – whether the donor wished it or not – effectively a gift in support of English royal authority. Neither the Welsh nor the Anglo/Cymro-Normans would have wished to give the king of England any more power than necessary within Wales. Those who wished to support the crusade therefore gave to the Hospitallers rather than to the Templars.

Hospitality

Another reason for choosing the Hospitallers in preference to the Templars was the Order's vocation. The Hospitallers were established to care for the poor and sick, and while the original establishment was specifically for the poor and sick in Jerusalem, Grand Prior Philip de Thame's report of 1338 reiterates that the houses of the Hospital in England and Wales were bound to help the poor, 'per ordinationem fundatoris dictorum locorum ex antiquo' (through arrangement by the founder of the said places, from antiquity), or 'prout ordinatum est per fundatores domus' (as was ordained by the founders of the house).[61] So a patron who set up a house of the Order of St John in Wales might expect to be establishing a hospice for poor travellers and the sick (Fig. 16.2). The Templars had no such obligation, although the report of 1338 mentions that eight former Templar houses did give hospitality, one of which was Garway, which in particular cared for many from Wales. Unlike other houses, there is no mention of the expense of caring for these travellers' horses, so presumably the Welsh travelled on foot. At Willoughton and Bruer in Lincolnshire, the obligation to supply hospitality was 'prout fundatores domus ordinaverunt' (as the founders of the house ordained), but not elsewhere.[62]

---

[60]   H. Nicholson, 'The Military Orders and the Kings of England in the Twelfth and Thirteenth Centuries', in A.V. Murray (ed.), *From Clermont to Jerusalem: the Crusades and Crusader Societies, 1095–1500* (Turnhout, 1998), pp. 203–18.

[61]   *Knights Hospitallers in England*, pp. 5, 8, 12, 14, 15, 18, 22, 25, 28, 39, 36, 43, 47, 50, 61, 76. Stanton in Hertfordshire may also have lodged travellers, as a possible hospice has been identified. R. Gilchrist, *Contemplation and Action: The Other Monasticism* (London, 1995), pp. 90–3; *Victoria County History, Hertfordshire*, ed. W. Page, 3 (London, 1912), p. 349.

[62]   *Knights Hospitallers in England*, pp. 137, 149 (Willoughton), 155 (Bruer), 158, 164, 186, 192, 198 (Garway).

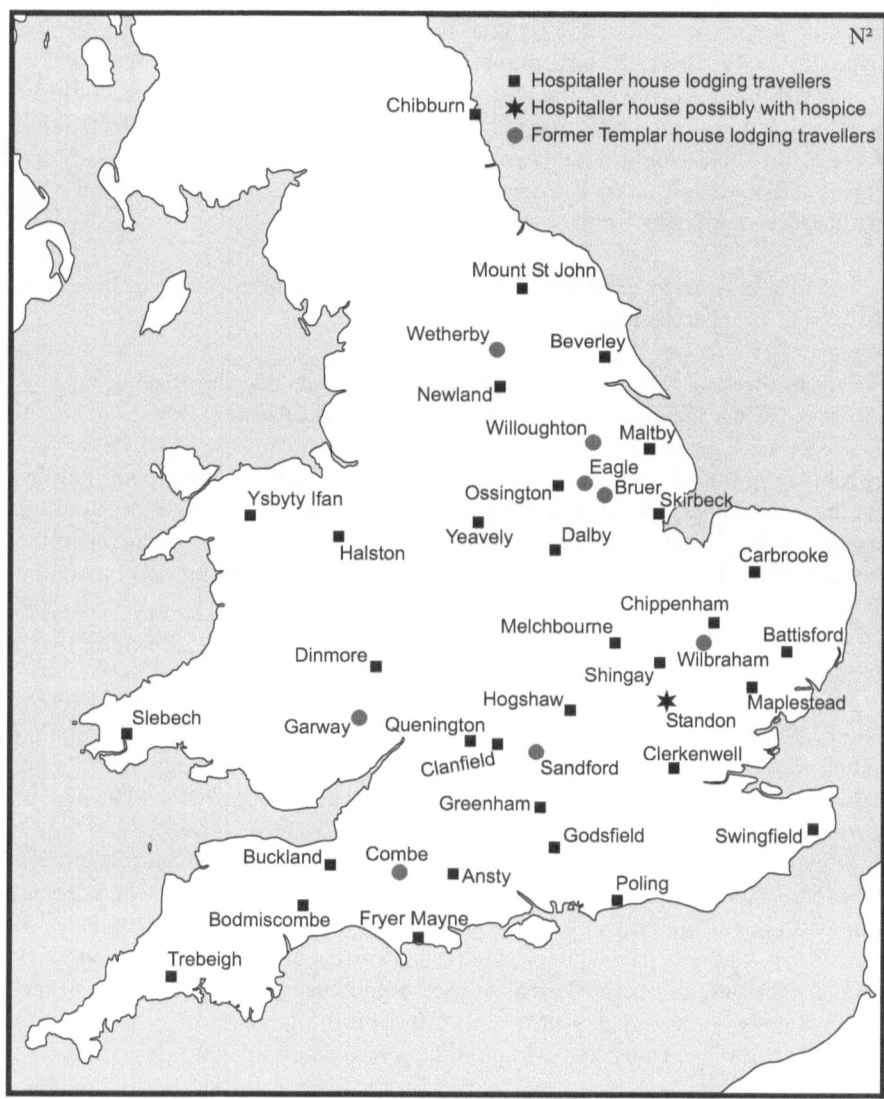

Figure 16.2   Hospitaller houses in England and Wales that lodged travellers in 1338, according to Philip de Thame's report (graphics: Helen J. Nicholson and N² Productions)

So, while a patron setting up a Templar house might expect the brothers to give hospitality to travellers, the obligation was not as strong as in the Hospital.

The Hospital received several gifts of small properties on roads where the most obvious function of the Order would be to provide hospitality for travellers. Some of these travellers would have been pilgrims on their way to St Davids, Strata Florida and other holy sites. The complaints of the Hospitallers of Slebech in 1338 that they were suffering from

> pluribus aliis supervenientibus de Wallia, qui multum confluunt de die in diem, et sunt magni devastatores, et sunt inponderosi.[63]

> (many others coming over from Wales, who rush in every day, and devastate the place, so that the expense that they cause cannot be calculated),

probably refer not to Welsh raiders, but pilgrims to St Davids.[64] The report implies that the commandery at Slebech had to cater for enormous crowds arriving at the house on a daily basis, eating the brothers out of house and home, taking anything that they needed on the excuse of taking charity, and disturbing the administration of the house and the running of its estates. For this reason, I have suggested elsewhere that the pilgrims were lodged on the south bank of the river, away from the main house.[65]

Pilgrims also travelled to the Order's houses on pilgrimage. The fifteenth-century Welsh poet Lewys Glyn Cothi (or Llywelyn y Glyn) wrote a poem stating that he was going to the house or Hospital of St John on pilgrimage, mentioning Slebech by name.[66] Dafydd Nanmor, writing in around 1460, cited the Hospital of St John as an example of generous hospitality.[67]

Other sites held by the Hospitallers in Welsh Wales may have cared for travellers, although precise evidence is lacking. Ystradmeurig in Ceredigion, donated to the Hospitallers by Roger de Clare and confirmed by Prince Rhys of Deheubarth, was only five kilometres west-south-west of the Cistercian monastery and pilgrim centre of Strata Florida, and so could have acted as a pilgrim hospice. Gwanas, now in Gwynedd, North Wales, lies on the junction of two important routes through the Cambrian Mountains, where it could have functioned as

---

[63]   Ibid., p. 35.

[64]   Rees, *History*, p. 33.

[65]   Nicholson, 'Sisters' House'.

[66]   Lewys Glyn Cothi, 'Mal Enlli amla' unllwybr', in *Gwaith Lewys Glyn Cothi*, ed. D. Johnston (Cardiff, 1955), no. 91, p. 208, and see also 'Am eni sacarïas' (in praise of St John the Baptist, patron of the Hospital of St John), in ibid., no. 2, pp. 17–18; cited in G. Hartwell Jones, *Celtic Britain and the Pilgrim Movement* (London, 1912), pp. 138–9.

[67]   'Megis ysbytau Ieuan / Yw i dai o fwyd i wan': Dafydd Nanmor, 'Rhys orau 'nhir Is Aeron', line 5, in T. Roberts and I. Williams (eds), *The Poetical Works of Dafydd Nanmor* (Cardiff, 1923), p. 1, and p. 122, n. on line 5; lines quoted in Evans, 'Ysbytty Ifan', p. 109.

a hospice for travellers. It is mentioned in Philip de Thame's report of 1338 as contributing forty shillings a year to the incomes of Halston commandery, but the report gives no further details of its function.[68] After 1338 no other sources mentioned it until the dissolution of the monasteries in England and Wales.[69]

Ysbyty Ifan ('St John's Hospital'), originally Dolgynwal in Gwynedd, comprised a hospice and church near a ford on the River Conwy on an important route from north to west Wales. Philip de Thame's record in 1338 that the Hospitallers at Halston and Dolgynwal had to bribe various local lords and officials 'to have their aid, and favour, and friendship', indicates that this was a dangerous area, where travellers would have valued a safe place to stay. During the revolt of Owain Glyndwr, the hospice was burned down and never rebuilt, but the location continued to be a place of refuge, including – according to local complaint in the fifteenth and sixteenth centuries – for bandits.[70]

William Rees and other scholars have also suggested that Ysbyty Cynfyn in Ceredigion belonged to the Hospitallers and lodged travellers, specifically pilgrims en route south to Strata Florida, but there is no documentary evidence of this.[71]

The Significance of the Military Orders' Welsh Properties to the Orders

Some of the churches that were entrusted to the Hospitallers and Templars, such as Slebech and Garway, became the centre of a commandery while also remaining the parish church. The brothers' responsibility was to keep the church in good repair and to appoint and support the priest. They could receive all the spiritual dues from the parish – in particular the tithes – that in 1338 were an important source of income for the Hospitallers' commanderies. Slebech, for example, received over £200 after expenses that year from its churches, their lands and associated revenues.[72]

However, the Hospitallers, with their interests extending across Christendom, did not always give priority to local parish responsibilities. In the late fourteenth century, when the Order's resources were focused on campaigns against the Ottoman Turks in the Balkans and Asia Minor, the bishop of St Davids noted that the Hospitallers' church of Llanrhidian in Gower was in such poor repair that 'for

---

[68]   *Knights Hospitallers in England*, p. 38.

[69]   Rees, *History*, pp. 66, 128 (quoting *Knights Hospitallers in England* and *Valor Ecclesiasticus*, 4, p. 455).

[70]   Davies, 'Church of St John the Baptist'; Evans, 'Yspytty Ifan'; Rees, *History*, pp. 63–7, 128.

[71]   Knowles and Hadcock, pp. 339, 409, citing W. Rees, [Ordnance Survey map of] *South Wales and the Border in the Fourteenth Century*, N.W. Sheet (Southampton, 1932). Rees does not repeat this suggestion in his *History*. I am very grateful to Paul Sambrook for his help with this point.

[72]   *Knights Hospitallers in England*, pp. 34–6.

a long time it had not been possible for the divine offices to be celebrated with due honour'. The bishop complained that at a previous visitation he had sternly warned the commander of the Hospital at Slebech to make the repairs, but as this had not been done, the bishop sequestrated the revenues of the rectory.[73] In 1397 the parishioners at Garway complained that although their priest worked hard, serving both Garway and Wormbridge with daily services, he could not perform his office properly because he could not speak Welsh and most of his parishioners could not speak English.[74]

> Item quod Ricardus, capellanus ibidem, celebrat bis in die, viz. hic apud Garwy et alibi apud Wormbrugge, et recipit duplex salarium. Item quod idem dominus est inhabilis ad gerendum curam animarum ibidem, quia nescit linguam Wallicanam et quia plures parochiani ibidem nesciunt linguam Anglicam.[75]

As the priest would have been appointed by the grand prior and his chapter at the Hospital's major English house at Clerkenwell near London, presumably they had not realized the importance of installing a Welsh-speaking priest.

It is interesting to note which brothers were promoted to being commanders of the Hospitallers' and Templars' houses in Wales and the March, as this gives some indication of the status of these houses within the orders as a whole. In particular, Slebech became a commandery held by Hospitaller brothers who were headed for high office in the Order, or who already held high office. Philip de Thame held the commanderies of Slebech and Sandford after he gave up the grand priory of England, until his death in 1358.[76] In 1338 John Frowyck was commander of Slebech; he was prior of Ireland 1356–59.[77] In 1358 Robert Hales, *socius* or personal aide of the grand master of the Order on Rhodes, was made commander of Slebech, an office he held until his death in 1381. After 1365 he also held Upleadon and Cardington. From 1372 he was grand prior of England.[78] John Weston was commander of Slebech in 1476, and continued to hold it after he

---

[73]   Williams, *Welsh Church*, p. 167, quoting *The Episcopal Registers of the Diocese of St Davids, 1397–1578*, ed. R. F. Isaacson, Cymru Record Series, 6 (1917–20), 1, pp. 174–6, 269.

[74]   L.B. Smith, 'A View from an Ecclesiastical Court: Mobility and Marriage in a Border Society at the End of the Middle Ages', in R.R. Davies and G.H. Jenkins (eds), *From Medieval to Modern Wales: Historical Essays in honour of Kenneth O. Morgan and Ralph A. Griffiths* (Cardiff, 2004), pp. 64–80, here p. 72, citing A.T. Bannister, 'Visitation Returns of the Diocese of Hereford in 1397', *English Historical Review*, 44 (1929), 279–89, 444–53 and 45 (1930), 92–101, 444–63, here 44, p. 289.

[75]   Ibid., p. 289.

[76]   NLM, AOM, 316, fols 199r–v.

[77]   *Knights Hospitallers in England*, p. 37; C.L. Tipton, 'The Irish Hospitallers during the Great Schism', *Proceedings of the Royal Irish Academy C*, 69 (1970), 38.

[78]   NLM, AOM, 316, fols 199r–v, 200r–v, 201r, 202r; NLM, AOM 319 fols 175r, 176r–v, 177r–178r, 179r; NLM, AOM 321, fols 145r–v.

became prior of England in 1477;[79] from 1485 until his death in 1513 it was held by Robert Evers, who was prior of Ireland from 1497.[80] From 1514 to 1540 it was held by Clement West, who became turcopolier of the Order in 1531, the highest office open to a brother of the English tongue.[81]

Conclusion

While an obvious motivation for donating to the military orders in Wales was to support the crusade and pilgrims to the Holy Land, it appears that donations were also made with regard to local patterns of power and authority, as donors within Wales – both the Anglo-Norman marcher lords and the Welsh – preferred not to give to the Templars, who were very close to the king of England, but gave instead to the more independent Hospitallers. The donations in Ceredigion by both the Anglo-Norman de Clare family and Prince Rhys combined piety with a strategic aim: the need to establish reliable men, on whose support they could rely, in a disputed area. There was also an immediate charitable aspect to donations: the Hospitallers were originally set up to care for Christian travellers to the Holy Land, and to judge from the report of 1338 they also took on this duty in England and Wales. So they were a suitable order for running hospices for travellers in remote areas. The Templars, in contrast, when they were given land in remote areas, did not develop it and sometimes gave it away. They did, however, lodge travellers at Garway on the Welsh/English border.

The military orders do not appear to have had a great deal of active power in Wales; in 1338 the Hospitallers complained about bandits and being imposed upon by numerous travellers, and having to pay protection money to local lords. On the other hand, they were clearly valued by those travellers and local lords. The Hospitallers' houses in Wales and the Welsh March were important to their Order, producing significant income towards the Order's work in the East. It is notable that, at least from the fourteenth century, the commandery of Slebech – the richest commandery in Wales – was usually entrusted to a brother who was headed for high office within the Hospital. Hence, although Wales and the Welsh March were geographically remote from the Hospital's centre of operations in the eastern Mediterranean, they were significant within the *langue* of England.

---

[79] O'Malley, pp. 45, 46.
[80] Ibid., pp. 50, 173, 311, 349.
[81] Ibid., p. 50.

Chapter 17

The Military Orders at the Court of King John[*]

Paul Webster

In the years either side of 1200, the military orders, in particular the Knights Templar, are considered to have occupied a place of considerable importance at the royal courts of medieval Europe, not least at that of the Plantagenet kings of England. Members of the orders served as close advisers to these kings, fulfilled administrative roles, acted as custodians of royal wealth, lent money, carried messages on the king's behalf, directed royal shipping, and served as royal almoners, to name just some of their roles.[1] It can reasonably be assumed that both sides benefited. The Plantagenets secured the services of an efficient body of men prepared to support their government. In return, the orders hoped to benefit from the opportunity both to pursue their interests without undue royal interference and to look for potential rewards, thereby contributing to their efforts to support the defence or recovery of Christian territory in the Holy Land.

This is how the relationship ought to have worked during the reign of a ruler who was friendly towards the church. This article, however, will consider the interaction between the crown and the military orders in the case of a king generally considered to have been an enemy of the organized religious, who has been described as being 'about as irreligious as it was possible for a man of his time to be': John, King of England from 1199 to 1216.[2] John's reign is defined by a series of interrelated crises. By 1204 large swathes of his inheritance in western France had been lost to the French king, Philip Augustus. John ruthlessly exploited his subjects in order to raise revenue to pay for attempts to win back these lands. His campaigns failed, most significantly when his army refused to fight at La-Roche-aux-Moines in Anjou in early July 1214 and when his allies were routed by Philip II at the Battle of Bouvines a few weeks later. The exploitation

[*]   I am grateful to Helen Nicholson and the participants at the 2009 conference, in particular Michael Heslop and Simon Phillips, for their helpful advice, questions and suggestions.

[1]   H. Nicholson, 'The Military Orders and the Kings of England in the Twelfth and Thirteenth Centuries', in A.V. Murray (ed.), *From Clermont to Jerusalem: The Crusades and Crusader Societies 1095–1500* (Turnhout, 1998), pp. 203–18; eadem, *The Knights Templar: A New History* (Stroud, 2001), pp. 160–80.

[2]   S. Painter, *The Reign of King John* (Baltimore, MD, 1949), p. 238.

during the years leading up to these defeats provoked the barons of England to combine against their king in 1215 and impose the celebrated charter of liberties, Magna Carta, which at the time provoked a civil war. Most critically in terms of John's posthumous reputation, the king fell into dispute with the church over the appointment of a successor to Hubert Walter as archbishop of Canterbury following the latter's death in 1205. John's intransigence resulted in the imposition of an interdict on England in 1207, which was to last until 1214, and the personal excommunication of the king between 1209 and 1213.

It might reasonably be assumed that established religious orders were therefore wary of cooperating with John or of becoming involved in government during his reign. However, the military orders certainly participated in royal service in this period and were prominent between c. 1212 and the conclusion of the reign, a period in which John lurched from one crisis to another. Royal allies were frequently thin on the ground, but the king regularly turned to the orders for support and men willing to carry out his instructions.

It is notable that, despite the scarcity of sources for the period from 1209 to mid-1212, the surviving evidence indicates that, during the interdict and the king's excommunication, the military orders did not distance themselves from the royal court. Prior to this, in 1207, the Templars and Hospitallers had already attracted papal attention for abusing their privileges during interdicts, effectively nullifying the effects of the sentence.[3] Innocent III clearly stated that the interdict on England was to apply to the military orders.[4] Yet following the royal seizure of church property in March 1208, the orders supported the king rather than the exiled archbishop and the pope. Within a week, the Hospitallers' possessions were restored.[5] Templar property passed to the king's justiciar, Geoffrey fitz Peter, an advocate of reconciliation between king and church.[6] It seems that restoration soon followed, with Geoffrey allowing the Templars to have their lands, perhaps on payment of a fine.[7] As there is no evidence for leading Templars joining the churchmen in exile, they are unlikely to have been the object of royal reprisals against the church for a

---

[3]   H. Nicholson, *Templars, Hospitallers and Teutonic Knights. Images of the Military Orders 1128–1291* (Leicester, 1993), p. 29; eadem, *The Knights Templar*, p. 179.

[4]   *Selected Letters of Pope Innocent III concerning England (1198–1216)*, eds C.R. Cheney and W.H. Semple (London, 1953), p. 96 (no. 31).

[5]   *Rotuli Litterarum Clausarum in Turri Londinensi asservati*, 1, 1204–1224, ed. T.D. Hardy (London, 1833), p. 108a.

[6]   Ibid., p. 110a; C.R. Cheney, 'King John's Reaction to the Interdict on England', *Transactions of the Royal Historical Society*, 4th ser. 31 (1949), 134; R.V. Turner, *Men Raised from the Dust. Administrative Service and Upward Mobility in Angevin England* (Philadelphia, PA, 1988), pp. 53–4; F.J. West, *The Justiciarship in England 1066–1232* (Cambridge, 1966), p. 133.

[7]   H. Nicholson, 'Aimery de Saint-Maur', in N. Bériou and P. Josserand (eds), *Prier et Combattre: Dictionnaire européen des ordres militaires au moyen âge* (Paris, 2009), p. 52. I am grateful to Helen Nicholson for supplying me with a copy of this entry ahead of publication.

sustained period. However, this did not mean that the military orders were immune from royal exactions, such as the heavy tallage of 1210.[8]

John's ability to withstand the sentences of interdict and excommunication depended on maintaining the support of the English barons. By mid-1212 his position seemed suddenly under threat when a proposed campaign in Wales was hastily abandoned amid rumoured plots to murder the king or hand him over to his enemies. Meanwhile, a threatened papal sentence of deposition raised the spectre of an invasion of England by Philip II of France.[9] The military orders responded to John's need for allies, playing an important role in negotiating and implementing a settlement with the church. One of the papal envoys to England, Durand, was a Templar, and military order houses and officials appear regularly in the sources relating to the negotiations.[10] In 1213, when John sought peace, his envoys included the Templar Alan Martel, who had served as preceptor of the New Temple in London in the early 1200s, and who was to become grand commander in England in 1219, and the Hospitaller William of St Ouen. These two, along with the abbot of Beaulieu, John's Cistercian foundation in Hampshire, were the only three of the six envoys to reach the papal curia, the remainder of the deputation having been taken captive in transit. John paid £500 each to the masters of the Temple and the Hospital to arrange for the ransoming of those captured.[11] Thereafter Alan and William were regular travellers on the business of the dispute, each returning to Rome on at least one further occasion, and William acting as messenger to the archbishop of Canterbury and the English bishops in exile.[12] In March 1214, while John was campaigning in Poitou, William of St Ouen delivered papal letters outlining the terms under which the interdict was to be lifted.[13]

In England, Temple Ewell, near Dover, provided the setting for negotiations concerning the resolution of the interdict. The papal envoy Pandulph was summoned to Dover from the continent having initially sent Templar messengers to John. At Ewell, on 15 May 1213, John issued a letter addressing some of the major elements on which the dispute hinged. This informed Stephen Langton, who was now finally to be accepted as archbishop of Canterbury, that the king had revoked the outlawry imposed on clerics. John acknowledged that he did

---

[8]   'Annales Monasterii de Waverleia', ed. H.R. Luard, *Annales Monastici*, RS 36 (London, 1864–69), 2, p. 264; 'Annales Monasterii de Bermundesia', ed. Luard, *Annales Monastici*, 3, pp. 451–2; *Rogeri de Wendover liber qui dicitur Flores Historiarum ab anno domini MCLIV annoque Henrici Anglorum Regis Secundi Primo*, ed. H.G. Hewlett, RS 84 (London, 1886–89), 2, p. 57.

[9]   W.L. Warren, *King John* (London, 1974), pp. 199–205.

[10]   For Durand, see *Selected Letters of Pope Innocent III*, p. 125 (no. 43), n. 1.

[11]   Ibid., pp. 130–31 (no. 45).

[12]   *Rotuli de Liberate ac de Misis et Praestitis regnante Johanne*, ed. T.D. Hardy (London, 1844), pp. 260, 264; *Rotuli Litterarum Patentium in Turri Londinensi asservati*, 1, pt 1, 1199–1216, ed. T.D. Hardy (London, 1835), p. 123b.

[13]   Ibid., p. 111b.

not have the authority to impose such a sanction.[14] Another major element of the submission, the surrender of England and Ireland to become fiefs of the papacy, was initially set out in a charter issued at Ewell on the same day. This document was later reissued, sealed with a golden bull, on 3 October 1213 at St Paul's, London, a day when the king's presence was also recorded at the New Temple. It referred to John's need for divine mercy, and noted his performance of liege homage to the pope at the hands of the papal legate, Nicholas cardinal bishop of Tusculum.[15] In addition, on 24 May 1213, John issued letters to Stephen Langton and the exiled bishops confirming that they would be allowed to exercise their offices without royal interference. The archbishop of Dublin, the bishops of Winchester and Norwich, and twelve of John's barons were named as guarantors.[16]

These events established the basis for a settlement. As the terms were gradually fulfilled or renegotiated, the military orders remained involved. Aimery de Saint-Maur, the Templar grand commander in England, provided a gold mark for John's offering on 20 July 1213, when the king was absolved from excommunication at Winchester cathedral by Archbishop Langton. In return, Aimery was reimbursed with the sum of nine silver marks.[17] The orders were also linked to the restoration of church property seized by the king. In October 1213, John was at the New Temple in London when he gave instructions for the restoration of some of the archiepiscopal lands and property.[18] Later that month, Pope Innocent III proposed that unclaimed church property should be requisitioned in aid of the Holy Land, and deposited in houses of the Templars and Hospitallers.[19] John also involved the two orders in handling payments of the tribute promised to the pope as part of the settlement.[20] In addition, in the winter of 1214–15 grants were issued from the New Temple to Archbishop Langton and the formerly exiled bishops. Although only one of these, to Hugh of Wells, bishop of Lincoln, referred to compensation for damages incurred during the interdict, there is a sense that the king was actively seeking to regain the loyalty of leading churchmen.[21] Meanwhile, the papal legate,

---

[14]   Ibid., p. 100a.

[15]   *Rogeri de Wendover*, 2, pp. 74–6; *The Letters of Pope Innocent III (1108–1216) concerning England and Wales: A Calendar with an Appendix of Texts*, eds C.R. and M.G. Cheney (Oxford, 1967), pp. 156 (no. 941), 160 (no. 962); *Rotuli Chartarum in Turri Londinensi asservati*, 1, pt 1, 1199–1216, ed. T.D. Hardy (London, 1837), p. 195a–b; *Rotuli Litterarum Clausarum*, pp. 153b–154a.

[16]   *Rotuli Litterarum Patentium*, pp. 98b–99a.

[17]   *Rotuli Litterarum Clausarum*, p. 148b.

[18]   Ibid., p. 152b.

[19]   *Selected Letters of Pope Innocent III*, pp. 159–60 (no. 58).

[20]   J. Piquet, *Les Templiers. Etude de leurs operations financiers* (Paris, 1939), p. 227; *Rotuli Litterarum Patentium*, p. 107a.

[21]   *Rotuli Litterarum Clausarum*, pp. 178b–179b, 182b, 187a; *Rotuli Litterarum Patentium*, p. 124a; *Rotuli Chartarum*, pp. 202b–204b; *The Cartae Antiquae Rolls 1–10*, ed. L. Landon, Pipe Roll Society 55, n.s. 17 (London, 1939), p. 4 (no. 11).

Nicholas of Tusculum, was provided with hospitality by both the Templars and Hospitallers for which they were reimbursed by the king.[22]

The other crises of the final years of John's reign follow a similar pattern: when the king needed support, the military orders responded. The continental campaign of 1214 was launched despite growing opposition in England. Yet the Templar Alan Martel accompanied the king to western France. For much of the expedition he was among those responsible for issuing royal letters, and he received orders concerning the queen and the royal children.[23] In September 1214, he was chosen as messenger along with other trusted advisors to the king, namely William earl Ferrers of Derby, Hubert de Burgh (who was destined for a long career in royal service in the reign of Henry III), and Falkes de Bréauté. These men were sent to Robert de Vieuxpont with verbal instructions relating to the king's nephew, the emperor Otto, and probably also concerning the failure of the continental campaign and John's intention that it should be abandoned.[24] At the same time, William Cadel, the Templar commander in the West, who was involved in negotiating peace between John and Philip II of France, was granted a safe conduct to go to England and return.[25] King John was back in England a month later.

During this campaign, preceptories in Aquitaine provided the king with administrative assistance. The Templars in Poitou were charged with the oversight of pensions established to maintain the loyalty of around twenty knights, including the Poitevin nobleman Ralph de Lusignan, count by marriage of the northern French province of Eu.[26] Ralph was exactly the sort of man whose loyalty, or at least neutrality, John had to secure in order to stand any chance of success. During this period, the Templars in Aquitaine regularly served as royal agents, notably Gerard Brochard, the grand commander in Aquitaine, and Brother Boes, preceptor of La Rochelle.[27] When John marched north into Anjou, he immediately took the Templars of the region into his protection, perhaps envisaging that they would perform a similar role on his behalf.[28]

The military orders also contributed to John's overseas expedition of 1214 through their diplomatic role. Their international structure was suited to aiding the construction of royal alliances such as that with the emperor Otto and the regional lords of northern France and the Low Countries. Robert, prior of the Hospitallers in

[22]   *Rotuli Litterarum Clausarum*, pp. 175b–176a.

[23]   Ibid., pp. 140b, 142b, 143a, 163b; *Rotuli Litterarum Patentium*, pp. 113b, 117a, 119a.

[24]   *Rotuli Litterarum Clausarum*, p. 172a; *Rotuli Litterarum Patentium*, p. 121b.

[25]   Ibid., p. 121b.

[26]   *Rotuli Chartarum*, p. 208b; *Rotuli Litterarum Patentium*, pp. 116b, 121b; A. Sandys, 'The Financial and Administrative Importance of the London Temple in the Thirteenth Century', in A.G. Little and F.M. Powicke (eds), *Essays in Medieval History Presented to Thomas Frederick Tout* (Manchester, 1925), p. 156.

[27]   *Rotuli Litterarum Patentium*, pp. 112b, 113b, 116b, 119a, 121b.

[28]   Ibid., p. 117a.

England, represented John in his negotiations with the emperor in 1209.[29] Ultimately such links provided the basis for the anti-Capetian coalition that underpinned the campaign that culminated in defeat at Bouvines. In March 1213 a pact of mutual assistance between John and Count William of Holland was finalized at the New Temple. Two months later, at Temple Ewell, the king received messengers from Count Ferrand of Flanders who sought a similar alliance.[30] John placed 20,000 marks at the New Temple for the use of his envoys to the Low Countries, John fitz Hugh and Falkes de Bréauté, in the summer of 1213.[31] Following the king's defeat, the Templar Hugh de Swaby was among those sent to recover money owed by the towns of the Low Countries, including a sum that was in turn to be paid to the Templar William Cadel.[32] The Templars also acted as peace-brokers. In September 1215, senior members of the order, including William Cadel, the commander in the West, Gerard Brochard, grand commander in Aquitaine, and Alan Martel, the future grand commander in England, were among those thanked by King John for negotiating a permanent truce with Philip II of France.[33]

The military orders continued to support John following his return to England, when large numbers of the English barons turned against him. In the months leading up to the agreement of Magna Carta, the king was a regular visitor to the New Temple in London, issuing a stream of documents when he did so. For example, in January 1215 he reissued an important charter granting freedom of election to the church in England.[34] Later, in May, he confirmed the liberties of the citizens of London, allowed them the right to elect their own mayor, and offered to put all grievances to arbitration, with the rebel barons being judged by their peers according to the law of the land.[35] These letters were to no avail, as the Londoners opened the city's gates to the king's opponents little more than a week later, paving the way for the issuing of Magna Carta. Aimery de Saint-Maur, the Templar grand commander in England, remained at the king's side in this period and was named as one of the advisers who counselled John to agree to Magna Carta in June 1215.[36]

---

[29]   Ibid., pp. 91b–92a; Piquet, p. 227.

[30]   *Rotuli Chartarum*, pp. 190b–191a; *Rotuli Litterarum Patentium*, p. 99a; Warren, p. 204.

[31]   *Rotuli Litterarum Clausarum*, p. 136b; J.E.A. Jolliffe, 'The Chamber and the Castle Treasures under King John', in R.W. Hunt, W.A. Pantin and R.W. Southern (eds), *Studies in Medieval History Presented to Frederick Maurice Powicke* (Oxford, 1948), p. 133.

[32]   *Rotuli Litterarum Patentium*, pp. 122b–124a.

[33]   Ibid., p. 140b. On William Cadel, see J. Burgtorf, *The Central Convent of Hospitallers and Templars. History, Organisation, and Personnel (1099/1120–1310)* (Leiden, 2008), pp. 672–5.

[34]   *Foedera, Conventiones, Litterae et cujuscunque generis Acta Publica inter Reges Angliae et alios quosuis imperatores, reges, pontifices, principes, vel communitates ab ingressu Gulielmi I in Angliam, A.D. 1066 ad nostra usque tempora habita aut tractate*, ed. T. Rymer, new edn, eds A. Clark and F. Holbrooke (London, 1816–69), 1, pt 1, pp. 126–7.

[35]   *Rotuli Chartarum*, pp. 207a, 209b; Warren, p. 234.

[36]   J.C. Holt, *Magna Carta*, 2nd edn (Cambridge, 1992), p. 448.

Although Magna Carta is of great historical significance, in 1215 it sparked a civil war in which the king's opponents invited Prince Louis of France to claim the English throne. Once again, King John called on his allies from the military orders. Roger the Templar oversaw royal shipping in this period. During the siege of Rochester between October and December 1215 he received a stream of orders relating to boats carrying men and supplies.[37] John's pressing need for mercenary troops is evident from the letters directed to Roger. This was another area in which the Templars came to John's aid. Even before Magna Carta was agreed, the Templars had provided money for Poitevin knights coming to England.[38] In 1216 the Templar grand commander in Aquitaine, Gerard Brochard, lent substantial sums to pay for mercenaries, money that was only repaid during the minority of Henry III.[39] Meanwhile Alan Martel remained active at the heart of government.[40] Aimery de Saint-Maur also remained loyal, and although it is unclear whether or not he was present at John's deathbed in October 1216, he was certainly listed among those who were to enact the terms of the king's final testament, issued in the last days of his life. In this document, John lavished praise on those appointed to oversee his wishes, men 'without whose counsel, even in good health, I would have by no means arranged my testament'.[41]

The regular presence of major figures from the ranks of the military orders among John's closest advisers raises the question of the nature of their involvement with the crown. Was this relationship based on the priorities of the orders themselves, or on personal links and ties of loyalty? This theoretical distinction is difficult to substantiate in practice. Royal letters were likely to be directed to high status individuals within the orders charged with carrying out the king's instructions. The recurrence of named individuals, such as Aimery de Saint-Maur, Alan Martel and Gerard Brochard, indicates those who held authority and to whom the king turned when he required the orders' involvement in government. Such interaction may have generated personal links, but these were not necessarily the initial stimulus. The issue of why John turned to the military orders is important here. He relied on their loyalty in times of crisis. Perhaps their international status, in which they had a place at princely courts across Europe, gave them something of a position of neutrality as agents of royal business, meaning that a Templar or Hospitaller was more likely to get through. This would help explain the role of the orders in

---

[37] *Rotuli Litterarum Clausarum*, pp. 230a–242b; Nicholson, 'The Military Orders and the Kings of England', pp. 211–12.

[38] *Rotuli Litterarum Clausarum*, pp. 194a, 198b, 221b.

[39] Sandys, p. 157. See also T.M. Parker, *The Knights Templar in England* (Tucson, 1963), pp. 66–7.

[40] *Rotuli Litterarum Clausarum*, pp. 231a, 240b, 261b–263a, 275b.

[41] S.D. Church, 'King John's Testament and the Last Days of his Reign', *English Historical Review*, 125 (2010), 515–18, 525–6.

John's diplomacy, and in collecting and transporting royal revenues and travelling between centres of royalist support in the civil war of the final years of the reign.[42]

While John could count on the military orders, there is also the question of what they expected in return. What were the rewards of serving the crown? In October 1214 Alan Martel, the Templar, was given the vill of Tolleshunt in Essex 'to sustain him in our service'. This was to be held by Alan himself, not the Templar Order, indicating the importance the king attached to providing specifically for Martel.[43] If anything happened to Alan, or if the king revoked the grant, the Templars were to be free of dues relating to grain and any items they had deposited there.[44] Similarly, in 1215 a land grant to the Templars specifically mentioned the king's regard for Alan Martel.[45] There is a sense in which he was treated first and foremost as a royal official and that his status as a Templar was of secondary importance to the king. However, this argument should not be taken too far. Alan's access to the higher reaches of government was surely useful to his Order, and in 1219 he succeeded Aimery de Saint-Maur as grand commander of the Templars in England.

Overall, the question of how extensively the military orders were rewarded by King John is difficult to answer. At the beginning of the reign, the king confirmed the charters of the Templars, the Hospitallers, and the leper Order of St Lazarus of Jerusalem.[46] However, that was not unusual at the start of a new reign. As John required the Templars and Hospitallers to pay substantial sums for these confirmations,[47] they were hardly rewards or acts of religious largesse. Although there was a steady pattern of further grants or confirmations across the reign, there is little indication of any particular royal favour to the military orders, beyond the grants to Alan Martel and occasional gifts from the royal parks.[48] That said, the king was keen to ensure that both the Templars and Hospitallers enjoyed protection across his territories, and that they should continue to trade.[49] John also confirmed his father's grant allowing a Templar knight to reside at the

---

[42]   N. Vincent, *Peter des Roches: An Alien in English Politics, 1205–1238* (Cambridge, 1996), pp. 129–30; Jolliffe, p. 133.

[43]   *Rotuli Litterarum Clausarum*, p. 173b; Nicholson, 'The Military Orders and the Kings of England', p. 206.

[44]   *Rotuli Litterarum Patentium*, p. 123a.

[45]   *Rotuli Chartarum*, p. 203a–b.

[46]   Ibid., pp. 1b–4a, 13a–b, 15a–16b, 67b–68a.

[47]   *Rotuli de Oblatis et Finibus in Turri Londinensi asservati, tempore regis Johannis*, ed. T.D. Hardy, Record Commission (London, 1844), pp. 13, 26. There is no record of the Order of St Lazarus paying for the confirmation of their charters.

[48]   *Rotuli Litterarum Clausarum*, pp. 28b, 125b–126a, 141a, 153b, 176a, 262a; Nicholson, 'The Military Orders and the Kings of England', p. 208; eadem, *The Knights Templar*, p. 121.

[49]   *Rotuli Litterarum Clausarum*, pp. 148b, 169b, 178a, 214a, 263a; *Rotuli Litterarum Patentium*, pp. 25b–27a, 30a, 66b, 85a, 86a, 104b, 111, 112, 117a, 121b, 138a, 142a, 159a, 162a, 164a, 167b, 173a, 192b.

royal court.[50] All this suggests that the orders were involved in royal affairs, and maintained a presence at court, in order to ensure that they could go about their business free from interference by the king.

Finally, the involvement of the military orders in King John's affairs was not confined to the realm of high-status politics. They also played a role in the king's devotional activity. James of the Temple served as a royal chapel clerk throughout the reign, including the period of the interdict and the king's excommunication. He was called upon to perform the laudatory chant *Christus Vincit* at the royal court on important religious occasions, such as Christmas, Easter, and Whitsunday.[51] Meanwhile, in a grant made 'for the sake of piety' (*pietatis intuitu*), the Hospitallers were given a gold crown from the royal regalia, the value of which was to be used for the benefit of the poor overseas.[52] The military orders acknowledged their relationship with the king after his death. A chaplain celebrated masses for John's soul at the New Temple, while the Hospitallers were granted 3d. a day in the late king's memory by the minority government of Henry III.[53] It is important, however, not to push this evidence too far. Roger the Templar, who was involved in the administration of shipping, is also described as the king's almoner. He was certainly responsible for gift giving, on one occasion being ordered to provide the king's illegitimate son Oliver with a tun of wine.[54] However, there is no evidence of him distributing royal alms, a role performed by a number of men of religion in the course of John's reign.[55] Nor did John regard the military orders as a source of prayers for his soul, as he did in the case of monastic orders and some cathedral communities.[56] One grant seems at first glance to contradict this conclusion: that made in the final days of John's life to Margaret de Lacy, daughter of William de Briouze, for her Hospitaller foundation at Aconbury (Herefordshire).[57] Between 1208 and 1211 John had hounded Margaret's father into exile and ordered the deaths of her mother, Matilda, and her brother (another William). He had confiscated the lands of Walter de Lacy (Margaret's husband), who had supported Briouze, and only restored them in the period 1213–15. John's grant to Margaret is more likely to have been motivated by desire to atone for his actions and regain the

---

[50]   Nicholson, *Templars, Hospitallers, and Teutonic Knights*, p. 104.

[51]   *Rotuli de Liberate*, pp. 14, 93; *Rotuli Litterarum Clausarum*, pp. 4a, 26b, 34b, 51b, 62b, 71a, 82a, 85b, 99a, 183b, 222a; *Rotuli Litterarum Patentium*, p. 150a; I.D. Bent, 'The Early History of the English Chapel Royal, ca. 1066–1327' (unpublished PhD thesis, University of Cambridge, 1969), 2, pp. 42–3, 142–3.

[52]   *Rotuli Litterarum Patentium*, p. 51b.

[53]   *The Register, or Rolls, of Walter Gray, Lord Archbishop of York*, ed. J. Raine, Surtees Society 56 (Durham, 1872), p. 24 (no. 115); Nicholson, 'The Military Orders and the Kings of England', p. 207; Vincent, p. 132.

[54]   *Rotuli Litterarum Clausarum*, p. 230b.

[55]   P. Webster, 'King John's Piety, c. 1199–c. 1216' (unpublished PhD thesis, University of Cambridge, 2007), pp. 144–7.

[56]   Ibid., pp. 31–116.

[57]   *Rotuli Litterarum Patentium*, p. 199b.

goodwill of the Lacys than by any attachment to the Hospitallers. In any case the house was refounded in 1237, for Augustinian canonesses, at Margaret de Lacy's instigation following a bitter dispute.[58]

Overall, it is clear that members of the military orders, in particular the Templars, were involved in conducting royal business in a series of events that proved central to the history of John's reign. This is particularly true in its final years, when the orders were instrumental in negotiating an end to the interdict, in organizing alliances against King Philip II of France, in arranging a truce when John's continental ambitions were thwarted, in the build-up to and agreement of the celebrated charter of liberties, Magna Carta, and through the Templar grand commander in England being named as an executor of the king's will. Members of the military orders were clearly among King John's closest confidants. This was surely a two-way arrangement: they must have had enough faith in John to continue to serve him, and have reckoned that the benefits of royal protection and support outweighed the lack of major benefactions received in return.

This article has presented a snapshot of a topic for which abundant evidence survives. There is more to be said about the role of the military orders at King John's court, for instance their relationship with prominent figures there, such as William Marshal, the servant of each of the first four Plantagenet kings until his death in 1219.[59] The relationship between the military orders and the queens of England is also worthy of further attention, with the Templars involved in the affairs of Eleanor of Aquitaine, Berengaria of Navarre, Isabella of Angoulême, and also Isabella's mother, Alice de Courtney, on various occasions.[60] It is also possible that the military orders played a part in the administration of royal territory in Ireland, similar to their role in Aquitaine.[61] Nonetheless, it may be argued that King John, a hitherto demonized ruler, renowned for falling out with the church and antagonizing his barons, found little difficulty in conducting business with the military orders in his kingdom. They found it equally possible to act on royal business on the king's behalf, without any apparent fear that 'Bad' King John would turn against them.

---

[58]   H.J. Nicholson, 'Margaret de Lacy and the Hospital of St John at Aconbury, Herefordshire', *Journal of Ecclesiastical History*, 50 (1999), 1–23.

[59]   Nicholson, *Templars, Hospitallers and Teutonic Knights*, pp. 57–8; S. Painter, *William Marshal: Knight-Errant, Baron, and Regent of England* (Baltimore, MD, 1933), pp. 56, 282, 284–5; D. Crouch, *William Marshal: Knighthood, War and Chivalry, 1147–1219* (London, 2002), pp. 56, 140, 155, 214.

[60]   *Rotuli de Liberate*, pp. 8–9; *Rotuli Chartarum*, pp. 128a–b, 213b–214a, 219b; *Rotuli Litterarum Patentium*, pp. 117a, 119a, 121b; Piquet, p. 227.

[61]   *Rotuli de Oblatis*, pp. 562–4; *Rotuli Chartarum*, p. 73a; *Rotuli Litterarum Patentium*, pp. 85a, 162a, 164a, 173a; *Rotuli Litterarum Clausarum*, p. 263a.

Chapter 18

Walking a Thin Line:
Hospitaller Priors, Politics and Power
in Late Medieval England

Simon Phillips

Politics and power in the medieval world went hand in hand. Monarchs needed military strength, political skills and authority in order to maintain power, and, when any of these were lacking, they found their position threatened, and that in turn would lead a realm into turmoil. The theme of this paper is how Hospitaller priors in England responded to crisis-situations. This discussion will focus on three examples: the crisis of the 1320s, with the rebellion against the Despensers and the overthrow of Edward II; the problems during the minority of Richard II that culminated in the Peasants' Revolt of 1381; and, finally, events concerning the reigns of Henry VI and Edward IV in the context of the Wars of the Roses. Power and politics concerned the relationship not just between the prior and the king, but also between the prior, the queen and important magnates, as for example Prior Archer and the Despensers and Mortimers, Prior Hales and John of Gaunt, and Prior Langstrother and Warwick the Kingmaker. Two of the aforementioned priors lost their lives during these crises, and so one has to ask how some were able to survive, whereas others did not. It is also crucial to discuss how their successors dealt with the aftermaths of these crises.

Underneath the Archers

The first crisis concerns the kingship of Edward II. The 1320s saw the baronial revolt against the Despensers, the overthrow of Edward II by Mortimer and Isabella, followed by their demise when Edward III asserted his right to rule. It was just as volatile a period in English history as those that Hales or Langstrother had to face, yet Archer and Tibertis managed to survive the political turmoil. Let us first look at the events surrounding the rebellion against the Despensers. Resentment arose due to Hugh Despenser the Younger's appetite for acquiring lands at others lords' expense and his exclusive control of access to the king. The earl of Hereford and Roger Mortimer of Wigmore took the lead in gathering a coalition of Marcher lords who felt threatened by the Despenser acquisitions. It was joined by Thomas earl of Lancaster, who in June 1321 held a meeting at which some sixty lords swore to bring about Despenser's downfall. At the July parliament, they demanded that Despenser be banished. The king resisted, but after the mediation of the earl of Pembroke he gave way. Yet within two months Despenser was back and, with the king, was planning reprisals. Lancaster held a further assembly at Doncaster on 29 November to draft a petition against the Despensers, appealing for the king to provide a remedy. However, Lancaster's association with the Scots had turned the majority opinion against him, and, apart from the Marcher lords, he had little cooperation. Edward began military preparations against the rebels in early November, rallying support in his Welsh lands. By mid-January 1322 Mortimer had surrendered, and Hereford, Mowbray and Clifford fled to join Lancaster. Less than two months later, with the battle of Boroughbridge (Yorks) and the deaths of both Hereford and Lancaster, the rebellion was over.

What evidence is there for Prior Archer's position during this revolt? He was present at the 1321 parliament, but appears to have been acting exclusively to gain the Templars' lands.[1] The prior remained loyal to the king, as did the majority of the lords, despite a general dislike of the Despenser hold on power. He received summonses to parliament, including the session called to debate the rebellion, he carried out his duties such as arranging a subsidy to finance the Scottish war, and throughout 1322 and 1323 he continued to receive the king's protection.[2] However, it is possible that a relative played a role in the disturbances, as, on 29 December 1322, at Prior Archer's request a Thomas Archer of Tanworth (Warwicks) received a pardon for his adherence to the rebels.[3] Perhaps by way of thanks, on 3 December 1325 a John Archer alienated three *selds* in London to Prior Archer for the maintenance of a chaplain in Clerkenwell church to pray for his soul and that of his brother, Thomas, while retaining his lands in Leicestershire and Warwickshire.[4] Others in the king's favour, such as Robert Baldock, the king's clerk, also requested pardons for people with whom they were associated.[5] The fact that Mortimer and his household had lodged overnight at Clerkenwell before the July parliament in 1321 might also have cast suspicion on Prior Archer himself, although, as London had refused Mortimer entry to the city, the prior could claim it was the Hospitallers' duty to offer hospitality.[6] That Prior Archer's request for a pardon was granted indicates that the prior still had the king's favour. His presumed relative's activities did not prevent him from gaining parliamentary confirmation in 1324 of the Hospitallers' right to Templar properties, although not without some losses to

---

[1]   C[alendar of]C[lose] Rolls] 1318–1323, p. 375; Rot[uli] Parl[iamentorum] (London, 1767–77), 1, p. 401.

[2]   CCR 1318–1323, pp. 527, 679, 687; C[alendar of] P[atent] R[olls] 1321–1324, pp. 101, 218, 222.

[3]   Ibid., p. 227.

[4]   T[he] N[ational] A[rchives] C 143/184/10; CPR 1324–1327, p. 199.

[5]   CPR 1321–1324, p. 101.

[6]   Annales Paulini in Chronicles of the Reigns of Edward I and Edward II, ed. W. Stubbs, RS (London, 1882–83), 1, p. 294.

the king and other lords, including the Despensers.[7] Nevertheless, parliamentary approval for the granting of the Templars' goods to the Hospitallers in 1324 was crucial, as it was a point of reference for all future confirmations.[8]

Apart from restrictions on sending money out of the realm,[9] the rest of Edward II's personal reign was advantageous for the Hospitallers, especially in gaining Templar property, as for example in 1325 in Warwickshire and Yorkshire, and in 1326 in Hertfordshire. The prior also continued to have the king's protection.[10] Yet in common with the majority of the lords, he did not come to the king's aid when Queen Isabella invaded in September 1326, possibly because at this stage she was intent only on the removal of Despenser and not the king's overthrow. Perhaps the resolution of London to support the queen's cause also influenced the prior's decision to follow a mainstream approach.[11] In any case, on 26 October 1326 Prince Edward was proclaimed keeper of the realm, with the king accused of deserting his kingdom. By December, Edward II was in custody, although proclamations were still being made in his name. It was only at the parliament of January 1327 that Mortimer set about having Edward II replaced. This took effect on 25 January, although at this stage Edward III was firmly under Isabella and Mortimer' control.

How did Archer respond to the new regime? He complied with it. He attended the January 1327 parliament, and was rewarded with grants of protection and further positive action on the Templar goods. For example, on 22 February 1327 the steward of Gower was ordered to take all former Templar possessions that were under the control of Stephen Barrett, knight, and deliver them to Archer or his attorney.[12] Archer had been careful to keep on good terms with the Despensers, for example, safekeeping sacks of wool for them in 1325, yet when the government changed, he did not resist passing these goods to Mortimer when requested.[13] Other signs that the Hospitallers were in favour were a licence to appropriate in mortmain

---

[7]   TNA E 40/1469; ibid., 42/49.

[8]   *Rot. Parl.* 2, pp. 21, 100; ibid., 6, p. 313. C. Perkins, 'The Wealth of the Knights Templars in England and the Disposition of it after their Dissolution', *American Historical Review*, 15 (1910), 252–63; H. Nicholson, 'The Hospitallers in England, the Kings of England and Relations with Rhodes in the Fourteenth Century', *Sacra Militia: Revista di Storia degli Ordini Militari*, 2 (2001), 25–45 at p. 30.

[9]   *CCR 1323–1327*, pp. 544–5.

[10]   Ibid., pp. 301, 481, 500–501; *CPR 1324–1327*, pp. 134, 295. For further details concerning the ex-Templar lands in England, see S. Phillips, 'The Hospitallers' Acquisition of the Templar Lands in England', in J. Burgtorf, P.F. Crawford and H.J. Nicholson (eds), *The Debate on the Trial of the Templars* (Aldershot, 2010), pp. 237–46.

[11]   A. Tuck, *Crown and Nobility: England 1272–1461*, 2nd edn (Oxford, 1999), p. 73.

[12]   *CPR 1327–1330*, 1, pp. 42, 114; *CCR 1327–1330*, p. 61.

[13]   *CCR 1323–1327*, pp. 490–91; *Calendar of Memoranda Rolls (Exchequer) preserved in the Public Record Office, Michaelmas 1326–Michaelmas 1327*, ed. R.E. Latham, RS 201 (London, 1968), p. 70.

the church of Brompton Bryan (Hereford),[14] the swift setting up of commissions of oyer and terminer to deal with attacks on Hospitaller servants and property,[15] and the appointment of the prior of Ireland to the chancery of Ireland.[16] This was all before Leonard de Tibertis (prior of Venice, procurator general of the grand master in the West, and future prior of England) arrived around 23 July 1327, when he was granted protection and safe conduct.[17] For the rest of Mortimer and Isabella's rule, the Hospitallers continued to remain on favourable terms with the government. The crown continued to recognize the Hospitallers' right to Templar goods, assisting them to recover these possessions and permitting them to lease them out, as for example in the cases of Balsall and Fletchampstead (Warwicks.), and Willoughton and Gainsborough (Lincs.). It also helped them to restore the English Province's finances.[18] This included instructing sheriffs throughout the realm to help Tibertis against rebel Hospitallers, some of whom were hiding goods so that he could not use them to clear Hospitaller debts.[19]

The third phase of the crisis concerned the overthrow of Mortimer and Isabella and the affirmation of Edward III's individual rule. Edward began to assert his authority in the spring of 1330 and this culminated on 19 October with the arrest of Mortimer in Nottingham Castle. Prior Tibertis had sworn an oath of fealty to Edward the previous month and remained loyal to the king.[20] Edward had summoned him to a meeting of the Great Council at Nottingham for 15 October, and we know he was there for on that day he received a licence to engage in land and rent transactions.[21] Tibertis was not one of those who took part in the arrest of Mortimer on the last day of the Great Council – that was done by Edward and a small group of household knights – but it is clear that he continued Archer's approach of consensus with the majority of lords. In any case, in the early 1330s the king continued to assist the prior in various ways that would have made a policy of non-cooperation unnecessary and unproductive. For example, Edward quickly restored the Hospitaller manors that had passed into the king's hands on Archer's death,[22] discharged the prior of debts to the exchequer, granted protection regarding the ex-Templar lands,[23] confirmed other grants,[24] and permitted free movement with safe conduct within and without the realm. This included, on 24

---

[14]   *CPR 1327–1330*, p. 19.

[15]   *CPR 1324–1327*, p. 347; *CPR 1327–1330*, pp. 84, 147, 152.

[16]   Ibid., p. 62.

[17]   Ibid., p. 138. C. Perkins, 'The Knights Hospitallers in England after the Fall of the Order of the Temple', *English Historical Review*, 45 (1930), 285–9.

[18]   *CCR 1327–1330*, pp. 153, 155–6, 234, 286; *CPR 1327–1330*, pp. 192, 223, 227, 247, 340, 354, 531; *CCR 1330–1333*, pp. 11–12.

[19]   *CCR 1327–1330*, pp. 220–21.

[20]   *CCR 1330–1333*, pp. 154–5.

[21]   Ibid., p. 153; *CPR 1330–1334*, p. 13.

[22]   *CCR 1330–1333*, p. 67.

[23]   Ibid., pp. 112–13; *CPR 1330–1334*, p. 244.

[24]   Ibid., pp. 88, 101.

November 1331, protection and safe conduct to Tibertis until Michaelmas 1332, so that he could visit Hospitaller preceptories in Scotland.[25] One might even describe relations between the Hospitallers and the crown as cordial at this time: the king asked that a Hospitaller knight named Adam de Cokerham be allowed to go to the Holy Land to fulfil a vow taken by Edward II, and he pardoned two non-Hospitaller knights, Thomas de Thornham and Thomas Wyneman, for actions against the king so that they could enter the Order, although neither appears in the 1338 survey.[26]

Minority and Revolting Peasants

Archer and Tibertis survived a difficult period in English History, but Prior Hales was not so fortunate. When Edward III died in 1377, Richard II was still a minor. It was going to be a long minority, and that meant intense political wrangling as magnates vied for influence. Furthermore, the decision-making process slowed down, as no one wanted to be accused of giving bad council once the king reached his majority. What should a Hospitaller prior do to ensure the smooth running of the Order's affairs in the realm? Hales became involved in government. In between 1377 and his murder in 1381, he served as admiral of the southern fleet, on the third 'continual council', as a diplomat in late 1380, and finally as royal treasurer. Was his decision a wise one, and did it have the effect he desired? His murder and the destruction of Hospitaller property in 1381 suggest not, yet in the years leading up to the Peasants' Revolt it paid dividends. For example, in 1379 he managed to get promises to repay his loans to the crown, such as the £100 lent on 6 March that year, and in 1380 he pushed through a legal action against as influential an official as the mayor of London, whose servants were accused of seizing Hospitaller goods and assaulting the prior's servants.[27] As king's treasurer, he did in fact secure repayment of the loans to the crown.[28] He also allied himself with the most powerful lord in England, John of Gaunt, under whom he served on diplomatic duty, and kept on friendly terms with others, such as William, Lord Latimer.[29] Hales appeared to take appropriate action to further Hospitaller interests. However, such influence attracted the disdain of those who resented his rise to power, and his execution at Tower Hill, as with the murder of lawyers,

---

[25] Ibid., pp. 221, 223; *CCR 1330–1333*, pp. 323, 325.

[26] Ibid., p. 315; *CPR 1330–1334*, p. 177.

[27] *CPR 1377–1381*, p. 635; ibid., p. 567.

[28] TNA E 403/484, mm.2, 6; warrant of 26 April 1381; TNA C 81/470/20; ibid., 81/470/94.

[29] Ibid., 71/60; *Foedera*, eds T. Rymer *et al.* (The Hague, 1737–45), 3, p. 105; *The Westminster Chronicle 1381–1394*, eds L.C. Hector and B.F. Harvey (Oxford, 1982), p. 2.

jurors and alien merchants, had more to do with settling old scores than with any evil counsel he had given.[30]

Just as interesting as Hales' tactics during this crisis is his successors' reactions. John Radington did not return to England immediately on his appointment as prior and perhaps was not in the realm before 23 September 1382, when he swore fealty to the king.[31] In the interim, Fr Hildebrand Inge dealt with the aftermath of the Peasants' Revolt. Inge, a prior's attorney since 1372 and a future turcopolier, was chosen by the English brethren to govern and preside over the Hospitallers in England during the vacancy after Hales' death.[32] He had two urgent tasks to perform, firstly to restore order within the Priory of England, which had broken down, and secondly to ensure the Hospitallers' safety. Regarding the restoration of order within the Hospital, Inge gained governmental assistance. On 29 July 1381 royal officials across the country were ordered to arrest all Hospitallers found wandering around, and take them to Inge to be reprimanded: some had renounced their vows and damaged Hospitaller goods, 'despising their profession and casting off their habit'.[33] It appears that, understandably fearing for their lives, some brethren wished to disassociate themselves from the Order. Concerning their safety within the realm, Inge managed to gain the king's protection for the Hospitallers on 20 July 1381.[34] He then set about promoting Hospitaller interests in areas where there had been particularly adverse reaction, and with some success. For example, over the summer of 1381 local officials in Lincoln, Leicester and Leicestershire had orders to instruct Hospitaller tenants to do their accustomed service,[35] to forbid unlawful assemblies, and to prevent attacks on Hospitaller brethren.[36] In the case of Wartnaby (Leics.), all men, whatever their estate, were not to be armed. The implication is that influential members of the local community had also been involved in the disturbances, and, indeed, there is evidence that the parson of Keadby (Lincs.) had been inciting violence against the Hospitallers.[37] By early August the situation had stabilized enough for Inge to delegate responsibility to local commanders, such as the commander of Skirbeck, who took action against tenants who had committed acts of violence against his preceptory.[38] By October Inge was performing duties for the state, as Hales had done before him, such as the delivery of jewels held as security for a loan to foreign merchants, and in

---

[30]   *The Great Chronicle of London*, eds A.H. Thomas and I.D. Thornley (London, 1938), p. 45; H. Nicholson, 'The Hospitallers and the "Peasants' Revolt" of 1381 revisited', MO 3, pp. 225–33.

[31]   *CCR 1381–1385*, p. 208.

[32]   *CPR 1381–1385*, p. 32.

[33]   *CCR 1381–1385*, p. 9.

[34]   *CPR 1381–1385*, p. 32.

[35]   Ibid., p. 75.

[36]   *CCR 1381–1385*, p. 5.

[37]   Ibid., p. 3.

[38]   Ibid., p. 7.

December he was one of the commissioners of the peace for Middlesex.[39] In other words, Inge did not shy away from crown service at a time of crisis, and neither did Prior Radington, once he returned. They had little choice, as crown service and the relative security it brought was one of the few options available to them.

Pick A Rose, But Beware, It Has Thorns

We have seen how the Hospitaller prior reacted during one minority. How does this compare with another, that of Henry VI, and the feeble leadership and periods of insanity that were to follow? This amounted to almost forty years between the death of Henry V and the seizure of power by Edward IV. Furthermore, despite Edward IV's strong leadership, as the 1460s progressed rivalries between the great magnates led to further instability and clashes. What did Hospitaller priors do during such a prolonged period of uncertainty? Robert Botyll was prior when Henry VI lost his senses in early August 1453. Hitherto Botyll had performed minimal duties to the crown,[40] but all this changed after Henry's incapacity, with a noticeable increase in Botyll's attendance at the king's council and involvement in political matters. Like Hales, it seems that Botyll thought that the best policy was to involve himself in the government of the country. Also like Hales, he used his position to further Hospitaller interests. For example, in connection with the granting of indulgences in aid of Rhodes, papal letters of 1 December 1454 noted that Henry VI was 'ready to risk the whole strength of his realm' to protect the Christian faith.[41] However, as Henry had been witless for sixteen months and did not recover until around 25 December 1454, it cannot have been the king's personal pledge. It is clear that this entry relates to the Great Council decision on 24 July, which echoes the decision in king's council on 14 July. Botyll was present at both the king's council and the Great Council on these days and one can hardly doubt that he promoted this resolution.[42]

Once the king had regained his senses, Botyll continued to attend the king's council, but his attendance was less intense and after July 1456 there is no record of his presence until 1461. There is one significant exception, and that is his attendance on 4 March 1458, which pre-empted his return to diplomatic duty in the forthcoming mission to Burgundy in May headed by the earl of Warwick.[43] This king's council took place soon after the reconciliation between Henry VI

---

[39]   *CPR 1381–1385*, pp. 46, 84.

[40]   *C[alendar of] F[ine] R[olls] 1446–1452*, pp. 122, 223–4, 444; *CCR 1447–1454*, p. 327; *CPR 1452–1461*, p. 42; *CCR 1447–1454*, p. 450.

[41]   *Calendar of Letters in the Papal registers relating to Great Britain and Ireland, 1447–1455*, pp. 263–6.

[42]   TNA E 28/85/29–34; *Foedera*, 5 (ii), pp. 57–8; *P[roceedings and Ordinances of the] P[rivy] C[ouncil of England]*, ed. H. Nicolas (London, 1834–37), 6, pp. 217–18.

[43]   TNA E 28/88/8; *PPC*, 6, pp. 294–5.

and the Yorkists, and Botyll's participation may indicate that he had sympathies with, or wished to remain on amenable terms with, the Yorkist camp.[44] Yet Botyll continued to tread carefully, maintaining his service to Henry VI – and Queen Margaret – after relations with the Yorkists once again broke down. One example of this is Botyll's service after the Coventry parliament of November 1459, the so-called 'Parliament of Devils', at which he was present and at which he with the other lords swore allegiance to Henry VI.[45] The top priority at this parliament was the passing of an act of attainder against the duke of York and his supporters. They had disobeyed a summons to a Great Council at Coventry in June the previous year, had gone on to fight against the Lancastrian army at Blore Heath (Staffs.) in September, and finally, in October, had only withdrawn at the last minute from a confrontation with an army headed by the king himself at Ludford Bridge (Salop.). For this, they were declared guilty of treason and sentenced to death, with their spouses and issue suffering disinheritance.[46] At this point Prior Botyll remained a loyal servant, even though the commercial embargo on Warwick's base at Calais hit the Hospitallers' wool trade, and therefore gave him reason to side with Warwick and the Staple.[47] On 12 February 1460, Botyll was appointed one of the commissioners to hear and determine treason in London and the suburbs relating to the aforementioned rebellion.[48] However, Henry's retraction of the confiscation of lands from the traitors' families and insistence on his right of pardon meant that none of the main culprits was executed. York was safe in Ireland and able to communicate with Warwick in Calais, where the garrison remained loyal to him. In June 1460, Botyll was one of those the king consulted about sending a deputation to the Yorkists, who had invaded the realm, yet by July he was present with the Yorkist army that defeated Henry VI at Northampton.[49] Botyll had been playing safe, but by July had switched allegiances. Gregory O'Malley has given credible reasons why Botyll changed sides.[50] However, surely a further consideration was that the Londoners, whose will the Hospitallers could ignore at their peril, especially with the Yorkist army camped just beyond Smithsfield, close to the Hospitaller headquarters at Clerkenwell, had converted to the Yorkist cause.[51] Yet even at this stage Botyll and the majority of the lords were not willing to retract their oath of loyalty by overthrowing Henry VI and replacing him with the duke of

---

[44]   See S. Phillips, *The Prior of the Knights Hospitaller in Late Medieval England* (Woodbridge, 2009), pp. 71–2.

[45]   *CCR 1454–1461*, p. 420; *Rot. Parl.*, 5, p. 352.

[46]   Tuck, pp. 276–7.

[47]   R.A. Griffiths, *The Reign of King Henry VI*, 2nd edn (Stroud, 1998), p. 827.

[48]   *CPR 1452–1461*, p. 565.

[49]   C.L. Scofield, *The Life and Reign of Edward the Fourth* (London, 1923), 1, pp. 78, 87.

[50]   G. O'Malley, *The Knights Hospitaller of the English Langue, 1460–1565* (Oxford, 2005), p. 125.

[51]   Tuck, p. 278. For a description of the position of London, see Griffiths, p. 860.

York, perhaps wary that Queen Margaret, who had orchestrated the Parliament of Devils, could act ruthlessly against her enemies.

Botyll's 'safety in numbers' approach meant that, once Edward IV took power, the Hospitallers faced no reprisals. It was still an unstable situation and his political and diplomatic service continued. He was summoned to the Westminster parliament due for July, where he performed his regular duty as a trier of petitions.[52] Good relations with the new regime had both commitments and advantages. In 1461, for example, apart from his parliamentary duties, Botyll was sent on a diplomatic mission to Burgundy in November,[53] and in December one of the Edward IV's servants was sent to the prior to receive a corrody.[54] On the other hand, in November licence was given for a visitation of the Hospitallers' English and Irish houses, and in December one of the prior's servants was exempted for life from being put on assizes or juries.[55]

We have seen how Prior Botyll managed to endure a dangerous crisis point, similar to that which Langstrother faced just over ten years later. What, if anything, did Langstrother do wrong? On the surface, it would appear that Botyll had backed someone who could provide strong government, whereas Langstrother chose to support a monarch that at least one historian has described as being 'a pathetic shadow of a king'.[56] However, the situation was more complex. The key to understanding Langstrother's actions is to recognize the necessary alliances that all priors needed with other powerful magnates in the realm. This was a successful policy for most priors, such as Prior Radington's association with John of Gaunt, Prior Hulles' with the bishop of Winchester, and even Botyll's not so obvious links with the duke of York and earl of Warwick. Botyll had nearly twenty years in office before he faced the crisis of the late 1450s and early 1460s, time in which to familiarize himself with the political scene in England. Langstrother had no such luxury. Additionally, he clearly had very strong links with Warwick, and that made a neutral approach difficult. At first, though, despite initial problems in getting possession of the priory, Langstrother was able to gain confirmation of the Hospitallers' rights in England.[57] Yet there were still serious doubts over his loyalties due to his suspected involvement in the Lincolnshire rebellion, and by Easter 1470 he had been arrested and was eventually imprisoned in the Tower of London.[58] It is worth noting that Henry VI was also confined to the Tower at this time, and perhaps this shared circumstance reinforced Langstrother's commitment

---

[52]   *CCR 1461–1468*, p. 59; *Rot. Parl.*, 5, p. 461.

[53]   *CPR 1461–1467*, p. 102; *Foedera*, 5, (ii), p. 106.

[54]   *CCR 1461–1468*, p. 99.

[55]   *CPR 1461–1467*, pp. 52, 82.

[56]   Griffiths, p. 775.

[57]   *CPR 1467–1477*, p. 189.

[58]   'Chronicle of the Rebellion in Lincolnshire', ed. J.G. Nichols, *Camden Miscellany*, 1 (Camden Society First Series, 39) (London, 1847), p. 8; *The Great Chronicle of London*, pp. 210–11.

to the usurped king. Certainly, his captivity made clear where Edward IV considered Langstrother's allegiance to be, and from this point on there was no possibility of remaining indifferent. Langstrother was freed from the Tower on 1 October 1470 at the same time as Henry VI.[59] His reappointment as treasurer, participation in the commission to bring Queen Margaret and Prince Edward back to England, and his involvement in arraying troops in the West Country further consolidated his position as a key Lancastrian figure and ensured that not even a promised pardon and sanctuary in Tewkesbury Abbey could save him from execution after the battle of Tewkesbury.[60]

Langstrother had fallen from the political tightrope. It is easy to accuse him of getting too involved in English politics, but one wonders whether, given the circumstances, he had any option.[61] He was, one might argue, carrying out the duty that his oath of fealty to Henry VI in October 1470 required, especially as parliament had sanctioned his retracted fealty to Edward IV.[62] Perhaps Edward IV understood this, for after Tewkesbury no other English Hospitallers were punished, and even Langstrother was allowed a noble death. This is exemplified by the treatment of his successor, Prior Tornay. As bailiff of Eagle, Tornay had been involved in the Lincolnshire rebellion, but had received a pardon in July 1470.[63] After the Readeption of Henry VI, he was appointed twice as a commissioner of the peace for Bedfordshire.[64] There were reasons, then, to suspect his loyalty to Edward IV, yet no action was taken against him, and he was not politically isolated. Instead, Edward IV wanted affirmation of allegiance to his reign. On 3 July 1471, Tornay along with other lords swore an oath of fealty to Edward IV's son, Edward Prince of Wales, and on 18 February 1472 Tornay received a general pardon for any involvement in the events of the previous year.[65] In the following years, he served on two governmental commissions and as a trier of petitions, though it is noticeable that he and future priors served as triers of foreign petitions, and not English petitions as had Priors Malory and Botyll. Whether this was a deliberate attempt to limit the priors' influence in internal English affairs or for other reasons is not clear.[66]

---

[59]   Ibid., p. 211.
[60]   Ibid., p. 218.
[61]   *CCR 1468–1476*, p. 161.
[62]   Ibid.; *CPR 1467–1477*, pp. 231–2.
[63]   Ibid., p. 217; *Foedera*, 7, p. 175.
[64]   On 7 November 1470 and on 28 February 1471. *CPR 1467–1477*, p. 607.
[65]   *CCR 1468–1476*, pp. 229–30; *Rot. Parl.*, 6, p. 234; *CPR 1467–1477*, p. 306.
[66]   *Rot. Parl.*, 6, pp. 3, 7, 42.

Conclusion

In summary, what can we learn from the above case studies about Hospitaller priors, politics and power in late medieval England? There are parallels between these crises points. Firstly, all took place under inept kings or those in their minority, yet Archer, Tibertis and Botyll survived, so further factors must have led to the deaths of Hales and Langstrother. The role of London in determining the priors' attitudes deserves mention: both Archer and Botyll followed the will of the Londoners and survived, whereas Hales and Langstrother did not and paid the price.[67] Finally, during a power vacuum, links with other powerful magnates became more important and all the priors discussed above followed this policy. The key, however, was to be aware of the will of the majority, not to align oneself too rigidly, and to walk a thin line.

---

[67]   For the wavering position of London in 1471, see *A Chronicle of the First Thirteen Years of the Reign of King Edward the Fourth by John Warkworth* (Camden Society 10) (London, 1839), p. 15; J.R. Lander, *Government and Community: England, 1450–1509* (Cambridge, MA, 1980), pp. 270–71; D. Loades, *Politics and Nation: England 1450–1660*, 5th edn (Oxford, 1999), p. 57.

Chapter 19

Procedure, Political Influence and Preceptorial Appointments in the Hospitaller Priory of England: The Templecombe Disputes of 1463–79

Greg O'Malley

On 2 July 1463 the master and council ordinary of the Order of St John in Rhodes called the members of the *langue* of England before them and asked if they had any interest in or claim to the preceptory of Templecombe in Somerset, regarding which King Edward IV of England had recently written to the master. In particular, the *langue* was asked to consider whether it claimed any right in Templecombe on behalf of Marmaduke Lumley, a conventual knight who had been given magistral license to return to England on his own business in the previous year. After two days' deliberation, the *langue*'s spokesman, Robert Pickering, reappeared in council and stated that both Lumley's visit to England and what he had done there had taken place without the *langue*'s consent and that as a body it therefore did not intend to say anything in the matter of Templecombe. In their own names, however, Pickering and Richard Sandford insisted that, if the preceptory was to be adjudged to any brother on grounds of *ancienitas*, it should not be to Lumley, because they were his elders.[1] Three days later the master wrote to Lumley saying that he had upheld the complaint of the lieutenant turcopolier, Brother John Weston, who had asserted that Lumley detained Templecombe unlawfully to the prejudice of the turcopolier William Dawney, who had been appointed its preceptor by magistral grace in 1456. In the usual manner, Lumley was summoned to Rhodes to account for himself.[2]

From the letter to Lumley, it would seem that the convent in Rhodes had come down on the side of Dawney. But this was not the end of the affair. Some months later the master dispatched the prior of Portugal, Vasco de Ataíde, to several western priories as his lieutenant, and on 5 July 1464 the latter presided over a provincial chapter at Clerkenwell in which commissioners were appointed on

---

[1]   Malta, Cod. 374, fols 139r–v. I have used the modern pencil foliation when referring to documents consulted in the archives of the Order of St John in the National Library of Malta.

[2]   Ibid., fol. 140r and Cod. 366, fol. 116v.

behalf of the parties to the dispute to draw up a concord that would be acceptable to both. Dawney, however, refused to accept the draft agreement and appealed to the convent, to which the provincial chapter agreed to refer the matter, ordering both parties to plead their case in Rhodes in person or by proctor within a year.[3] Accordingly, the master and council appointed further commissioners in Rhodes to examine the disputants and determine justice. John Weston appeared before these on behalf of Dawney, but Lumley refused to substantiate his claims to Templecombe and appealed instead to a future chapter general. His appeal was rejected as frivolous, and the commissioners adjudged the house to Dawney and insisted he be put into possession.[4]

Once again an apparently definitive conventual sentence only served as the prelude to further deliberations in England. On 23 July 1466 Dawney and Lumley appeared before the prior of England, the bailiff of Eagle, and the receiver of the conventual common treasury in England at Clerkenwell, and came to an agreement over the future of Templecombe. Dawney was to retain the title of preceptor for life, and was to receive full possession of the house from Lumley. Following this the turcopolier, in contemplation of the (wishes of the) king and of other nobles and worthies of the kingdom, and at the instance of the prior, bailiff and receiver, but nonetheless of his own free will, conceded Templecombe to Lumley in return for a pension of fifty marks sterling per annum. Lumley was to support the charges of the house, with help from Dawney if they proved to be excessive, and was to receive the turcopolier and his household for three days and nights once a year. After Dawney's death, Lumley was to become preceptor of Templecombe by title of *cabimentum* as if by election of the *langue* in Rhodes. At least initially, this arrangement appeared to be acceptable to the convent. It was confirmed by chapter general in Rome on 22 December 1466, and early in 1467 Lumley, still described as a conventual knight, was licensed to go to England for two years, presumably to administer Templecombe and its estates.[5] The agreement was still regarded as valid in August 1468, when Lumley, by now styled preceptor, was given licence to let Templecombe out at farm for three years once he had received the turcopolier's consent.[6]

---

[3]   Malta, Cod. 73, fol. 152v; J. Sarnowsky, *Macht und Herrschaft im Johanniterorden des 15. Jahrhunderts: Verfassung und Verwattung der Johanniter auf Rhodos (1421–1522)* (Münster, 2001), p. 143. As well as England, Ataíde's visitation and magistral lieutenancy encompassed the Iberian Peninsula and Ireland. Malta, Cod. 73, fols 130r–131r and Cod. 374, fol. 98r; J. Sarnowsky, 'The Convent and the West: Visitations in the Order of the Hospital of St John in the Fifteenth Century', in K. Borchardt, N. Jaspert and H.J. Nicholson (eds), *The Hospitallers, the Mediterranean and Europe: Festschrift for Anthony Luttrell* (Aldershot, 2007), pp. 151–62, at p. 155n.

[4]   Malta, Cod. 375, fols 100r–v.

[5]   Malta, Cod. 376, fols 155r–156r, 154v.

[6]   Malta, Cod. 377, fol. 141v.

There are some interesting elements in the manoeuvrings outlined in these documents. First, it was highly unusual for an English conventual knight to seek to remove an established preceptor without an accompanying accusation that he had either been improperly appointed or had mismanaged the house. Second, while kings of England often showed an interest in the prior of England and his doings, they did not make a habit of intervening in disputes over individual preceptories. If the nature of Edward IV's involvement in this case does not come out very clearly from the documents bearing directly on the government of Templecombe, political circumstances, the personal and family connections of the two brethren, and further documents go some way to suggest that the king was acting as an advocate for Lumley rather than as an impartial arbiter. Finally, the to-ing and fro-ing between Clerkenwell and Rhodes, while hardly out of the ordinary, highlights the differences between the interests and concerns of the prior and preceptors in England and of the *langue* in the Dodecanese.

To turn to the first point, it is not surprising that Lumley was not willing to question the circumstances of Dawney's appointment in convent or seek to remove him for mismanagement. Not only had Dawney been appointed by magistral grace, but brother Thomas Damport, who in 1456 had felt that Templecombe should have been in the *langue*'s gift, and was then evidently the most likely candidate for promotion to it, had been compensated with a pension on Dawney's appointment, so that any claim the *langue* might have had to appoint could be said to have been satisfied.[7] Moreover, Dawney was an extremely experienced and, as far as the convent was concerned, much esteemed senior knight-brother. Having made his profession by 1437, he had subsequently served as turcopolier, captain of Bodrum, captain of the galleys, and as a diplomat and visitor in east and west.[8] He was also noted for his financial acumen, having held the office of procurator of the common treasury and advanced the convent thousands of ducats in the 1450s and having

---

[7]   Malta, Cod. 366, fol. 119r.

[8]   Malta, Cod. 353, fol. 140r (1437); Cod. 361, fols 237v–8r (appointed turcopolier); Cod. 359, fol. 213v (Bodrum); Cod. 363, fol. 259r (galleys); Cod. 363, fol. 260r; Cod. 364, fol. 195r–v (visitor); Cod. 364, fol. 117v; Cod. 367, fols 118v–119r (embassies); some editions in Z.N. Tsirpanlis, *Anekdota eggrapha gia te Rodo kai te Noties Sporades apo to archeio ton Ionniton Ippoton* (Unpublished Documents concerning Rhodes and the south-eastern Aegean Islands from the Archives of the Order of St John) [in Greek] (Rhodes, 1995), pp. 486–7, 629–30, 649–51, 740. A panel of Dawney's arms, quartered with those of Newton of Newton, Yorkshire, was placed on a little tower at the south-west corner of the castle of St Peter accompanied by his motto, *Drede Shame*. Echoing, if somewhat more dramatically, Dawney's ejection from his preceptory by Lumley, the turcopolier's arms were blasted from their setting by a French shell during the First World War, ending up in a ditch. I am grateful to Jean-Bernard de Vaivre for pointing out the locations of this panel, and informing me of its vicissitudes, and to Michael Siddons for establishing the blazon, which differs somewhat from that given for the family in the *Visitation of Yorkshire* (n. 10 below).

paid the *soldea* of the English brethren in Rhodes out of his own pocket.[9] Given the circumstances of Dawney's appointment, and the trust in which he was held, it is not surprising that Lumley should have been reluctant to challenge the turcopolier in convent. Given the weakness of his own claims, it is also not surprising that Lumley sought to secure royal support for his attempt to wrest Templecombe from Dawney. All this, however, leaves the further questions of why Lumley should have sought to relieve such a highly regarded officer of one of his appointments and why the king appears to have been amenable to his attempts to do so.

The answers might lie partly in the English connections of both knights. From an established Yorkshire knightly family, Dawney had been granted the preceptory of Dinmore in Herefordshire in 1439 and recommended to the convent in the warmest terms by Henry VI of England in the following year.[10] Although not appointed to Templecombe until 1456, Dawney had been connected to a sister at the Hospitaller nunnery at Buckland in Somerset by 1451 and in 1458 had arranged confraternity with the Order for James Butler, earl of Wiltshire and Ormond, an important landowner in both Herefordshire and Somerset and one of the leading members of the Lancastrian court.[11] At the same time Dawney had also secured confraternity with the Order for members of two important Somerset gentry families, the Sydenhams and Carents, the latter also being influential at court.[12] Although Dawney was still in Rhodes when he made these arrangements, it is quite possible that the grant was born out of longer-standing relationships. Both the Sydenhams and Carents were to receive grants from the Order after Dawney's death,[13] and on the occasions when they were both resident in England, Butler might have become acquainted with Dawney either at court or in Herefordshire.[14]

---

[9]   e.g. Malta, Cod. 364, fols 129r–129v; Cod. 366, fols 163v–164v; Cod. 367, fols 222r, 222v.

[10]   *The Visitation of Yorkshire in the Years 1563 and 1564, made by William Flower, Esquire, Norroy King of Arms*, ed. C.B. Norcliffe, Harleian Society Publications 16 (London, 1881), p. 94; Malta, Cod. 354, fol. 201r; T. Bekynton, *Official Correspondence*, ed. G. Williams (London, Rolls Series, 1872), 1, p. 87.

[11]   Malta, Cod. 363, fols 155r–v; Cod. 367, fol. 118r.

[12]   Malta, Cod. 367, fol. 118r. Nicholas Carent, or Caraunt, was the queen's secretary and received several important benefices through her influence. R.A. Griffiths, *The Reign of King Henry VI* (Stroud, 1998), p. 258.

[13]   London, British Library, MS Lansdowne 200, fol. 45v; MS Cotton Claudius E. vi, fols 47r–v, 288r–v.

[14]   As the crow flies, Butler's castle of Kilpeck, held in right of his wife Eleanor, is within five miles of the Hospital's former preceptory at Garway, which had been amalgamated with Dinmore earlier in the century. Dawney would certainly have been expected to visit Garway and possibly resided there for short periods. C. Robinson, *A History of the Mansions and Manors of Herefordshire* (London, 1872; repr. Almeley, 2001), p. 179, and map p. 365.

Perhaps Butler's interest in the Hospital was given some added weight by his heir Sir John's pilgrimage to the Holy Land in 1478.[15]

In view of Dawney's longstanding ties with the Lancastrian regime, it is difficult to imagine Lumley attempting to remove him from Templecombe while that government endured, not least because no branch of the Lumley family is known to have had estates in Somerset. The Lancastrians, however, had effectively been removed from England south of the Tyne by their defeat at Towton in March 1461, and Butler had been captured and executed by the Yorkists shortly afterwards. As royal treasurer and an intimate of queen Margaret of Anjou, Butler had been singled out by Edward IV's father as one of the evil counsellors who had surrounded the king and controlled access to him. If Dawney was associated with Butler or his supporters there may have been an argument for his removal from a position of influence in the vicinity of the Butler estates. This, however, remains no more than a suggestion given that only one proven link between Dawney and Butler has been found, and that gentry families such as the Sydenhams and Carents, who probably had much closer ties with Butler, survived his fall undamaged.[16] In any case, Dawney's associations with the new dynasty's enemies were certainly not considered to be deep enough to require his complete removal, as he was able to retain two further preceptories after the loss of Templecombe without difficulty.[17]

Probably more telling is the position enjoyed by the Lumley family in the early Yorkist polity. Having forfeited their baronial rank through rebellion against Henry IV in 1401, the Lumleys remained one of the wealthiest knightly families in the realm, holding lands worth more than £450 per annum in the four northern counties of Durham, Northumberland, Westmoreland and Yorkshire,[18] and their relative importance was greatly increased by deaths among the magnate families of northern England in 1460 and 1461 and by the continued threat to north-eastern

---

[15]   C. Costello, *Ireland and the Holy Land: An Account of Irish Links with the Levant from the Earliest Times* (Alcester, 1974), p. 57.

[16]   The Sydenhams and Carents seem likely to have followed the lead of their relative, Lord Stourton, who managed to remain neutral during the conflict. History of Parliament Trust, London, unpublished articles by H.W. Kleineke on John Sydenham of Bossington and Brimpton, Somerset (c.1410–68) and by L.S. Clarke on John Carent of Ash in Stourpaine, Dorset and Toomer in Henstridge, Somerset (c.1418–83) and William Carent of Toomer in Henstridge (d. 1476). I am grateful to the History of Parliament Trust for allowing me to see these articles in draft. See also R.E. Stansfield, *Political Elites in South-West England, 1450–1500* (Lampeter, 2009), p. 194.

[17]   These were Willoughton in Lincolnshire, to which he had been appointed in 1449, and the amalgamated houses of Battisford (Suffolk) and Dingley (Northamptonshire), which the convent in Rhodes had permitted him to exchange for Dinmore in May 1461. The exchange is unlikely to have been prompted by politics, as it must have been initiated before the Yorkists were in a position to demand Dawney's removal from Herefordshire, and the beneficiary was his nephew John Weston. Malta, Cod. 360, fol. 88r and Cod. 371, fol. 141r.

[18]   *Calendar of Inquisitions Miscellaneous, 1399–1422* (London, 1968), nos. 54–7.

England from Henry VI and his followers, who had taken refuge at the Scottish court and continued to hold castles in Northumberland until 1464. The loyalty of the Lumleys was therefore critical to the stability of the new regime in the North-East, as is evidenced by the promotion of Sir Thomas Lumley, the head of the family, to the peerage within a few months of Edward IV's accession, and the fact that in 1461 the frontier garrison of Tynemouth was entrusted to his son George.[19] While we cannot be sure that the Hospitaller Marmaduke Lumley belonged to the main line of the family based at Lumley castle, evidence in Malta shows him to have been all too convinced of his own nobility, and Marmaduke was a recurring name in the family, serving to remind its members of their ancestor Marmaduke Thweng, the head of a knightly and Hospitaller-producing lineage, the Thwengs of Thweng, whose seat the Lumleys had inherited.[20] The timing of Fr Marmaduke Lumley's decision to return to England on private business so soon after the elevation of Sir Thomas might also suggest wider family involvement in the affair.

Edward IV's involvement in the dispute is in any case likely to have been occasioned only by one of three circumstances. Either the contending candidates for Templecombe were disturbing the public peace, in which event he might indeed have acted as an impartial arbiter, or it suited the king to reduce Dawney's influence, or persons of significance had advocated Lumley's cause. As discussed above, it is unlikely that Lumley was in a position to disturb Dawney, or the peace, without the help of the king or one of his supporters in the South-West, so the idea of the monarch's impartiality can probably be dispensed with. And there is also a further piece of evidence that suggests that the king was acting as an opponent of Dawney or advocate for Lumley. In January 1465 Dawney's proctor in Rhodes, his nephew John Weston, learned that he had been summoned to England to answer a charge of disloyalty to the king, who had written to the convent personally to demand that Weston come home. Given that it was a matter of importance, it was decided 'for the honour of the religion' that the lieutenant turcopolier should return, although he did not do so until nearly a year later. Whether he was afraid of royal agents, personal enemies or the Venetians, when he did set out for home, Weston was sufficiently concerned about his safety to request licence to put off his habit and make the journey incognito.[21] While there is no mention of the nature

---

[19]   C. Ross, *Edward IV* (London, 1974), pp. 73n, 46.

[20]   The pedigree given in R. Surtees, *The History and Antiquities of the County Palatine of Durham* (London, 1816–40), 2, p. 162, states that the Sir Marmaduke Lumley who married Margaret née Thweng in the mid-fourteenth century was 'said to be Prior of St John of Jerusalem, at Kilmainham in Ireland', a post the Hospitaller was granted in 1482. This suggests that there might have been a tradition that a member of the family had held the post of prior of Ireland, but that lacking more suitable candidates the family had settled on the fourteenth century knightly head of the family as the most likely holder. I would suggest that the Hospitaller might be an otherwise unrecorded son of Sir Thomas Lumley, who was himself the nephew of Marmaduke Lumley, bishop of Carlisle.

[21]   Malta, Cod. 73, fols 157v, 158r; Cod. 375, fols 101r, 101v; G.J. O'Malley, *The Knights Hospitaller of the English Langue, 1460–1565* (Oxford, 2005), p. 126. Weston,

of Weston's offence, the only other recorded interventions in the Order's internal affairs by Edward IV before 1466 all concerned Templecombe, and by supporting Dawney's position and undermining Lumley's in July 1464, probably in despite of the king's letters, Weston had undoubtedly delayed any resolution of the affair.

The king's decision to support Dawney's removal from Templecombe while ignoring his retention of two further preceptories perhaps also takes us back to the circumstances of the turcopolier's appointment in 1456. Given the right presentation, it would not have been difficult to turn the master's decision to overturn the *langue*'s rights to appoint into a matter of both national honour and personal grievance demanding royal redress on behalf of Lumley as the self-proclaimed injured party. Edward IV's suspicion of foreign agencies and determination in his first years on the throne to uphold justice with an Arthurian integrity might have rendered him sympathetic to such claims even without any advocacy the Lumley family might have exercised on behalf of their presumed relative.[22]

Finally, the Templecombe dispute casts some light on the differing roles of the senior brethren resident in England and of the *langue* in Rhodes. The chief concern of the prior, the bailiff of Eagle, and of the receiver in this affair was not to uphold the strict letter of the Order's statutes, according to which Lumley's position was untenable, but to come to some kind of arrangement that might preserve the dignity of both disputants and serve the royal will. Their desire to do so is understandable, as the priory required continuous royal support to send men and money abroad, and to defend its properties at law.[23] The primary consideration of the *langue*, in contrast, was that its members should be justly rewarded for their conventual service, and in the right order. It did not especially object to the attempt to remove Dawney from Templecombe *per se*, as Dawney had been appointed by the master rather than by the *langue*. Rather it disagreed with Lumley going to England to arrange his own promotion, insisted that if the house was to be given to a conventual brother it should be the *langue* that allocated it, and demanded that the grant of Templecombe to Lumley should not serve as a precedent. Admirable consistency was shown in adherence to these principles across the years. Despite the confirmation of the agreement over Templecombe in the chapter general of 1466–67, the same assembly saw the *langue* protest lest

---

'whose journey we much desire should be fortunate and safe', was granted a safe conduct as well as licence to leave. Malta, Cod. 375, fol. 188r.

[22]   J. Hughes, *Arthurian Myths and Alchemy: the Kingship of Edward IV* (Stroud, 2002), chapters 5–6.

[23]   The leading members of the priory also had personal reasons to support Lumley. Robert Botyll had thrown his weight very publicly behind the Yorkist regime in 1460–61, and is unlikely to have been willing to upset the king over the collation to a single preceptory. The bailiff of Eagle, John Langstrother, was a native of the lordship of Kirkby Kendall, of which the Lumleys held an eighth. O'Malley, p. 125; S. Phillips, 'Walking a Thin Line: Hospitaller Priors, Politics and Power in Late Medieval England', above, pp. 225–6; *Calendar of Patent Rolls, Henry VI* (London, 1910–11), 3, *1436–41*, p. 284.

any bull of expectancy or confirmation granted Lumley be to its prejudice.[24] In November 1471, perhaps because it was thought possible to remove Lumley in the light of the political upheavals in England over the previous year, the *langue* persuaded the chapter general to overturn the agreement of 1466 as contrary to the statutes, and to collate a conventual knight, John Borough, instead.[25] While Borough was evidently unable to unseat Lumley, and was granted the preceptory of Carbrooke instead in 1475, in November 1478 Walter Fitzherbert was granted Templecombe in turn, and the agreement of 1466 once more declared null and void.[26] Although Fitzherbert complained shortly afterwards that John Weston, now prior of England, was detaining Templecombe from him, the complaint was not repeated, and it seems that Fitzherbert was able to secure possession.[27] The *langue* had its victory, and having never been granted another house in addition to Templecombe, Lumley was returned to the status of conventual knight, from which he was elevated to the priorate of Ireland in place of the disgraced James Keating in December 1482. Once again, and this time with the support of the convent and of his *langue*, Lumley set out to unseat an incumbent senior brother, but this time with even less success.[28]

---

[24]   Malta, Cod. 376, fol. 154r.

[25]   Malta, Cod. 380, fol. 136r. The chapter general and *langue* must have been aware that Edward IV had regained the throne well before this date, for they had been informed of the death of prior John Langstrother on 6 May, in the aftermath of the decisive battle of the war at Tewkesbury, by 28 August at the latest.

[26]   Malta, Cod. 382, fol. 141r and Cod. 386, fols 128v–129r.

[27]   Malta, Cod. 386, fols 129v–130r.

[28]   O'Malley, pp. 243–4. s

PART 4
Italy

Chapter 20

Sta Maria in Carbonara in Viterbo: History and Architecture of a Templar Preceptory in Northern Lazio

Nadia Bagnarini

The architectural complex of Sta Maria in Carbonara[1] consists of two buildings located inside the city walls[2] and in the shadow of the dome of the cathedral of San Lorenzo in the south-western area of *Castrum Viterbii*. The church and its conventual wing, placed at right-angles to each other, are to be found in the present-day Via di Sant'Antonio. This street, once named Via di Valle, represents a branch of the Roman consular road, the Via Cassia, which, starting from the hill on which the cathedral now stands, ran through the centre of the city and left at the Porta Vallia to meet the Via Cimina.[3] The church owes its name 'de Carbonaria' to its construction on so-called *carbonare*:[4] large ditches fitted with strong wooden piles, dug for defence in the plain beside the Via di Valle.[5]

Architectural analysis suggests that the church building may already have existed when the Knights Templar arrived in Viterbo. Its walls are mostly composed of blocks of longitudinal *peperino* stone, laid horizontally and occasionally interrupted by headers – a typical building technique at Viterbo in the second

[1]   G. Romalli, 'La domus templare di Santa Maria in Carbonara', in L.P. Bonelli and M.G. Bonelli (eds), *Dal castrum Viterbii alla Civitas Pontificum: Arte e architettura a Viterbo dall'xi al xiii secolo: Atti del Convegno di Studi, Viterbo Aula Magna del Rettorato dell'Università della Tuscia 21–22 aprile 2005* (Viterbo, 2005), pp. 37–68; N. Bagnarini, 'La domus templare di Santa Maria de Carbonaria a Viterbo', in *Il Tesoro della città: strenna dell'Associazione Storia della Città* (Rome, 2003), 1, pp. 40–48.

[2]   M. Miglio, 'Riflessioni sulle mura di Viterbo', in E. Guidoni and E. de Minicis (eds), *Le mura medievali del Lazio: studi sull'area viterbese* (Rome, 1993), pp. 11–15; S. Valtieri, *La genesi urbana di Viterbo* (Rome, 1977).

[3]   M.G. Bonelli, 'Viterbo, Urbanistica', in *Enciclopedia del'arte medievale* 11 (Rome, 2000), p. 705; A. Mosca, *Via Cassia: un sistema stradale romano tra Roma e Firenze* (Florence, 2002).

[4]   S. del Lungo, *La toponomastica archeologica della Provincia di Viterbo* (Tarquinia, 1999), p. 127.

[5]   I. Ciampi, *Cronache e statuti della città di Viterbo* (Florence, 1872), pp. 499, 525 no. 109; A. Scriattoli, *Viterbo nei suoi monumenti* (Viterbo, 1915–20), pp. 158–60.

half of the twelfth century.[6] Furthermore, architectural ornament is limited to the outer parts of the apse of the church: two ornamental corbel tables with geometric, zoomorphic and anthropomorphic decoration, the upper one supporting small blind arches – similar to those found in Viterbo and dating from the last quarter of the twelfth century or the beginning of the thirteenth on the outer sides of the apses of the churches of San Sisto and San Carlo a Pianoscarano.[7] As for written sources, the first mention is on 12 December 1232, when a certain Spiccagonella granted his son Fuscum a house with a turret 'posita in hora Sancti Peregrini' and a vineyard 'positam in contrada Vallis Olmetuli', bordering the vineyard of Sta Maria in Carbonara: 'iuxta quoque vineam Sancte Marie in Carbonarie'.[8] In a second archival source, dated 4 April 1236,[9] Bishop Matteo, at the instigation of Pope Gregory IX (1227–41), attempted to resolve the conflicts that had arisen among the dignitaries and canons of the minor churches of Viterbo concerning the appointment and ordination of heads of houses. The document was composed 'de comuni voluntate pro pace conservando' in the presence of representatives of the chapters of these churches, one of whom was 'presbyter Bartolomeus' of Sta Maria in Carbonaria.[10] A few years later a contract was drawn up in the same church whereby forty-two houses next to Sta Maria in Poggio and San Giovanni in Zoccolo were sold to the emperor, Frederic II,[11] to allow him to build his imperial residence there.[12] According to Romalli, the document, drawn up on 3 February 1243 in the presence of 'magistro preposito Ulfreducio', testifies to the importance

---

[6]   D. Andrews, 'L'evoluzione della tecnica muraria nell'Alto Lazio', *Biblioteca e società*, 4 (1982), 1–16; R. Chiovelli, *Tecniche costruttive murarie medievali: la Tuscia: storia della tecnica edilizia e restauro die monumenti* (Rome, 2007); L.P. Bonelli, 'Le murature della chiesa di San Sisto come testimonianza dello sviluppo tecnico-strutturale dell'edilizia viterbese medievale', in *Informazioni: Periodico del Centro di Catalogazione dei Beni Culturali della Provincia di Viterbo*, n.s. 3 (1994), pp. 11, 93–100.

[7]   Romalli, 'La domus templare', pp. 50–51.

[8]   *Il 'Liber Quatuor Clavium' del Comune di Viterbo* 1.2, ed. C. Buzzi, Istituto Storico Italiano per il Medio Evo, Fonti per la Storia dell'Italia medievale, Regesta Chartarum 46–7 (Rome, 1998), pp. 9–11.

[9]   Archivio Comunale di Viterbo, perg. 1124.

[10]   G. Signorelli, *Viterbo nella storia della chiesa* (Viterbo, 1907–1969), 1.3, p. 195 no. 22.

[11]   E. Bentivoglio, 'Il "bello et grande palazzo" di Federico II a Viterbo: strategia politica, processo e tecniche di realizzazione', in (ed.), *Federico II: cultura istituzioni arti, Atti del seminario di studi, Reggio Calabria, 20–21 maggio 1995*, Quaderni PAU 9 (Catanzaro, 1996); A. Spina, 'Il Palazzo di Federico II a Viterbo: un progetto sul Poggio del Tignoso', *Biblioteca e Società* 25 (2006), 18–20.

[12]   '… in volta Santa Maria in Carbonara die tertio febrarii sub annis domini millesimo duecentesimo quadragesimo tertio frasumundus 9 imper auctoritate judex ordinaries et notarius'. Archivio di Stato di Viterbo, perg. 60.

of the church in the twelfth century; but it makes no mention of the presence there of any members of a military order.[13]

Although written sources are far from comprehensive for this period, of particular interest is one of the records of the proceedings against the Templars in Tuscia, in the territory of the Papal State, which offers the only certain proof of a Templar community's presence in the preceptory of Viterbo. The case is recorded in a parchment roll (*rotulus*) preserved in the Vatican *Archivio Segreto* and published in 1982 by Anne Gilmour Bryson.[14] On 20 December 1309 the inquisition held in the bishop's palace in Viterbo decided to send messengers to serve a summons 'in ecclesiarum Sancte Marie de Carbonara'. The mission was discharged the next day, when papal legates exhibited a 'cartam sive membranam continentem dicte ditationis edictum in hostiis ecclesie Sanct Marie de Carbonaria dicti ordinis militiae Templi'.[15] In the following months, pending its deliberations, the inquiry's commission ordered the monastery's sequestration and handed over its temporary management to Giovanni, prior of the nearby church of San Giovanni in Valle.

The most interesting information is to be found in the depositions of three *servientes* among the seven witnesses called to testify in the bishop's palace in Viterbo between 8 and 10 June 1310. These three were Gerardo da Piacenza, Pietro Valentini and Vivolo *de Sancto Iustino*.[16] Gerardo da Piacenza, when asked about the preceptors who had governed the Order in recent times, recalled,

> Et post dictum fratrem Gulielmum fuit magnus preceptor frate Artusio de Pocopalgia qui ut audivit mortuus fuit Viterbio et sepultus in Sancta Maria de Carbonaria de Viterbio dicti ordinis.[17]

Pietro Valentini[18] responded similarly, whereas Vivolo *de Sancto Iustino, familiaris* of the preceptor of San Bevignate at Perugia,[19] confirmed that he had been present at a meeting of the Order's chapter 'in ecclesia Sancte Marie de Carbonaria de Viterbio dicti ordinis', presided over by Brother Guglielmo Cernerio 'qui gerebat

---

[13]   Romalli, 'La domus templare', p. 42.

[14]   A. Gilmour Bryson, *The Trial of the Templars in the Papal State and Abruzzi*, Studi e Testi 303 (Vatican City, 1982).

[15]   Ibid., p. 93.

[16]   Ibid., p. 91. On Gerardo da Piacenza, see A. Gilmour-Bryson, 'Italian Templar Trials: Truth or Falsehood', in N. Housley (ed.), *Knighthoods of Christ: Essays on the History of the Crusades and the Knights Templar, Presented to Malcolm Barber* (Aldershot, 2007), p. 221. The other witnesses included Guillelmus, priest of Verduno, and Fr Henricus de Balnoregio.

[17]   Gilmour-Bryson, *Trial of the Templars*, pp. 188–9.

[18]   On Pietro Valentini, see ibid., p. 201

[19]   F. Tommasi, 'L'Ordine dei Templari a Perugia', *Bollettino della Deputazione di Storia Patria per l'Umbria*, 78 (1981), 5–79.

se pro magno preceptore in Patrimonio beati Petri in Tuscia' in the presence of 'preceptores locorum dicti ordinis de dicto Patrimonio'.[20]

A study of these depositions reveals that the Grand Preceptor Artusio de Pocapalgia had succeeded Guglielmo Provincialis, who is mentioned in 1286,[21] and preceded Guglielmo de Canellis, who was in office in 1292.[22] By then Artusio de Pocapalgia must already have been dead, and his body buried in the church of Viterbo. The year 1292 is consequently a certain *terminus ante quem* for the establishment of the Templars in Sta Maria in Carbonara. What we learn from these documents is that Sta Maria in Carbonara must have been held in high prestige at that time among the Order's houses in the northern borderlands of the Papal State,[23] as they had chosen this church for the burial of a grand preceptor and had even held a chapter of the Order in it.

The Templars established at Viterbo certainly had to manage an immense landed estate in the territory of Viterbo as far as Lake Bolsena, including the possessions of Sta Maria in Capita at Bagnorea,[24] San Benedetto di Burlegio,[25] Sta Maria di Castell'Araldo, San Savinio near Tuscania,[26] San Giulio and Sta Maria at

---

[20]   Gilmour-Bryson, *Trial of the Templars*, p. 221. This dignitary has been identified by Tommasi, pp. 12, 17, as a French knight and preceptor of San Giustino d'Arno in Pilonico Paterno, near Perugia.

[21]   Gilmour-Bryson, *Trial of the Templars*, p. 188.

[22]   B. Capone and L. Imperio, 'I cavalieri templari della nobile famiglia de Canellis', in *Atti del X Convegno di Ricerche Templari (Poggibonsi (Si), 12–13 settembre 1992* (Florence, 1994), pp. 54–5; M.-L. Bulst-Thiele, *Sacra domus militiae Templi Hierosolymitani magistri: Ricerche sulla storia dell'ordine dei Templari 1118/9–1314: I Grandi Maestri Templari*, trans. R. Pardi (Perugia, 2004), p. 279; E. Bellomo, *The Templar Order in North-West Italy (1142–c.1330)* (Leiden–Boston, 2008), pp. 101–3.

[23]   A. Luttrell, 'Two Templar–Hospitaller Preceptories North of Tuscania', *Papers of the British School at Roma* 39 (1971), 90–124; G. Romalli, 'La Magione di Bagnoregio: una precettoria templare nella Tuscia Romana', in C. Ciammaruconi (ed.), *L'Ordine templare nel Lazio meridionale, Atti del Convegno (Sabaudia, 21 ottobre 2000)*, Biblioteca Casaemariensis 7 (Casamari, 2003), pp. 157–200; N. Bagnarini, 'Architettura templare nel Lazio', in *Atti del XXIII Convegno di Ricerche Templari, Cervia 24–25 settembre 2005* (Latina, 2006), pp. 110–34.

[24]   N. Bagnarini, 'Santa Maria in Capita: la casa dei cavalieri templari in località Bagnorea', in *Atti del XXII Convegno di Ricerche Templari, Trieste 25–26 settembre 2004* (Latina, 2005), pp. 29–44.

[25]   Archivio Segreto Vaticano: Armadio XXXV (14), fol. 85r: 'frater Joanne de Briscio preceptore Mansionis de Burlea ordinis Templi'; G. Silvestrelli, 'Le chiese e i feudi dell'Ordine dei Templari e dell'Ordine di Gerusalemme nella regione romana', in *Rendiconti della Reale Accademia dei Lincei classe di scienze morali storiche e filologiche* 26 (Rome, 1917), pp. 496, 503–4.

[26]   Ibid., pp. 496, 501–2.

Civitavecchia,[27] San Biagio at Vetralla,[28] Sta Maria at Valentano,[29] San Benedetto at Montefiascone,[30] and Sta Maria del Tempio at Sutri;[31] but we are not at all certain about the Templars' role in the papal curia, located nearby in the palace of San Lorenzo,[32] where they may perhaps have acted as papal guards.[33]

It was largely papal activities that conditioned the nature of Templar life at Viterbo, ranging from periods of prosperity to periods of abandonment. In good times the Templars built their first residential wing, spread over three levels including the basement; but even this was provided with defensive elements: a wooden gallery around the whole structure, a door[34] on the building's east side, and a very small number of slit openings. The Templars' residence was originally one storey lower than the present building. Slits are still visible in the masonry of the southern façade. A distinct horizontal break, with brackets underneath to support the gallery runs around the exterior of the walls. The structure was topped by a flat terrace and enclosed on three of its sides by a projecting wooden gallery to guarantee protection and defence from above. Furthermore, the bulwark-like character of the residence was emphasized by a door, fitted into the eastern wall at first-floor level, as well as by the very small number of slits, the only openings that pierce the imposing peperino walls. Masonry analysis has also identified another break, which runs at a higher level across the eastern face of the building, next to the corner towards the church; this feature confirms that the structure had been enlarged, expanding the superstructure beside the church. For a better understanding of the different building phases, it is worth studying the plan attached to the *cabreo* of 1629 preserved in the Archives of the Order of the Knights of Malta in Rome.[35] The building, designed to house a limited number of knights, was connected to the church through a square tower between the southern

---

[27]   E. Valentini, *I Templari a Civitavecchia e nel territorio fra Tarquinia e Cerveteri* (Tuscania, 2008).

[28]   Silvestrelli, p. 495.

[29]   Ibid., p. 495.

[30]   Ibid., pp. 503–4.

[31]   E. Susi, 'Culti e agiografia a Sutri tra tardoantico e alto medioevo', in *Sutri Cristiana. Archeologia, agiografia e territorio dal IV all'XI secolo* (Rome, 2006), p. 177 nos. 366–7.

[32]   M.T. Gigliozzi, 'Il ruolo del Palazzo Pontificio nella Viterbo duecentesca', in Bonelli and Bonelli (eds), pp. 89–104; M.T. Gigliozzi, *I palazzi del papa: architettura e ideologia: Il Duecento*, La Corte dei papi 11 (Rome, 2003).

[33]   A. Luttrell, 'Templari e Ospitalieri in Italia', in *Templari e Ospitalieri in Italia: la chiesa di San Bevignate a Perugia* (Milan, 1987), p. 22.

[34]   M.R. Giordani, 'Ricognizioni delle torri medievali di Viterbo', in E. de Minicis and E. Guidoni (eds), *Case e torri medievali: Atti del convegno di studi la città, le torri e le case. Indagini nei centri dell'Italia comunale (sec. XI–V). Toscana, Lazio, Umbria, Città della Pieve, 8–9 novembre 1996* (Rome, 1996, 2001), 2, p. 155.

[35]   Archivio del Sovrano Militare Ordine di Malta (Rome): Cabrei, 192, fols 19v–20r. L. Bartolini Salimbeni, '"I Cabrei" e i "Processi di miglioramento" dell'Ordine di Malta:

side of the church and the northern wall of the *domus*; it contained a staircase connecting the different floors and at the same time giving access to the church's crypt. It could also act as a lookout tower, thus contributing to the complex's defence and ability to control the surrounding area.

In the 1270s the number of knights residing in the *domus* had probably increased; hence the decision to add another floor and enlarge the entire residential wing. Three striking architectural elements mark this second building period: the trilobate arch of evidently gothic pattern, produced in the nearby Cistercian workshop at San Martino at Cimino,[36] inserted into the right-side wall of the presbytery as a connection to the staircase-tower to the right of the church; the transverse arches[37] raised inside the church to support the roof-trusses and at the same time to offer a counter-buttress to the pressure on the southern walls caused by the same tower; the two-light windows on the three visible sides of the elevated *domus*. Of these, only the two-light windows on the main façade show decorative motifs: a blazon (its heraldic figures illegible) and a rose on the tympana of the northern one and a two-barred cross and the outline of a knight in those of the southern one, but they do serve to emphasize that the building now functioned as an administrative centre rather than as a place of defence.[38] A similar decoration with a five-petalled rose appears on the tympana of the windows of the nearby tower of Messer Braimando, mentioned in the city of Viterbo's statutes of 1251 as a prescriptive style for civic fortified architecture;[39] we find comparable decoration on the windows of the tower of the Palazzo degli Alessandri in the San Pellegrino quarter,[40] and on the former palace of Cardinal Raniero Capoci near the castle.

The architecture of this second building period can be rightly compared with that of the nearby papal palace, thus allowing a precise stylistic differentiation of the original nucleus of the Templar residence, finished at latest in 1260s, from the enlargements undertaken in the 1270s. There too the outer walls were relieved on the side towards Faul by a row of three two-light windows, very similar in architectural

---

una fonte per la storia dell'architettura fra XVI e XVIII secolo', *Architettura: storia e documenti* 3 (1987), 165–83.

[36]   M. de Paolis and M.C. Oberti, 'L'abbazia di S. Martino al Cimino', in A.M. Romanini (ed.), *I Cistercensi e il Lazio. Atti delle giornate di studio dell'Istituto di Storia dell'Arte dell'Università di Roma, Roma 17–21 maggio 1977* (Rome, 1978), pp. 169–75; D. Borghese and S. Santolini, 'Note aggiuntive su di un cantiere cistercense: l'abbazia di San Martino al Cimino', *Strenna dei romanisti* 57 (1996), 75–108.

[37]   M.E. Savi, 'Archi diaframma: contributi per una tipologia architettonica', in *Arte medievale* 3 (Rome 1987), 163–81.

[38]   A. Cadei, 'Gli Ordini di Terrasanta e il culto per la Vera Croce e il Sepolcro di Cristo in Europa nel XII secolo', *Arte medievale*, n.s. 1 (2002), 51–69.

[39]   Romalli, 'La domus templare', p. 61 no. 68.

[40]   N. Cucu, 'La casa medievale nel Viterbese', *Ephemeris Dacoromana, Annuario della scuola romena di Roma* 8 (Rome, 1938), pp. 7–20; M.R. Giordani, 'Ricognizione delle torri medievali di Viterbo', in de Minicis and Guidoni (eds), 2, p. 159.

form to those on the east side of the preceptory, topped by an architrave composed of two peperino blocks, placed side by side and sustained by a pillar.

Studies of Italian Templar architecture have identified two famous earlier preceptories with similar ground-plans, in which the church and the residential building are joined. One is the preceptory of Sta Maria on the Aventine Hill in Rome[41] and the other the preceptory of San Bevignate in Perugia.[42] Sta Maria on the Aventine hill was studied in depth by Pio Francesco Pistilli, who tried to reconstruct the appearance of the medieval site. As authorities he studied the drawings of du Perac and the famous record of the Vatican archivist Giacomo Grimaldi of September 1619, but he also carried out a detailed archaeological analysis of the existing buildings. These studies established the preceptory as the earliest example of a Templar site, of a type that was to be developed in the following decades in two Templar foundations in central Italy, San Bevignate and Sta Maria in Carbonara. In contrast to those two foundations, where the conventual building is placed at right-angles to the presbytery of the church,[43] the residential building of the Roman house is placed alongside and slightly oblique to it; its planners may have incorporated parts of the pre-existing Cluniac monastery, but they may also have taken into account the building's strategic and panoramic position.[44] They had in fact the chance of taking advantage of the River Tiber, to which the house was connected by a ramp from the Via Marmorata and leading to a natural landing place. Moreover, its principal façade was built with an open gallery, placed in an open space between the apse of the church and the eastern side of the little palace.

The house of San Bevignate, built from 1256 onwards, does not seem to have shown a principal façade. The structure was built in the eastern suburbs of Perugia at the end of the Via Spargente and consists of a conventual building of two wings placed at two right-angles to one another, the church building, and a massive defensive tower inserted between the right side of the church and the northern convent wing.

---

[41] P.F. Pistilli, 'Due tipologie templari: la domus romana sull'Aventino e il locus fortificato di San Felice Circeo', in *L'Ordine templare nel Lazio meridionale* (as n. 23 above), pp. 157–200.

[42] P. Raspa and M. Marchesi, 'Note sull'architettura di San Bevignate', in *Templari e Ospitalieri in Italia: la chiesa di San Bevignate a Perugia* (Perugia 1987), pp. 79–86; R. Pardi, *Architettura templare e crociata 1118/9–1314* (Perugia 2004), pp. 35–6, 47–50; idem, 'Analisi comparata delle testimonianze architettoniche templari', in S. Merli (ed.), *Milites Templi: il patrimonio monumentale e artistico dei Templari in Europa* (Perugia, 2008), pp. 335–6; G. Casagrande, 'San Bevignate: una chiesa per la città', in *Milites Templi: Il patrimonio monumentale e artistico dei Templari in Europa*, ed. S. Merli (Perugia 2008), pp. 191–202.

[43] A. Cadei, 'La chiesa di San Francesco a Cortona', in *Storia della città* 9 (Rome, 1978), pp. 16–23.

[44] A. Ilari, 'Il granpriorato Giovannita di Roma: ricerche storiche e ipotesi', *Militesia* 4 (Taranto, 1998).

Extensive researches into Templar residences in Europe have traced other architectural structures where a similar constructive typology to Sta Maria in Carbonara is found. Two preceptories in particular should be mentioned: one in England, in the county of Kent, and one in Spain, in the western part of Catalonia. The English preceptory of Strood is at Temple Manor,[45] located near Rochester on the western bank of the River Medway; it fits into the typology of residential structures, built with the capacity to house just a few knights or sergeants. The property at Strood, granted the Templars by King Henry II,[46] was a rural holding but included some urban incomes and a watermill on the Medway. Unfortunately lack of documents prevents us knowing about the different building periods, but architectural details such as the use of Purbeck marble for the inner wall coverings of the first-floor rooms has led experts to date the structure to the first half of the thirteenth century. Some precise information is preserved in documents of 1308, composed at the time of the seizure of Templar goods; there are four distinct reports about Strood, forwarded to the sheriff of Kent by John of St Denis, who had to administer all the properties in the area. The surviving Templar building at Strood, with its two stories and simple rectangular ground plan, has similarities with the house at Viterbo. The ground floor is spanned by three bays of rib-vaults on low clustered pillars, while the first floor, provided with an external staircase and a timber roof, is constructed with arcades of Purbeck marble, following patterns commensurate with the Westminster Abbey workshop. The easternmost bay, lightly separated from the others, was probably used as a chapel. This first-floor level, known as the *camera*, would have been the residential part of the preceptory. The word *camera* can mean various things: it can mean a room next to a hall, an autonomous two-storied building, or even a subordinate room in a *domus*; in a wider sense it can be any room or series of rooms used to accommodate a privileged resident and his *familiares*. The hypothesis put forward by the late S. Rigold is that Strood's structures were the *camerae* of normally secular and occasionally residential use, comprising a private inner room and an outer antechamber for staff or guests.[47]

Unlike Temple Manor, the Catalan preceptory of Gardeny,[48] placed on a promontory of the same name to the south of Lerida, was built to lodge a large

---

[45]   S.E. Rigold, 'Two Camerae of the Military Orders: Strood Temple, Kent, and Harefield, Middlesex', *Archaeological Journal*, 122 (1965), 86–132; P. Ritook, 'The Architecture of the Knights Templars in England', in *MO* 1, pp. 167–78.

[46]   B.A. Lee, *Records of the Templars in the Twelfth Century: The Inquest of 1185 with Illustrative Charters and Documents*, British Academy Records of the Social and Economic History of England and Wales 9 (London, 1935), p. xcviii.

[47]   Rigold, 'Two Camerae', p. 121.

[48]   J. Fuguet Sans, *L'arquitectura dels Templers a Catalunya* (Barcelona, 1995); A. Cadei, 'Templari', *Enciclopedia dell'arte medievale* 9 (Rome, 2000), p. 101; L. Dalliez, 'Les Templiers dans la peninsule iberique', *Archeologia*, 27 (1968), 36–41; J. Miret y Sans, *Les cases da Templers y Hospitalers en Catalunya* (Barcelona, 1910), pp. 67, 88; M. Bonet

number of knights. Although the form and composition of the different buildings seem to be typical of those found on rural agricultural sites, their monumental effect reveals the site's strategic importance as the administrative centre of the Order's immense estates on the River Segre. The preceptory at Gardeny is built on two levels: the lower level was mostly incorporated into the eighteenth-century defence works, while the upper level comprises three Templar buildings, most probably of the late twelfth century, namely the church Sta Maria of Gardeny, the treasure tower and the residence (or *donjon*).

The church, built facing east, has a single nave covered by a pointed vault and a polygonal choir; the original roofing is lost. The conventual wing, the so-called residence, is placed at right-angles to its east end. It is a spacious structure, built to a rectangular ground plan, with a small square tower at its eastern corner. The convent building is attached to the church's choir by a long passage, added afterwards. Old drawings provide details of the building's interior, which has been altered by restorations in the 1970s. Both floors were covered with pointed vaults. The lower level, a basement without windows, was presumably used for the staff, whereas the upper level, a spacious hall lit by slits, probably accommodated the brothers and the commander. On the northern part of the west side, towards the chapel, is a minor structure containing the staircase and bedchambers. The whole building was covered by a terrace with battlements, nowadays completely remade.

These comparisons have shown how in geographically distant areas different preceptories follow similar pattern to the house in Viterbo; such comparisons also highlight the preceptory's high strategic and defensive significance for the surrounding area, always an important consideration in the organisation of Templar estates. Both Gardeny and Sta Maria on the Aventine Hill were built on promontories, which, besides offering sites defended by nature, allowed their owners to overlook the River Segre and the city of Lerida on the one hand and the River Tiber on the other. The same consideration applies to the house of Strood. Even without any defensive elements in its architecture, as it was exclusively used to manage the Templars' landed properties in Kent, it was built near to the River Medway, which offered a privileged access to the Thames estuary and was exploited to work a system of watermills.

The preceptories of Sta Maria di Carbonara and of San Bevignate fit into the same model. Both were built next to important main roads (the Via Cassia and the Via Spargente), even if they show different types of defensive architecture: at Viterbo, an imposing construction of two levels, with a first-floor door and wooden gallery; at San Bevignate, the raising of a real defensive tower.

Donato, 'Historiografia e investigacion sobre el Temple en la Corona de Aragon', in *Milites Templi* (as n. 42 above), pp. 71–9.

Chapter 21

The Spanish Military Orders in Italy: Initial Remarks on Patronage and Properties (Twelfth–Fourteenth Centuries)

Elena Bellomo

The presence of the Spanish military orders in Italy is a field of research that still requires exhaustive investigation (see Fig. 21.1).[1] Existing studies have focused mainly on reconstructing the history of a few specific properties, and there is no comprehensive catalogue of the houses or a reconstruction of the development and organization of these orders in the Peninsula, Sicily and Sardinia during the Middle Ages. Moreover, evidence from Italian primary sources has rarely been set alongside documents from the Spanish archives. This paper offers some initial remarks on the presence of the Spanish military orders in Italy, concentrating on the kind of patronage they enjoyed and the particular character and development of their presence. As this research has not long started, only preliminary results can be presented here. These first findings, however, can serve to illustrate possible lines of future research and emphasize the original and valuable contribution that a comprehensive investigation into this subject will make to military orders' studies.

Founded in the Iberian Peninsula with the shared intent of fighting against the Muslims, but differing from one another in structure and observance, the Spanish military orders made a crucial contribution to the *Reconquista*. They were also involved in the struggles among and within the Christian Iberian kingdoms. Once established, they were subject to increasing influence from the various royal dynasties. By the end of the fifteenth century, the Spanish military orders were controlled by Iberian kings and formed an important tool in royal domination over both the aristocracy and the land.[2]

[1]   I would like to thank Martín Alvira Cabrer, Carlos Barquero Goñi, Natalia Fernández Casado, Alan Forey and Philippe Josserand for their help in finding some of the primary sources and secondary works mentioned in this article. I am also especially grateful to Cadw: Welsh Historic Monuments for its financial support for early-career scholars living outside the UK and presenting a communication at the Cardiff conference.
[2]   On these orders see D.W. Lomax, *Las Órdenes Militares en la Peninsula Ibérica durante la Edad Media* (Salamanca, 1976); C. de Ayala Martínez, *Las Órdenes Militares hispánicas en la Edad Media. Siglos XII–XV* (Madrid, 2003); *Las Órdenes militares en la Península Ibérica*, eds R. Izquierdo Benito, F. Ruiz Gómez and J. López Salazar-Pérez

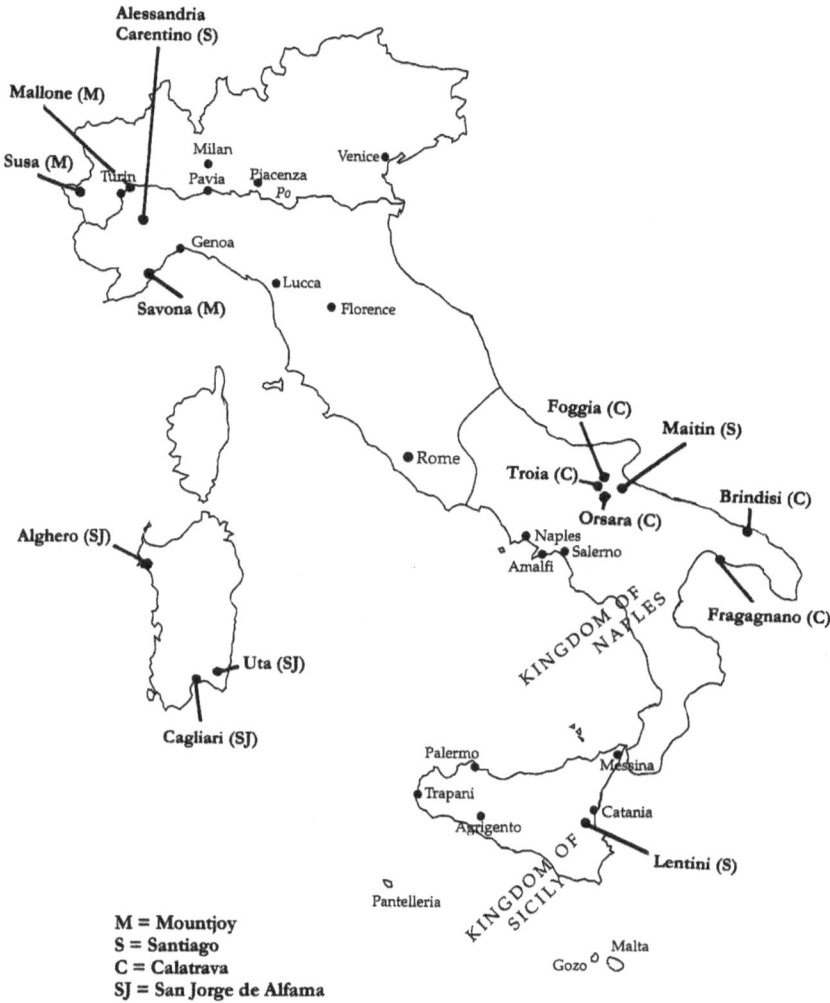

Figure 21.1   The Spanish Military Orders in Italy (graphics: E. Bellomo)

This picture would suggest that the Spanish military orders were strangers to the international dimension that is a basic feature of the crusades. However, a close scrutiny of the primary sources shows that in the twelfth and thirteenth centuries they were present in France, Germany and Italy, and that their involvement in the Holy Land and Latin Romania was repeatedly sought. In fact the international vocation and activities of the Spanish military orders are still a matter for debate. Some scholars consider such activity to be the result of sporadic ecclesiastical attempts to involve them in new theatres of war, attempts that were doomed to failure because of their predominantly national character.[3] Others claim that, at least until the mid-thirteenth century, the Spanish military orders did indeed demonstrate an international vocation, which stemmed from their dependence on the papacy and which was modelled on the international structure and activities of the first military orders, the Temple and the Hospital.[4] It is true, however, that the Iberian monarchies limited the expansion of the Spanish military orders outside the Iberian Peninsula when it did not coincide with their own agendas, and the increasing royal control over these institutions heavily conditioned their development, forging an exclusive association with the Iberian territories and then bringing their activities into conformity with royal aims and strategies.

A study of the presence of the Spanish military orders in Italy can contribute substantially to this debate. Italy was a crucial crossroads, and any order aiming at Mediterranean expansion had almost necessarily to settle in this area. Moreover, in the Italian Peninsula and its islands the Spanish military orders enjoyed two different types of patronage. In the twelfth and thirteenth centuries, the presence of these orders gained the support of the local nobility, the cardinals and the papacy, while from the fourteenth century onwards the establishment of Aragonese control – first on Sicily and Sardinia, and then on continental southern Italy – extended

---

(Cuenca, 2000); P. Josserand, *Église et pouvoir dans la Péninsule ibérique. Les ordres militaires dans le Royaume de Castille 1252–1369* (Madrid, 2004); E. Rodríguez-Picavea Matilla, *Los monjes guerreros en los reinos hispánicos. Las órdenes militares en la Península Ibérica durante la Edad Media* (Madrid, 2008).

[3]   Ayala Martínez, pp. 8–16, 529–32.

[4]   J.M. Rodríguez García, 'El internacionalismo de las órdenes militares "hispanas" en el siglo XIII', *Studia historica*, 18–19 (2000–2001), 187–209; A. Mur i Raurell, 'Relaciones europeas de las Órdenes Militares hispánicas durante el siglo XIII', in K. Herbers, K. Rudolf and J. Valderón Baruque (eds), *España y el Sacro Imperio. Procesos de cambio, influencias y acciones recíprocas en la época de la 'europeización'. Siglos XI–XIII* (Valladolid, 2002), pp. 179–272; P. Josserand, 'L'Ordre de Santiago en France au Moyen Age', in *Saint Jacques et la France* (Paris, 2003), pp. 451–68; A. Mur i Raurell, 'Relaciones europeas de las Órdenes Militares hispánicas durante el siglo XIV', in K. Herbers and N. Jaspert (eds), *Spanien und das römisch–deutsche Reich vom 14. Jahrhundert bis zum Beginn des habsburgischen Großreiches: Konstruktionen des Eigenen und des Fremden* (Münster and Berlin, 2004), pp. 135–84; A. Mur i Raurell, 'Gracia la inspiración y a pesar de la desconfianza: la Orden de Calatrava en el área germánica y este de Europa', *Cistercium*, 238 (2005), 213–52.

royal patronage to these areas as well. A study covering both periods can provide interesting insights into these orders' activities beyond the Iberian Peninsula, particularly when their presence in the kingdoms of Sicily and Naples before and after the establishment of Aragonese supremacy is compared.

A first significant case for consideration is the Order of Mountjoy, founded in the 1170s and soon, in the decades that followed, to be merged with other orders. Its founder, Rodrigo Álvarez from Leon, made a pilgrimage to the Holy Land in 1176–77 and named it Mountjoy after the *Mons Gaudii*, the hill from which the pilgrims could first glimpse the Holy City.[5] The Order also received some property in the Latin East, where Rodrigo himself promised to fight.[6] These facts imply that the founder of the Order was already planning for international expansion. However, the real focus of this short and, in the end, unsuccessful Mediterranean development was not the Holy Land but Italy. In 1180 Rodrigo was in Italy in order to secure the papal confirmation of the property already acquired by his Order.[7] That year he also received a house close to a bridge on the River Mallone in Piedmont as a gift from the marquis William V of Montferrat.[8] William was well connected to the Latin East thanks to his participation in the Second Crusade and the marriage of his son William Longsword to Sybilla, the heiress to the throne of Jerusalem.[9] But perhaps William's gift was not intended simply to establish a connection with an order that was supposed to fight in the East. His family's patronage of religious institutions in Piedmont also had clear political aims, since the monasteries and hospitals supported by the marquises were usually located in

---

[5]   On this Order see A.J. Forey, 'The Order of Mountjoy', *Speculum*, 46 (1971), 250–66, repr. idem, *Military Orders and Crusades*, Variorum Collected Studies Series (London, 1994), XI; N. Jaspert, 'Transmediterrane Wechselwirkungen im 12. Jahrhundert. Der Ritterorden von Montjoie und der Templerorden', in R. Czaja and J. Sarnowsky (eds), *Die Ritterorden als Träger der Herrschaft: Territorien, Grundbesitz und Kirke* (Toruń, 2007), pp. 257–78; A. Demurger, 'Belchite, le Temple et Montjoie: la Couronne d'Aragon et le Temple au XIIᵉ siècle', in N. Housley (ed.), *Knighthoods of Christ: Essays on the History of the Crusades and the Knights Templar, Presented to Malcolm Barber* (Aldershot, 2007), pp. 123–35.

[6]   *Codice diplomatico del sacro militare ordine gerosolimitano*, ed. S. Paoli (Lucca, 1733–37), 1, no. 63, p. 63; J. Delaville Le Roulx, 'Inventaire de pièces de Terre Sainte de l'Ordre de l'Hospital', *ROL*, 3 (1895), no. 119, p. 61; *RRH*, nos. 553, 553a; H.E. Mayer, *Die Kanzlei der lateinischen Könige von Jerusalem*, MGH Schriften 40 (Hannover, 1996), 2, pp. 383–5, 875–6.

[7]   *PUTJ*, 2, no. 122, pp. 309–11 and nos. 125–6, pp. 315–21.

[8]   Ibid., 2, no. 125, p. 317; Forey, 'The Order of Mountjoy', p. 254.

[9]   WT, p. 760. On William Longsword see G. Ligato, 'Guglielmo Lungaspada di Monferrato e le istituzioni politiche dell'Oriente latino', in L. Balletto (ed.), *Dai feudi monferrini e dal Piemonte ai nuovi mondi oltre gli oceani* (Alessandria, 1993), pp. 153–85; B. Hamilton, *The Leper King and his Heirs: Baldwin IV and the Crusader Kingdom of Jerusalem* (Cambridge, 2000), pp. 109–18; G. Ligato, *Sibilla regina crociata. Guerra, amore e diplomazia per il trono di Gerusalemme* (Milan, 2005), pp. 57–88.

strategic areas in their dominions.[10] The gift to Mountjoy might therefore have also had a local relevance, being targeted at keeping the bridge close to the Mountjoy house under the marquises' indirect influence.

In 1181 Fr Fralmo of Lucca, on behalf of the Order of Mountjoy, received some lands close to Susa from the sons of the local castellan who were also interested in possibly donning the Order's habit in the future.[11] The same year Fralmo was entrusted with running a hospital in Savona.[12] He then succeeded Rodrigo as head of the Order, becoming the only non-Spanish master of a Spanish military order. At least one other member of Mountjoy, Hildebrand, was from Lucca.[13]

In the end, problems in recruitment, a lack of precise identity, and, above all, internal schisms led Mountjoy to an early and troubled disappearance. No traces of its Italian houses survive in the sources after the references to their establishment. However, the attempt at recruiting internationally, even though it was most probably very limited, and the settlements in places such as Susa, Savona and Montferrat, which were connected to the Latin East and not to Spain, can be interpreted as evidence for an embryonic but clear effort to expand beyond the Iberian Peninsula and attain international significance.[14]

Shortly after the appearance of Mountjoy in Piedmont, another Spanish military order is attested to in the same area. Before 1184 the Order of Santiago received *terram et possessiones* in Carentino, a village close to Alessandria, from Marquis Alberto of Incisa (1157–c.1188).[15] It has been convincingly argued that representatives of Santiago travelled through France in the 1180s looking for donations, alms and recruits.[16] In 1184 Pope Lucius III confirmed the Order's

---

[10]   R. Bordone, 'I marchesi di Monferrato e i cavalieri di San Giovanni di Gerusalemme durante il XII secolo', in G. Soldi Rondinini (ed.), *Il Monferrato: crocevia politico, economico e culturale tra Mediterraneo e Europa* (Ponzone, 2000), pp. 73–4, 85; A.A. Settia, '"Postquam ipse marchio levabit crucem". Guglielmo V di Monferrato e il suo ritorno in Palestina (1186)', in Soldi Rondinini (ed.), pp. 102–7.

[11]   AHN OO.MM, carpeta 582, no. 34; Forey, 'The Order of Mountjoy', p. 255.

[12]   *Il Cartulario di Arnaldo Cumano e Giovanni di Donato (Savona, 1178–1188)*, eds L. Balletto, G. Cencetti, G. Orlandelli and B.M. Pisoni Agnoli (Rome, 1978), part 2, no. 869, pp. 454–5.

[13]   Forey, 'The Order of Mountjoy', pp. 255ff.

[14]   For a more detailed discussion on the international ambitions of the Order of Mountjoy, see E. Bellomo, 'Fulfilling a Mediterranean Vocation: The *Domus Sancte Marie Montis Gaudii de Jerusalem* in North-west Italy', in H.J. Nicholson (ed.), *On the Margins of Crusading: The Military Orders, the Papacy and the Christian World* (Farnham and Burlington, VT, 2011), pp. 13–29..

[15]   *Bullarium equestris Ordinis Sancti Iacobi de Spatha*, eds A.F. Aguado de Córdova, A.A. Alemán y Rosales and J. López Agurleta (Madrid, 1719), no. 5, p. 31; D.W. Lomax, *La Orden de Santiago. 1170–1275* (Madrid, 1965), pp. 21, 116; Mur i Raurell, 'Gracia la inspiración', p. 236.

[16]   Lomax, *La Orden de Santiago*, p. 160; J.L. Martín Rodríguez, *Orígines de la Orden Militar de Santiago. 1170–1195* (Barcelona, 1974), pp. 89–90; Josserand, 'L'Ordre

property, and it is possible that members of the Order passed through Alberto's dominions on their way to the papal curia. A personal acquaintance with the Order might provide a sufficient justification for his gift. Moreover, it seems that the marquis's example was later followed by others. A document issued in 1240 deals with the Order of Santiago's house at Alessandria, whose revenues were being handed over to a lay affiliate of the Order. The charter does not mention Carentino (which is 21km from Alessandria) and refers only to one house. In all likelihood, other donations enabled the Order to establish itself in Alessandria, and this convent became the Order's most important settlement in the area. What is significant, however, is that in 1240 Santiago was not interested in administering its possessions there directly.[17]

The features of Santiago's appearance in southern Italy are markedly different. It only began in 1251 when the Order was entrusted with the church of S. Spirito de Maitin in the diocese of Salpi (close to Foggia). A highly influential personality, Pietro Capocci, cardinal deacon of San Giorgio al Velabro and papal legate, assigned this church and all its possessions to the Spanish military order, specifying that the grant had been solicited by Santiago itself.[18] Just a few years earlier, Santiago had been involved in planning a military expedition to the Latin Empire of Constantinople (eventually unrealized), which was conceived during the council of Lyons in 1245.[19] That council could also have been the occasion for a meeting between Pietro Capocci and the master of Santiago, Pelayo Pérez Correa (1243–72).[20] It has been argued that Santiago's arrival in Apulia was planned in order to support the military aid promised to Baldwin II of Constantinople.[21] Furthermore, in the same period Bohemond V of Antioch had probably asked for the military intervention of Santiago in the Latin East.[22] In this context, the acquisition of a convent in Apulia could have been the first step in a plan for Mediterranean expansion. However, even if the idea of settling in Apulia was initially linked to military campaigns in the eastern Mediterranean that eventually

---

de Santiago', pp. 456–7.

[17] AHN OO.MM, carpeta 58, no. 6; Lomax, *La Orden de Santiago*, p. 38.

[18] *Bullarium equestris Ordinis Sancti Iacobi de Spatha*, no. 2, p. 187; A. Javierre Mur, 'Un contacto de la Orden de Santiago con la Puglia en el tempo de Corrado de Scavia', *Archivio Storico Pugliese*, 13 (1960), no. 1, pp. 93–4.

[19] R.L. Wolff, 'Mortgage and Redemption of an Emperor's son: Castile and the Latin Empire of Constantinople', *Speculum*, 29 (1954), 82–4; E. Benito Ruano, *Estudios Santiaguistas* (León, 1978), pp. 31–60; Josserand, *Église*, pp. 271, 607, 611.

[20] A. Paravicini Bagliani, *Cardinali di curia e 'familiae' cardinalizie dal 1227 al 1254* (Padua, 1972), 1, pp. 302–3; idem, 'Capocci Pietro', in *Dizionario Biografico degli Italiani* 18 (Rome, 1975), p. 605; Benito Ruano, pp. 42–3; Josserand, *Église*, p. 611.

[21] A. Mur i Raurell, 'Gli Ordini di Calatrava e di Santiago in Puglia nel secolo XIII', in *Sulle orme dei Calatrava. I rapporti tra la Spagna e l'abbazia Sancti Angeli di Orsara di Puglia nel XII e XIII secolo* (Orsara, s.d.), p. 34.

[22] *Primera historia de la orden de Santiago*, ed. A. de Vargas Zúñiga (Badajoz, 1978), p. 369; Josserand, *Église*, p. 604.

did not materialize,[23] the church of S. Spirito was granted to Santiago at a time when it was apparent that the Order was not going to fulfil its military obligation on the Greek side of the Adriatic Sea or to fight the infidels in the Holy Land.

Bartholomew, the officer who received the church of S. Spirito on the Order's behalf in 1251, was its master in Italy, and the appointment of a dignitary in charge of overseeing the Order's Italian convents could imply that Santiago's plan to settle in Apulia was part of an emergent scheme of expansion in the Peninsula. According to the *Kalendarium* de Uclés, Pelayo Pérez Correa also 'established new convents in Lombardy', and, even though this information is not confirmed by any contemporary evidence, his mastership was definitely marked by a positive approach towards opportunities to recruit, settle and fight outside the Iberian Peninsula.[24]

The Order of Santiago persisted in its interest in acquiring S. Spirito even though the takeover proved to be difficult. In 1266 the Order was not yet serving the church, and popes Alexander IV (1254–61) and Clement IV (1265–68) were both asked to intervene and confirm its transfer to the Spanish Order.[25] A charter of 1251 mentions the possibility that Emperor Frederick II might already have granted the church of S. Spirito to another assignee, and the invalidity of any other transfer was reinforced in later papal letters relating to S. Spirito. In all likelihood it was competition with another assignee, chosen by the Hohenstaufen, that prevented the Order from obtaining S. Spirito. Since Frederick II had built one of his palaces in Salpi, it seems plausible that he had also tried to control this neighbouring church.[26] This reconstruction suggests that Pietro Capocci's action was mainly related to the contemporary strife against Frederick's heirs.[27] Pietro had been personally involved in the papal military campaign against the emperor shortly before promoting the arrival of the Order of Santiago in Apulia.[28] By handing S. Spirito over to the Order, he could both fulfil the request of the Spanish Order to settle in the region and try to inhibit Hohenstaufen rule in the Salpi area. Unfortunately, further thirteenth-century information on S. Spirito does not survive. According to Derek Lomax, Santiago succeeded in securing this church in the following century, and, just as happened in Piedmont, it remained an isolated acquisition, administered together with the Order's Sicilian houses and

---

[23]   Javierre Mur, p. 93.

[24]   Lomax, *La Orden de Santiago*, p. 21; Rodríguez García, pp. 191–209; Ayala Martínez, pp. 530, 633; Josserand, *Église*, pp. 615–16; Mur i Raurell, 'Gracia la inspiración', pp. 213–52.

[25]   *Bullarium equestris Ordinis Sancti Iacobi de Spatha*, no. 3, pp. 187–8; nos. 1–2, pp. 191–2; no. 1, p. 206; Javierre Mur, nos. 2–5, pp. 94–6.

[26]   R. Licinio, *Castelli medievali. Puglia e Basilicata: dai Normanni a Federico II e Carlo I d'Angiò* (Bari, 1994), pp. 128–9.

[27]   Josserand, *Église*, p. 608 n. 152.

[28]   Paravicini Bagliani, 'Capocci Pietro', pp. 604–7.

property.[29] What is significant, however, is that Santiago demonstrated a persistent determination to get hold of this church. This obstinacy is consistent with the enduring interest in international openings shown by Master Pelayo Pérez Correa, and it gained approval in papal circles. Nevertheless, in the end the Order was unable to establish a major presence in continental southern Italy.

The foundation of the Order of Santiago's first Sicilian convent in Lentini took place in a completely different context. The creation of this house is linked to Riccardo da Passaneto da Lentini, a prominent member of the local Norman nobility. After the Sicilian Vespers, Riccardo had promptly joined the Aragonese party. He had actively supported the Aragonese conquest of Sicily, fighting alongside King Peter III, and it was perhaps in this period that he met Ruy Himenez de Luna, later master of Montalbán, who played a noteworthy role in the Aragonese conquest of the island. Afterwards Riccardo acted as a public officer in the newly established kingdom of Sicily. Riccardo's familiarity with the Aragonese royal family is also borne out by the fact that he borrowed a book from the young King James II, who solicited its return in 1293. For their services and loyalty towards Frederick III of Sicily, the Passanetos were eventually rewarded with the title of Count of Grasiliato.[30] In 1309 Riccardo wrote to James II asking the king to commend him to the master of Santiago, Juan Osórez. In his letter he qualified himself as *preceptor ordinis militie sancti Iacobi in regno Sicilie*.[31] The first attestation to the Santiago house of Lentini, most probably founded by Riccardo, dates to 1317 when, probably shortly after Passaneto's death, the new master of the house asked for the endorsement of his election.[32] Later evidence shows that the convent was housed in the former palace of the Passaneto family, who also controlled the appointment of its master. Thanks to the generosity of his founder, the influential connections with local nobility and the crown's favour, the house of Lentini became the Order's most important property in Sicily.[33] In this case it was the enforcement of Aragonese control on the island, which concurred with the Passanetos' political choices and family strategies, that decisively paved the way for the expansion of Santiago's presence beyond the Iberian Peninsula.

In the thirteenth century the Order of Calatrava also settled in Italy. Its most important house in the Peninsula was at Orsara, close to Foggia, which had belonged to the Order since 1229. Even before being assigned to the military order, the monastery of S. Angelo di Orsara had strong links with the Iberian Peninsula. Its monks and superiors were most probably recruited in Leon and Asturias; it had a priory at Bamba in the diocese of Zamora and it also enjoyed royal patronage in

[29]   Lomax, *La Orden de Santiago*, p. 21.
[30]   L. Sciascia, 'Riccardo Passaneto e la Commenda dei Cavalieri di Santiago di Lentini', in G. Arlotta (ed.), *Santiago e la Sicilia* (Perugia, 2008), pp. 146–50.
[31]   Ibid., p. 149.
[32]   Ibid., pp. 151–2.
[33]   Ibid., pp. 151–4.

Leon.[34] Nevertheless, in the first decades of the thirteenth century, the monastery was facing a severe crisis that was to lead Pope Honorius III to entrust it to the Order of Calatrava in hope of improving its critical situation.[35]

Two Iberian cardinals, Pelayo (or Pelagius) Gaitán (c.1165–1230), cardinal bishop of Albano, and Gil Torres (1206–1254), cardinal deacon of SS. Cosmas and Damian, who had already overseen the sale of Orsara's Spanish priory at Bamba to the bishop of Zamora in 1225, acted as intermediaries in the transfer of S. Angelo to Calatrava.[36] The activities of Pelayo Gaitán are particularly interesting since some of them are connected to southern Italy and the crusades. A renowned canonist, Pelayo had a brilliant career in the papal entourage. Moreover, he had been on a diplomatic mission to Constantinople in 1213 and later on was offered the patriarchate of Antioch. Appointed papal legate during the Fifth Crusade (1217–21), he was then entrusted with the difficult task of persuading Frederick II to fulfil his crusading vow. Later on Pelayo was called to guide in person the papal forces in the struggle against the emperor.[37]

It was also true in the case of Calatrava that Apulia could form the perfect base for future expansion into the eastern Mediterranean. As early as 1206 the Order had considered engaging in military activities in the Latin East.[38] Perhaps Pelayo Gaintán, whose career was closely linked to the Holy Land, hoped that Calatrava's possession of Orsara could foster its involvement in the East. Four years after the arrival of the Knights of Calatrava in Apulia, Gregory IX tried to further their presence in Antioch.[39] Unfortunately we cannot prove that these events were connected.

---

[34] R. Hiestand, 'San Michele in Orsara. Un capitolo dei rapporti pugliesi-iberici nei secoli XII–XIII', *Archivio Storico Pugliese*, 44 (1991), 67–79; Mur i Raurell, 'Gli Ordini di Calatrava', pp. 10ff.

[35] *Bullarium Ordinis militiae de Calatrava*, eds I.J. de Ortega y Cotes, J.F. Álvarez de Baquedano and P. de Ortega Zúñiga y Aranda (Madrid, 1761), no. 3, pp. 60–61; *Les registres de Gregoire IX*, ed. L. Auvray (Paris, 1890–1955), no. 286.

[36] *La documentación pontificia de Honorio III (1216–1227)*, ed. D. Mansilla (Rome, 1965), no. 570, pp. 422–4. On Gil Torres, see P. Linehan, *The Spanish Church and the Papacy in the Thirteenth Century* (Cambridge, 1971), pp. 276ff.

[37] On Pelayo, see S. Kuttner, *Repertorium der Kanonic (1140–1234): Prodromus corporis glossarum* (Vatican City, 1937), pp. 53–4; J.P. Donovan, *Pelagius and the Fifth Crusade* (Phildelphia, PA, 1950); D. Mansilla, 'El cardinal hispano Pelayo Gaitán (1206–1230)', *Anthologica annua*, 1 (1953), 11–66; Linehan, pp. 279ff; B. Hamilton, *The Latin Church in the Crusader States: The Secular Church* (London, 1980), pp. 182, 224–7, 254, 301–2, 317; A. García y García, 'La canonística ibérica (1150–1250) en la investigación reciente', *Bulletin of Medieval Canon Law*, 11 (1981), 41–75; W. Maleczek, *Papst und Kardinalkolleg von 1191 bis 1216: die Kardinäle unter Coelestin III. und Innocenz III.* (Vienna, 1984), pp. 166–9; J.M. Powell, *Anatomy of a Crusade, 1213–1221* (Philadelphia, PA, 1986), *passim*; Paravicini Bagliani, *Cardinali*, 1, pp. 11–12.

[38] *Bullarium Ordinis militiae de Calatrava*, no. 8, pp. 39–40.

[39] Ibid., no. 10, p. 67; Mur i Raurell, 'Gli Ordini di Calatrava', pp. 21–4.

It has also been hypothesized that Calatrava's settlement in Orsara was supported by the papacy in order to establish a military presence that could control the neighbouring Muslim community of Lucera, created by Frederick II.[40] The truth is that Calatrava received the monastery of S. Angelo in a successful phase of the papal campaign against the Hohenstaufen – led by Pelayo Gaitán among others. S. Angelo lay on the road connecting Benevento and Gargano,[41] and it is possible that the cardinal backed the establishment there of an Order with which he was familiar with a view to controlling this area. It cannot therefore be ruled out that Calatrava's arrival in Apulia was bound up with anti-Hohenstaufen endeavours.

In 1295 Pope Boniface VIII granted S. Angelo and all the properties of Orsara to Archbishop Filippo of Trani. Boniface himself had received the Order's possessions from the master of Calatrava, Ruy Pérez Ponce (1285–96), before becoming pope.[42]

Then, in 1300, following the death of Filippo of Trani, the new master of Calatrava, García López de Padilla (1297–1329, 1329–36 in Alcañiz), exchanged S. Angelo di Orsara and all its possessions for some property located in Castile that belonged to Ferdinand IV of Castile (1295–1312).[43] Most probably this transaction did not have a positive conclusion, and four years later the master handed over all Calatrava's Italian possessions to John, the son of James II of Aragon.[44] By the beginning of the fourteenth century Calatrava had acquired churches and property in Apulia; there were possessions, whose exact location has still to be traced, in Sicily and Calabria.[45] However, these scattered properties were probably unable to produce considerable profit and were not easy to run in an effective way. Moreover, this transfer has to be seen against both the positive relationship between the master of Calatrava and the king, and the hegemonic ambitions of the Aragonese Crown in the central Mediterranean. The handover of these possessions to the infante could provide a useful footing for the Aragonese in southern Italy and at the same time strengthen the ties between the master of Calatrava and King James.[46]

Papal documents issued at the turn of the thirteenth and fourteenth centuries further mention Calatrava with reference to Italy. The Order is listed among the addressees of some papal letters directed to several areas of Italy and Latin

---

[40]   L. Cotugno, *Orsara di Puglia. Notizie storiche* (Foggia, 1999), p. 22.

[41]   C. Raimondi, 'Sulle tracce dei Cavalieri di Calatrava: cenni sulla presenza dell'ordine in età medievale con particolare riferimento alla Puglia', *Rivista cistercense*, 11 (1994), 213.

[42]   *Les registres de Boniface VIII*, eds G. Digard *et al.* (Paris, 1889–1939), no. 696.

[43]   *Bullarium Ordinis militiae de Calatrava*, no. 3, pp. 154–6.

[44]   P.C. Picatoste, 'Intereses translapinos de Jaime II en la época de la conquista del Reino de Murcia. La donación de los Calatravos al infante Juan en 1304', in *Actas del Congreso Internacional Jaime II 700 años después, Anales de la Universidad de Alicante. Historia Medieval*, 11 (1996–97), pp. 463–4.

[45]   *Les registres de Boniface VIII*, no. 696; Picatoste, p. 463; Hiestand, 'S. Michele', pp. 72–3, 77; Raimondi, pp. 210–12; Mur i Raurell, 'Gli ordini di Calatrava', pp. 24–5.

[46]   Picatoste, pp. 457–62; Josserand, *Église*, pp. 531–4.

Romania. A letter from 1290 was addressed to several military orders present in Italy, including Santiago and Calatrava. In 1301 another papal missive was dispatched to, among others, members of Calatrava in Italy, Sicily, Sardinia, Corsica, the principality of Achaia, the duchy of Athens and the neighbouring islands. Another document issued in 1304 was similarly sent to the brethren of several military orders, among which Calatrava, *per provincias Tuscie, Romaniole, Marchie Tarvisine ac partes circumadiacetes constituti.*[47]

This information has to be used with great caution since the presence of Calatrava among the military orders listed in these documents could simply be attributable to the erroneous use of a standard list for these letters rather than to the actual presence of the Order in these regions.[48] Moreover, it could be that Calatrava was present in only one of the regions mentioned in a letter. Further evidence would therefore be needed to confirm Calatrava's ownership of these places. It is worth noting that Calatrava's property in Romania is only mentioned in Boniface VIII's 1295 grant of S. Angelo's possessions and in the transfer to John of Aragon in 1304. Unfortunately, these references are so vague that it is not possible to ascertain whether the Order was actually present in both central Italy and Latin Romania, the two regions mentioned in the papal documents. We can hope that future research will clarify whether any expansion towards the East had actually taken place, with the Calatrava house of Orsara as a point of reference.

A final Spanish military Order with significant connections with Italy is San Jorge de Alfama. Between 1355 and 1363 the Order received from Aragonese authorities in Sardinia some property and a house in Alghero, the monastery of San Saturnino in Cagliari, and the church of S. Maria in Uta. Presumably Peter IV of Aragon wanted to reward the Order for its support in the pacification of the island. Some local scholars believe that he also hoped to improve the situation of a monastery – San Saturnino – and an area – Uta in the county of Quirra – that were both experiencing a period of acute crisis. In addition, S. Saturnino was strategically located between Cagliari castle and the hill of Bonaria.[49] However, San Jorge de Alfama never settled in Sardinia. The Order lacked a precise structure and, despite its foundation dating from 1201, formal papal approval was only given much later, in 1373. By 1400 its few members had been integrated into the Order of Montesa.[50]

---

[47] *Les registres de Boniface VIII*, nos. 4316–17, 5127; *Les registres de Bénoit XI*, ed. C. Grandjean (Paris, 1905), nos. 1171–2.

[48] Josserand, *Église*, p. 619.

[49] On these houses see L. D'Arienzo, 'San Saturno di Cagliari e l'ordine militare di S. Giorgio di Alfama', in *Actas del Congreso Internacional hispano-portugués sobre 'Las Órdenes militares hispánicas en la Península Ibérica durante la Edad Media'*, Anuario de Estudios medievales, 11 (1981), 823–52, also published in *Archivio Storico Sardo*, 34 (1983), 43–80.

[50] On this Order see R. Sáinz de la Maza Lasoli, *San Jorge de Alfama. Aproximación a su historia* (Barcelona, 1990); Ayala Martínez, pp. 130–31.

To conclude: effective patronage and local connections always played a crucial role in the successful establishment of a religious order in a new environment. Obviously this applies to the Spanish military orders in Italy, whose presence in the Peninsula, Sicily and Sardinia was also conditioned by other significant factors. The international aspirations of Mountjoy were frustrated by the weakness of this short-lived Order. Likewise, San Jorge de Alfama disappeared shortly after receiving houses and lands in Sardinia. Santiago and Calatrava were solid institutions when they tried to build a sizeable presence in Italy. Nevertheless, even the powerful patronage of cardinals and popes, possibly tinged with crusading and anti-Hohenstaufen considerations, proved insufficient to secure a reliable local footing for these orders. Italian evidence clearly demonstrates that the Spanish military orders had international ambitions but that in the end only the expansion of the Aragonese dominions could and did create the essential conditions for their enduring presence in Italy. This international development was therefore only successful as an aspect of national expansionism.

Chapter 22

The Hospitallers in Southern Italy: Families and Power

Mariarosaria Salerno

The defeat of Manfred, the natural son of Frederick II, at the hands of Charles of Anjou, the brother of the king of France, at Benevento in 1266 represented a victory for the papacy over the empire and the beginning of a new dynasty in the kingdom of Sicily.[1] Once he had become king, Charles put an end to the seizure of Hospitaller lands, and on 9 August 1268 he ordered the return of properties occupied, or encroached upon, by Frederick and his sons to their lawful owners.[2] The Hospital, which on behalf of the pope supported Charles against Conradin and his followers, regained its properties, but the conflict had left both the kingdom and the papacy in debt. Pope Clement IV appealed to all his loyal followers to raise money, and in 1268 he instructed Rodolphe de Chevrières to collect the tithe for Charles of Anjou from all exempt orders such as the Cistercians, the Templars and the Hospitallers.[3]

In an economic and financial context characterized by an oppressive tax system, by the debts accumulated by the Angevins, who had to deal with the long and disastrous War of the Sicilian Vespers, and by the difficulties the dynasty had in guaranteeing the loyalty of the aristocracy, the Hospitallers gained the confidence of the sovereigns, and profited from and returned their favours.

Southern Italy, especially the region of Apulia, was the nearest part of western Europe to the Latin East, and so it was able to provide the necessary imports, especially cereals, at the lowest transport cost. The royal favours therefore took the form not only of guarantees for their territorial possessions, but also of the frequent intervention of the kings or their officers to allow the Hospitallers a free hand to export, without paying tax, their merchandise, animals and weapons, first to Syria and later to Cyprus and Rhodes. This authorization was fundamental for Hospitaller economic and commercial development. The Order's great ships left

---

[1]   E.G. Léonard, *Gli Angioini di Napoli* (Varese, 1967); S. Runciman, *I Vespri siciliani* (Milan, 1986), p. 124.

[2]   *I Registri della Cancelleria angioina*, eds R. Filangieri *et al.* (Naples, 1963–), 1, no. 349, pp. 182–4.

[3]   *CH*, 3, nos. 3122, 3279, 3318, 3321; Runciman, p. 140.

from the Apulian ports of Brindisi and Manfredonia with aid for the brethren, their retainers and the poor.[4]

In his native France, Charles of Anjou was overlord of Provence and the county of Folcalquier to the east of the Rhône. A large number of French families came to southern Italy in his wake. At first they supported Charles's military actions and then, after he became king of Sicily, they were rewarded with fiefs or benefices, settled in the kingdom and became 'Italians'.[5] There was a strong French–Provençal component in the Hospital, making up a majority in the Convent, with the result that in the kingdom of Sicily there was a special relationship between the Hospitallers and the Angevin monarchy.[6] Charles I pursued an ambitious Mediterranean policy, and in 1277 he became king of Jerusalem by acquiring rights to that kingdom from Mary of Antioch.[7]

In southern Italy the Hospital was organized into three priories – Capua, Barletta and Messina – and the so-called 'bayllis par chapitre general'. The latter were independent entities within the priories, directly dependent on the General Chapter of the Convent, composed primarily of French–Provençal brethren, which could use them directly or could assign them to the Master.[8] The 'bayllis par chapitre general' were the commanderies of Naples, the county and *excambium* of Alife, Sant'Eufemia in Calabria, SS. Trinità of Venosa in the Basilicata and S. Stefano of Monopoli in Apulia.

This situation enhanced the position of French–Provençal Hospitallers in the *domus* of southern Italy, especially the officers who controlled the priories and capitular commanderies. However, it is not easy, particularly as we come into the fourteenth century, to establish simply through a surname whether the *fratres* in question were Provençal or whether they were simply of Provençal origin, having settled in the kingdom of Sicily and become 'Italians'. Between 1266 and the end of the fourteenth century, at least five out of the fifteen priors of Capua were French–Provençals, and the Italian origin of the others is not always certain. However,

---

[4]   J.M. Martin, 'Fiscalité et économie étatique dans le Royaume angevin de Sicile à la fin du XIIIe siècle', in *L'Ètat angevin. Pouvoir, culture et société entre XIIIe et XVIe siècle* (Rome, 1998), p. 614; *Registri*, 3, p. 189 and 29, p. 81; *CH*, 3, nos. 3401, 3423.

[5]   S. Pollastri, 'La noblesse provençale dans le royaume de Sicile (1265–1282)', *Annales du Midi*, 184 (1988), 404–34; S. Kelly, 'Noblesse de robe et noblesse d'esprit à la cour de Robert de Naples. La question d'italianisation', in N. Coulet and J.M. Matz (eds), *La noblesse dans les territoires angevins à la fin du Moyen Âge* (Rome, 2000), pp. 347–61.

[6]   A. Luttrell, 'Change and Conflict within the Hospitaller Province of Italy', in *Studies on the Hospitallers after 1306: Rhodes and the West*, Variorum Collected Studies Series (Aldershot, 2007), XV, p. 189.

[7]   'Les Gestes des Chiprois', *RHC Arm.*, 2, p. 777; J.L. LaMonte, *Feudal Monarchy in the Latin Kingdom of Jerusalem* (Cambridge, MA, 1932), pp. 77–9.

[8]   M. Salerno, *Gli Ospedalieri di San Giovanni di Gerusalemme nel Mezzogiorno d'Italia (secc. XII–XV)* (Taranto, 2001), pp. 110–13; A. Luttrell, 'Le origini della Precettoria Capitolare di Santo Stefano di Monopoli', in *Studies*, XIV, p. 90; M. Salerno, *Le precettorie capitolari degli Ospedalieri di San Giovanni di Gerusalemme (secc. XIII–XIV)* (Bari, 2009).

only Italians can be found after 1365. At the priory of Barletta, at least nine out of seventeen of the documented priors were French–Provençal, but only Italians held office after 1375. Before the War of the Sicilian Vespers, at the priory of Messina three out of twenty-two were French–Provençal. At Alife, all the known commanders bore Provençal names. At Naples, when the *domus* became a capitular commandery, four out of seven were Provençal. At Sant'Eufemia, there were seven out of the fourteen documented and, from 1347 onwards, they were all Provençal. At Venosa, five out of the fourteen mentioned by name certainly had a French–Provençal background. At Monopoli, the same was true of eight out of eleven.[9]

Some *fratres* received support from the Convent in the form of high-ranking positions, and, from the Angevins, through political office. In the early years of the reign of Charles I, some men in his entourage, or their relatives, were also prominent Hospitallers. First among these was Jacques de Taxi, Charles' counsellor and *familiare*, seconded by the master, Hughes Revel. De Taxi arrived in Sicily with the king in 1265 and became prior of Messina. Between 1268 and 1269, he obtained significant funding for the Hospital from the king and was able to discourage the harassment of the *fratres* and the Order's properties; in 1271 he took command of the castle of Reggio Calabria after the capitulation of the city that had fallen into the hands of Conradin's supporters. After the disastrous end to King Louis IX's crusade, Charles I sent De Taxi on a diplomatic mission to the king of Tunis. He was entrusted with the royal treasure to take back to Trani, and he sold the king a new, fully equipped ship. De Taxi also continued to hold prestigious offices in the Hospital: he was probably prior of Messina until 1275; he was prior of Barletta from 1277 to 1281; and, in 1284, grand preceptor of Acre.[10] From the list compiled by Paul Durrieu of the 'French' who came to Italy as followers of Charles I, another de Taxi emerges, Jean, who, between 1278 and 1284, was a knight, member of the king's *familia*, and lord of the castle of Valona.[11] In the same period, Philippe d'Egly was the ruthless enforcer of papal directives in repressing Conradin's Sicilian partisans.[12] In 1269–70 Fr Peter, Hospitaller in Barletta, was the king's almoner and was carrying out repairs to royal ships; in 1271 Fr Simone de Letto (Lettre), almoner and member of the king's *familia*, was commander of

---

[9]   M. Salerno and K. Toomaspoeg, *L'inchiesta pontificia del 1373 sugli Ospedalieri di San Giovanni di Gerusalemme nel Mezzogiorno d'Italia* (Bari, 2008), pp. 33, 72–3; K. Toomaspoeg, *Templari e Ospitalieri nella Sicilia Medievale* (Bari, 2003), pp. 247–52; Salerno, *Le precettorie*; F. Tommasi, 'L'ordinamento geografico-amministrativo dell'Ospedale in Italia (secc. XII–XIV)', in A. Luttrell and F. Tommasi (eds), *Religiones militares* (Città di Castello, 2008), pp. 102–3.

[10]   *CH*, 2, nos. 3347–8, 3473, 3483; 3, nos. 3498, 3503, 3717; Toomaspoeg, pp. 70–72; *Registri*, 15, pp. 37, 41; 16, p. 138; 20, p. 182; 24, pp. 82, 122–3; 37, p. 365; J. Delaville Le Roulx, *Les Hospitaliers en Terre Sainte et à Chypre (1100–1310)* (Paris, 1904), p. 419.

[11]   P. Durrieu, *Les archives angevines de Naples* (Paris, 1887), 2, p. 388; *Registri*, 25, p. 35; 27, pp. 365, 367, 369.

[12]   Tommasi, p. 103.

Aversa.[13] Among the clerics of the royal *hospicium* or *familia*, the almoner had preferential treatment, as had the cleric-doctors and the chaplain. In 1280, the Hospitaller, Simone de Breban, was Charles I's chaplain, and was involved in a mission to Acre for the king.[14] Two other de Brebans – Gossequin, a knight, and Henri, a royal squire – are also documented as followers of Charles I at the same period.[15] In 1277 Fr Raymond from Avignon was overseeing the excavation of the mines in Calabria on the king's behalf.[16]

In 1283, in the middle of the War of the Vespers, when the Angevins and Aragonese fought for control of Sicily, the priors of Capua, Sant'Eufemia and Barletta, along with many bishops and representatives of other religious orders within the kingdom, were summoned to court to discuss the Sicilian situation. The following year, on 29 April, Prince Charles, the future Charles II, asked the Hospitallers of Barletta and Capua for financial and humanitarian aid.[17] In the years 1278–92 Guillaume de Villaret, prior of St Gilles and later master of the Order, was successively counsellor to Charles I and Charles II.[18] In those years, Pierre de Musac (or Moysac), the preceptor of Sant'Eufemia, was summoned to court, and the king entrusted him with the *capitania* of Nicastro, Maida, Tiriolo and Castiglione, and with the defence of the Lametian coast, mainly against the danger of the Aragonese and the Sicilian rebels. He may be the *Petrus de Meysac* (or *de Moysac*) who was a royal squire in 1278–79; if so, that would prove the close connection between the Hospital and the Angevins. The preceptor, probably taking advantage of his 'political' role, committed outrages in Calabria against properties and knights loyal to the dynasty. In 1284, however, he was replaced by the king because he had to go to Acre to support his *confrères*.[19]

Other significant brethren arrived in the kingdom of Naples during the following century when the papacy was established at Avignon, and we know of Provençal Hospitallers who made careers for themselves thanks to the patronage of the popes or the masters of the Order. Among them was a Provençal official named Isnard du Bar, already commander of Aix-en-Provence, who moved to southern Italy where

---

[13]   *Registri*, 3, p. 79; 7, pp. 197, 268.

[14]   Ibid., 23, p. 279; A.M. Voci, 'La cappella di corte dei primi sovrani angioini di Napoli', *Archivio Storico per le Province Napoletane*, 113 (1995), 69–126.

[15]   Durrieu, p. 292.

[16]   *Registri*, 18, pp. 93, 98.

[17]   Ibid., 26, p. 151; 27, p. 443.

[18]   L. Villari, 'Giovanni de Villers, Gran Maestro dell'Ordine Gerosolimitano, combattente angioino al tempo del Vespro siciliano', *Archivio Storico messinese*, 17–19 (1968), 179–83.

[19]   Durrieu, p. 348; *Registri*, 27, no. 147, p. 229; 26, nos. 302, 344, p. 156; C. Carucci, 'Le operazioni militari in Calabria nella guerra del Vespro Siciliano', *Archivio Storico per la Calabria e la Lucania*, 2 (1932), 10; M. Mafrici, 'Calabria Ulteriore', in *Storia del Mezzogiorno*, 7 (Rome, 1986), p. 100. M. Salerno, 'Templari ed Ospedalieri di San Giovanni in Calabria in età medievale: risultati ed ipotesi', in Luttrell and Tommasi (eds), p. 231.

he was prior of Capua from 1336 to 1365.[20] He held a variety of offices: in 1336 he was lieutenant master in the kingdom of Sicily and rector on his behalf in the county of Alife, becoming preceptor there in 1340.[21] He may also have been commander of Sant'Eufemia in Calabria from 1347. Between 1361 and 1363, he presided over the process whereby the Order disposed of the county of Alife, which had become difficult to maintain because of war. The Hospital decided to exchange the county for properties, mostly located around Naples and Aversa, that were more useful and better suited to its needs.[22] Isnard du Bar behaved with questionable integrity, not least in his dealings with important men of the kingdom including Goffredo Marzano, count of Squillace and admiral of the kingdom, who had occupied Alife for many years; Cristoforo of Costanzo, a Neapolitan knight; and Matteo Capuano, another Neapolitan knight and *magister rationalis* of the royal court, supervisor of the management of royal officials. At the same time Isnard held political office in the kingdom of Naples. He was advisor to and *familiare* of both King Robert and Joan I. Matteo Camera reported that Joan I recommended Isnard to the pope for promotion to the highest dignities. In 1347 Isnard held *ad interim* the office of seneschal of Provence in the absence of Filippo Sangineto; between 1346 and 1347 he was present in the East as captain of the papal and Hospitaller galleys. Later the master, Déodat de Gozon, granted him, in view of 'all his merits', the right to move freely through the priory of Capua and Provence and go to Rhodes.[23] The 'exchanges' were probably among the last acts carried out by Isnard, who, in the meantime, had also added the Neapolitan commandery to his dignities and resided in the capital of the kingdom; he died before 18 December 1365, when the master and the convent designated a successor.[24] Isnard assisted his family members in their entry into the kingdom; his nephew, Monréal du Bar, was preceptor of Naples

---

[20] N. Coulet, 'La noblesse provençale dans l'entourage du roi René', in Coulet and Matz (eds), p. 321; M. Hébert, 'La noblesse et les Etats de Provence', in Coulet and Matz (eds), p. 341. M. Salerno, 'Da domus a sede priorale: l'evoluzione della fondazione giovannita capuana nei suoi aspetti giurisdizionali ed economici', in A. Pellettieri (ed.), *Il Gran Priorato giovannita di Capua* (Matera, 2008), pp. 77–8.

[21] Biblioteca Nazionale di Napoli, MS XV, D, 15, *Reassunto de' Diplomi esistenti nell'Archivio della Regia Zecca appartenenti all'abolito Ordine de' Templari, ed all'attuale S.M. Ordine de' Cavalieri di S. Giovanni di Gerusalemme, compilato sotto gli ordini del signor Balio frà Francesco Antonio Cedronio, ricevitore e ministro dell'Ordine presso S.M. Siciliana, per opera dell'avvocato Felice Parrilli* (1803), XV, XXXVII; Tommasi, pp. 112–14.

[22] Malta, cod. 317, fols 191v, 196r; cod. 280, fol. 46r; F. Russo, *Regesto Vaticano per la Calabria* (Rome, 1974), 1, nos. 7590, 7595, 7635, 7709; G. Bosio, *Dell'istoria della sacra religione et ill.ma militia di San Giovanni Gerosolimitano* (Rome, 1630), 2, p. 62; Luttrell, 'Le origini', p. 93; idem, 'Change', p. 193; Salerno, *Le precettorie*.

[23] M. Camera, *Annali delle due Sicilie* (Naples, 1860), 2, p. 394; Russo, 1, no. 6995; Malta, cod. 317, fol. 191v; K. Setton, *The Papacy and the Levant, 1204–1571* (Philadelphia, PA, 1976), 1, pp. 205, 208–10.

[24] Salerno, *Le precettorie*; Malta, cod. 319, fol. 243v. Salerno, 'Da domus', p. 78.

in 1343. He was also a loyal royal advisor, but documents from the reign of Joan I refer to him as 'Fr Moriale', a brutal leader who became involved in the dynastic struggles and was executed in 1354.[25]

Bertrando of Malobosco, probably another Provençal, is recorded as having been the prior of Barletta between 1307 and 1315 and then between 1322 and 1329, and was described as 'advisor and *familiare* of Robert of Anjou'. In 1312, as prior of Barletta and lieutenant of Master Helion de Villeneuve in the Kingdom of Sicily *citra Farum* and in the priory of Sant'Eufemia, he obtained funds from Robert of Anjou but stole from a shop in Molfetta belonging to the Bonaccorsi family.[26] After 1324 it would seem that he administered the capitular commanderies of Naples and Venosa, firstly as the master's procurator and then, transferred from the Priory of Barletta, as preceptor.[27] His career indicates the assimilation in terms of rank of a capitular commandery to a priory. He was dead by 14 May 1347.[28] We know nothing about the beginning of his career, and so we cannot say whether Bertrando came from Provence or was a descendant of the Malbois who were followers of Charles I; in 1273 there had been a royal knight named Raymond *de Malbosco*.[29]

Garin de Châteauneuf, who had recently come to Italy, had the support of Pope Innocent VI. In 1347 he was preceptor of Jalles and Valdrôme in the priory of St-Gilles in Provence, and prior of Barletta (1347–58). The priory and the capitular commandery of Monopoli were in a 'statum miserabilem', and the General Chapter of the Hospital assigned them to Fr Garin for ten years with the task of recovering properties and the responsions that his predecessors had not paid to the treasury. In 1356 Pope Innocent VI asked the master and the convent to transfer him to his own land or to grant him the commandery of S. Stefano and other lands belonging to the priory. However, the Master ordered him to remain in Monopoli.[30]

---

[25]   Domenico da Gravina, *Chronicon*, ed. L.A. Muratori, *RIS*, 12, col. 557; M. Camera, *Elucubrazioni storico-diplomatiche su Giovanna I regina di Napoli e Carlo III di Durazzo* (Salerno, 1889), p. 149.

[26]   Salerno and Toomaspoeg, pp. 72–3; G. Crudo, *La SS. Trinità di Venosa* (Trani, 1899), p. 350; *Reassunto*, XLV. R. Caggese, *Roberto d'Angiò e i suoi tempi* (Florence, 1922), 1, p. 583.

[27]   Camera, *Annali*, 1, p. 298; 2, p. 283; B. Del Pozzo, *Ruolo generale de' cavalieri gerosolimitani della Lingua d'Italia* (Turin, 1714), s.a. 1340; M. Radogna, *Monografia di S. Giovanni a Mare baliaggio del S.M.O. Gerosolimitano in Napoli* (Naples, 1873), p. 21; M. Gattini, *I priorati, i baliaggi e le commende del sovrano militare ordine di S. Giovanni di Gerusalemme nelle province meridionali d'Italia* (Naples, 1928), pp. 20, 77. C. Tipton, 'The 1330 Chapter General of the Knights Hospitallers at Montpellier', *Traditio*, 24 (1968), 308.

[28]   Crudo, pp. 352–3. Malta, cod. 317, fol. 191v.

[29]   Durrieu, p. 342.

[30]   Malta, cod. 317, fol. 190r; cod. 316, fols 267r–v. Del Pozzo, s.a. 1355; Innocent VI, *Lettres secrètes et curiales*, ed. P. Gasnault (Rome, 1959–66), no. 1897. A. D'Itollo, *I più antichi documenti del libro dei Privilegi dell'Università di Putignano, 1107–1434* (Bari, 1989), pp. 50–51.

In 1365 Fr Bertrand Flotte, of S. Salvatore in the county of Nice, was commander of Monopoli, and in 1373 he was commander of Naples but resident in Rhodes.[31] Gregory XI recommended him to Master Raymond Berenger, granted him the Hungarian Priory, and gave the Neapolitan commandery to Jean Flotte, Bertrand's nephew. Bertrand, though, remained as commander of Naples at least until the end of December 1374. At that time the master, Robert de Juilly, was in Naples; Flotte was ready to leave for the *passagium*, but he needed money. He distinguished himself in the East, becoming Grand Commander in Rhodes, and in 1382 he was granted the commandery of Monopoli for life.[32] His nephew, Jean Flotte, was also rewarded by the master in southern Italy and in March 1382 became commander of Venosa.[33] However, two months later, at a time of political confusion, King Charles III of Anjou-Durazzo, with the support of the Roman pope, Urban VI, gave the commandery of Venosa to the Neapolitan Enrico Dentice, described as a 'devoted follower of the king'.[34]

Another Provençal, Bertrand de Boyson, arrived in the kingdom of Sicily in the 1360s with the backing of Master Raymond Berenger. Already prior of Barletta, he was given the capitular commandery of Sant'Eufemia in 1365. Boyson, who resided in Rhodes, was a leading figure in relations between the Convent and the *domus* of the kingdom of Sicily and was one of the master's trusted men. He was allowed to leave the Convent to perform tasks in Sicily, choose men, arms and horses to send to the East, and organize his preceptory before returning to Rhodes. It seems he was initially opposed in his commandery by Raymo de Sabran, a brother to whom we shall return and who for some time refused to leave the Calabrian house.[35] On 16 June 1370, with Boyson dead and his commandery vacant, Pope Urban V, by an irregular act, designated Manuel Chabaud, Boyson's nephew, his successor at Sant'Eufemia. Chabaud had been in the kingdom of Naples for at least ten years, as is evidenced by the fact that in 1361, as prior of Marigliano, he was among the *fratres* who supported Isnard du Bar in the acts by

---

[31]   Malta, cod. 319, fols 1r–v, 2v; J. Delaville Le Roulx, *Les Hospitaliers à Rhodes jusqu'à la mort de Philibert de Naillac, 1310–1421* (Paris, 1913), pp. 149–51, 185, 197, 205–6; D'Itollo, pp. 60–63; Luttrell, 'Le origini', p. 98; idem, 'The Hospitallers at Rhodes, 1306–1421', *HC*, 3, pp. 299–304; idem, 'Intrigue, Schism and Violence among the Hospitallers of Rhodes, 1377–1384', in *The Hospitallers in Cyprus, Rhodes, Greece and the West, 1291–1440*, Variorum Collected Studies Series (London, 1978), XXII, pp. 33, 37, 39–40; Gattini, pp. 14, 77; Luttrell, 'Change', p. 197.

[32]   Gregory XI, *Lettres secrètes et curiales ... les pays autres que la France*, ed. G. Mollat (Paris, 1962–63), nos. 1502–3, 2622–3, 2768, 2994; Malta, cod. 320, fol. 57r; cod. 321, fols 205v–206r; Delaville Le Roulx, *Les Hospitaliers à Rhodes*, p. 150.

[33]   Malta, cod. 321, fol. 205v.

[34]   Crudo, pp. 354–6.

[35]   Malta, cod 319, fols 246r–v, 247r, 251r–v; M. Salerno, 'Dipendenze e dignità dell'Ordine di San Giovanni di Gerusalemme nel XIV secolo: il caso di Bertrand de Boyson, precettore di Sant'Eufemia in Calabria', *Archivio Storico per la Calabria e la Lucania*, 71 (2004), 55–75.

which the Hospital divested itself of the county of Alife. Pope Gregory XI, Urban's successor, cancelled the irregularity and wrote to Master Raymond Berenger 'ut Manuelem Chabaudi ... commendatum habeat'. The Convent awarded Chabaud the commandery for ten years, and he continued to enjoy the support of Pope Gregory, who in 1373 asked the master to assign him Sant'Eufemia for life.[36]

Boyson was supported by the Master, and his nephew Chabaut by the pope; both are remembered for misappropriations perpetrated in Calabria. Manuel Chabaut was excommunicated, and in 1376 Gregory XI asked the master to assign Sant'Eufemia to Geraud de la Roche, the pope's nephew. [37] In 1382, in the midst of the confusion surrounding the papal schism, Chabaut was declared an 'enemy of the Church' by the Avignonese pope Clement VII for having favoured Charles of Anjou-Durazzo. In 1384 he was recognized as the legitimate commander of Sant'Eufemia by Anti-Master Riccardo Caracciolo, but he paid no responsions. Caracciolo proposed that Chabaud relinquish Sant'Eufemia in exchange for priory of Petite Provence, with the *camere* of Aquis, Sta Croce, Nice and Comis left vacant by the death or disobedience of the previous commanders. Chabaut's decision is unknown.[38]

These foreign *fratres milites* were engaged in activities overseas and often did not reside in their assigned commanderies; they abused their power and appropriated the profits without paying the responsions to the treasury.

It was not only French–Provençal Hospitallers who held political office in the kingdom of Sicily; in 1289–90 Fr Matteo di Ruggero of Salerno held the position of *magister rationalis* of the *magna curia*. The office represented the peak in his career because in 1270–71, before professing, he had been justiciar of Sicily. In 1272–74 and again in 1282, he was justiciar of Calabria, and between 1278 and 1283 he was vice-admiral of the Principato and Terra di Lavoro, one of the few Italians to hold public offices in those years.[39]

Sometimes the Angevins had a role in the allocation of Hospitaller commanderies. According to Matteo Camera, in 1330 the commander of Naples was Fr Ruffo de Marinis of Genoa, a brother who was very close to King Robert, who had been admiral of the kingdom before professing, and who had a son named Ambrogio who was appointed *ciambellano* by the king that year.[40] So when in 1330 the commandery of Naples was at the disposal of the master, it was probably the king who was able to nominate his own loyal supporter as its administrator.

---

[36]   Urbain V, *Lettres communes*, eds M. and A.M. Hayez (Rome, 1983), nos. 25896, 26657; Russo, 2, nos. 7895, 7909, 8083.

[37]   Ibid., 2, nos. 8110, 8312, 8356. Salerno, 'Dipendenze', pp. 66–8; A. Luttrell, 'Introduzione', in Salerno and Toomaspoeg, p. 26.

[38]   Russo, 2, nos. 9010–11. Malta, cod. 281, fols 3r, 4v, 42v–43r, 76r.

[39]   *Registri*, 30, p. 88; 32, pp. 100, 145, 174, 190, 236–7, 240, 242; Durrieu, 2, pp. 191, 211, 213.

[40]   Camera, *Annali*, 2, pp. 359–60.

Other *fratres* from the nobility of the kingdom acquired important offices in the Order. In 1373 Domenico de Alamania was commander of Monopoli, and two years later he also became preceptor of Naples. He notionally retained office as preceptor until 1399,[41] but his administration was interrupted in 1383 by the presence of Anti-Master Riccardo Caracciolo, who was resident in Naples. In fact Alamania, although himself a Neapolitan, remained loyal to the Convent of Rhodes. He later conducted negotiations for the ransom of Master Juan Fernandez de Heredia, who had been captured in an Albanian ambush in 1378. From 1379 he was lieutenant in Italy and was designated as commander of Cyprus.[42] A few years later, in 1386, he became prior of Monopoli once again, but he was probably unable to gain effective control of the revenue of the *domus* until after the death of Caracciolo. Naples was granted to him for life by Master Philibert de Naillac on 6 October 1399,[43] and he also retained S. Stefano until 1411. In 1386 he had also been assigned the island of Nysiros, which belonged to the Hospital and which until then had been enfeoffed to the Assanti brothers of Ischia, subjects of the kings of Naples.[44]

At the time of the papal inquiry into the Hospitallers in 1373, the priories of Capua and Barletta were both held by members of titular families of counties in the kingdom. Berardo of Aquaviva, son of the count of San Valentino,[45] was prior of Capua from 18 December 1365, succeeding Isnard du Bar. Earlier he had been preceptor of Bari and Monopoli, and so acquiring a priory was a promotion from the master and the Convent, and it also brought him closer to his family's domains.[46] He was a *frater sacerdos* of between thirty-one and thirty-eight years of age in 1373. By December 1373, when Riccardo Caracciolo was illegally occupying the priory of Capua, Aquaviva was probably already dead, or perhaps a prisoner in Greece; he was certainly dead by 23 August 1381.[47]

The prior of Barletta, Raymo de Sabran, belonged to an illustrious Provençal family that had come to Italy with the followers of Charles I and was by now 'Italian'. In 1373 members of the family held the counties of Ascoli, Ariano and Anglona, and there was also a saint in the family, Count Elzear of Ariano, a model

---

[41]   Delaville Le Roulx, *Les Hospitaliers à Rhodes*, p. 190; Bosio, 2, p. 96; Malta, cod. 321, fol. 201r; cod. 326, fols 112v, 113v; cod. 330, fol. 107r.

[42]   Luttrell, 'Intrigue', p. 41. A. Luttrell, 'Sugar and Schism: The Hospitallers in Cyprus from 1378 to 1386', in *The Hospitaller State on Rhodes and its Western Provinces, 1306–1462*, Variorum Collected Studies Series (Aldershot, 1999), IV, p. 163. Radogna, pp. 21–2; Gattini, p. 77

[43]   Malta, cod. 323, fol. 215v; cod. 324, fols 135v–136r, 137v–138v; cod. 326, fols 112v, 113v; cod. 330, fol. 107r. Delaville Le Roulx, *Les Hospitaliers à Rhodes*, p. 190.

[44]   A. Luttrell, 'Feudal Tenure and Latin Colonization at Rhodes', in *The Hospitallers in Cyprus*, III, pp. 761–2.

[45]   Gregory XI, no. 1496.

[46]   Malta, cod. 319, fols 247v, 248r–v; cod. 320, fol. 56r–v.

[47]   Archivio Vaticano, *Reg. Vat.* 292, fols 35r–v. Malta, cod. 321, fol. 200v.

of lay and noble holiness.[48] Fr Raymo was at Sant'Eufemia around 1366, when he refused to leave the commandery to the Provençal Boyson; in 1372 he became prior of Barletta, where he remained at least until 1375, and in the inquiry of 1373 he was described as a *frater miles* of about thirty-five years of age.[49]

In 1366 Venosa was assigned to an Italian, Ruggero de Sansonisiis, for ten years with the intention that he ameliorate the desolate state of the preceptory. Sansonisiis did not disappoint the master and proved to be a good preceptor; he was excused payment of the responsions on condition that he built a fortress at his own expense in Corneto within five years.[50] He was also an important *frater* in the Order: as preceptor of Venosa, he took part in the assembly of Avignon in 1373 and in the 1379 Epirot crusade. Perhaps he died following that exploit, because his *dispolia mortuaria* are listed in a document approved at Rhodes in 1381. When he died, he possessed a large silver service, with a value of 624 ducats, and wheat worth around 24.5 gold ounces.[51]

By the time of King Robert at the beginning of the fourteenth century the Angevin dynasty of Naples had already become 'Italian'. An indigenous nobility had re-emerged, as ultramontane families were by then assimilated with their local counterparts. The Order of St John, on the other hand, directly dependent as it was on the papacy in Avignon and influenced by the French–Provençal component, continued to use the *domus* of southern Italy for the ambitions of ultramontane *fratres* who enjoyed the support of popes and masters. But some Italians did rise to acquire important positions within the Hospital.[52]

[48]   Delaville Le Roulx, *Les Hospitaliers a Rhodes*, p. 52; Gregory XI, no. 1496. G. Klaniczay, 'La noblesse et le culte des saints dynastiques sous les rois angevins', in Coulet and Matz (eds), pp. 511–26.

[49]   Malta, cod. 319, fol. 251r; Archivio Vaticano, *Reg. Aven.* 187, fol. 526r; Malta, cod. 320, fol. 59r.

[50]   Malta, cod 319, fols 247v, 250r; 252v–253r. A. Luttrell, 'The Hospitaller Commandery of the Morea, 1366', in *Studies*, XXI, p. 299.

[51]   Crudo, pp. 353–4; Gattini, p. 20. Malta, cod. 320, fols 55v, 56r; cod. 321, fol. 202v; R. Iorio, 'Un priorato medievale del Mezzogiorno: geografia economica e assetti amministrativi', *Studi Melitensi*, 6 (1998), 56. Salerno, 'Da domus', pp. 93–7.

[52]   Kelly, pp. 347–8; S. Pollastri, 'La présence ultramontaine dans le Midi italien (1265–1340)', *Studi storici meridionali*, 1–2 (1995), 4; G. Vitale, 'Nobiltà napoletana della prima età angioina. Elite burocratica e famiglia', in *L'État*, pp. 535–76.

Chapter 23

The Teutonic Order in Italy: An Example of the Diplomatic Ability of the Military Orders

Kristjan Toomaspoeg

In the closing centuries of the Middle Ages, the military orders often found themselves in complex political situations, where they had to preserve their interests and remain on good terms with sovereigns who were at war with one another. To give just a few examples among many: the Templars and Hospitallers sought to maintain their presence and favoured position in both France and England despite the frequent warfare between the two monarchies;[1] the Templars managed to keep their possessions in the Genoese territories, even though they were allied to Venice in the war of St Sabas (1256–58);[2] the Hospitallers were the allies of the papacy and the Angevin dynasty during the crusade against Aragonese following the Sicilian Vespers of 1282, but they also supported Peter III of Aragon;[3] while in the Iberian Peninsula, it was not rare to find Hospitallers fighting on both sides in the wars between Aragon and Castile.[4] It is clear that the orders developed diplomatic skills that allowed them to be friends of opposing powers, even in very difficult situations.

---

[1]   M.-L. Bulst-Thiele, 'Templer in königlichen und päpstlichen Dienste', in P. Classen and P. Scheibert (eds), *Festschrift Percy Ernst Schramm* (Wiesbaden, 1964), 1, pp. 289–308; H. Nicholson, 'The Military Orders and the Kings of England in the Twelfth and Thirteenth Centuries', in A.V. Murray (ed.), *From Clermont to Jerusalem: The Crusades and Crusader Societies (1095–1500)* (Turnhout, 1998), pp. 203–18; J. Sarnowsky, 'Kings and Priors: The Hospitaller Priory of England in the Later Fifteenth Century', in J. Sarnowsky (ed.), *Mendicants, Military Orders and Regionalism in Medieval Europe* (Aldershot, 1999), pp. 83–102.
[2]   E. Bellomo, *The Templar Order in North-west Italy (1142–c.1330)* (Leiden, 2008), pp. 52–3.
[3]   N. Housley, *The Italian Crusades: The Papal–Angevin Alliance and the Crusades against Christian Lay Powers (1254–1343)* (Oxford, 1982), p. 77; A. Luttrell, 'The Aragonese Crown and the Knights Hospitallers of Rhodes (1291–1350)', *English Historical Review*, 76 (1961), 4 (repr. in Luttrell, *The Hospitallers in Cyprus, Rhodes, Greece and the West: 1291–1440*, Variorum Collected Studies Series (London, 1978), XI); P. Bonneaud, *Le prieuré de Catalogne, le couvent de Rhodes et la couronne d'Aragon, 1415–1447* (Millau, 2004), p. 247.
[4]   See for example P. Josserand, *Église et pouvoir dans la Péninsule Ibérique. Les ordres militaires dans le royaume de Castille (1252–1369)* (Madrid, 2004), p. 581.

This paper sets out to show that the same is true of the Teutonic Order, especially in its Italian provinces. The Teutonic Order in Italy has been the subject of my research for several years. The Order's history in the Italian Peninsula began shortly before its official foundation as military order in 1198–99,[5] as by then the brothers had already established houses in some southern Italian cities.[6] During the first half of the thirteenth century, the Order managed to expand into almost all regions of Italy with the exception of Umbria, the Marches, Liguria, Piedmont and Lombardy, and created its own regional structures.[7] These comprised four provinces, Sicily,[8] Apulia,[9] 'Lombardia' (in fact, its possessions in Friuli, the Veneto and Emilia-Romagna)[10] and An der Etsch (in the southern Tirol),[11] and a group of houses in Rome, Latium and Tuscany administered by the Order's procurator at the papal court.[12]

---

[5] For the foundation of the Teutonic Order, see M.L. Favreau, *Studien zur Frühgeschichte des Deutschen Ordens* (Stuttgart, 1974); U. Arnold, 'Entstehung und Frühzeit des Deutschen Ordens', in J. Fleckenstein and M. Hellmann (eds), *Die Geistlichen Ritterorden Europas* (Sigmaringen, 1980), pp. 81–108; and most recently N.E. Morton, *Teutonic Knights in the Holy Land 1190–1291* (Woodbridge, 2009), pp. 9–30.

[6] In Brindisi, Messina and probably also in Gela (southern Sicily). K. Toomaspoeg, *Les Teutoniques en Sicile (1197–1492)* (Rome, 2003), pp. 73, 103; idem, 'I cavalieri teutonici tra Sicilia e Mediterraneo', in A. Giuffrida, H. Houben and K. Toomaspoeg (eds), *I Cavalieri Teutonici tra Sicilia e Mediterraneo. Atti del Convegno Internazionale (Agrigento, 24–25 marzo 2006)*, Acta Theutonica, 4 (Galatina, 2007), pp. 75–90, here p. 76.

[7] See B. Schumacher, 'Studien zur Geschichte der Deutschordensballeien Apulien und Sizilien', *Altpreußische Forschungen*, 18 (1941), 187–230 and 19 (1942), 1–25; K. Forstreuter, *Der Deutsche Orden am Mittelmeer*, QuStDO 2 (Bonn, 1967).

[8] Toomaspoeg, *Teutoniques*.

[9] H. Houben, 'Zur Geschichte der Deutschordensballei Apulien. Abschriften und Regesten verlorener Urkunden aus Neapel in Graz und Wien', *Mitteilungen des Instituts für österreichische Geschichtsforschung*, 107 (1999), 50–110.

[10] G.B. Altan, *Precenicco, i conti di Gorizia, i cavalieri teutonici e la sua comunità* (Udine, 1981); idem, *Precenicco, i cavalieri teutonici, le sue vicende e la sua comunità* (Udine, 1992); M. Fanti and G. Roversi, *S. Maria degli Alemanni in Bologna* (Bologna, 1969); G. Cagnin, 'La controversa donazione del Castello di Stigliano ai cavalieri teutonici (Acri, 15 dicembre 1282)', in F. Tommasi (ed.), *Acri 1291. La fine della presenza degli ordini militari in Terrasanta e i nuovi orientamenti nel XIV secolo* (Perugia, 1996), pp. 99–119; K. Toomaspoeg, 'La fondazione della provincia di "Lombardia" dell'Ordine dei Cavalieri Teutonici (secoli XIII–XIV)', *Sacra Militia*, 3 (2003), 111–59; P. Cierzniakowski, 'L'Ordine Teutonico nell'Italia settentrionale', in H. Houben (ed.), *L'Ordine Teutonico nel Mediterraneo. Atti del Convegno internazionale di studio. Torre Alemanna (Cerignola)– Mesagne–Lecce 16–18 ottobre 2003*, Acta Theutonica, 1 (Galatina, 2004), pp. 217–35.

[11] *Der Deutsche Orden in Tirol. Die Ballei an der Etsch und im Gebirge*, ed. H. Noflatscher, QuStDO 43 (Marburg, 1991).

[12] J.-E. Beuttel, *Der Generalprokurator des Deutschen Ordens an der römische Kurie. Amt, Funktionen, personelles Umfeld und Finanzierung*, QuStDO 55 (Marburg, 1999); T. Frank, 'Der Deutsche Orden in Viterbo (13.–15. Jahrhundert)', in F.J. Felten and N.

While establishing its estates, the Teutonic Order entered into diplomatic and political relations with widely differing powers that meant it enjoyed the patronage of the kingdom of Sicily, the Western Empire and the Papacy, and also the Republic of Venice, the city of Padua, the county of Gorizia and the patriarchate of Aquileia. All these institutions developed differently as a result of their own dynastic, political or ideological vicissitudes, so that the groups with whom the Order had to deal did not always remain the same. A good relationship with those powers would help create and maintain the Order's patrimony and obtain privileges such as tax exemptions or right to control the administration of justice, while conflict could end with confiscations or other forms of discrimination. What follows are some brief observations about the ways in which the Teutonic Order achieved a diplomatic and political coexistence with the various Italian polities.

The Empire

The first institution to support the Teutonic Knights in the peninsula was the imperial court. It is difficult to make a distinction between the imperial and Sicilian policies before 1250, as from 1194 to 1197 and 1220 to 1250 the emperor was also king of Sicily. The Order greatly benefited from the patronage of Emperor Henry VI, who intended to use it as a specifically German institution that would contribute to the conquest of the kingdom of Sicily and to the crusade.[13] Henry, and before him his brother Frederick of Swabia, can be seen as the real creators of the Teutonic Order. The relationship between the Order and the imperial power remained close,[14] despite the problems caused by Henry VI's death in 1197, the minority of his successor, Frederick II, who until 1213 was officially under the tutelage of Pope Innocent III,[15] and the fact that the power in the Empire passed to Otto IV and it was only conferred on Frederick in 1220.

In this situation, the Teutonic Knights proved their diplomatic abilities for the first time. They did not support the papal regency in Sicily and allied themselves with Otto in Germany and with the Sicilian usurpers. These usurpers, Markward of Anweiler, Dipold of Vohburg and Wilhelm of Capparone, governed Sicily in

---

Jaspert (eds), *Vita religiosa im Mittelalter. Festschrift für Kaspar Elm zum 70. Geburtsdag* (Berlin, 1999), pp. 321–43; B. Bombi, 'L'Ordine Teutonico nell'Italia centrale. La casa romana dell'Ordine e l'ufficio del procuratore generale', in Houben (ed.), pp. 197–216.

[13]   Toomaspoeg, *Teutoniques*, pp. 21–34. See also the 'classic' monography by W. Cohn, *Das Zeitalter der Hohenstaufen in Sizilien* (Breslau, 1925); F. Giunta, 'Il furor Theutonicus', in F. Giunta and U. Rizzitano (eds), *Terra senza crociati* (Palermo, 1967), pp. 99–110.

[14]   K. Militzer, *Von Akkon zur Marienburg. Verfassung, Verwaltung und Sozialstruktur des Deutschen Ordens 1190–1309*, QuStDO 56 (Marburg, 1999), pp. 16–17.

[15]   F. Baethgen, *Die Regentschaft Papst Innocenz III im Königreich Sizilien* (Heidelberg, 1914); K. Wieser, 'Gli inizi dell'Ordine Teutonico in Puglia', *Archivio storico pugliese*, 26 (1973), 475–87.

the name of the young Frederick II; they each had their own political stance, but all of them were in some way following the line taken by Henry VI.[16] Until 1209 the Teutonic Order had excellent relations with these men and as a result received many properties and privileges in the kingdom.[17] At the same time, this attitude damaged diplomatic relations between the knights and Pope Innocent, but it was obviously worth it.

After 1209, power in Sicily passed to noblemen of Norman origin including Pagan of Parisio who continued to control royal policy and also supported the Teutonic Order.[18] The upshot was that, when in 1213 Frederick II began to rule independently, he had good reason to repress the Order that had usurped control of some territories in the royal domain. The Order's continuing presence in southern Italy was saved only in 1216, thanks to an alliance agreed in Germany between Frederick II and the master of the Order, Hermann of Salza.[19] This pact was intended to help Frederick secure the German and imperial crowns, but the Teutonic brothers in southern Italy went on to receive many privileges and possessions between 1216 and 1225.[20]

As is well known, Hermann of Salza became one of Frederick II's closest confidants and a mediator between the papal and imperial courts. Scholars often therefore present us with the image of an order that was completely subservient to the Empire until the master's death in 1239.[21] In reality, the topic is more complex. On the one hand, Hermann, though master, did not represent or control the entire Order, many of whose brothers were opposed to Hermann's imperial obedience

---

[16]   For the historical background, see Baethgen, and N. Kamp, 'Die deutsche Präsenz im Königreich Sizilien (1194–1266)', in T. Kölzer (ed.), *Die Staufer im Süden. Sizilien und das Reich* (Sigmaringen, 1996), pp. 141–85.

[17]   See Wieser; Toomaspoeg, *Teutoniques*, pp. 36–43; idem, 'Gli insediamenti templari, giovanniti e teutonici nell'economia della Capitanata medievale', in *Federico II e i cavalieri teutonici in Capitanata: recenti ricerche storiche e archeologiche. Atti del Convegno internazionale, Foggia–Lucera–Pietra Montecorvino 10–13 giugno 2009*, Acta Theutonica (Galatina, forthcoming).

[18]   These nobles were also great supporters of Templars and Hospitallers. See L. Villari, *Templari in Sicilia* (Latina, 1993), pp. 26–8; K. Toomaspoeg, 'Templari e Ospitalieri nella Sicilia Medievale, Gran Priorato di Napoli e Sicilia del Sovrano Militare Ordine di Malta', *Melitensia*, 11 (2003), 60–63.

[19]   A. Koch, *Hermann von Salza* (Leipzig, 1884); E. Caspar, *Hermann von Salza und die Gründung des Deutschordensstaats in Preussen* (Tübingen, 1924); H. Kluger, *Hochmeister Hermann von Salza und Kaiser Friedrich II.: ein Beitrag zur Frühgeschichte des Deutschen Ordens*, QuStDO 37 (Marburg, 1987); H. Houben, 'Alla ricerca del luogo di sepoltura di Ermanno di Salza a Barletta', *Sacra Militia*, 1 (2000), 165–77.

[20]   See Houben, 'Deutschordensballei'; Toomaspoeg, *Teutoniques*, pp. 46–52.

[21]   See for example J.M. Powell, 'Frederick II, the Hohenstaufen, and the Teutonic Order in the Kingdom of Sicily (1187–1230)', in *MO* 1, pp. 236–44.

and wanted him to spend more time in the Holy Land.[22] On the other, Frederick II was a pragmatic ruler and his favourable policy towards the Teutonic Order had its limits. At first, before 1225, Frederick gave the knights privileges in the kingdom of Sicily, but after that, there were no more grants there; instead, in the 1230s, the emperor gave the Order new privileges elsewhere in the Italian peninsula. The new possessions gained by the knights in Tuscany and Padua came in the context of Frederick II's military expeditions.[23]

With the death of Hermann of Salza and the second excommunication of Frederick in 1239, relations between the emperor and the Teutonic Knights suffered. Internal dissentions divided the Order into two camps, the supporters of the emperor and those of the pope, and in 1249 there was even a short schism with election of two opposed masters.[24] In Italy the Order received no new privileges or possessions, and at some point in 1248–49 Frederick II took back into the royal domain lands that the knights had usurped in northern Apulia.[25]

After the Frederick's death in 1250, the Empire was no longer of importance for the development of the Italian provinces of the Teutonic Order, although in the first half of the fourteenth century, when the Sicilian Aragonese court was allied to the emperors Henry of Luxemburg and Ludwig of Bavaria, the Order, recalling its imperial origin, was able to derive some benefit from the situation.[26]

The Papal Court

Officially the Teutonic Order was answerable to no one but the pope, and so its relations with the Roman Church had to be positive. It was not always, however, so simple, especially at the beginning of the thirteenth century, when the knights supported Innocent III's enemies in Sicily, and during the conflict between the papal and imperial courts that lasted from 1229 until 1250. There was in fact a difference between the papal policies towards the Teutonic Order as a whole with its headquarters in the Holy Land, and towards the Italian provinces of the Order. The warmest relations between the knights in Italy and the papacy dated from the

---

[22]   H. Cleve, 'Kaiser Friedrich II. und die Ritterorden', *Deutsches Archiv für Erforschung des Mittelalters*, 49 (1993), 39–73, here pp. 63, 73; U. Arnold, 'Der Deutsche Orden zwischen Kaiser und Papst im 13. Jahrhundert', in Z.H. Nowak (ed.), *Die Ritterorden zwischen geistlicher und weltlicher Macht im Mittelalter*, Colloquia Torunensia Historica, 5 (Toruń, 1990), pp. 57–70, here p. 60; K. Militzer, 'From the Holy Land to Prussia: The Teutonic Knights between Emperors and Popes and their Policies until 1309', in *Mendicants*, pp. 71–81, here pp. 72–3.

[23]   Toomaspoeg, 'La fondazione', pp. 121–3.

[24]   Arnold, 'Kaiser und Papst', p. 63; Militzer, 'Holy Land', p. 73.

[25]   Toomaspoeg, 'Gli insediamenti'.

[26]   S.V. Bozzo, 'Giovanni Chiaramonte II nella discesa di Ludovico il Bavaro, saggio critico', *Archivio Storico Siciliano*, 3 (1878), 155–85; Toomaspoeg, *Teutoniques*, p. 177.

time of Honorius III (1216–27) when the pope gave the Order's houses privileges in the ports such as Messina or Brindisi that were used by crusaders.[27]

When in 1220 he granted them the church of Sta Maria in Domnica at Rome, Honorius became the only pope to give the Teutonic Order property in the Papal State. It was only a temporary concession, and at the beginning of the fourteenth century, when the relations between the Teutonic Order and the papal court at Avignon were not so cordial, Clement V took the church back and gave it to one of his nephews.[28] Another pope with a very positive attitude towards the Order was Alexander IV (1254–61), who gave many privileges to the houses of the Order in Padua and elsewhere in Italy. In 1260, when the Order was defeated at the battle of Durbe in Livonia, Alexander IV gave the knights the monastery of San Leonardo di Siponto in Apulia.[29] Papal support for the Order was in large part due to the diplomatic activities of its envoys in Rome, especially John of Capua, notary of the papal court and procurator of the Order.[30]

In the fourteenth century the Teutonic Knights also managed to retain papal favour even when they were supporting the pope's enemies. In the island of Sicily under the domination of the excommunicated Aragonese dynasty, Nicholas IV allowed the knights to celebrate Mass in their churches,[31] and in 1307–1311 Clement V helped them against the bishop of Cefalù.[32] During the Great Schism, the Teutonic Knights on the island gave their obedience to the Avignonese pope, while those in the rest of Italy give theirs to the pope at Rome.

---

[27]   See H. Houben, 'Friedrich II., der Deutsche Orden und die Burgen im Königreich Sizilien. Eine unbekannte Urkunde Honorius III von 1223', *Deutsches Archiv für Erforschung des Mittelalters*, 56 (2000), 585–91; P.-V. Claverie, 'Les relations du Saint-Siège avec les ordres militaires sous le pontificat d'Honorius III (1216–1227)', in N. Bériou and P. Josserand (eds), *Élites et ordres militaires au Moyen Âge. Rencontre en l'honneur d'Alain Demurger, Lyon, 21–23 octobre 2009* (Lyon, forthcoming).

[28]   K. Toomaspoeg, 'Die Deutschordenskirche Santa Maria in Domnica im Licht eines unbekannten Inventars von 1285', *Quellen und Forschungen aus italienischen Archiven und Bibliotheken*, 83 (2003), 83–101.

[29]   See *San Leonardo di Siponto. Cella monastica, canonica, domus Theutonicorum. Atti del Convegno internazionale (Manfredonia, 18–19 marzo 2005)*, ed. H. Houben, Acta Theutonica, 3 (Galatina, 2006).

[30]   *Berichte der Generalprokuratoren des Deutschen Ordens an der Kurie*, eds K. Forstreuter and H. Koeppen, Veröffentlichungen der Niedersächsischen Archivverwaltung, 12, 13, 21, 32, 37 (Göttingen, 1960–76), 1, pp. 52–62; Beuttel, p. 5ff; F. Delle Donne, 'Giovanni da Capua', in *Dizionario Biografico degli Italiani*, 55 (Rome, 2000), pp. 759–61.

[31]   Toomaspoeg, *Teutoniques*, no. 284, p. 652 (29 Jan. 1291); no. 285, p. 652 (5 Feb. 1291).

[32]   Ibid., no. 499, p. 722; no. 520, p. 729; no. 535, pp. 734–5.

The Two Kingdoms of Sicily

The situation was the most complicated in Sicily. After 1250 the Teutonic Order supported King Conrad and King Manfred, the sons of Frederick II, but that was a political necessity, and it is not possible to speak of the knights' fidelity to the Hohenstaufen dynasty.[33] After the fall of Manfred in 1266, they allied themselves immediately to the Angevin dynasty. From the Order's point of view, Charles I of Anjou was much better than Manfred. He was the papal candidate for the Sicilian throne, was interested in the war in the Holy Land, and had a special regard for the military orders. Although not as much as the Templars and Hospitallers, but nevertheless to a considerable extent, the Teutonic Knights served the new king and greatly benefited from their relations with the dynasty.[34]

In 1282, however, the island of Sicily passed into Aragonese control.[35] From then on the possessions of the Teutonic Order in southern Italy were split between the two separate kingdoms, but the knights had excellent relations with the royal courts in both. Their adherence to the Aragonese in Sicily was immediate, and the new dynasty gave them privileges explicitly to thank them for their support.[36] Here again the Order seems not to have been guided by fidelity or ideology but simply accepted the new rulers. In its relations to the Aragonese, the Order made use of its imperial origins, especially during the reign of Frederick III, who was actively involved in the Ghibelline cause.[37]

---

[33]   M. Hellmann, 'König Manfred von Sizilien und der Deutsche Orden', in K. Wieser (ed.), *Acht Jahrhunderte Deutscher Orden in Einzeldarstellungen. Festschrift zu Ehren Sr. Exzellenz P. Dr. Marian Tumler O.T. anläßlich seines 80. Geburtstages*, QuStDO 1 (Bad Godesberg, 1967), pp. 65–72.

[34]   For Charles I of Anjou and the military orders in the kingdom of Sicily, see K. Toomaspoeg, 'Le ravitaillement de la Terre Sainte: l'exemple des possessions des ordres militaires dans le royaume de Sicile au XIIIe siècle', in *L'Expansion occidentale (XIe–XVe siècles). Formes et conséquences. XXXIIIe Congrès de la Société des Historiens Médiéviste de l'Enseignement Supérieur Public (Madrid, Casa de Velázquez, 23–26 mai 2002)* (Paris, 2003), pp. 143–58; idem, 'Templari', pp. 69–75; D. Carraz, 'Christi fideliter militantium in subsidio Terre Sancte. Les ordres militaires et la première maison d'Anjou (1246–1342)', in I.C. Ferreira Fernandes (ed.), *As Ordens Militares e as Ordens de Cavalaria entre o Occidente e o Oriente. Actas do V Encontro sobre Ordens Militares, 15 a 18 de Fevereiro de 2006* (Palmela, 2009), pp. 549–82; K. Toomaspoeg, 'Les ordres militaires dans les royaumes de la Méditerranée centrale (XIIIe–XIVe siècles)', in Ferreira Fernandes (ed.), pp. 435–50, here p. 442.

[35]   S. Runciman, *The Sicilian Vespers* (Cambridge, 1958); I. Peri, *La Sicilia dopo il vespro: uomini città e campagne, 1282–1376* (Rome–Bari, 1981).

[36]   Toomaspoeg, *Teutoniques*, pp. 65–8.

[37]   C.R. Backman, *The Decline and Fall of Medieval Sicily: Politics, Religion, and Economy in the reign of Frederick III (1296–1337)* (Cambridge, 1995). I use here the updated Italian edition of this book, *Declino e caduta della Sicilia medievale. Politica, religione ed economia nel regno di Federico III d'Aragona Rex Siciliae (1296–1337)* (Palermo, 2007), pp. 47, 67–9, 71–3, 76–7.

In Sicily in the second half of the fourteenth century, royal authority gave way to feudal anarchy and the island came to be dominated by some important local noble families, notably the Chiaramonte and the Ventimiglia,[38] both friends and allies of the Teutonic Order. But when at the end of the century the island was conquered by Martin of Barcelona, the knights abandoned their former allies and gave financial support to Martin with the result that they became closely associated with the new regime.[39] However, their relation to the Sicilian court moved in the direction feudal subordination, and the commander of the Order's Sicilian province was later forced to serve in the king's army.[40]

Other Powers

Elsewhere in Italy the Teutonic Order usually reaped the benefits of its association with the international powers such as the Empire or the Church. Thus in 1240 the knights were established in Padua at a time when the local despot, Ezzelino III of Romano, was a close ally of Frederick II,[41] and the acquisition of possessions in Friuli and the Tirol was also related to the imperial politics.[42]

The case of Venice was different. Since the Fourth Crusade the Order was an ally of the Republic of Venice, and, like the Templars, the knights collaborated with the Serenissima in many fields. In consequence, they were automatically opposed to the Genoese and could never establish themselves in Liguria. Nevertheless, despite their good relations with the Teutonic Order, the doges of Venice only ever gave it one privilege: in 1258 the Order received a church on the Isola della Dogana.[43] In the fifteenth century Venice conquered a large part of the Veneto, including the territories of Padua and Treviso where the Teutonic Knights had

---

[38] See I. Peri, *La Sicilia; Restaurazione e pacifico stato in Sicilia, 1377–1501* (Rome and Bari, 1988).

[39] S. Fodale, *Il clero siciliano tra ribellione e fedeltà ai Martini (1392–1398)* (Palermo, 1983); Toomaspoeg, *Teutoniques*, pp. 275–80.

[40] *La contabilità delle Case dell'Ordine Teutonico in Puglia e in Sicilia nel Quattrocento*, ed. K. Toomaspoeg, Acta Theutonica, 2 (Galatina, 2005), p. lxxxii.

[41] C. Cantù, *Ezzelino da Romano; Storia di un ghibellino* (Milan, 1901); *Nuovi studi ezzeliniani*, ed. G. Cracco, Comune di Romano d'Ezzelino, Nuovi studi storici, 21 (Rome, 1992); C.F. Polizzi, 'Comune, signoria ezzeliniana e Chiesa di Padova nel secolo XIII attraverso le carte dei frati alemanni. Con due appendici: 1. Edizione di 57 documenti inediti (1232–1463), 2. Regesto e regesto-estratto di 316 documenti inediti (1219–1399)', unpublished thesis, University of Padua, 1981–82; Toomaspoeg, 'La fondazione', p. 123.

[42] Forstreuter, pp. 137–8.

[43] F. Corner, *Veneta Ecclesiae Illustratae*, 5 (Venice, 1749), p. 1; Forstreuter, pp. 139–41; U. Arnold, 'Der Deutsche Orden und Venedig', in E. Coli, M. De Marco and F. Tommasi (eds), *Militia Sacra. Gli ordini militari tra Europa e Terrasanta* (Perugia, 1994), pp. 145–65, here p. 146.

many properties. The Republic thereby became one of the Order's most important landlords in Italy.

In Italy the links between the establishment of the Order and its relations with the crusaders in the Holy Land were less evident than in the rest of Europe, and only in the case of the patriarchs of Aquileia is there a direct connection between Italy and the East. At the beginning of the thirteenth century, the patriarch was Wolfger of Ellenbrechtskirchen, one of the founders of the Teutonic Order; he himself gave landed properties to the knights and persuaded the counts of Gorizia to do likewise.[44]

The diplomacy of the Teutonic Order in Italy acted at two different levels: on the one hand, the Order as a whole could influence the papal and imperial courts in obtaining privileges for its Italian provinces, while on the other, the local provinces pursued their own diplomacy so as to maintain good relations with the powers who were officially allied to the Order and at the same time with whom the Order was not supposed to have dealings because of papal opposition. Taking the lead in this diplomacy were the masters of the Order and also the provincial commanders and its procurators at the papal court. Many of the thirteenth-century masters, and not only Hermann of Salza, spent quite a lot of time in Italy and acted to safeguard the Order's local possessions. From 1291 to 1309 the masters resided in Venice; afterwards they governed from Prussia, and the officer in charge of Italy was the so-called master of Germany, the *Deutschmeister*.[45]

The provincial commanders[46] were often men of important origin, such as Florence of Holland, a relative of German king William of Holland, who was commander in Sicily in the 1260s and 1270s,[47] and who acted as the link between the Order and Charles I of Anjou. The other military orders employed the same strategy: the Hospitallers had as delegates at the court of Charles I two officers, Philip d'Egly and Jacques de Taxi, who were also provincial commanders in Messina and Barletta.[48] The procurators at the papal court had the task of furthering by every means possible the Order's initiatives throughout the Christian world. Very often they also had a role in papal administration.[49]

We have to ask whether or not the Teutonic Order's diplomacy was a success. There were some territories in Italy where the knights could never

---

[44]   Altan, *Gorizia*, pp. 64–5.

[45]   H.H. Hofmann, *Der Staat des Deutschmeisters* (Munich, 1964); Militzer, *Von Akkon*, pp. 213–23, 307–32.

[46]   See K. Toomaspoeg, 'Les premiers commandeurs de l'Ordre Teutonique en Sicile (1202–1291). Évolution de la titulature, origines géographiques et sociales', *Mélanges de l'École française de Rome, Moyen Âge*, 109 (1997), 443–61 (Sicily); H. Houben, 'Die Landkomture der Deutschordensballei Apulien (1225–1474)', *Sacra Militia*, 2 (2001), 116–54 (Apulia); Toomaspoeg, 'La fondazione' (Lombardy).

[47]   Toomaspoeg, *Teutoniques*, no. 46, p. 462.

[48]   Toomaspoeg, 'Templari', pp. 70–72; Carraz, pp. 553–9.

[49]   See Beuttel and the introductions of the volumes of *Berichte* (as n. 30).

establish themselves. They had no point of entry into Liguria while they were allies of Venice. The Templars, who similarly were allied to the Venetians, did hold properties there,[50] but these were old grants from the time of St Bernard of Clairvaux.[51] In Piedmont the Teutonic Knights never managed to foster relations with the local nobility, as for example with the marquis of Montferrat, as had the Templars and Hospitallers. Finally, the Order had very few possessions in the Papal States, from Latium to the Marches. These failures can be explained by the later foundation of the Teutonic Order, which was created some eighty years after the emergence of the Hospital as religious order and seventy years after the Temple. Both older orders had already acquired properties in Italy in the twelfth century:[52] the Teutonic Order simply arrived too late.

When reflecting on the Order's diplomatic activities, it is evident that there was no dynastic or political fidelity. Thus in Sicily, the knights moved from the Hohenstaufen to the Angevins and from the Angevins to the Aragonese with a good intuition for guessing the winner of the game. They presented themselves as a papal Order to the Angevins and as an institution of imperial origin to the Aragonese. Even at a more local level, they showed the same skills. In Padua, they were protected by the despot Ezzelino da Romano and also by his successors, the Carraresi dynasty. The Order's diplomacy was also very pragmatic and could be adapted to quite different circumstances. Even when under Hermann of Salza the Teutonic Knights were allied to Frederick II, their adherence to the imperial politics was not total, and later, when the emperor and the pope were at daggers drawn, the Order kept on the side of both thanks to its own internal divisions.

The pope, the emperor, the Angevins, the Aragonese and the Venetians – all of them supported the Teutonic Order, which could pursue its projects using a two-level diplomatic system, based on its political status and the personal relations between the Italian powers and the local commanders.

---

[50]   Bellomo, pp. 52–3.

[51]   Ibid., p. 18.

[52]   For the settlement of Templars and Hospitallers in Italy, see Bellomo; F. Tommasi, 'L'ordinamento geografico-amministrativo dell'Ospedale in Italia (secc. XII–XIV)', in A. Luttrell and F. Tommasi (eds), *Religiones Militares. Contributi alla storia degli Ordini religioso-militari nel Medioevo* (Città di Castello, 2008), pp. 61–130; F. Bramato, *Storia dell'Ordine dei Templari in Italia. Fondazioni* (Rome, 1991).

PART 5
Northern and Eastern Europe

Chapter 24

The Counts of Brienne and the Military Orders in the Thirteenth Century[1]

Karol Polejowski

The origins of the Brienne family date back to the tenth century. Two centuries later the counts of Brienne were considered as ranking among the 'old aristocracy' of the county of Champagne alongside the counts of Bar-sur-Seine, Arcis, Dampierre, Traînel, Chappes and Plancy.[2] The real founder of the family was Walter I (?1035–89). As the count of Brienne he took control of the county of Bar-sur-Seine through his marriage to Eustache, the daughter of Milo count of Tonnerre and Bar-sur-Seine. After his death in 1089 his heritage was divided between his two sons with the elder, Erard, taking over the county of Brienne and the younger, Milo, becoming count of Bar-sur-Seine.[3] There is no doubt that both Erard of Brienne and Milo of Bar-sur-Seine knew Hugh of Payns, the founder of the Templars, as even before 1118 their names had appeared together on charters issued by Count Hugh I of Champagne.[4] Starting with Erard I, members of the family participated in each of the crusades to the Holy Land. We know that Erard joined the ranks of the First Crusade in 1097 and that he returned from Palestine and died before 1125.[5] He was a benefactor of the Benedictines of Montier-en-Der and the Norbertines of Beaulieu.[6] Erard's son Walter (II) was the founder of the Norbertine monastery of Basse-Fontaine in 1143.[7] In the second half of

---

[1]   I would like to express my gratitude to the Cardiff Centre for the Crusades, to Dr Helen J. Nicholson for her help and support, and to Cadw: Welsh Historic Monuments for financial support for my participation in the Cardiff conference.

[2]   T. Evergates, *Feudal Society in the Baillage of Troyes under the Counts of Champagne, 1152–1284* (Baltimore, MD, 1975), p. 104. On the formation of the county of Champagne, see M. Bur, *La formation du comté de Champagne (v.950–v.1150)* (Nancy, 1977).

[3]   T. Evergates, *The Aristocracy in the County of Champagne, 1100–1300* (Philadelphia, PA, 2007), p. 169; A. Roserot, *Dictionnaire historique de la Champagne méridionale (Aube) des origines à 1790*, 1 (Langres, 1942), p. 108.

[4]   Archives Départementales de l'Aube (Troyes) (henceforth ADA) 6H38. Charter of Hugh I of Champagne (1113).

[5]   'Catalogue d'actes des comtes de Brienne, 950–1356', in H. d'Arbois de Jubainville (ed.), *Bibliothèque de l'École des Chartes*, 33 (1872), no. 20 (hereafter CAB).

[6]   CAB, no. 28; Roserot, p. 243.

[7]   ADA 1H1; CAB, no. 40.

1147 he left France and took part in the Second Crusade. He returned in 1151 and died probably ten years later.[8] His eldest son Erard II (1161–89) was heir to the county of Brienne and took part in the Third Crusade along with his younger brother Andrew of Ramerupt. Both died fighting the Muslims in the Holy Land.[9] Erard, like his ancestors, favoured the Norbertine houses at Basse-Fontaine and Beaulieu, where his younger brother John was the prior.[10]

Thus far little can be said about the policy of the counts of Brienne towards the Templars and Hospitallers, who in the twelfth century established commanderies in southern Champagne. On one occasion, in 1133, Count Walter II was listed as a witness confirming a grant made by Andrew of Baudement, the chamberlain of Champagne, to the Templars.[11] No other twelfth-century document gives even the slightest indication that the counts of Brienne showed any interest in the military orders.

This lack of interest is astonishing once one realizes that the Brienne family seat lay just a short distance from Payns, the home of the Templars' founder. Moreover, both the Templars and Hospitallers had large estates in southern Champagne and their presence there must have been perfectly evident at the castle of Brienne.[12] This attitude could be explained by the political absence of the family in Palestine, its participation in donations to the Norbertines and Benedictines in southern Champagne and its rivalry with the counts of Troyes, who usually held the Templars in favour. Apart from that, the counts of Brienne actively supported hospitals and other charitable institutions in their county, elsewhere a Hospitaller field of activity.

This situation changed in the beginning of the thirteenth century, when two of Erard II's sons, Walter III and John, became involved in papal policy directed against the Germans in southern Italy after the death of the emperor Henry VI (Innocent III's crusade against Markward of Anweiler and Diepold of Vohburg). It began with the marriage (1199/1200) between Walter III and Elvira of Lecce, daughter of Tancred of Sicily and Sybil of Apulia. This marriage gave the count of Brienne rights to lands in southern Italy, in particular the county of Lecce, and obliged him to organize a military campaign against the German forces. By the end of the year 1204 the conquest of southern Italy by Walter's army, with the blessing of Pope Innocent, was almost complete; the Germans, who after the death of Markward were led by Diepold of Vohburg, retained control over

---

[8]   CAB, nos. 45–6, 59.

[9]   *HC*, 2, pp. 50–52.

[10]   CAB, nos. 60, 62.

[11]   Ibid., no. 37. Walter's second wife, Adelaide of Baudement, was Andrew's daughter. See Evergates, *Aristocracy*, p. 240.

[12]   For the Templars and Hospitallers in southern Champagne in the twelfth century, see J. Richard, 'Les Templiers et les Hospitaliers en Champagne méridionale (XIIe–XIIIe siècles)', in J. Fleckenstein and M. Hellman (eds), *Die geistlichen Ritterorden Europas* (Sigmaringen, 1980), pp. 231–42.

only small parts of the country. In June 1205 Walter of Brienne, while besieging Diepold at Sarno in Campania, was ambushed by a sally and mortally wounded. It was probably after his death that Elvira gave birth to his son, Walter IV, but his Italian inheritance was confiscated by Frederick II of Hohenstaufen.[13] Alberic of Trois-Fontaines, a French annalist of the thirteenth century and a supporter of the Brienne family, left in his chronicle a description of these dramatic events:

> Walter, the count of Brienne, son of Erard, married a daughter of Tancred, king of Sicily and Sybil queen of Apulia, and came with her to Apulia, where, in course of several years, he conquered this land that belonged to him by right of his wife, until he was killed by a German called Diepold. However, he left a little boy named Walter, whom he [Diepold] feared and gave himself up to the Teutonic hospital where he spent there the rest of his life.[14]

It is clear that Alberic of Trois-Fontaines ends his account with a reference to the Teutonic Knights, but we are not sure exactly when and where Diepold of Vohburg joined the Teutonic Order – probably not until 1221, and not in Italy. But the French annalist, who began his *Chronica* after 1232, brought together information concerning the role of the Brienne family in the political situation in southern Italy and material about the Teutonic Knights and Diepold. In so doing, Alberic, who was certainly in possession of very good information so far as the Brienne family was concerned and knew about their relationship with the Teutonic Knights, had created a narrative in which the German military Order provided shelter for the murderer of the father of Walter IV, the founder of the Teutonic commandery of Beauvoir in southern Champagne. It is not exactly clear why Alberic should have constructed his account in this way, but it is definitely not favourable to the Germans in general and the Hohenstaufen in particular.[15]

According to Alberic's revelation, Walter IV's first contact with the Teutonic Knights would not have made him enthusiastic, but during the next twenty years the relations between the Briennes and the Teutonic Order changed for the better and became close, especially during the first half of the thirteenth century. First, John of Brienne, as king of Jerusalem, issued a few charters in which he granted (or confirmed) lands and properties in Acre and Tyre to the Teutonic Order.[16] These documents relate to the political situation in the Holy Land, where the military

---

[13] For the political and military situation in southern Italy after the death of Emperor Henry VI (1197) and the intervention of Walter III of Brienne, see D. Abulafia, *Frederick II: A Medieval Emperor* (Oxford, 1988), pp. 89–102; F. de Sassenay, *Les Brienne de Lecce et d'Athènes* (Paris, 1869), pp. 30–86.

[14] 'Chronica Albrici Monachi Trium Fontium, a monacho Novi Monasterii Hoiensis interpolata', *MGH SS*, 23 (Hannover, 1874), p. 879; see also *Eracles*, pp. 234–8.

[15] For Alberic's origins and sources, see M. Schmidt-Chazan, 'Aubri de Trois-Fontaines: historien entre la France et l'Empire', *Annales de l'Est*, 36 (1984), 163–92.

[16] *Tabulae ordinis Theutonici*, ed. E. Strehlke (Berlin, 1869), nos. 49, 50, 51, 57.

orders were accepted as an important factor in the defensive structure of the kingdom, and John was behaving in the same way towards the orders as had his predecessors, Henry of Champagne[17] and Aimery of Lusignan.[18]

However, in the first decades of the thirteenth century the Teutonic Knights were very active on the political scene, and, as the newest of the military orders, they had to find their own place in Europe and the Latin East. By the middle of the thirteenth century they had established commanderies in Germany, Italy and Sicily, Palestine, Lesser Armenia, Spain, Prussia, Poland, Frankish Greece and France.[19] On the one hand, their growth was a result of the diplomatic activity of Herman of Salza, the grand master of the Order in the years 1209–1239; on the other, it was thanks to the bravery and charitable activities of these 'New Maccabees', especially at Damietta during the Fifth Crusade (1218–21).[20] Their courage and medical care in Egypt would explain why some French noblemen, mostly from southern Champagne and northern Burgundy, made the first grants to the Teutonic Order in France (1218–19).[21] The most important among them were Milo of Bar-sur-Seine, count of Le Puiset, and his son Walter, the leading members of a separate branch of Brienne family. They granted some lands in the county of Le Puiset to the Teutonic Order. However, both Milo and Walter were probably killed in July 1219 near Damietta, and this line of the Brienne family became extinct. The county of Bar-sur-Seine was absorbed into the county of Champagne, but the county of Le Puiset was inherited by Simon of Rochefort-en-Brevon, Milo's son-in-law. Simon, like his father-in-law, made donations to the Teutonic Knights in the county of Le Puiset in 1225.[22] This was the beginning of the small Teutonic properties in Neuvy-en-Beauce and St Michel de l'Hermitage (near Chartres). It is certain that the grants made by Milo and Walter of Bar-sur-Seine and Simon of Rochefort were connected with the events of the Fifth Crusade. About the

---

[17]   *Tabulae*, nos. 28–32; M.-L. Favreau, *Studien zur Frühgeschichte des Deutschen Ordens* (Stuttgart, 1976), pp. 59–60.

[18]   *Tabulae*, nos. 34–6, 38; Favreau, pp. 73, 123–5.

[19]   M. Tumler, *Der Deutsche Orden im Werden, Wachsen und Wirken bis 1400. mit einem Abriß der Geschichte von 1400 bis zur neusten Zeit* (Vienna, 1955); K. Militzer, *Die Entstehung der Deutschordensballeien im Deutschen Reich*, QuStDO 16 (Bonn-Godesberg, 1970); K. Forstreuter, *Der Deutsche Orden am Mittelmeer*, QuStDO 2 (Bonn, 1967).

[20]   On Herman of Salza's diplomacy, see H. Kluger, *Hochmeister Hermann von Salza und Kaiser Frierdrich II. Ein Beitrag zur Frühgeschichte des Deutschen Ordens*, QuStDO 37 (Marburg, 1987), pp. 9–20. For the Teutonic Knights as 'Novi Machabei', *Tabulae*, nos. 72, 321. For the Fifth Crusade and the military orders, see J.M. Powell, *Anatomy of a Crusade 1213–1221* (Philadelphia, PA, 1994); K. Polejowski, *Geneza i rozwój posiadłości zakonu krzyżackiego na terenie Królestwa Francji do połowy XIV wieku (Genesis and Development of Property of the Order of Teutonic Knights in the Kingdom of France to the Middle of the XIVth Century)* (Gdansk, 2003), pp. 22–60.

[21]   For more details, see ibid., pp. 42–56; H. d'Arbois de Jubainville, 'L'Ordre Teutonique en France', *Bibliothèque de l'École des Chartes*, 32 (1871), 65.

[22]   ADA, 3H3530.

same time, in January 1224, another member of the Brienne family, Walter IV, granted the Teutonic Knights some lands and titles in his county, but the political circumstances surrounding this action were different.

Walter was born in 1205 in southern Italy, but before 1221 his life is virtually unknown. We do know that, at the time Walter arrived in the Holy Land (probably after 1216), John of Brienne was acting as his legal guardian. In April 1221 and in May 1222 Count Thibaut IV of Champagne and his mother Blanche of Navarre were asked by John of Brienne to accept Walter's homage.[23] In March 1223 John arrived at Ferentino in Italy to take part in a meeting between Pope Honorius III, the emperor Frederick II and the master of the Teutonic Order, Herman of Salza. According to *Eracles* Herman of Salza proposed that Frederick should wed Isabella, John's daughter and heiress of Jerusalem.[24] No certain information is available as to whether the young Walter IV participated in that meeting, but we cannot rule out the possibility that he was indeed present. Nonetheless, in January 1224 at Brèvonne, near Brienne-le-Château, he issued his first known charter in the county of Brienne in which he gave the Teutonic Knights a manorial farm at Bugney, near Brienne-la-Vieille, with meadows, pastures, livestock with the rights of pasturing in his forests, and the right to receive a quantity of wine pressed between Brienne and Brienne-la-Vieille.[25] Even though the charter gives religious reasons for Walter's grant, the donation's political context is unquestionably bound up with the settlement at Ferentino. It is my opinion that the donation of January 1224 was a mark of gratitude to Herman of Salza for his initiative, but perhaps Walter (and maybe his uncle John) was deluding himself into believing that, thanks to the marriage between Frederick and Isabella, it might prove possible to recover the county of Lecce.

In spite of the violent severance of familial relations between Frederick II and the Brienne family just after the emperor's marriage to Isabella, which took place at Brindisi in November 1225, Walter IV continued to favour the Teutonic Knights. In January 1231 the Teutonic Order was granted Walter's rights over the hospital in Brienne-le-Château, a donation he confirmed at Acre in 1237.[26] Five months later, in June 1231, Walter IV issued charters in favour of the Teutonic Knights, the Hospitallers and the Premonstratensians. The Teutonic Order received 300 acres of forest near Chaumesnil with permission to grub out the trees and to build a house or manorial farm. The Order was allowed to use the meadows situated *citra Albam* ('on this side of Aube river') to pasture

---

[23] *Catalogue des actes des comtes de Champagne depuis l'avènement de Thibaud III jusqu'à celui de Philippe le Bel*, ed. H. d'Arbois de Jubainville (Paris 1863–66) (hereafter *CAC*), nos. 1330, 1407.

[24] *Eracles*, p. 358.

[25] ADA, 3H3529.

[26] Ibid., 3H3541; CAB, no. 170. This donation was contested by the Benedictines from Montier-en-Der. Polejowski, *Geneza*, pp. 184–8.

their livestock.[27] This donation of June 1231 also marked the foundation of the Teutonic commandery of Beauvoir, which, during the next hundred years, was to become the main Teutonic commandery in France.

At the same time Walter also granted the Hospitallers 500 acres of *li Bateiz* forest where they might grub out the trees and build a house or chapel, but without fortifications. He added the right to pasture livestock in his Orient forest (excluding goats). He confirmed this donation in Acre in 1235.[28] It was the beginning of the Hospitaller commandery of Orient (later Bonlieu). The Premonstratensian monastery of Basse-Fontaine obtained 200 acres of Walter's forest at Wévre.[29] Between 1224 and 1231 the Templars received nothing directly from the count of Brienne. But in 1228 he granted Bernard of Montquc, chamberlain of Thibaut IV of Champagne, 400 acres of his *li Bateiz* forest with permission to sell it to the Hospitallers, Templars or Cistercians from Clairvaux if necessary.[30] We know that between 1232 and 1233 Bernard sold 120 acres of the forest to the Templars, a sale that was confirmed by Erard II of Chacenay, Walter's governor in the county of Brienne. In 1238 Walter allowed the Templars to take over the remaining 280 acres of this forest, an act that allowed the Templars to found their commandery at Bonlieu.[31]

Walter's activity towards the other orders in June 1231 was related to his project to leave France and settle in the Latin East, and by April 1232 he was absent from his French county. Erard of Chacenay, who was appointed governor in Brienne, was a relative and one of the French nobles who had made donations to the Teutonic Order in Egypt in 1219. We cannot exclude the possibility that, before Walter left Champagne, he wanted to gain the military orders' favour and cooperation in the Holy Land in the near future. After leaving France, he went to the Latin East and married Maria of Lusignan, the sister of King Henry of Cyprus (after June 1231 but before 1235). Probably as a direct result of this marriage Walter acquired the county of Jaffa, the dowry of his mother-in-law, Alice of Champagne. Between 1235 and 1244 he spent most of the time in Palestine. According to Alberic of Trois-Fontaines, in 1237 a large company of 120 Templars attacked Muslim bands operating in the region between Acre and Atlit. Despite Walter's warnings, the Templars ran up against a larger than expected force and were defeated. Only the master, Armand of Périgord, and nine Templars managed to escape.[32] This event probably deepened the animosity between the count and the Templars in the Holy Land, an animosity that, if we believe the chronicler's description, was transformed into outright hostility before the battle of Forbie (October 1244).

---

[27]   ADA, 3H3538.

[28]   *CH*, nos. 1985, 2123.

[29]   CAB, no. 166.

[30]   Ibid., no. 158. The forest was situated between Piney, Gerosdot and Brèvonne.

[31]   Ibid., nos. 168, 171. For the Bonlieu commandery, see M.T. Boutiot, 'Les Templiers et leurs établissement dans la Champagne méridionale', *Annuaire de l'Aube* (1866), p. 36.

[32]   'Chronica Albrici', p. 942.

Perhaps the situation was more complex. Four charters have survived from the time Walter was at Jaffa. Three of them, all issued at Acre, were concerned with the military orders' presence in his French county. Among other things, the Templars received several hundred acres of *li Bateiz* forest. Is it possible that the chronicler's reports are over-coloured?[33]

From his marriage with Maria of Lusignan Walter had three sons, John, Hugh and Aimery. In 1244 the children were still minors – the eldest, John, can have been born no earlier than in 1235 or 1236 – and they probably remained with their mother's family in Cyprus or Antioch. John, as the eldest, inherited his lands and titles, although not the county of Jaffa, which, after the death of Alice of Champagne (1246), eventually passed to John of Ibelin.[34] In 1247 King Henry conferred all the rights that he possessed in Champagne and Brie to the young John of Brienne. This grant was confirmed in 1258 in Nicosia by Henry of Antioch and his wife Isabella, the daughter of Hugh I of Cyprus and Alice of Jerusalem and sister of Maria.[35] These charters should be linked to the situation in Champagne, where in July 1247 Thibaut IV, count of Champagne, uncertain of Walter's death, temporarily leased Alice's properties to her sister Philippa (the wife of Erard of Ramerupt, himself a member of a cadet branch of the Brienne family). In March 1250 the count of Champagne transferred the rights to Onjon, Luyères and Ville-sur-Terre to Walter of Reynel, a relative of the Briennes, ignoring the rights of the sons of Walter IV.[36] In 1254 Walter, as governor in the county of Brienne, confirmed the Templars' acquisition of 600 acres in the *li Bateiz* forest, between Brèvonne and their house at Bonlieu.[37] Clearly, during the absence of Walter IV's heir from the county, parts of his inheritance had been taken over by distant relatives with the approval of the counts of Champagne. It would seem that the young John of Brienne was appointed by his mother's family to set the affairs of their French domain to rights. John arrived in France in about 1259 but survived no longer than

---

[33]   Walter was one of the most important barons in the Kingdom of Jerusalem and one of the commanders during the crusades of Thibaut of Champagne and Richard of Cornwall (1239–41). He took part in the battle of Forbie, in October 1244 where he was captured; he died in an Egyptian prison before June 1247. His policy in Outremer (1233/5–1244) required further research.

[34]   The heirs of Walter IV made no claim to the county of Jaffa. For the controversy surrounding the county of Jaffa in the first half of the thirteenth century and after 1247, see H.E. Mayer, 'Ibelin vs. Ibelin: The Struggle for the Regency of Jerusalem 1253–1258', *Proceedings of the American Philosophical Society*, 122 (1978), 25–57; P.W. Edbury, 'John of Ibelin's title to the county of Jaffa and Ascalon', *English Historical Review*, 98 (1983), 115–33; H.E. Mayer, 'John of Jaffa, His Opponents and His Fiefs', *Proceedings of the American Philosophical Society*, 128 (1984), 134–63. Despite his extensive arguments concerning Walter IV as count of Jaffa, Mayer is not entirely persuasive.

[35]   CAB, no. 178.

[36]   Ibid., no. 176.

[37]   This land was sold by Guy of Milly and his wife Agnes for 1700 *livres tournois*; *CAC*, no. 3066.

two years, a charter issued in Champagne in January 1261 confirming his death.[38] During his presence in southern Champagne[39] he made decisions concerning the Templars and, indirectly, the Teutonic Order. On 18 April 1260 John and Guy de Basainville, the Templars' visitor in the West, appointed arbitrators to adjudge a dispute concerning the Templars' acquisitions between their houses at Bonlieu and Ville-sur-Terre.[40] This dispute had probably arisen as a consequence of Walter of Reynel's decision of 1254. In turn, in September 1260, John issued a charter in which he restored full rights to the Benedictines from Montier-en-Der over the hospital in Brienne.[41] This decision abrogated Walter IV's grant to the Teutonic Order in 1231. However, John of Brienne probably died at the end of 1260, and both the Templars and the Teutonic Knights kept their rights and lands.

At the same time Aimery, John's youngest brother, died, and the sole heir to the lands and titles was now Hugh, Walter IV's second son. Hugh remained in the Holy Land for several years, but his French overlord, Thibaut V of Champagne, was dissatisfied with his absence from France. In May 1267 the magnates of the kingdom wrote a letter to Thibaut explaining Hugh's long absence from Champagne.[42] Shortly afterwards he left Outremer for ever and went to France, where his presence was noticed in April 1268.[43] A few weeks later he went to Rome and probably took part in the battle of Tagliacozzo (23 August 1268) as a follower of Charles of Anjou. The French victory and the death of Conradin of Hohenstaufen allowed him to recover the county of Lecce. From this moment Hugh of Brienne and his successors were involved in Angevin policy in Italy and the Latin East. Hugh spent most of his time in Italy and then in Frankish Greece, and his visits to France were rare. It seems that after 1267 he lost interest in Palestinian (but not Cypriot) affairs yet retained an interest in the military orders present in his French domain.

In July 1269 Hugh, during a stay in France, approved Templar acquisitions in the county of Brienne to the sum of 1,000 *livres tournois* and in May of the following year the Templars' house Bonlieu obtained the right to draw 70 *livres* as a pension from Brienne.[44] On the other hand, in August 1269 the Teutonic Order's commandery of Beauvoir received the exclusive right to use the forest near Ville-sur-Terre and ownership of a quarter of this forest. Hugh also gave his consent

---

[38]   His first known charter issued in Champagne is from April 1259, the last one from September 1260. CAB, nos. 179, 182.

[39]   A certain John of Brienne (probably a different person) was a *bailli* of Thibaut V of Champagne in Troyes (June 1260). *CAC*, no. 3201.

[40]   CAB, no. 180.

[41]   Ibid., no. 182.

[42]   *CAC*, no. 3407.

[43]   Roserot, pp. 245–6.

[44]   CAB, nos. 186, 191; Polejowski, *Geneza*, p. 87.

for the Order to build a new enclosure at Beauvoir.[45] In May 1270 the count of Brienne approved the possession of a mill that was built near the Teutonic Order's commandery and exempted it from taxes. The Teutonic Knights also gained the right to grind grain from Chaumesnil, Morvilliers and La Chaise and from between Morvilliers and Soulaines.[46] In the end, the Hospitallers received Hugh's confirmation of his brother John's last will, in which he bequeathed them 20 *livres* from Brienne to build St Mary's Chapel in their Orient commandery.[47]

As had been the case in the activity of his father towards military orders in June 1231, Hugh's activity in this matter was connected with his preparations to leave France. In June 1270 Hugh of Brienne visited Marseille and probably took part in King Louis IX's second crusade.[48] We know that Hugh returned to France, because his presence was noted in southern Champagne between April 1272 and January 1273.[49] But after this date he left his French domain and went to Italy, where he married Isabel of La Roche, widow of Geoffrey of Bruyères, lord of Karyteina and Thebes. She was a daughter of Guy, duke of Athens, and gave her new husband two children, son Walter and daughter Agnes. It seems certain that this marriage was an element in Charles of Anjou's political ambitions to build an empire for himself in the eastern Mediterranean.[50]

Hugh's grants to the military orders in the spring of 1270 were one of the last signs of the Brienne family's patronage of these crusading institutions. Only once again, in 1288 during a short stay in France, did he give proof of his interest in the Teutonic commandery of Beauvoir. In August 1288 Hugh was a mediator between the Teutonic Knights and Walter d'Ecot and his wife. According to the documents, the Beauvoir commandery had illegally seized some rights and properties belonging to Walter and his wife in Morvilliers. Hugh, as Walter's feudal overlord, negotiated a settlement on the basis that the d'Ecot family would approve the Beauvoir acquisitions in return for compensation amounting to 10 *livres*.[51] This charter is similar to another document issued in 1286 by a certain knight called John of Brienne, guardian of the fairs in Champagne and Brie (we do not know to which branch of the Brienne family he belongs). He too recognized a number of illegal acquisitions made by Beauvoir near Brienne-la-Vieille and remitted the debts of the commandery.[52] Shortly afterwards Hugh left France and went to Italy,

[45]   ADA, 3H3538; Polejowski, *Geneza*, p. 89; K. Polejowski, 'Les comtes de Brienne et l'ordre teutonique (XIII–XIV siècle)', *La Vie en Champagne*, 32 (Oct.–Dec. 2002), 7.

[46]   ADA, 3H3538; Polejowski, *Geneza*, p. 89.

[47]   CAB, no. 189.

[48]   Ibid., no. 193; Roserot, p. 246.

[49]   CAB, nos. 194–5.

[50]   For Charles of Anjou's East Mediterranean policy before 1282 see S. Runciman, *The Sicilian Vespers* (Cambridge, 1958), pp. 171–200.

[51]   ADA, 3H3562. Note also that in October 1288 Hugh confirmed his father's grant to the Teutonic Order of June 1231. Polejowski, 'Brienne', p. 7. For Walter of Ecot see Roserot, p. 975.

[52]   ADA 3H3531; Polejowski, *Geneza*, pp. 126–7.

where in 1291 he took the widow Helena Komnena as his second wife.[53] He died in 1296, and his heir was his son Walter V, who was born after 1272 and who was to die on 15 March 1311.

Like his father, Walter spent most of the time in his Italian inheritance, fighting as one of the Angevin commanders against the Aragonese, and also in Greece, where he inherited the duchy of Athens after the death of Guy II of La Roche in 1308. Walter's interest in the military orders was insignificant. Only the Teutonic Knights featured in his activities during his short stays in France. In June 1303 Walter was present in Beauvoir, where he confirmed the grant in favour of the Teutonic Knights made by Margaret of Chaumesnil.[54] In April 1304 he granted the Beauvoir commandery the right to acquire property in the county up to the sum of 40 *livres*.[55] With these two charters the interest of the Brienne family in the military orders in southern Champagne came to an end.[56] There is just one final echo of their interest in the military orders: when in 1347 Walter VI of Brienne drew up his last will, he bequeathed for *l'abbée* of Beauvoir 10 *livres tournois*.[57]

It is self-evident that in the second half of the thirteenth century the Brienne interest in the religious orders steadily declined. Only Hugh of Brienne between 1269 and 1270 showed a strong interest in them. It is possible that he wanted to strengthen the position of the military orders in France because he was aware of the situation in the Holy Land and, as one of the most important Latin nobles there, was conscious of the orders' importance.

During the twelfth and thirteenth centuries the Templars and the Hospitallers built a dense and rich system of commanderies in southern Champagne. For example, in the county of Brienne, the Templar commandery at Beaulieu owned more than 2,000 hectares of land, not to mention other rights and properties. Similarly, the Hospitallers commandery of Orient was vast and rich. For those two military orders the kingdom of France was always one of the most important

---

[53]   She was the heiress of Lamia and Larissa in Greece and mother of Guy II de La Roche, duke of Athens.

[54]   ADA, 3H3531. Margaret granted to Beauvoir five *mansi* and some lands near Chaumesnil.

[55]   ADA, 3H3531; 3H3528.

[56]   It is worth mentioning that, after the defeat at the Battle of Cephissus (15 Mar. 1311), Pope Clement V supported the Brienne efforts for a military recovery of Athens (1312–14). The main element in the struggle against the Catalans was to be the Hospitallers. See *Codice diplomatico del sacro militare ordine gerosolemitano oggi di Malta*, ed. S. Pauli (Lucca 1733–37), 2, p. 395; *Regestum Clementis papae V*, eds cura et studio monachorum ordinis S. Benedicti (Rome, 1885–92), nos. 7891, 10166–8; *Acta Aragonensia*, ed. H. Finke (Leipzig/Berlin, 1908–1922), 2, no. 466.

[57]   C. Paoli, 'Nuovi documenti intorno a Gualtieri VI di Brienne, duca d'Atene e signore di Firenze', *Archivio Storico Italiano*, serie 3, 26 (1872), p. 42. From the same document (p. 47) we learn that Walter VI had, for financial reasons, sequestered three *casaulx* (*de Locorotonde, de Potignan, de Casable*) belonging to the Hospitallers in Apulia, which, after his death, were to return to the Order.

fields for recruiting new members and a major source of income for the brothers fighting in the Holy Land. The Teutonic Order, evidently favoured by the Brienne family in the thirteenth century, remained weak in France. The Beauvoir commandery possessed only about 200 hectares of land.[58] It must be added that, in the thirteenth century in southern Champagne and northern Burgundy (in the Orbec commandery),[59] the Teutonic Knights, like the Hospitallers, were mainly perceived through their charitable vocation, but, as an imperial and German institution, they were regarded as outsiders. We have no information about any French nobles donning the habit of the Teutonic Order and joining the Beauvoir or Orbec commanderies in the thirteenth century. The association between the counts of Brienne and the Teutonic Order was therefore quite exceptional, especially in view of the rather cool and distrustful relations between the French nobility and the Teutonic Knights in both Palestine and France in the thirteenth century.

---

[58]   Polejowski, *Geneza*, pp. 97–8, 213–16.

[59]   On the origins of the Teutonic commandery in Orbec, see K. Polejowski, 'Sur l'origine de la maison de l'ordre teutonique à Orbec. Un document de Renaud, évêque de Nevers (avril 1224)', *Bulletin de la Société Nivernaise des Lettres, Sciences et Arts*, 52 (2003), 253–7.

Chapter 25

A Geography of Power: The Hospitallers in the Territorial Policies of the Bishops of Strasbourg in Lower Alsace in the Thirteenth Century

Nicolas Buchheit

The commanderies of the Hospital of St John of Jerusalem in Lower Alsace were founded during the thirteenth century (see Fig. 25.1). The first one was Dorlisheim, which dates from the beginning of the century, while Rhinau and Sélestat only appeared in the early 1260s. As elsewhere, these establishments had to support the Order's mission of war and charity in the East. However, the Hospitallers' appearance in Lower Alsace is relatively late by comparison with the beginning of their development in the south and the west of Europe, directly after the founding of the Order in the early twelfth century. It was only in the thirteenth century that they benefited from the interest of local society and especially of the bishops of Strasbourg.

These bishops took an interest in the Hospitallers from the moment they appeared in Lower Alsace. However, after the initial phase, documentation is lacking until the mid-thirteenth century. When the materials resume, they reveal a continuing relationship between the bishops of Strasbourg and the Hospitallers. Relations became closer at the beginning of the 1260s, during the episcopacy of Walther of Geroldseck, but then they appear to have fallen into abeyance until the end of the century. The Order's establishments in Lower Alsace therefore coincided with the period of sustained relations with the bishops lasting from the beginning of the thirteenth century to the 1260s.

They are in fact conterminous with the decades in which the bishops pursued an intensive policy of domination in the region, a policy that began after the death of Emperor Henry VI in 1197.[1] Until the end of the twelfth century the bishops of Strasbourg had been under the control of the German monarchy, but, at this point,

---

[1]   A. Hessel, 'Die Beziehungen der Straßburger Bischöfe zum Kaisertum und zur Stadtgemeinde in der ersten Hälfte des 13. Jahrhunderts', *Archiv für Urkundenforschung*, 6 (1918), 266–75; A. Hessel and M. Krebs, *Regesten der Bischöfe von Straßburg*, 2 (Innsbruck, 1928); F. Rapp, *Le château-fort alsacien dans la vie médiévale et dans la politique territoriale* (Strasbourg, 1968).

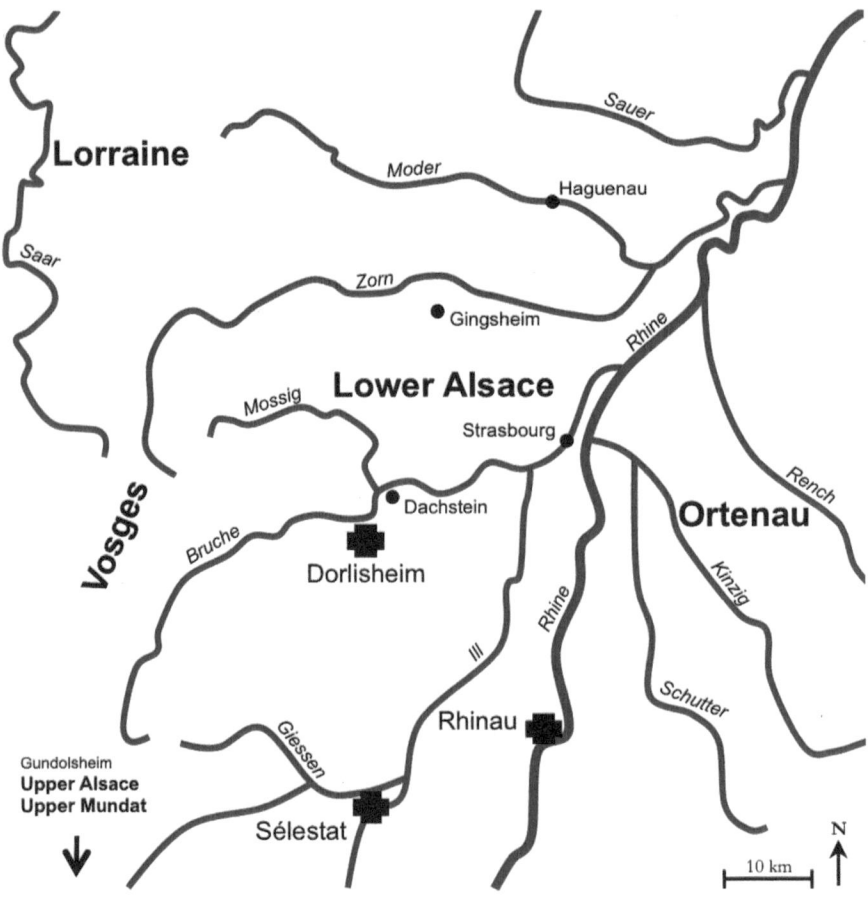

Figure 25.1   The Hospitallers in Lower Alsace (graphics: N. Buchheit)

Bishop Conrad of Huneburg decided to assert a greater degree of independence. He was an opponent of the Staufen and an ally of Otto of Brunswick in his struggle for the crown against Philip of Swabia. The bishop was in competition with the Staufen family, who had made Alsace one of its centres of power.[2] By contrast, Henry of Veringen, who was elected in 1202, followed the instructions of Pope Innocent III and supported the accession of Frederick II in 1212. The situation changed again with the episcopate of Berthold of Teck in 1223. On the one hand, the extinction of the Dagsburg family in 1225 exacerbated the disputed claims to

---

[2]   J.-Y. Mariotte, 'Les Staufen en Alsace au XIIᵉ siècle d'après leurs diplômes', *Revue d'Alsace*, 119 (1993), 43–74; A. Meister, *Die Hohenstaufen im Elsass. Mit besonderer Berücksichtigung des Reichsbesitzes und des Familiengutes derselben im Elsass, 1079–1255* (Mainz, 1890).

their important territorial inheritance in the region. The bishop was to obtain a substantial part of it after the defeat of King Henry (VII), who was administering Germany on behalf his father Fredrick II, and his allies. On the other hand, since Frederick's excommunication in 1227, Berthold of Teck had the support of Pope Gregory IX in his territorial policy against the Staufen. The submission of the emperor in 1230 resulted in the suspension of hostilities, but the struggle restarted in 1238. From immediately after his election in 1244, Bishop Henry of Stahleck followed his predecessor's aggressive policy. In the context of the Great Interregnum, he supported Henry Raspe from 1246, and then, after Frederick's death in 1250, William of Holland and Richard of Cornwall; in the process, he became the most powerful territorial prince in Lower Alsace. When Walther of Geroldseck acceded to the episcopate in 1260, he inherited an extensive territorial base, but he wanted to bring Strasbourg, which aspired to greater independence, under his control. Walther waged war against the town, but he was defeated at the battle of Hausbergen, near Strasbourg, in 1262. This event put an end to the bishops of Strasbourg's attempts to turn Lower Alsace into an episcopal lordship. They remained the most important power in the region but henceforth would have to compromise with the imperial towns.

The fact that between the early thirteenth century and the 1260s successive bishops of Strasbourg both supported the development of the Hospitallers and pursued a policy of territorial expansion in Lower Alsace raises the question of whether the Hospitallers contributed to this policy. The foundation of the commandery of Dorlisheim had implications for territorial control that interested the bishops of Strasbourg. There is evidence for Hospitaller involvement in episcopal territorial ambitions in the middle of the thirteenth century, and it became a matter of deep significance during the episcopate of Walther of Geroldseck, who had close ties with Dorlisheim and Rhinau. However, the failure of this policy at the beginning of the 1260s put an end for a time to the connections between the bishops and the Hospitallers, who then had to find support elsewhere.

The earliest document to provide evidence for the presence of the Hospitallers in Lower Alsace is an agreement between the bishop of Strasbourg, Henry of Veringen, and Dorlisheim in 1217.[3] The commandery could have been founded by the counts of Dagsburg, because it was established on land that had belonged to them, or alternatively by Duke Theobald I of Lorraine, who had possession of the Dagsburg patrimony thanks to his marriage to Gertrude, the heiress of Count Albrecht II who died in 1212.[4] The duke of Lorraine would have been following a family tradition of supporting the Order of St John that went back to the mid-twelfth century.[5] Establishing a Hospitaller presence in Dorlisheim could have

---

[3]   Arch. dép. Bas-Rhin, H1392/1; *CH*, 2, no. 205; Hessel and Krebs, no. 837, p. 19.

[4]   F. Legl, *Studien zur Geschichte der Grafen von Dagsburg-Egisheim* (Sarrbrücken, 1998), pp. 344–56, 414–15.

[5]   G. Giulato, 'La commanderie Saint-Jean du Vieil Aître de Nancy', in *Retour aux sources. Textes, études et documents d'histoire médiévale offerts à Michel Parisse* (Paris,

been a way of expressing his power within his wife's inheritance and unifying it
with his own. The foundation would also have been connected with his territorial
policy, which was directed towards the Rhine valley and which brought him into
competition with Frederick II whose opponents he had joined. But the duke's
defeat in 1218 put an end to his policy in Lower Alsace.[6] Henry of Veringen's
interest in the commandery of Dorlisheim does not necessarily mean he had good
relations with the count of Dagsburg or the duke of Lorraine, but it did accord
with the anti-Staufen stance initiated by his predecessor, Conrad of Huneburg.[7]
After his election in 1202, Henry of Veringen, in keeping with Pope Innocent III's
instructions, had supported Otto IV of Brunswick against Philip of Swabia in the
struggle for the throne. Similarly, when he moved closer to Frederick II in 1212,
he was obeying papal orders. Even so, his attitude to the contest between Frederick
and the duke of Lorraine was equivocal. His support for the commandery of
Dorlisheim, close to an area of opposition to the Staufens, could indicate a latent
anti-Frederick position, but it did serve to further a shrewd territorial policy.

In this respect, the commandery of Dorlisheim had an interesting location.
Placed at the mouth of the valley of the River Bruche in the Vosges, it controlled an
important Roman crossroads. One route followed the Rhine valley from north to
south. The other connected Alsace to Lorraine from east to west through the Bruche
valley.[8] Trade seems to have been increasing along this road at the beginning of the
thirteenth century.[9] The commandery was established on one of the few possible
crossing points on the River Bruche,[10] and so control of the commandery meant
control over a strategic line of communication, something that had been a goal of

---

2004), pp. 253–64; M. Henry, *Les ordres militaires en Lorraine* (Metz, 2006), pp. 117–18.

[6]   M. Parisse, *La Noblesse lorraine (XIe–XIIIe s.)* (Paris, 1976), pp. 731–52; idem,
Noblesse et chevalerie en Lorraine médiévale. Les familles nobles du XIe au XIIIe siècle
(Nancy, 1982), pp. 93–4, 99–100.

[7]   F.J. Fuchs, 'Veringen, Heinrich von', in *Nouveau dictionnaire de biographie
alsacienne* 38 (Strasbourg, 2002), p. 3987; Hessel and Krebs, no. 838, pp. 19–20.

[8]   J. Braun, 'Les voies romaines autour de Molsheim', *Annuaire de la Société
d'histoire et d'archéologie de Molsheim et environs* (1967), 20–27.

[9]   J.-L. Fray, 'Sarrebourg und der obere Saargau im Lichte der Zentralitätsforschung.
Ein Beitrag zur Geschichte der mittelgroßen lothringischen Städte im Mittelalter', in
H.-W. Herrmann (ed.), *Die alte Diözese Metz – L'ancien diocèse de Metz. Referate eines
Kolloquiums in Waldfischbach-Burgalben vom 21. bis 23. März 1990* (Saarbrücken, 1993),
pp. 152–8; idem, *Villes et bourgs de Lorraine. Réseaux urbains et centralité au Moyen Âge*
(Clermont-Ferrand, 2006), pp. 433–6; F. Rapp, 'Routes et voies de communication à travers
les Vosges du XIIe au début du XVIe siècle', in *Les pays de l'Entre-Deux au Moyen Âge:
questions d'histoire des territoires d'Empire entre Meuse, Rhône et Rhin. Actes du 113e
Congrès national des Sociétés savantes de Strasbourg, Section d'histoire médiévale et de
philologie, 1988* (Paris, 1990), pp. 197–8.

[10]   G. Oswald, *Molsheim à la fin du Moyen Âge (1308–1525). Essai d'histoire politique,
économique, sociale et religieuse d'une cité épiscopale de Basse-Alsace* (Strasbourg, 1993–
94), pp. 135–6.

the bishops of Strasbourg since the early thirteenth century. They wanted access to their territories, but above all they wanted to assert their territorial claims in this part of Lower Alsace. That explains why Henry of Veringen built the castle of Dachstein in 1214, whose role was to keep watch over the mouth of the Bruche and the Mossig valley.[11] The relationship established with the Hospitallers of Dorlisheim was connected with this policy, as were the first possessions of the commandery elsewhere.

The agreement between Bishop Henry of Veringen and the commandery of Dorlisheim in 1217 was an exchange of an annual rent on a property he had given the Hospitallers '*pro divino respectu et amore domus predicte*', in return for another annual rent on vines located near the mill of *Gundoltesheim*. It is difficult to know whether this village is Gingsheim in the north of the Kochersberg, or Gundolsheim in Upper Alsace. In any case, both are well away from Dorlisheim. Gundolsheim is situated in the Upper Mundat, an enclave around Rouffach and Soultz in the diocese of Basle that belonged to the bishop of Strasbourg. He could control the north–south road that followed the Rhine valley and earn some income from the vineyard and the trade, but also be one of the main powers in the Oberrhein.[12] As for Gingsheim, the bishops of Strasbourg had had a presence there since the end of the twelfth century; it was in a region that they controlled through their castles and vassals, and, in the fourteenth century, they owned the village itself.[13] In both places, it is difficult to explain the episcopal land ownership. But the allocation of these possessions to the Hospitallers of Dorlisheim could have been a way to increase allies in areas where the bishop could gain income and control through the commandery. The development of Dorlisheim is to be seen as a further expression of Bishop Henry of Veringen's territorial ambition.

However, the lack of sources after the 1217 charter makes it impossible to trace the influences on the commandery of Dorlisheim until the middle of the thirteenth century. But on 6 January 1256, a knight named Ulrich of Ittenheim gave the commandery lands and buildings in Dachstein, where he was *burgensis*, in the presence of the bishop of Strasbourg, Henry of Stahlek, and a group of episcopal officers, *ministeriales* and vassals.[14] This donation shows that, even if the connection between the commandery and the bishops of Strasbourg had been interrupted earlier in the thirteenth century, it was certainly present in the context of the Great Interregnum. Like his predecessor, Bishop Berthold of Teck, Henry of Stahleck was engaged in the struggle against the Staufen. He supported the

---

[11]   Rapp, *Le château-fort alsacien*, pp. 74–5.

[12]   O. Kammerer, *Entre Vosges et Forêt-Noire: pouvoirs, terroirs et villes de l'Oberrhein (1250–1350)* (Paris, 2001), pp. 70–85.

[13]   J. Fritz, *Das Territorium des Bisthums Strassburg um die Mitte des XIV. Jahrhunderts und seine Geschichte. Ein Beitrag zur deutschen Territorialgeschichte* (Köthen, 1885), p. 6; J. Burnouf (ed.), *Le Kochersberg. Histoire et paysage* (Strasbourg, 1980), p. 53; *Das Reichsland Elsass-Lothringen*, 3 (Strasbourg, 1903), p. 343.

[14]   Arch. dép. Bas-Rhin, H1382/2; Hessel and Krebs, no. 1482, p. 154.

claims to the throne of Henry Raspe in 1246, William of Holland in 1247 and then Richard of Cornwall in 1254. In return, he was given responsibility for the imperial territories in Alsace. The weakening of the Staufens and royal power gave the bishops the opportunity to dominate Lower Alsace.[15] By acquiring properties in Dachstein, where there was a castle held by the bishops of Strasbourg against the Staufen and the centre of their power in this part of the Bruche valley, the commandery of Dorlisheim was bound up with this process.[16]

Walther of Geroldseck, who was elected in 1260, inherited this connection and developed even closer links with the commandery of Dorlisheim. He buried his brother Hermann in front of the great altar of the commandery church after he was killed during the battle of Hausbergen in 1262. According to the chronicle written by the Benedictine Richer of Senones during this period, other warriors killed in the fight were buried in the commandery as well.[17] Walther of Geroldseck himself was buried next to his brother,[18] probably having taken the Order's habit: the bishop's tombstone has vanished, but the epitaph revealed that he was 'frater ordinis nostri', and the *Pappenheim-Chronik*, written c. 1535, claimed that he was an *Ordensbruder*.[19] Their father, also named Walther, made a donation in 1265 for the remission of his sins and the memory of his sons.[20] He confirmed this donation in 1266, which was also for his wife Elisabeth's salvation.[21] So the commandery church served as a burial place for those who fought with the bishop during the battle of Hausbergen, before becoming a place where the Geroldseck family was commemorated. The commandery had thus passed from episcopal influence to that

---

[15]   F. Rapp, 'Stahleck, Heinrich III von', in *Nouveau dictionnaire de biographie alsacienne* 35 (Strasbourg, 2000), pp. 3725–6.

[16]   B. Metz, 'Essai sur la hiérarchie des villes médiévales d'Alsace (1200–1350). 1ère partie', *Revue d'Alsace*, 128 (2002), 73–5; Rapp, *Le château-fort alsacien*, p. 75.

[17]   D. Dantand, 'La chronique de Richer de Senones. Présentation, édition et traduction', thèse, Université de Nancy II, 1996.

[18]   'Annales Maurimonasterienses', *MGH SS*, 17, p. 182; Arch. dép. Bas-Rhin, H1382/3; 'Bellum Waltherianum', *MGH SS*, 17, p. 113; C. Bühler, *Geroldsecker Regesten. Regesten der Urkunden des Hauses und der Herrschaft Geroldseck*, at http://www.buehler-hd.de/reg/regesten1.pdf, no. 161 (accessed 23 November 2011); *CH*, 3, no. 3202; Hessel and Krebs, no. 1719, p. 220; P.C. von Planta (ed.), *Adel, Deutscher Orden und Königtum im Elsaß des 13. Jahrhunderts. Unter Berücksichtigung der Johanniter* (Frankfurt/Main, 1997), no. 20, p. 304; 'Richeri Gesta Senoniensis ecclesiae', *MGH SS*, 25, p. 343.

[19]   P.A. Grandider, *Œuvres historiques inédites*, 5 (Colmar, 1867), p. 354; J.J. Reinhard, *Pragmatische Geschichte des Hauses Geroldseck wie auch derer Reichsherrschaften Hohengeroldseck, Lahr und Mahlberg in Schwaben* (Frankfurt/Main-Leipzig, 1766), no. 1, p. 18.

[20]   Arch. municipales de Haguenau, GG206/1; Bühler, *Geroldsecker Regesten*, no. 157; *Cartulaire de l'église S. George de Haguenau*, ed. C.A. Hanauer (Strasbourg, 1898), no. 19, pp. 12–13.

[21]   Arch. dép. Bas-Rhin, H1382/3; Bühler, *Geroldsecker Regesten*, no. 161; *CH*, 3, no. 3202; Hessel and Krebs, no. 1719, p. 220; P.C. von Planta, no. 20, p. 304.

of the Geroldsecks, a characteristic consequence of the interpenetration of episcopal and family interests that at this point marked the height of Geroldseck power.[22]

The same is true for the foundation of the commandery of Rhinau in the early 1260s. Rhinau was an episcopal town located near one of the few fords across the Rhine linking Lower Alsace and the Ortenau, an area where both episcopal and Geroldseck territory was located.[23] The presence of a hospital in the commandery indicates the strategic significance of this place that Walther of Geroldseck and his family wanted to control.[24] However, the first mention of this establishment dates to 19 November 1264, more than a year after Walther of Geroldseck's death, when Theodoric, the bishop of Wierland, gave an indulgence to help the Hospitallers to finish building the commandery.[25] At this time, the foundation appears to have been recent and so it could date from the episcopate of Walther of Geroldseck. Moreover, on the same day, Thedoric issued an indulgence for the commandery of Dorlisheim, where the Geroldseck family retained close relations.[26] The foundation of this commandery enabled Rhinau to grow in the second half of the thirteenth century, so that by then it was the largest town in Lower Alsace between Strasbourg and Sélestat.[27] It illustrates the tendency for the Hospitallers to be established at strategic locations for communication and also in towns, the new places of power.

If the Order of St John could grow until the early 1260s because it allowed Walther of Geroldseck and his family to use it as an instrument for their territorial ambitions, it also had to pay the prince when their policy failed. After Bishop Walther of Geroldseck's defeat at the hands of the citizens of Strasbourg at Hausbergen in 1262, his territories were devastated by the victorious troops and the village of Dorlisheim was set on fire. We do not know, however, whether the commandery was affected.[28] Walther of Geroldseck withdrew to the nearby castle of Dachstein from where he pursued his struggle against Strasbourg. It was at this

---

[22]   C. Bühler, *Die Herrschaft Geroldseck. Studien zu ihrer Entstehung, ihrer Zusammensetzung und zur Familiengeschichte der Geroldsecker im Mittelalter* (Stuttgart, 1981), pp. 33–40; B. Metz, 'Geroldseck, über Rhein von', in *Nouveau dictionnaire de biographie alsacienne*, 13 (Strasbourg, 1988), pp. 1170–71.

[23]   J. Braun, 'Les voies romaines de l'arrondissement d'Erstein', *Revue d'Alsace*, 98 (1959), 32–3; P. Claus, 'Rhinau', in *Encyclopédie de l'Alsace*, 11 (Strasbourg, 1985), p. 6410; O. Kammerer, 'Le Haut-Rhin entre Bâle et Strasbourg a-t-il été une frontière médiévale?', in *Les pays de l'Entre-Deux au Moyen Âge*, p. 75.

[24]   Arch. Ville et Communauté urbaine de Strasbourg, 1AH353/1 and 1AH854, fol. 138; *Urkundenbuch der Stadt Straßburg*, ed. A. Schulte, 3 (Strasbourg, 1881), no. 108, pp. 38–9.

[25]   Arch. dép. Bas-Rhin, G4213/8.

[26]   Arch. dép. Bas-Rhin, H1360/3; *CH*, 3, no. 3109; Hessel and Krebs, no. 1769, p. 236.

[27]   B. Metz, 'Essai sur la hiérarchie des villes médiévales d'Alsace (1250–1350). 2e partie', *Revue d'Alsace*, 134 (2008), pp. 155–7.

[28]   Hessel and Krebs, no. 1678, p. 209.

time that the commandery became the place where the supporters of the episcopal party and the members of the Geroldseck family who had been killed were commemorated. When King Richard of Cornwall decided on 3 November 1262 to confer the parish of St George in Haguenau, an imperial town, on the Hospitallers, it was not granted to the nearby commandery of Dorlisheim but to the Order in general.[29] In fact relations between the king and Walther of Geroldseck had deteriorated,[30] and the burghers of Haguenau wanted to break free of the episcopal domination.[31] Richard's motivation could well have been more to do with his good relations with the Order – something that had existed since his crusade of 1240–42[32] – and with the prior of Germany, Fr Henry of Boxberg,[33] rather than as the consequence of episcopal influence, even if the bishop approved this grant of a parish that lay within his diocese.[34]

The relations with the Geroldseck continued after Walther's death. This is shown by his father's donation in 1265 and the 1266 confirmation, as he was still at war with Strasbourg.[35] However, after that there is no documentary evidence to show that they persisted beyond the end of the thirteenth century, nor whether Walther of Geroldseck senior's death in 1275/77, which prefigured the division of his legacy, put an end to them. But during the 1260s the commandery of Dorlisheim remained in the Geroldseck fold. That is why Bishop Walther's successor seems to have had no consistent relations with Dorlisheim and Rhinau. After his election in 1263, the new bishop of Strasbourg, Henry of Geroldseck (am Wasichen), wanted to restore peace and good relations with Strasbourg.[36] The commandery of Dorlisheim was too much involved with the aggressive interests of the Geroldsecks and so could not be expected to work with the new bishop. When he acted on the commandery's behalf, he did so in accord with his everyday episcopal prerogatives. Thus, in 1264, he gave his consent to the indulgence granted the commandery by Bishop Theodoric of Wierland,[37] and, when in 1266 the abbey of Schwarzach sold the commandery a house in Strasbourg, he approved the transfer of this piece of

---

[29]   *Acta imperii inedita*, ed. E. Winckelmann, 1 (Innsbruck, 1880), no. 573, pp. 459–60; *Cartulaire de l'église S. George de Haguenau*, no. 18, pp. 11–12; *CH*, 3, no. 3041; *Regesta imperii*, ed. J.F. Böhmer et al., 5, (Innsbruck 1881–1901), 1, 2, no. 5411.

[30]   Hessel and Krebs, nos. 1696–7, pp. 215–16.

[31]   Ibid., nos. 1604, 1645, pp. 183, 197.

[32]   J. Riley-Smith, *The Knights of St. John in Jerusalem and Cyprus, c. 1050–1310* (London, 1967), pp. 177–8.

[33]   *Alsatia diplomatica*, ed. J.D. Schoepflin, 2 (Mannheim, 1775), no. 612, p. 441; *Regesta imperii*, 5, 1, 2, no. 5412.

[34]   *Cartulaire de l'église S. George de Haguenau*, no. 21, pp. 13–14; Hessel and Krebs, no. 1696, p. 215.

[35]   Bühler, *Geroldsecker Regesten*, no. 162; *Urkundenbuch der Stadt Straßburg*, ed. W. Wiegand, 1 (Strasbourg, 1879), no. 615, pp. 463–4.

[36]   B. Metz, 'Geroldseck am Wasichen von, Heinrich', in *Nouveau dictionnaire de biographie alsacienne*, 13 (Strasbourg, 1988), p. 1169.

[37]   Arch. dép. Bas-Rhin, H1360/3; *CH*, 3, no. 3109; Hessel and Krebs, no. 1769, p. 236.

church property within his diocese.[38] This sale is the commandery's only known acquisition in Strasbourg, and that suggests that, because Strasbourg had defeated Walther of Geroldseck and was still at war with his father, its development within the town was very restricted at this period.

Bereft of the support of the bishop of Strasbourg, in 1273/74 the commandery of Dorlisheim found a new patron in German-speaking Lorraine in the person of Renaud, count of Blieskastel and lord of Bitche.[39] He was the youngest son of the Duke Ferry II of Lorraine, the brother of Theobald I who may have founded the commandery of Dorlisheim, and the uncle of the Duke Ferry III (1251–1303). This could explain why Conrad of Lichtenberg, who was elected bishop of Strasbourg in 1273,[40] had no interest in Dorlisheim or in any commanderies in Lower Alsace during the first years of his episcopate. He was frequently at war with the duke of Lorraine before becoming bishop and until 1286. But in 1287 he appended his seal on the charter of the donation of one of his vassals, Cuno of Geispolsheim, to the commandery of Dorlisheim.[41] The development of the commandery in Lorraine was disrupted in the late 1270s by the war of succession for the county of Blieskastel,[42] with the result that the Hospitallers focused anew on Lower Alsace. The episcopal gesture might be considered an expression of favour for the commandery, but it is the only one known from the entire episcopate of Conrad of Lichtenberg until his death in 1299. On the other hand, it shows that the Hospitallers had established a new network of relationships at the end of the thirteenth century with members of the local minor nobility, often close to the bishop.

The development of the Hospitallers in Lower Alsace in the thirteenth century had depended on the support of the bishop of Strasbourg up to and including the episcopate of Walther of Geroldseck. However, the failure of his territorial policy and the influence of his family during the 1260s interrupted this special relationship between the Hospitallers and the bishops of Strasbourg. They were to resume later, in other forms, through the minor nobility and the urban burghers who supported the military orders in Lower Alsace at the end of the century. The commanderies of Dorlisheim and Rhinau were dependent on aristocratic strategies in the region, which aimed to assert, enhance and perpetuate their social domination. These

---

[38] Arch. dép. Bas-Rhin, H1514/2; Hessel and Krebs, no. 1795, p. 242; *Urkundenbuch der Stadt Straßburg*, ed. Wiegand, no. 606, p. 457.

[39] N. Buchheit, 'Une commanderie alsacienne en Lorraine: les Hospitaliers de Dorlisheim à Puttelange au XIIIe siècle', *Revue d'Alsace*, 136 (2010), 47–59.

[40] J.-M. Rudrauf, 'Lichtenberg, Conrad', in *Nouveau dictionnaire de biographie alsacienne*, 24 (Strasbourg, 1994), p. 2349.

[41] Arch. Ville et Communauté urbaine de Strasbourg, CH272; *Corpus der altdeutschen Originalurkunden bis 1300*, eds F. Wilhelm *et al.* 5 (Lahr, 1986), no. 325, p. 247; Hessel and Krebs, no. 2189, p. 332.

[42] J. Gayot, 'Histoire de la seigneurie de Bliescastel', *Bulletin de la Société des amis des pays de la Sarre*, 2 (1925), 59–344; H.-W. Herrmann, 'Die Grafen von Blieskastel', in H.-W. Herrmann and K. Hoppstätdter (eds), *Geschichtliche Landeskunde des Saarlandes*, 2 (Saarbrücken, 1977), pp. 254–61.

establishments were instruments of territorial policies. Their exploitation signalled the evolution of a geography of power in Lower Alsace, from the control of strategic places of communication to the control of the expanding towns, the new seats of power. Nevertheless the question remains of the specific role of the Order of St John in the aristocratic developments and territorialisation in Lower Alsace, as compared with other religious establishments, and, in particular, in the light of the logistical and the propagandist functions of the Hospitallers' commanderies in their support of the Latin East and crusading values.

Chapter 26

The Priors of the Knights Hospitaller from the Piast Dynasty in the Province of Bohemia: Hereditary Princes or Ecclesiastical Dignitaries?

Maria Starnawska

It was customary for medieval rulers to intend one of their sons for the priesthood, a practice that was embraced by the Piast dynasty in medieval Silesia (see Fig. 26.1).[1] This area, which had been ruled by a separate line of the Piasts since Poland split into five major duchies in 1138, had, in the thirteenth century, been divided further into even more duchies as the dynasty expanded. In the fourteenth century Silesia was not incorporated into the reunified Polish kingdom, but instead it broke up into several duchies, some of which included only a few towns. In the first half of the fourteenth century most of the Silesian dukes became vassals of the Bohemian rulers of the Luxemburg dynasty. Many of them served the Bohemian kings and participated in Bohemia's political life. After Charles IV became emperor, they also became involved in the politics of the Empire. Some of the Silesian dukes constituted a separate élite, superior in prestige to the Bohemian magnates, as was evidenced by the marriages of Silesian Piast duchesses to Emperor Charles IV as well as to Polish and Hungarian kings.[2]

---

[1]   My thanks are due to Marcin Lewandowski for translating this paper into English. The following abbreviations have been used:

APWr – The State Archive in Wrocław

CDS – Codex diplomaticus Silesiae (Breslau, 1857–1933)

LB – Lehns- und Besitzurkunden Schlesiens und seiner einzelnene Fürsthethümer im Mittelalter, eds C. Grünhagen, H. Markgraf, 1–2 (Leipzig, 1881–83)

RML – Prague, The National Archives, Řad Maltézsky – Listiny. Also availabe at www.monasterium.net (accessed 24 November 2011).

RMS – Prague, The National Archives, Řad Maltézsky – Spisy

Smitner 3, 4 – Prague, The National Archives, Řad Maltézsky – Spisy, nos. 74, 75, kn. 3, 4: F.P. de Smitner, Diplomatarium Ordinis S. Johannis Hierosolymitani Magni Prioratus Bohemiae ..., vols 3, 4.

[2]   M. Czapliński, E. Kaszuba, G. Wąs and R. Żerelik, Historia Śląska (Wrocław, 2007), pp. 50–120; L. Bobková, 'Slezští Piastovci na dvoře Karla IV', in A. Barciak (ed.), Piastowie śląscy w kulturze i europejskich dziejach (Katowice, 2007), pp. 168–78.

Alternative Names

Brzeg	Brieg
Bytom	Beuthen
Chojnów	Hainau
Cieszyn	Teschen Těšín
Głogów	Gross Gogau
Lubin	Lüben
Mała Oleśnica	Klein-Öls
Niemcza	Nimptsch
Oława	Ohlau
Ścinawa	Steinau
Strzelin	Strehlen
Wrocław	Breslau
Ziębice	Münsterberg
Żory	Sohrau

Figure 26.1   Silesia in the Late Middle Ages (graphics: N² Productions)

Many descendants of these Silesian dukes pursued clerical careers, thereby guaranteeing their relatives who ruled the neighbouring duchies influence in the Church. Since the late twelfth century at least twenty-four members of the Silesian Piast ducal house had become clergy. Most of them were members of cathedral chapters, usually in Wrocław. Sometimes they became bishops.[3] Five dukes joined military orders. Three entered the Order of St John: two, the Teutonic Knights.[4] Most of these clerical dynasts did not rule in their home duchies. There were, however, at least eight Silesian Piasts who as ecclesiastics ruled or co-ruled hereditary principalities (see Fig. 26.2). Five were bishops or canons (Jarosław, bishop of Wrocław and duke of Opole; Władysław, archbishop of Salzburg and duke of Wrocław; Konrad the Humpbacked, provost of Wrocław and duke of Ścinawa and Żagań; Bolesław, archbishop of Esztergom and duke of Toszek; John the Aspergillum, bishop of Włocławek, Kamień, Chełmno and Poznań, archbishop of Gniezno and duke of Opole).[5] Three dukes, in turn, belonged to the military orders. These were Conrad VIII, ruler of Ścinawa in the first half of the fifteenth

---

[3]   K. Jasiński, *Rodowód Piastów śląskich* (Kraków, 2007), pp. 78, 127–8, 134–5, 172, 181–2, 184–5, 336–7, 355, 437–8, 441, 526, 533, 561, 573–4, 580–81, 600–601, 613, 618, 660–61; R. Samulski, *Untersuchungen über die persönliche Zusammensetzung der Breslauer Domkapitels im Mittelalter bis zum Tode Bischofs Nanker (1341)* (Weimar, 1940), pp. 81–2, 103, 131, 147, 152, 154–6; B. Zientara, 'Bolesław Wysoki – tułacz, repatriant, malkontent. Przyczynek do dziejów politycznych Polski XII wieku', *Kwartalnik Historyczny*, 62 (1971), 383; idem, *Henryk Brodaty i jego czasy* (Warsaw, 1975), pp. 89, 351; K. Dola, 'Piastowie śląscy na europejskich stolicach biskupich', in Barciak (ed.), pp. 182–5; J. Gottschalk, 'Auswärtige auf dem fürstbischöflichen Stuhl zu Breslau von 1456–1945 und Schlesier als Bischöfe von 1204–1903', in E. Brzoska (ed.), *Neunhundertfünfzig Jahre Bistum Breslau. Vorträge zur 950-Jahrfeier gehalten in der Universität Frankfurt m Main vom 9. –15. Oktober 1950* (Königstein/Ts., 1951), pp. 54, 59–61, 63–4; T. Jurek, 'Konrad I głogowski. Studium z dziejów dzielnicowego Śląska', *Roczniki Historyczne*, 44 (1988), 115; J. Chrząszcz, 'Herzog Boleslaw von Tost, nachmals Erzbischof von Gran (+1329)', *Zeitschrift des Vereins für Geschichte und Alterthum Schlesiens*, 37 (1903), 334–5; J. Horwat, *Książęta górnośląscy z dynastii Piastów. Uwagi i uzupełnienia genealogiczne* (Ruda Śląska, 2005), pp. 46, 93–4, 102, 106–7; C. Kuchendorf, *Das Breslauer Kreuzstift in seiner personlicher Zusammensetzung von der Gründung (1288) bis 1456* (Breslau, 1937), pp. 93–4; S. Sroka, 'Bolesław – arcybiskup ostrzyhomski (1321–1328)', *Nasza Przeszłość*, 79 (1993), 120–48; idem, *Z dziejów stosunków polsko-węgierskich w późnym średniowieczu. Szkice* (Kraków, 1995), pp. 49–101.

[4]   Dola, p. 185; Sroka, *Z dziejów*, pp. 84–101; Jasiński, pp. 199, 336, 444–5, 533, 618–19; J. Mitáček, 'Ziemovit těšínský – generální převor řádu johanitů a slezský kníže', *Sborník Prací Filozofické Fakulty Brněnské Univerzity*, 46 (1999), 17–40; M. Starnawska, 'Siemowit', in *Polski Słownik Biograficzny*, 37 (Warsaw and Kraków, 1996), pp. 75–8; A. Wędzki, 'Rupert (Ruprecht) II', in *Polski Słownik Biograficzny*, 33 (Wrocław, Warsaw and Kraków, 1991–92), p. 107.

[5]   Jasiński, pp. 77–8, 133–5, 354–5, 526, 573–4; Dola, pp. 182–5; Gottschalk, pp. 59–61, 63–4; Sroka, 'Bolesław', pp. 120–48; idem, *Z dziejów*, pp. 49–83; Chrząszcz, pp. 334–5; Horwat, p. 102.

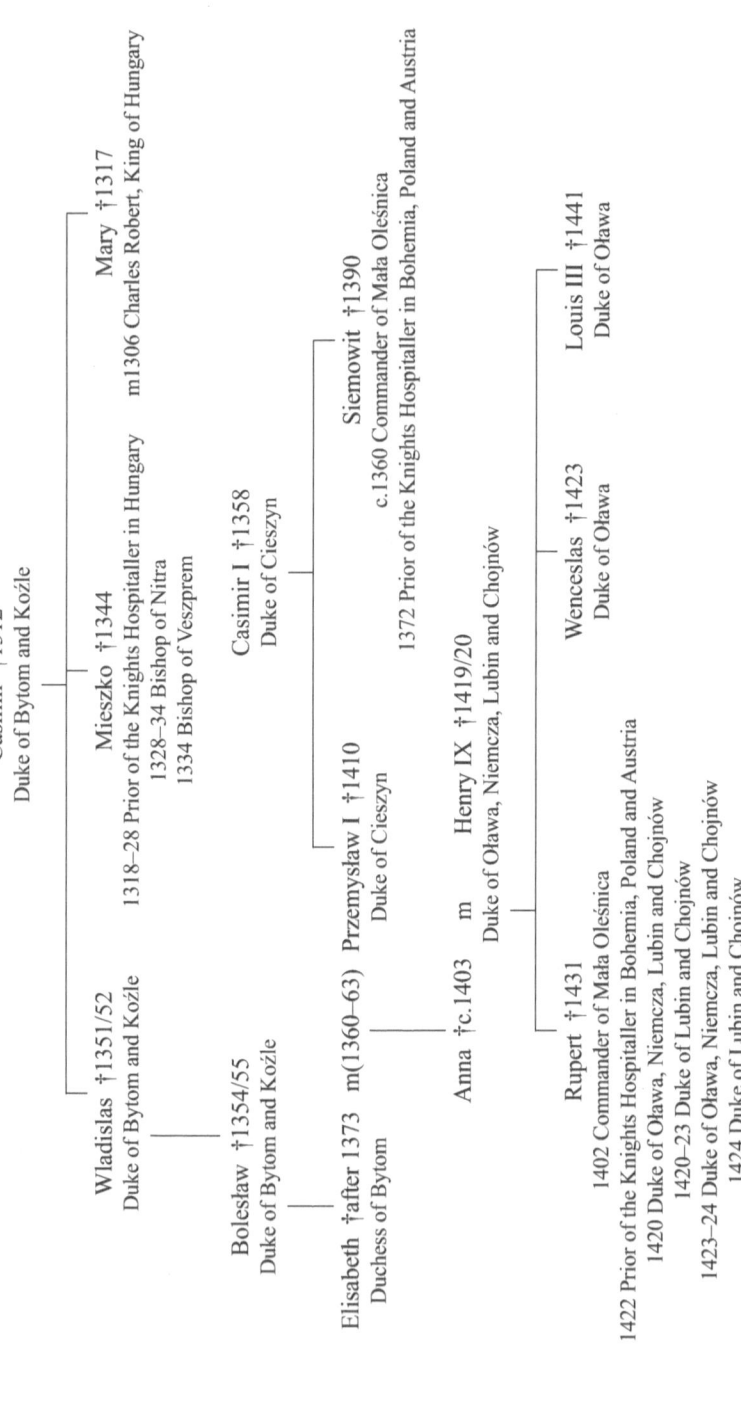

Figure 26.2   Hospitaller Priors of the Piast dynasty (graphics: N² Productions)

century and head of the Teutonic Order in Bohemia and Moravia,[6] as well as two priors of the Knights of St John, Siemowit and Rupert.

The lives of these two dukes will serve as examples of how the Silesian Piasts combined the roles of secular ruler and ecclesiastic. Born c. 1340 Siemowit was the son of Casimir I, duke of Cieszyn. On his death in 1358, Casimir was succeeded by Siemowit's elder brother, Przemysław I called *Noszak*, who had links with Charles IV's court as a courtier and diplomat.[7] Between 1359 and 1362 (most likely in 1361) Siemowit became commander of Mała Oleśnica in Silesia. From 1372 until his death in 1390, he was prior of Bohemia, Poland and Austria. Together with his brother he also co-ruled the duchy of Cieszyn, although without having his own district.[8]

Born c. 1390 Rupert, the son of Duke Henry IX of Oława, Niemcza, Lubin and Chojnów and Anna, the daughter of Duke Przemysław I of Cieszyn, was Siemowit's great-nephew.[9] In 1402 he became commander of Mała Oleśnica. On his father's death at the turn of 1420, he briefly co-ruled with his brothers Louis and Wenceslaus. In the same year he received a separate district of the duchy including the towns of Chojnów and Lubin. Following Wenceslaus's death in 1423, Rupert and Louis co-ruled their hereditary duchy; however, in 1424 Rupert again became sole ruler of Lubin and Chojnów. By June 1422 he had also become prior of the Hospitallers in Bohemia. He held both of these posts until his death in 1431.[10]

The careers of both dukes were almost identical, yet they took place against a background of differing political circumstances in both Silesia and Bohemia. The priors' status among the Silesian dukes was also different. The duchy of Cieszyn, Siemowit's patrimony, was one of the largest in Silesia and, as a result, flourished

---

[6]   Jasiński, pp. 444–5.

[7]   Ibid., p. 145; Horwat, p. 61; I. Panic, *Książę cieszyński Przemysław Noszak (*ok. 1332/1336 - + 1410). Biografia polityczna* (Cieszyn, 1996), pp. 50–60; idem, 'Działalność polityczna i dyplomatyczna księcia cieszyńskiego Przemysława Noszaka w czasach panowania Karola IV jako cesarza (1355–1378)', *Watra* (1988), 29–30, 34–45; idem, *Księstwo cieszyńskie w średniowieczu. Studia z dziejów politycznych i społecznych* (Cieszyn, 1988), pp. 45, 48–9; R. Žáček, 'K počátkům politické kariéry vévody Přemysla I. Těšínského', *Acta historica et museologica Universitatis Silesianae Opaviensis*, 3 (1997), 44–53.

[8]   R. Heś, *Joannici na Śląsku w średniowieczu* (Kraków, 2007), p. 494; RML 787–8; Mitáček, 'Ziemovit', pp. 18, 21–40; Starnawska, 'Siemowit', pp. 75–6. Mitáček's date for Siemowit's death (1390) is more accurate than mine.

[9]   Jasiński, pp. 195, 198–9, 629 put the date of his birth as c. 1396. However, given his appointment as commander in 1402, it seems that Rupert must have been born a few years earlier. The dates of his parents' births and of their marriage are unknown, but Rupert may have been born c. 1390.

[10]   RML 839; Heś, p. 496; APWr, Rep. 132a, Chojnów, no. 92; K. Klose, *Beiträge zur Geschichte der Stadt Lüben* (Lüben, 1924), p. 35; Wędzki, p. 107; Czapliński *et al.*, p. 96.

under his brother's rule.[11] The area ruled by Rupert and his brothers, which was composed of several districts with no territorial links, was much smaller.

Siemowit's entry into priesthood was a natural step. However, the duke of Cieszyn's younger brother was only the second Silesian Piast to have joined the Hospitallers and the first in the Bohemian priory. The first knight of St John among the Piasts, Mieszko, duke of Bytom, prior of Hungary in 1318 and then bishop of Nitra and, from 1334, bishop of Veszprem, held these posts because he was the brother of Mary, the first wife of Charles Robert, king of Hungary. He was also the great-uncle of Elisabeth, duchess of Bytom, who became engaged to Przemysław I in 1356. She married him around 1360 at the same time Siemowit became commander of Mała Oleśnica. Przemysław, who sought the succession after his father-in-law died without leaving a male heir, must have been familiar with the history of that line of the Piast dynasty.[12] He was hopeful that, thanks to his position at the king of Bohemia's court, his brother could follow a career in the Hospitallers' Bohemian priory similar to the one Mieszko had pursued in the Hungarian priory.

There is no documentation that would confirm the duke of Cieszyn's influence on his brother's decision. However, it appears that Siemowit, in his early twenties at the time, can only have become commander of Mała Oleśnica thanks to his brother's support. We do not know the minimum age requirement for becoming a commander in the fourteenth century, although, according to the records, the youngest commanders in that period were about thirty-three years of age. The average age of commanders in the Prague diocese mentioned in the inspection documents barely exceeded forty-seven.[13] It would have been impossible to appoint a 20-year-old duke to head one of the richest commanderies in Silesia without the powerful support of his brother, the duke of Cieszyn. This nomination was also beneficial to the Order itself since it gained prestige by admitting a member of the Piast dynasty into its ranks.

In the fourteenth century the Hospitallers in Silesia and Bohemia offered great career opportunities. The Bohemian priory consisted of twenty-six commanderies in Bohemia, Moravia and Austria, about ten in Silesia and two in Greater Poland, in the area of the Polish kingdom.[14] The young duke's career started in a rural

---

[11]   Panic, *Księstwo*, pp. 24–30, 44–54, 77–9.

[12]   Sroka, *Z dziejów*, pp. 84–101; Jasiński, pp. 527–30, 533, 548, 567.

[13]   *Inquisitio domorum Hospitalis S. Johannis Hierosolimitania per Pragensem Archidioecesem facta anno 1373*, ed. V. Novotný, Historický Archiv 19 (Prague, 1901), pp. 21, 32, 37, 41, 44, 51, 57, 61, 63, 68.

[14]   M. Starnawska, 'Crusade Orders on Polish Lands during the Middle Ages. Adaptation in e Peripheral Environment', *Quaestiones Medii Aevi Novae*, 2 (1997), 137–8; eadem, 'Der Johanniterorden und Schlesien im Mittelalter', *Würzburger medizinhistorische Mitteilungen*, 22 (2003), 405–15; Heś, pp. 55–83, 123–231; J. Mitáček, 'Převorství Havla z Lemberka (1337–1366)', *Acta Musei Moraviae. Scientiae sociales*, 90 (2005), 201–10; idem, 'Čeští johanité v předvečer husitských válek', *Acta Musei Moraviae. Scientiae sociales*, 91 (2006), 143–5.

commandery with a huge estate. The commandery of Mała Oleśnica – the only rural commandery founded by the Piasts – was particularly suited to this role. It had been created in the 1220s by Henry the Bearded, duke of Lower Silesia, for the Templars and was taken over by 1314 by the Hospitallers.[15] The dukes of Cieszyn were not descendants of Henry the Bearded, but of his paternal uncle, Mieszko the Stumbling.[16] Yet because the post of commander was held by two later Silesian dukes, it could be concluded that the Piasts regarded Mała Oleśnica as a family estate and claimed the right to appoint their heirs as head of that commandery.

The commandery of Mała Oleśnica consisted of several villages located in the duchies of Strzelin-Ziębice and Oława.[17] Siemowit's appointment had two consequences: first, he managed to gain an income from the estate, and, second, his brother's influence was extended and came to embrace these duchies. Until 1370 Siemowit's activity as commander was restricted to managing the estate. He would have spent most of the time in Mała Oleśnica or its neighbourhood.[18] There are no confirmed traces of his activity in the priory apart from his participation in the general chapter, most probably at Svetla (Bohemia), held in 1370.[19] Nor are there any indications of his involvement in ruling the duchy of Cieszyn.

His lifestyle differed, however, from that of other rural commanders. Siemowit repeatedly used the title of duke of Cieszyn and commander of Mała Oleśnica even when he called himself a *frater*. The inscription on his seal provides information about his rank.[20] That seal was impressed on red wax, which in the late medieval period was a privilege granted to rulers. Other commanders impressed their seals on black wax.[21] The seal featured an eagle, the coat of arms of the Piasts, with Maltese crosses over the wings. Siemowit's dynastic background was also

---

[15] E. Burzyński, 'Templariusze na Śląsku. Komandoria w Oleśnicy Małej', *Słupskie Studia Historyczne*, 13 (2007), 157–72; P. Hope, 'Kwestia sprowadzenia templariuszy do Polski. rozwój uposażenia zakonu w Wielkopolsce', *Poznański Rocznik Archiwalno-Historyczny*, 1 (1993), 15–21; M. Goliński, 'Uposażenie i organizacja zakonu templariuszy w Polsce do 1241 roku', *Kwartalnik Historyczny*, 98 (1991), 11–13; K. Eistert, 'Der Ritterorden der Tempelherren in Schlesien', *Archiv für schlesische Kirchengeschichte*, 14 (1956), 1–24; R. Stelmach, 'Komenda joannitów w Oleśnicy Małej w świetle zachowanych dokumentów z Centralnego Archiwum w Pradze', *Sobótka*, 57 (2002), 211–23.

[16] Jasiński, pp. 63–8, 501–2, 507–8, 515–16, 595–6.

[17] Heś, pp. 158–65; Mitáček, 'Ziemovit', pp. 18–19.

[18] RML 788, 790, 792, 794–6, 1716; APWr, Akta Wrocławia 369; Mitáček, 'Ziemovit', pp. 18–19.

[19] *CDS*, 9, ed. C. Grünhagen (Breslau, 1870), p. 249.

[20] RML 790, 792, 794, 1716; E. Šefčik, *Pečeti téšinských Piastovců* (Ostrava, 1982), pp. 7, 22; M. Kaganiec, *Heraldyka Piastów śląskich 1146–1707* (Katowice, 1992), p. 165; S. Mikucki, 'Heraldyka Piastów śląskich do schyłku XIV wieku', in W. Semkowicz (ed.), *Historia Śląska od najdawniejszych czasów do roku 1400*, 3 (Kraków, 1936), 3, pp. 548, table CXXXV.

[21] M. Gumowski, M. Haisig and S. Mikucki, *Sfragistyka* (Warsaw, 1960), pp. 138–9; RML 646, 2212, 2239.

something to which other Silesian dukes often referred. In documents addressed to
him, they called him duke and commander and often referred to their relationship
with him, though sometimes wrongly. For example, Louis, duke of Brzeg, called
Siemowit 'sororio nostro karissimo', although he was not his mother's brother.
Other dukes did not perceive Siemowit simply as an ecclesiastic, but as one of
themselves.[22]

Siemowit's position changed in 1371 on the death of John of Zvířetice, the
prior of Bohemia. In June of that year Siemowit was acting as governor of the
priory, and, after 19 April 1372, he himself became prior of Bohemia thanks to
Charles IV's support and his appeal to the Pope.[23] The grand master of the Order,
Raymond Béranger, had actually appointed Hesso Schlägelholz, who was highly
influential on Rhodes, as John's successor.[24] As prior of Bohemia, Siemowit not
only ruled his province determinedly, but also became involved in his brother's
policies. He was able to combine these roles because he was a frequent visitor to
Mała Oleśnica and the other Silesian commanderies.[25] He often celebrated such
feasts as Whitsunday, St Hedwig's Day, or Trinity Sunday in Silesia.[26] Once a
year or once every two years, he left for Bohemia for a few months to inspect
the local commanderies. Sometimes he would celebrate feasts there: for example,
in 1387 he was at Strakonice for Whitsunday.[27] He also resolved to enlarge the

---

[22]   RML 793, 795.

[23]   Mitáček, 'Ziemovit', 21–2; idem, 'Čeští johanité 1367–1397 – správci a dipomaté.
Česká johanitská provincie za správy Jana ze Zvířetic (1367–1371), knížete Ziemovita
Těšínského (1372–1390) a Markolta z Vrutice (1391–1397)', *Časopis Národního Muzea*,
174 (2005), 1–2; Starnawska, 'Siemowit', p. 75; *Libri confirmationum ad beneficia
ecclesiastica pragensem per archidioecesim*, 2, ed. F.A. Tingl (Prague, 1868), p. 53;
Smitner, 3, no. 501; *Monumenta Vaticana res gestas Bohemica illustrantia*, ed. C. Stloukal,
4 part 1 (Prague, 1949), nos. 413, 456, 457.

[24]   A. Luttrell, 'The Hospitallers at Rhodes, 1306–1421', in *HC*, 3, pp. 301, 303; idem,
'Intrigue, Schism and Violence among the Hospitallers of Rhodes 1377–1384', *Speculum*,
41 (1966), 40.

[25]   *CDS*, 9, reg. 405, 526, 1634; *LB*, 1, pp. 348–9; RML 107–8, 204, 246, 355, 549,
577, 801, 804, 808–10, 812, 814–16, 818–24, 828, 830, 832, 900, 1717, 1720; Smitner, 3,
no. 543; APWr, rep. 12, no. 12, rep. 68, no. 69, rep. 123, no. 48, rep. 132a, Złotoryja, no. 62;
Mitáček, 'Ziemovit', pp. 23–6, 28–30, 34–5, 37–8; Stelmach, pp. 222–3.

[26]   RML 107, 204, 806, 809, 822, 824.

[27]   *Codex diplomaticus et spistolaris Moraviae*, ed. V. Brandl, 10 (Brünn, 1878), p. 213,
no. 190; *Libri confirmationum ad beneficia ecclesiastica pragensem per archidioecesim*,
3–4, ed. J. Emler (Prague, 1879), p. 55, 58, 86, 96, 109–10, 123, 134, 145, 148, 152–5,
190–92, 197; *Soudní akta konsistoře pražské – Acta iudiciaria consistorii Pragnesis*, ed.
F. Tadra, 2 (Prague, 1893), p. 411; *Regesta Bohemiare et Moraviae aetatis Venceslai IV.
[1378 dec.–1419 aug. 16.]*, 1, ed. V. Jenšovská (Prague, 1979), no. 2848; RML 554, 2009,
2138, 2240, 2451–2; APWr, Akta Wrocławia, 25.07.1379; Mitáček, 'Ziemovit', pp. 25–9,
32, 34, 36–7.

prior's residence in Prague.[28] Since Mała Oleśnica was just one day away from Wrocław on the road to Prague,[29] he was able to keep in contact with the Bohemian commanderies.

Sometimes Siemowit had to attend ceremonies or events in Silesia in the company of his brother Przemysław. So for example, in 1378, he came to Żory when the town was purchased from the dukes of Opawa. In 1385, together with his brother, he received an oath of fealty from the councillors of Głogów when that town came under the rule of the dukes of Cieszyn. Along with other Silesian dukes, in 1389 he was involved in concluding a treaty with Jobst, margrave of Moravia, and Nicholas, bishop of Olomouc, regarding the security of their lands. Finally, in 1390 in Cieszyn, Siemowit, his brother, nephews and some magnates swore to repay a debt incurred from the Jews.[30] Przemysław and Siemowit's agreement to purchase the town of Strzelin from Bolesław, duke of Ziębice, was signed in 1385 in Mała Oleśnica, even though the deal did not concern the Order's affairs.[31] Siemowit was involved with his brother's political activity in Silesia partly because of his dynastic background but also because of his affiliation to the Order. It was largely due to this dual position that the agreement to purchase Strzelin was signed in Mała Oleśnica. Apparently the deal could be signed thanks to Siemowit's contact with the duke of Ziębice, in whose duchy part of the commandery of Mała Oleśnica was located.

While Siemowit's status within the Order rose as he pursued his career, his brother also derived political benefits. That was largely thanks to the fact that the Hospitallers acted as diplomats and courtiers for the rulers of Bohemia.[32] Siemowit's subsequent promotions in the Order's hierarchy went hand in hand with his brother's close contacts with the Luxemburg court. At the time Siemowit became commander of Mała Oleśnica in about 1360, his brother Przemysław was greatly involved as a diplomat at the court of Charles IV. Siemowit's career then remained at a standstill because the duke of Cieszyn had fallen into disfavour. He became prior in 1372 thanks to Charles IV's support now that Przemysław had returned to the emperor's diplomatic service. Siemowit received another

---

[28]   A. Breycha-Vauthier de Baillemont, 'Das Grosspriorat von Böhmen-Österreich', in A. Wienand (ed.), *Der Johanniter-Orden. Der Malteser-Orden. Der ritterliche Orden des hl. Johannes vom Spital zu Jerusalem* (Cologne, 1970), p. 352.

[29]   J. Nowakowa, *Rozmieszczenie komór celnych i przebieg dróg handlowych na Śląsku do końca XIV wieku* (Wrocław, 1951), pp. 59–60, 69–76.

[30]   *Codex diplomaticus et spistolaris Moraviae*, ed. V. Brandl, 11 (Brünn, 1885), no. 536 and ed. B. Bretholz, 15 (Brünn, 1903), no. 312; *LB*, 1, pp. 197–8 and 2, p. 389; Mitáček, 'Ziemovit', pp. 26, 35, 38; K. Orzechowski, *Ogólnośląskie zgromadzenia stanowe* (Wrocław and Warsaw, 1979), pp. 117–18.

[31]   *LB*, 1, pp. 348–9; Mitáček, 'Ziemovit', p. 35.

[32]   I. Hlaváček, 'Zur Rolle der geistichlichen und ritterlichen Orden am Hofe der böhmischen Luxemburger', in *Die Ritterorden zwischen geistlicher ud weltlicher Macht im Mittelalter*, ed. Z.H. Nowak, Ordines militares. Colloquia Torunensia Historica, 5 (Toruń, 1990), pp. 157–8.

promotion in 1384 when Riccardo Caracciolo, Hospitaller grand master of the
Rome Obedience, appointed him treasurer and governor of the Empire. The
nomination of a *frater* who had close links with the Luxemburg family was
indicative of the good relations between Wenceslaus IV, king of Germany and
Bohemia, and the Roman pope. This promotion coincided with Przemysław's
I increased involvement in imperial politics at Wenceslaus's court.[33] Thus did
Siemowit's promotions complement his brother's political activities.

Siemowit's rise in status as prior of Bohemia and co-ruler of the duchy of
Cieszyn strengthened his ducal aspirations. After 1378 he ceased altogether
calling himself *frater*, a term he had tended to omit beforehand.[34] From 1376 he
used a new and larger seal that depicted a shield divided into four fields, featuring
the Piast eagle and the Greek cross (a Hospitaller emblem) quartered. According
to the official record, it was the seal of the priory; however, its inscription and
heraldic symbols indicate that it belonged to the prior, not the priory as an
institution.[35] So Siemowit strove to reinforce his prestige both as a duke and as
an ecclesiastic. What mostly concerned him were the Order's affairs. His secular
activity complemented his brother's policies as duke of Cieszyn, and it was
thanks to the prosperity of Bohemia and Silesia in the fourteenth century and his
brother's outstanding personality that the prior could pursue such a successful
career.

Rupert's career as a secular ruler and a senior officer in the Order took a
different course. He was no more than twelve years of age when he became
commander, and that was quite exceptional and contrary to church law. This
unusual nomination must have stemmed from the lofty aspirations of his maternal
grandfather Duke Przemysław of Cieszyn, whose ambitions apparently included
regaining the influence with the Hospitallers he had lost twelve years earlier
on his brother's death. Przemysław's involvement appears central to Rupert's
appointment as the boy's father, Henry IX, who ruled a small duchy, did not have
the necessary contacts to obtain commandery for his son.

Thanks to the complex political situation in 1402, the duke of Cieszyn could
achieve his ambitions. Rupert became commander in the spring of 1402 when
Wenceslaus IV was imprisoned in Bohemia by his brother, King Sigismund of
Hungary. In July 1402 Przemysław established the league of Silesian dukes,
which included his son-in-law Henry IX, in defence of the Bohemian king.
However, Henry of Hradec, who had been elected prior of Bohemia in 1402,
was opposed to the king and supported the appointment of his follower, Bohuš
the White.[36] In this context, the appointment of a duke from Wenceslaus's camp

---

[33]   RMS, no. 77, kart. 41, k. 337v, 338; Mitáček, 'Ziemovit', p. 33; idem, 'Čeští
johanité 1367–1397', pp. 10–14; Panic, *Książę*, pp. 54–65, 74–6.

[34]   He called himself *frater* at least on 4 October 1377. RML 808.

[35]   Ibid., 204, 549, 554, 577, 806, 824, 2138.

[36]   J. Spěváček, *Václav IV. 1361–1419. K předpokladům husitské revoluce* (Prague,
1986), pp. 336–49; Orzechowski, p. 118; Mitáček, 'Čeští johanité v předvečer', pp. 139–41.

to the post of commander of Mała Oleśnica would have been perceived as an attempt by this faction to gain influence in the Order. What contributed to the success of this plan was the fact that Mała Oleśnica was located in the duchy of Oława, which belonged to Rupert's father. Apparently, this commandery was viewed as the Silesian Piasts' appanage. Rupert was the eldest son of Henry IX, yet in 1402 his younger brothers could not assume the post of commander. They were destined to pursue secular careers and ensure the continuity of the dynasty.

Rupert's activities as commander of Mała Oleśnica from 1402 to 1419 were similar to those of Siemowit in the same capacity. Rupert was concerned with the commandery' estates and was not involved in the affairs of the priory. He did not forget his ducal background. He used the title of Silesian prince, sometimes *dominus* of Lubin, commander and *frater*. The inscription on Rupert's seal, modelled on Siemowit's, indicated his ducal background. Unlike Siemowit's seal, however, Rupert's was habitually impressed on black wax, a sign of diminished status.[37] Only once in that period, in 1409, did Rupert become involved in the affairs of his father's duchy by joining him in confirming the sale of a rent to the councillors of Chojnów. In the document Rupert is referred to as *dominus* of Lubin and Chojnów – his clerical office is not mentioned.[38] This is indicative of Rupert's future resolve to determine the role he would assume at any particular time: whether secular ruler or senior officer in his Order.

In the early years of Rupert's career, some problems may have been caused by his age. He issued documents at Mała Oleśnica, and that might have meant that he had resided there permanently since 1402. However, the village is not far from Oława, one of Henry IX's estates, and Rupert could have stayed at his father's court, visiting the commandery a few times a year to confirm the transactions conducted there in the meantime. There are no records to show that the young Rupert was accompanied by a guardian or a counsellor. The only indication of his father's discreet guidance is suggested by a few grants that Henry IX bestowed on two of the Silesian Hospitaller commanderies.[39]

In 1422, or perhaps 1421, Rupert was elected prior of Bohemia, succeeding Henry of Hradec who had died from the wounds inflicted by the Hussites.[40] There are no precise details of his appointment, but the main reason for his election was that he was the only commander who could ensure protection for the Order's interests against the Silesian dukes during the Hussite revolution. In the first stage of the conflict, the Hussites had destroyed the priors' estate in Prague and most of

---

[37]   RML 838–9, 841–4, 846–54, 857–9, 861, 863, 865–6, 868–70; Kaganiec, p. 165.

[38]   APWr, Rep. 132a, Chojnów, no. 71.

[39]   RML 840, 845, 871.

[40]   APWr, Rep. 132a, Chojnów, no. 92; J. Mitáček, 'Češká provincie řádu sv. Jana Jeruzalémského za vlády Lucemburků (1310–1419)', Autoreferát disertační práce (Brno, n.d.), p. 8; Heś, p. 352; Wędzki, p. 107.

the Bohemian commanderies.[41] In 1420–22 Rupert had become an independent ruler of a small duchy and the head of a disordered province, and he had to combine both functions to the best of his ability.

Since the Bohemian estates of the Hospitallers were destroyed and the war was under way, Rupert could not remain in Bohemia. Instead he took up permanent residence in Silesia, primarily in Mała Oleśnica. He could only exercise authority over the commanderies based in Greater Poland, Silesia and Lusatian Zittau.[42] In 1425, as the sole Bohemian prior, he visited the house in Kościan (Greater Poland). There he met the king of Poland, Ladislaus II Jagiello, who confirmed the privileges of the commanderies in Kościan and Poznań. In return, Rupert relinquished to the king's benefit his right to appoint commanders in Poznań. Such an easy cession of this right was indicative of the weakness of the prior's position, even though the king courteously addressed Rupert as duke.[43] In 1431 he convened a chapter in Wrocław in order to decide whether some revenues could be channelled into the war against the Hussites.[44] Rupert's performance as prior consisted of only minor actions, as making war on the Hussites hampered the normal functioning of the Bohemian commanderies.

As the secular ruler of a small duchy, Rupert could not pursue ambitious policies. His political activity was restricted to endorsing his subjects' grants and transactions.[45] In 1423, together with his brother Louis, he granted the town of Lubin the right to coin money.[46] The most important theme in his political activity was his involvement in the anti-Hussite alliance of the Silesian dukes. However, all he did was in 1428 to command the defence of the towns he owned, Lubin and Chojnów. For a short while in that year he participated in a military expedition led by Bishop Conrad.[47] He was also prepared to donate part of his income from the

---

[41]   F. Skřivánek, *Heraldické památky kostela Panny Marie pod Řetězem českého velkopřevorstvi maltézských rytirů v Praze* (Prague, 2002), pp. 147–8.

[42]   Mitáček, 'Češká provincie', p. 8; *LB*, 1, p. 373; RML 700, 872–4; Smiter, 4, no. 584; KDS 194, APWr, Rep. 3, no. 23; J. Prochno, 'Regesten zur Geschichte der Stadt und des Landes Zittau 1234–1437, zweiter Teil 1378–1437', *Neues Lausatisches Magazin*, 114 (1938), nos. 1450, 1629, 1649.

[43]   *Kodeks dyplomatyczny wielkopolski*, eds A. Gąsiorowski and T. Jasiński, 8 (Warsaw and Poznań, 1989), no. 1038; RMS, inw. 2560, kart. 1030, sygn. 5.

[44]   Smiter, 4, no. 584.

[45]   KDS 146, 167, 168, 181, 188, 206, 209, 214, 218; APWr, Rep. 132a, Chojnów, nos. 92, 99; Rep. 135, sygn. 129, Lubin, nos 16–17; RML 206; G. Thebesius, *Liegnitzische Jahr-Bücher*, 2 (Jauer, 1732), p. 275; *CDS*, 9, reg. 860

[46]   *CDS*, 12, ed. F. Friedensburg (Breslau, 1887), pp. 45–6; Klose, p. 35.

[47]   *Scriptores rerum silesiacarum*, ed. C. Grünhagen, 6 (Breslau, 1871), pp. 10–11, 63–4, 69, 74, 78–9; C. Grünhagen, *Hussitenkämpfe der Schlesier 1420–1435* (Breslau, 1872), pp. 149–53, 157, 165, 170; Klose, p. 36; T. Scholz, *Chronik der Stadt Haynau* (Haynau, 1869), pp. 38, 41; P. Kouřil, D. Prix and M. Wihoda, *Hrady českého Slezska* (Brno-Opava, 2000), p. 536; F. Szafrański, *Ludwik II brzesko-legnicki, feudał śląski doby*

Zittau commandery for military purposes.[48] He did not adopt a prominent role, and his approach was basically similar to that of other dukes facing the Hussite threat. Rupert did not call upon the few Silesian Hospitaller brothers to fight the Hussites. His involvement in the conflict had nothing to do with his Order.

The duke made sure that his two roles, prior and duke of Lubin and Chojnów, remained distinct. The titles he used varied depending on the content of the document he was signing. In documents concerned with the Order's affairs, besides the titles of Silesian duke and 'lord of Chojnów and Lubin', he would invariably indicate his position in the Order.[49] However, his ecclesiastical title was often omitted in the documents concerning the duchy's affairs.[50] The latter were usually produced in Lubin, Chojnów or on the estates of the dukes with whom Rupert signed treaties.[51] Certificates regarding the Order's affairs were typically produced at the seats of the commanderies.[52] Evidently Rupert dealt with the Order's affairs in religious houses, tackling the duchy's problems on its own territory. Sometimes, however, he invited his brothers to the commanderies to make joint decisions about dynastic affairs.[53] Depending on the role he assumed, he surrounded himself with different groups of people. There are no records to show that he appeared in the company of other Hospitallers outside religious houses. However, his chamberlain, George Falkenheim, accompanied Rupert both when he assumed the role of duke and also when he visited religious houses.[54]

It seems, then, that Rupert tried to keep his functions separate even though it was hardly possible to prevent the two fields of activity from interacting. Rupert's attitude distinguished him from Siemowit, who tolerated the permeation of the Order's affairs with those of the duchy of Cieszyn. However, we must remember that Siemowit did not have a separate district and cooperated closely with his brother. Both brothers complemented each other's actions when necessary. Rupert ruled as a sovereign, and his activity both as a prior and a ruler was restricted. Hence he followed the procedures necessary to exercise power both as a senior officer in his Order and as a secular ruler in an attempt to raise the prestige of the two posts he held.

---

późnego średniowiecza (Wrocław, Warsaw, Kraków and Gdańsk, 1972), p. 73; Wędzki, p. 107; Heś, p. 353.

[48]   Prochno, no. 1649.

[49]   RML 700, 872–4.

[50]   R. Żerelik, *Katalog dokumentów przechowywanych w archiwach państwowych Dolnego Śląska*, 9 (Wrocław, 1998), nos. 146, 167, 176, 181, 188, 206, 209; *CDS*, 4, ed. A. Meitzen (Breslau, 1863), pp. 153–4, 9, reg. 860, 12, pp. 45–6; *LB*, 1, pp. 371–3; RML 206, APWr, Rep. 135, sygn. 129, Lubin, no. 17.

[51]   *LB*, 1, pp. 371–2; *CDS*, 9, reg. 860; Żerelik, nos. 146, 176, 209; RML 206, APWr, Rep. 132a, Chojnów, nos 92, 99; Rep. 135, sygn. 129, Lubin, nos 16, 17.

[52]   RML 700, 869–70, 872–4, *Kodeks dyplomatyczny wielkopolski*, 8, no. 1038; Prochno, no. 1649.

[53]   *LB*, 1, pp. 373; Żerelik, no. 194.

[54]   RML 206, 700, 872–3.

Destroyed by the Hussite wars, the Bohemian priory of the Knights Hospitaller lost its appeal for Silesian dukes with ambitions to pursue ecclesiastical careers. Combining a career in the Order with that of a secular ruler was no longer attractive. Siemowit's and Rupert's careers represented one phase in the much longer history of the Bohemian priory.

Chapter 27

Royal Power and the Hungarian–Slavonian Hospitaller Priors before the Mid-fifteenth Century

Zsolt Hunyadi

This paper focuses on the ways in which the rulers of Hungary influenced the activity of the Hungarian–Slavonian priory of the Hospital from the mid-fourteenth century. It should be said at the outset that it is not possible to construct a full narrative of the events in this period as the disparate nature of the sources prevents us from tracing developments in their entirety, but there are certain moments in the history of the Hospital in medieval Hungary for which there are sufficient records at our disposal for scholarly investigation. Secondly, there are a number of features that reveal that the affairs of the Hungarian–Slavonian priory diverged from general Hospitaller policy and, from the mid-fourteenth century onwards, were increasingly influenced by the Hungarian rulers. Thirdly, it is very likely that not all the questions raised will be answered satisfactorily in the course of this survey, as the history of the priory in the fifteenth and early sixteenth centuries has not been studied in depth since the 1920s.

On the basis of many years of research into the history of the priory,[1] it seems that until the 1340s the Hospitallers in the Hungarian–Slavonian priory were not involved in any major political issues apart from the support that, at the behest of Pope Boniface VIII, they apparently provided the Neapolitan Angevin Charles Robert (1301/09–1342) against his rival, Wenceslas, in the succession struggle for the Hungarian throne.[2] With the accession of his son, Louis I 'the Great' (1342–82), a new era was inaugurated that brought changes to the life of the priory in many respects, as became apparent in the course of the wars Louis waged against Venice and the kingdom of Naples. The tension with Venice, however, gave rise to a peculiar triangular confrontation between Venice, Louis and the Hospital in their rivalry over control of the Dalmatian coast. It should also be noted that Venice and the Hospital collaborated in the crusade to Smyrna. This situation changed when the Croatian Mladen Subić (III), himself since 1343 a Venetian

---

[1]   See Z. Hunyadi, *The Hospitallers in the Medieval Kingdom of Hungary, c.1150–1387* (Budapest, 2010), passim.

[2]   *Vetera Monumenta Historica Hungariam Sacram Illustrantia, 1216–1352*, ed. A. Theiner (Rome, 1859–60), 1, pp. 401–2.

citizen, besieged Vrana, one of the most important Hospitaller castles – it was their effective headquarters – during the winter of 1345.[3] It is not clear whether the prior of Hungary, Pierre Cornuti (1335–48), or his lieutenant, Giovanni Latini of Perugia, attempted to escape from Vrana during the siege. King Louis led his army to Zara against Venice, but his tactical errors contributed to an unexpected fiasco in 1346. The king's response was delayed as, in the meantime, Prince Andrew, his younger brother, had been assassinated in Naples, and the main orientation of Hungarian foreign policy was immediately altered. Accordingly, much less attention was paid to the report sent by the Hungarian–Slavonian vice prior, Baudoin Cornuti, to Nicholas of Lendva, Ban(us) of Slavonia, concerning Venetian military preparations for the defence of Zara in the spring of 1348.[4] The Hungarian struggle with Venice over the coastal areas was finally settled in February 1358 with the signing of the Treaty of Zara. Prior Baudoin Cornuti was among the witnesses present in Venice on behalf of the Hungarian king, while the prior of Venice, Napoleone de Tibertis, witnessed the oath of the Venetian party two days later.[5]

Until he was able to re-establish friendly relations with Venice, Louis had to face a very hard and busy decade. As far as the Neapolitan wars were concerned, he encroached on the autonomy of the priory (and that of the Order) when he conferred the office of prior on the Provençal Hospitaller, Montreal du Bar, also known as Frà Moriale: in 1354 this man was to be executed in Rome for his earlier crimes.[6] Moriale's behaviour is illustrated by several charters from 1353 which show that he had removed part of the archives as well as the seal of the priory and had used the seal in alienating many of the priory's properties.[7] Baudoin was able to persuade the king to invalidate Moriale's charters of alienation. All this meant that by the middle of the fourteenth century the Hospitallers were finding themselves involved in high-level internal politics.

After several years of tranquillity, royal power again interfered in the affairs of the Order in the early 1370s. After Baudoin Cornuti, King Louis wanted Raymond de Beaumont, another Provençal, as prior of the Hungarian–Slavonian priory. Pope

---

[3] *Diplomácziai emlékek az Anjou-korból. Acta extera Andegavensia*, ed. G. Wenzel (Budapest, 1874–76), 2, p. 124.

[4] *Codex diplomaticus regni Croatiae, Dalmatiae ac Slavoniae*, eds M. Kostrenčić and T. Smičiklas (Zagreb, 1904–1998), 11, pp. 444–5.

[5] *Acta extera Andegavensia*, 2, pp. 501–4, 513–18.

[6] E. Reiszig, *A jeruzsálemi Szent János lovagrend Magyarországon* (Budapest, 1925–28), 1, pp. 100–112. See also L. Gessi, 'Fra Moriale', *Rivista del Sovrano Ordine Militare di Malta*, 2/3 (1938), 13–23; I. Miskolczy, *Magyar-olasz összeköttetések az Anjouk korában. Magyar-nápolyi kapcsolatok* (Budapest, 1937), pp. 250–52; A. Bárány, 'The Communion of English and Hungarian Mercenaries', in K. Papp and J. Barta (eds), *The First Millennium of Hungary in Europe* (Debrecen, 2002), p. 129; M. Dupuy, 'The Master's Hand and the Secular Arm: Property and Discipline in the Hospital of St. John in the Fourteenth Century', in D.J. Kagay (ed.), *Crusaders, Condottieri, and Cannon* (Leiden, 2003), pp. 329–33.

[7] *Codex diplomaticus regni Croatie*, 12, pp. 165–7.

Gregory XI, however, wanted to give the priory to the Provençal Bertrand Flotte, but the master eventually appointed Giovanni Rivara, prior of Venice.[8] Such papal interventions were not unparalleled: in 1372 Gregory XI wanted Fr Hesso Schlegelholtz as prior of Bohemia, but King Charles IV (1316–78) was able to prevail with his own local nominee, Ziemovit Těšín.[9] The Hungarian king, nonetheless, objected to both candidates on the basis of the agreement between the *langues* of Italy and Provence enacted in November 1373.[10] By the terms of this agreement it was the master who would appoint the Hungarian prior at the next vacancy, and subsequently the prior would be chosen alternately from the *langues* of Italy and Provence. Despite papal intervention, the post was given to Raymond de Beaumont, who was in possession by the beginning of autumn 1374.[11]

The anarchic period following the death of Louis I on 10 September 1382 saw the rise of an ambitious Hospitaller of local origin who was well aware of the importance of both politics and power. This was John of Palisna, who had assumed the title of prior of Hungary by April 1379 when a local preceptory styled him thus in a charter though Raymond de Beaumont bore the title until 1381.[12] The master, Juan Fernández de Heredia, only appointed him prior officially as late as July 1382.[13] Palisna soon found himself embroiled in the political turmoil that followed the king's death.[14] The crown immediately passed to Louis' daughter Mary, with her mother, Queen Elisabeth, acting as regent. In the autumn of 1383 Palisna joined a revolt that had broken out against them.[15] The exact reasons for his rebellion are not known; he may have objected to female rule, as Neven Budak

---

[8]   *Vetera Monumenta Historica Hungariam*, 2, p. 197; J. Delaville Le Roulx, *Les Hospitaliers à Rhodes jusqu'à la Mort de Philibert de Naillac: 1310–1421* (Paris, 1913), p. 197; A. Luttrell, 'The Hospitallers in Hungary before 1418: Problems and Sources', in Z. Hunyadi and J. Laszlovszky (eds), *The Crusades and the Military Orders: Expanding the Frontiers of Medieval Latin Christianity* (Budapest, 2001), p. 275.

[9]   J. Mitáček, 'Ziemovit Těšínský – generální převor řádu johanitů a slezský kníže', *Studia Historica Brunensia*, 46 (1999), 21–3.

[10]   On 22 November 1373 (the extant copy was prepared in 1427). Malta, Cod. 347, fols 51r–v. Edited in Hunyadi, *The Hospitallers*, pp. 321–2.

[11]   *Codex diplomaticus Hungariae ecclesiasticus ac civilis*, ed. G. Fejér (Buda, 1829–44), 9/4, pp. 614–16.

[12]   National Archives of Hungary, Budapest, Collectio Antemohacsiana, original charters (henceforth Dl.), Dl.7550 (1379). BAV Ottoboni lat. 1769. fol. 323v (1381).

[13]   Malta, Cod. 322, fol. 251r. See Z. Hunyadi, 'Adalékok a johannita magyar-szlavón (vránai) perjelségre kirótt rendi adók kérdéséhez', *Acta Universitatis Szegediensis, Acta Historica*, 116 (2002), 42–3. Edited in Hunyadi, *The Hospitallers*, pp. 329–30.

[14]   For the period, see most recently S. Süttő, *Anjou-Magyarország alkonya. Magyarország politikai története Nagy Lajostól Zsigmondig, az 1384–1387. évi belviszályok okmánytárával* (Szeged, 2003), pp. 17–35.

[15]   Ibid., pp. 36–9.

has suggested,[16] but in any case he presumably aimed at gaining power in the southern part of Hungary. There is no sign of Palisna having any political dealings with the Angevin claimant to the throne, Charles of Durazzo, who was at that time in Italy. By November 1383 the queen mother had taken steps to depose Palisna; the prior surrendered the Order's castle of Vrana and found shelter for a while at the court of the Bosnian king Tvrtko. Queen Elisabeth immediately appointed ecclesiastical *gubernatores* to administer the priory's goods.[17] By the autumn of 1384 Raymond de Beaumont, the former prior, had returned to Hungary where he was apparently acting as the legitimate prior and using the priory's seal.[18] Since early 1381 Raymond had been in the service of Charles of Durazzo, for whom he and Palisna had both fought in alliance with Padua back in 1372. In August 1385 Charles was invited to take the Hungarian throne, and he was crowned as Charles (Parvus) II at the end of that year. Palisna did not play an active role in the making of the new king; that was the achievement of the rebels known as the Horvats' party.[19] Palisna's ambitions might have given rise to his own permanent occupation of the priory for decades to come under Charles's rule, but there was an uprising against Charles in February 1386, and he was murdered later that month. The Angevin party had plans for revenge and persuaded Palisna to side with them in the spring of 1386.[20] He joined the rebellion against the queens and perhaps even mobilized Hospitaller troops in July 1386 when the Angevin party attacked and imprisoned the queens and their entourage on their way to Gora. Although it is certain that Palisna did not take part in the strangulation of the queen mother in November 1386, he was clearly an outlaw from that time onwards. He was not in any immediate danger of arrest, especially as the rebels were still strong enough in Slavonia to oppose the troops of the new claimant, Sigismund of Luxemburg. In January 1387, as Sigismund was beginning to overcome the rebels' resistance, John of Palisna was besieging Darnóc.[21] Several weeks later Sigismund was crowned king, and he took further steps to consolidate his power by force and by promises, partly against the Hospitaller prior.

After the deposition of Palisna, Emeric Bwbek became prior of the Hungarian–Slavonian priory in 1392,[22] but was only fully effective from 1394. There is no sign that he owed his appointment to either the convent or the master. He was a member of a middling noble family that had risen quickly and had begun his

---

[16]   N. Budak, 'John of Palisna, the Hospitaller Prior of Vrana', in Hunyadi and Laszlovszky (eds), p. 286.

[17]   Dl.35269. E. Mályusz, 'A szlavóniai és horvátországi középkori pálos kolostorok oklevelei az Országos Levéltárban', *Levéltári Közlemények*, 8 (1930), 66–7.

[18]   *Codex diplomaticus Hungariae*, 10/2, p. 179 (Dl.7111).

[19]   Süttő, p. 127.

[20]   Cf. Budak, p. 286.

[21]   *Codex diplomaticus Hungariae*, 10/1, p. 135.

[22]   The first charter to refer to him as prior was issued on 13 November 1392. Dl.7811.

career as a royal official in a Hungarian county thanks to the influence of his father, Detric Bwbek. Detric had first appeared as the queen's master of the Stewards in 1379 and between 1389 and 1392 as well as in the period from 1394 to 1397 was acting as ban(us) of southern region.[23] In 1397 King Sigismund elevated him to the office of the palatine of the kingdom that he held until 1402. Through his family connections his son Emeric became the member of the royal court, and in this capacity he followed his king to Nicopolis in 1396. His loyalty, however, soon ceased. In 1401 Emeric joined the league of aristocrats that conspired against Sigismund in 1401; they succeeded in imprisoning the king and made Emeric the ban(us) of Slavonia. On his release, Sigismund deprived Emeric of the office of ban(us) in 1402,[24] but he did not attempt to depose him as prior. Emeric Bwbek never pretended to be an adherent of the king, as is clear from the fact that he participated in the subsequent rebellion against Sigismund. The baronial rebels invited Ladislas of Naples to the Hungarian throne and crowned him at Zara in Dalmatia in 1403 in the presence of, among others, Emeric and Detric Bwbek. Bwbek, moreover, gave Ladislas Vrana and its dependencies which he sold to Venice in 1409; it thus passed out of the Order's possession.[25] By the autumn of 1403, however, the majority of the rebels had surrendered and sought the king's pardon. Sigismund pardoned Emeric, but he deposed him as prior and appointed lay administrators to the priory in November 1405.[26]

The vacancy in the prior's office became widely known, but the papal appointee, Bartholomew Carraffa, died unexpectedly before taking his office.[27] The master thereupon appointed Michael Ferrand in 1405,[28] who bore the title of prior until 1417, but there is no record to show that he ever came to Hungary or even attempted to exercise this position. The reason for his absence was that sometime around 1408 Sigismund wanted to appoint his supporter, Albert of Nagymihály, to govern the priory.[29] Albert's ancestors were the retainers of members of the upper nobility in the time of the Angevin kings.[30] He himself was a married layman and Sigismund's confidant. He grew up in the royal court and was invited to join the chivalric Order of the Dragon, the circle of the ruling elite, as *homo novus* in 1408. Albert participated in wars against the Turks (1411–12) as

---

[23]   For the career of the Bwbek kindred, see *Magyarország világi archontológiája, 1301–1457*, ed. P. Engel (Budapest, 1996): CD-ROM edition: Arcanum, Budapest, 2001.

[24]   Cf. *Zsigmondkori oklevéltár. 1387–1423*, eds E. Mályusz *et al.* (Budapest, 1951–2009), 2. no. 2424.

[25]   *Monumenta spectantia historiam slavorum meridionalium, 960–1527*, ed. S. Ljubić (Zagreb, 1868–75), 5, pp. 181–202.

[26]   *Codex diplomaticus Hungariae*, 10/8, p. 470.

[27]   *Vetera Monumenta Slavorum meridionalium Historiam Illustrantia*, ed. A. Theiner (Rome and Zagreb, 1863–75), 1, pp. 344–5.

[28]   Malta, Cod. 333. fol. 93v.

[29]   T. Botka, *Bars vármegye hajdan és most* (Pest, 1868), 1, pp. 85–6.

[30]   For his career, P. Engel, *A nemesi társadalom a középkori Ung megyében* (Budapest, 1998), pp. 158–60.

well as against Venice. Having been widowed, he entered the Hospital. Sigismund strongly lobbied Philibert de Naillac on Albert's behalf, notably at the Council of Constance where all three of them were present. Sigismund was successful, and Philibert de Naillac eventually appointed Albert as Hungarian–Slavonian prior in February of 1417. Moreover, the Hungarian king conferred on him the office of ban(us) of Croatia and Dalmatia, thus making Albert one of the most powerful aristocrats of the realm until his death in 1434. In return for effective royal backing, Albert fought not only against the Turks but also against the Hussites in 1420–22 when he recruited troops at his own expense.

The rule of Sigismund placed the Hungarian–Slavonian priory in a new situation as the king now openly intervened in its affairs. In addition, as with other ecclesiastical establishments, he often allowed the priory to remain vacant, or, rather, kept it vacant, and appointed secular governors (*gubernatores*) to collect its revenues. These years also proved to be a turning point since, from this time onwards, local men were appointed as priors of Hungary, irrespective of whether or not they were adherents of the current ruler. Between the early 1380s and the 1410s the preponderance of the foreign office-holders ceased. Most of the late fourteenth- and early fifteenth-century priors, or their families, had risen during the Angevin period to such a height that some of them belonged in the upper ranks of the nobility that was to turn against Sigismund of Luxemburg at the beginning of the fifteenth century. Moreover, from this time onwards the Hungarian rulers laid claim to the right to appoint the prior, and they sometimes deposed priors as in the case of John Palisna and Emeric Bwbek following their rebellions. By the beginning of the fifteenth century, many of the Hospitallers' landed properties were occupied or usurped by powerful noblemen, and, as mentioned, Vrana was sold the Venetians in 1409.[31] Some scholars have asserted that such changes were closely connected to the Great Schism, but that is still unproven. Certainly both the grand master of the Hospital and the popes protested against Sigismund's interventions, especially when he entrusted laymen – often the current Slavonian wardens – with the administration of the estates of the Hungarian priory, even though the grand master had appointed priors. Recent investigation has proved, however, that all this was part of Sigismund's new policy in the face of the Turkish threat.[32]

By the end of Sigismund's rule, several political and administrative changes had taken place, partly in connection with the growing Ottoman menace. The king managed to consolidate his power during the second half of his long reign, and there was no longer any baronial threat to his authority. In addition, he conferred aristocratic titles on those of his adherents who had remained loyal to him for

---

[31]   Z. Hunyadi, 'The Hungarian Nobility and the Knights of St John', in N. Coulet and J.-M. Matz (eds), *La noblesse dans le territoires angevins a la fin du Moyen Âge* (Rome, 2000), pp. 607–18.

[32]   Z. Hunyadi, 'Entering the Hospital: A Way to the Elite in the Fifteenth Century?' in N. Bériou, P. Josserand and L.F. Oliveira (eds), *Élites et ordres militaires au Moyen Âge* (Madrid, 2012) (forthcoming).

decades. In order to offset the power of the major office-holders, Sigismund introduced a system of 'twin' office-holding: he appointed two officers to the same office, not seldom brothers or relatives.[33] The practical importance of this new system manifested at the frontier region, where both officers always had to be present in order to provide prompt reaction for the Turkish operations.[34] This system outlasted Sigismund as it functioned well until the end of the fifteenth century. It may have affected the office of the Hospitaller prior as well. Following the death of Albert of Nagymihály, Sigismund appointed Matkó of Thallóc as administrator of the priory's properties. Since he was already a royal officer and the captain of Belgrade, Sigismund expected him to play a crucial role in the maintenance of the defensive system on the southern frontiers.[35]

Matkó of Thallóc originated from the island of Curzola, and he had arrived in Hungary via Dubrovnik accompanied by his brothers, Jovan and Frank.[36] In 1438, directly after the death of Sigismund (1437), Jovan was appointed to the office of the prior that he was exercising by 1439. His main duty was the same as his brothers': to organize the fight against the Turks. In 1440 when the Mamluks besieged Rhodes, the troops of the Ottoman Sultan Murad attacked Belgrade. Jovan of Thallóc successfully defended the castle, but after the assault he resigned as captain of Belgrade. He continued fighting the Turks and in 1444 took part in the crusade of Varna together with his brother Matkó. Both men managed to escape from this disaster. That same year, Jovan asked for a dispensation from his duties as prior in order to go to the convent at Rhodes.[37] But this plan never came to pass as he was killed in the civil war in 1445.[38]

It has long been debated whether the Hungarian–Slavonian priory frequently failed to send the payments to the convent. From this time onwards, the Hungarian–Slavonian priors spent a major part of their incomes on the military operations against the Turks. They are also known to have levied extra taxes from the tenants and leaseholders of the landed estates of the priory.[39] Later on they even pledged lands as happened during the tenure of both Thomas Szekler of Szentgyörgy (1453–61) and Bartholomew Berislavić (1475–1512). The more important role

---

[33]   For the Hungarian society in the Sigismund era, see P. Engel, *The Realm of St Stephen: A History of Medieval Hungary, 895–1526* (London and New York, 2001), pp. 211–13.

[34]   F. Szakály, 'The Hungarian–Croatian Border Defense System and its Collapse' in J.M. Bak and B.K. Király (eds), *From Hunyadi to Rákóczi: War and Society in Late Medieval and Early Modern Hungary* (New York, 1982), pp. 141–58.

[35]   Dl. 12641: *Codex diplomaticus Hungariae*, 10/7, p. 564; *Monumenta Historica Episcopatus Zagrabiensis*, ed. I. Tkalčić (Zagreb, 1873–74) 2, p. 108.

[36]   For the career and activity of the Thallóci brothers, see E. Mályusz, 'The Four Tallóci Brothers', *Quaestiones Medii Aevi Novae*, 3 (1998), 137–75.

[37]   *Vetera Monumenta Slavorum*, 1, p. 383.

[38]   Mályusz, 'The Four Tallóci Brothers', pp. 169–70.

[39]   *Codex diplomaticus partium regno Hungariae adnexarum*, eds L. Thallóczy, A. Hodinka and A. Áldásy, 33 (Budapest, 1907), p. 134.

the priors played in the Ottoman wars, the higher the prestige they enjoyed in aristocratic circles. From the reign of King Matthias Corvinus (1458–90), the priors were regarded as prelates who had the right to lead their troops under their own banner (*domini banderiati*). From the mid-fifteenth century they also had the right to be present at the royal council. Eventually, by virtue of article 22 of the law of 1498,[40] the aristocratic status of the Hungarian–Slavonian prior was secured; in return, the Jagellonian kings, Wladislas and Louis II, reserved the right to have a say in the prior's appointment.

---

[40] A new, bilingual edition is in press. *The Laws of the Medieval Kingdom of Hungary*, 4, *1490–1526*, eds J.M. Bak *et al*. (Budapest and Idyllwild, CA, 2012), pp. 102–4.

Chapter 28

Politics, Diplomacy and the Recruitment of Mercenaries before the Battle of Tannenberg–Grunwald–Žalgiris in 1410

Sven Ekdahl

In 2010 the decisive battle of Tannenberg will be commemorated in many countries, above all in Poland (as the battle of Grunwald) and Lithuania (as the battle of Žalgiris), but also by other nations whose soldiers had fought in 1410 in the victorious armies and so contributed to the defeat of the Teutonic Order in Prussia, hitherto the dominant power in the east of Central Europe.[1] This paper will reflect on power politics in the months preceding the battle and their consequences for the all-important recruitment of mercenaries and thus for the outcome of the conflict. Despite the enormous amount published on this famous battle over the course of the last century-and-a-half, it is still worth examining more closely old stereotypical theses, some of which have proved to be wrong,[2] and to look for new or unconsidered sources, especially in the archives of the Teutonic Order in the Geheimes Staatsarchiv in Berlin.[3]

---

[1] S. Ekdahl, 'Tannenberg, Battle of (1410)', in A.V. Murray (ed.), *The Crusades: An Encyclopedia* (Santa Barbara *et al.*, 2006), 4, pp. 1145–6; idem, 'Tannenberg/Grunwald – ein politisches Symbol in Deutschland und Polen', *Journal of Baltic Studies*, 22 (1991), 271–324; idem, 'The Battle of Tannenberg–Grunwald–Žalgiris (1410) as Reflected in Twentieth-Century Monuments', *MO*, 3, pp. 175–94. For the sources, see S. Ekdahl, *Die Schlacht bei Tannenberg 1410. Quellenkritische Untersuchungen*, 1: *Einführung und Quellenlage*, Berliner Historische Studien, 8 (Berlin, 1982). Polish translation: *Grunwald 1410. Studia nad tradycją i źródłami*, trans. M. Dorna (Cracow, 2010). As yet there is no second volume. See the author's bibliography at www.ekdahl.de (accessed 24 November 2011).

[2] See for instance S. Ekdahl, 'Aufmarsch und Aufstellung der Heere bei Tannenberg/ Grunwald (1410). Eine kritische Analyse', in J. Gancewski (ed.), *Krajobraz grunwaldzki w dziejach polsko-krzyżackich i polsko-niemieckich na przestrzeni wieków. Wokół mitów i rzeczywistości* (Olsztyn, 2009), pp. 31–103 (and three maps at the end of the book).

[3] Geheimes Staatsarchiv Preussischer Kulturbesitz Berlin, XX. Hauptabteilung Historisches Staatsarchiv Königsberg (GStA PK, XX. HA StA Kbg.): Ordensbriefarchiv (OBA), Pergamenturkunden (Perg.-Urk.), Ordensfolianten (OF). The documents on paper in the OBA are listed in *Regesta Historico-Diplomatica Ordinis S. Mariae Theutonicorum*

During the months preceding the battle the political and diplomatic activity on and behind the scenes was intense. The main actors in this drama were Ulrich von Jungingen, grand master of the Teutonic Order; Władysław II (Jagiełło, in Lithuanian: Jogaila), king of Poland; Vytautas (Polish: Witold), grand duke of Lithuania; Venceslas IV, king of Bohemia; and Sigismund (or Sigmund) of Luxembourg, king of Hungary and *vicarius generalis* of the Holy Roman Empire. We may also add the Teutonic Order's master in Livonia, the three dukes in Pomerania and many more dukes in Silesia and other parts of the Reich as well as sovereigns in neighbouring countries and the Catholic Church.

Both sides in this battle for power between Prussia and Poland–Lithuania depended on reinforcing their own military forces by using the possibilities offered by the mercenary recruiting markets, especially in Silesia, Bohemia, Moravia, Lausitia, Saxony and Thuringia, but also in other countries and German provinces. During the first half of the 'The Great War' of 1409–1411, in other words the war between the Teutonic Order and Poland–Lithuania in 1409, the grand master Ulrich von Jungingen had engaged about 2,500 mercenaries who fought in the Order's field armies and were used to garrison their own or conquered castles.[4] According to the Treasurer's book, the costs were 46,000 marks, a considerable sum, which, however, was only one-fifth of the money that had to be spent on mercenaries the following year, when the rule of the Knights in Prussia was in danger following the defeat at Tannenberg.[5]

I

After the successful campaign of the Knights in the late summer and autumn of 1409, an armistice was concluded on 8 October between Prussia and the kingdom of Poland, and the war was suspended until the end of St John's Day (24 June) 1410. The truce did not include the grand duchy of Lithuania. An arbitration was to be announced by King Venceslas by, at the latest, 9 February the following year. To make sure that his judgement would be favourable, he was promised 60,000 florins by the grand master, on condition that he respected the Order's wishes.[6] One may regard this as a perfect example of bribery.[7]

1198–1525, eds E. Joachim and W. Hubatsch, Pars I, 1–3 (Göttingen, 1948, 1950, 1973). The parchments are listed in Pars II (Göttingen, 1948). There is an index of Pars I, 1–2, and Pars II (Göttingen, 1965). Pars I, 3 also contains an index.

[4]   S. Ekdahl, 'The Teutonic Order's Mercenaries during the "Great War" with Poland–Lithuania (1409–11)', in J. France (ed.), *Mercenaries and Paid Men: The Mercenary Identity in the Middle Ages* (Leiden and Boston, 2008), pp. 345–61.

[5]   Ibid., pp. 353–4.

[6]   OBA, nos. 1629 and 1630.

[7]   Cf. K. Neitmann, *Die Staatsverträge des Deutschen Ordens in Preußen 1230–1449. Studien zur Diplomatie eines spätmittelalterlichen deutschen Territorialstaates*,

At the beginning of 1410 the political situation of Poland was indeed difficult, whereas in Prussia things seemed very good. The Order could now attack Lithuania without fear of Polish intervention. What was more, on 20 December 1409, Ulrich von Jungingen had concluded a treaty with Sigismund of Hungary against Poland, which therefore threatened to encircle the enemy and make him more vulnerable.[8] Sigismund's brother Venceslas of Bohemia would surely also support the Order. Both kings were eager to obtain money from the wealthy Order. Despite these developments, Jagiełło and his advisers among the Polish nobility and clergy tried to form a coalition that might withstand the pressure from their mighty neighbours. As early as at the beginning of December the king and his vice-chancellor, Nicolaus Trąba, had met Grand Duke Vytautas at Brest-Litowsk in order to discuss the political situation and prepare a campaign against Prussia after the end of the armistice. Vytautas was accompanied by Dżelal-ed-Din, who was later to command the Tatars at Tannenberg.[9] From then on not only Prussian but also Polish propaganda was reaching Western Europe, including France and England, with the intention of getting support or at least preventing the enemy from receiving military aid from crusaders or mercenaries from those regions.

According to the Prussian *Annalista Thorunensis* Venceslas's arbitration was pronounced in Prague on 8 February 1410.[10] As Ulrich von Jungingen had anticipated, it was in his favour. The Polish delegates demurred and left the Bohemian capital in anger. It was now obvious that Venceslas could no longer play an active role as a mediator. One clause in the arbitration provided for a further meeting to be held in Breslau (Polish: Wrocław) on 11 May, but it was unlikely that the Poles would change their minds in any new negotiations with the Bohemian king. There remained the possibility that Sigismund might replace his brother as mediator, intervene in the conflict and thus prevent a war in which he would otherwise be involved because of his treaty with the Teutonic Knights. Such a war would reduce his capacity to deal with other important problems, in particular the Schism and the political situation in the Balkans. He therefore arranged a meeting with Vytautas in the Hungarian town of Käsmark (the German name; Hungarian: Késmárk; Polish: Kieżmark; now Kežmarok in Slovakia) in mid-April.[11] Sigismund is said to have offered the Lithuanian grand duke a royal

---

Neue Forschungen zur Brandenburg-Preußischen Geschichte, 6 (Cologne and Vienna, 1986), p. 606.

[8]   *Lites ac res gestae inter Polonos ordinemque Cruciferorum*, 2nd edn, 2, ed. Z. Celichowski (Posnaniae, 1892), Additamentum, no. 53, pp. 443–4 (Latin); *Die Staatsverträge des Deutschen Ordens in Preußen im 15. Jahrhundert*, 1 (1398–1437), 2nd rev. edn, ed. E. Weise (Marburg, 1970), no. 77, pp. 78–9 (German).

[9]   *Joannis Dlugossii Annales seu Cronicae incliti Regni Poloniae, Liber decimus et Liber undecimus*, eds M. Plezia *et al.* (Warsaw, 1997), pp. 43–4.

[10]   'Franciscani Thorunensis Annales Prussici (941–1410)', ed. E. Strehlke, *SRP*, 3, p. 312.

[11]   *Joannis Dlugossii Annales*, liber XI, pp. 53–6.

crown during this meeting, and it may well be that he was thereby, albeit without success, trying to put an end the Polish–Lithuanian alliance.[12] He would also have got in contact with Jagiełło, who was residing in the nearby town of Nowy Sącz on the Polish side of the border.

The most important result of the Käsmark negotiations was the decision to meet again in the Prussian city of Thorn (Polish: Toruń) in June, before the truce between Prussia and Poland expired. It was intended to involve the grand master of the Teutonic Order, the Lithuanian grand duke, and the kings of Poland and Hungary in person.[13] This move has to have been at Sigismund's insistence and not that of the Teutonic Order, who, however, could not oppose its powerful ally. We know this because Ulrich von Jungingen was planning to begin the war with Poland again on 1 June.[14] He regarded a surprise attack as legitimate, because the Poles had rejected King Venceslas's arbitration of 8 February and were hardly likely to change their position at 11 May meeting in Breslau. About 1,800 mercenaries (600 *Spiesse*, 'spears'; Latin: *lanceae*) were now secretly recruited by the Order and brought to Prussia.[15] The allied dukes of Pomerania-Stettin (Polish: Szczecin) and Pomerania-Wolgast were also to bring their forces to join the Order's army. At the meeting in Breslau the grand master merely waited for a message that the Poles had not accepted Venceslas's arbitration of February. His plan seemed to be perfect. He would attack Poland from the north, and his treaty with Sigismund would guarantee an attack from Hungary in the south. No Polish delegation arrived in Breslau, and that served to confirm his assumption that they had rejected the arbitration.

Sigismund was informed of these plans, but, as they went against his own interests, he insisted that the high-level meeting take place in Thorn in June. Ulrich von Jungingen received this message from Käsmark by royal emissary on 11 May and immediately stopped his preparations for the surprise attack on Poland on 1 June.[16] His war machinery had thus been brought to a standstill by his mighty ally.

For Jagiełło and Vytautas, the decision in Käsmark was most opportune: the Teutonic Order was now under an obligation to respect the truce, and they could

[12]   G. Mickūnaitė, *Making a Great Ruler: Grand Duke Vytautas of Lithuania* (Budapest and New York, 2006), pp. 66–9.

[13]   Letter of Ulrich von Jungingen to the dukes of Stettin and Wolgast, 11 May 1410. OBA, no. 1276.

[14]   S. Ekdahl, 'Die Söldnerwerbungen des Deutschen Ordens für einen geplanten Angriff auf Polen am 1. Juni 1410. Ein Beitrag zur Vorgeschichte der Schlacht bei Tannenberg', in B. Jähnig (ed.), *Beiträge zur Militärgeschichte des Preußenlandes von der Ordenszeit bis zum Zeitalter der Weltkriege* (Marburg, 2010), pp. 89–102; idem, 'Werbowanie żołnierzy zaciężnych ...', in J. Cygański (ed.), *Rocznik Olsztyński* (Olsztyn, 2012) (forthcoming).

[15]   Letter from the deputy *Komtur* of Thorn to Ulrich von Jungingen, 8 June 1410, which has usually been misdated to 22 January 1410. OBA, no. 1248. Cf. Ekdahl, 'Die Söldnerwerbungen'.

[16]   OBA, no. 1276. See n. 13.

continue their own preparations for an offensive campaign against Prussia without fear of intervention. It can therefore be claimed that one important reason for the Knights' defeat at Tannenberg can be traced back to Käsmark. It was now necessary for the Polish king and the Lithuanian grand duke to pretend that they agreed to the meeting and would come to Thorn. Therefore on 27 April Jagiełło issued a letter of safe-conduct for Sigismund to travel through Poland to Thorn with 1,500 horses.[17] Grand Duke Vytautas had indicated his consent. In Prussia the matter was taken seriously. Much preparation was needed to take care of the expected guests and their many horses. Among other things quantities of food and fodder had to be taken to Thorn, where the meeting was scheduled for 17 June.[18] Ulrich von Jungingen arrived on time with high-ranking knights, some important guests from abroad (among which were the heralds of the margrave of Meissen, the landgrave of Thuringia, and the duke of Brunswick),[19] a part of the Order's army, and the 1,800 mercenaries, who had been secretly recruited for the planned but suspended attack on Poland on 1 June.

However, the situation looked more like 'Waiting for Godot', because neither the kings of Hungary and Poland, nor the grand duke of Lithuania appeared. Sigismund only sent a delegation led by the magnates Nicolaus of Gara and Stibor of Stiboritz with 200 horses.[20] It was humiliating for Ulrich von Jungingen and the Teutonic Order and did not bode well for the general situation before the battle of Tannenberg that was to take place on 15 July. The Polish and Lithuanian diplomacy had proved superior to that of the Knights, but only thanks to King Sigismund's interference. Within a few weeks the political situation and the respective roles of the players on the political, diplomatic and military scene had totally changed: the Order was no longer on top, but heading for a disastrous defeat.

II

The most fateful consequence of the decision at Käsmark and the aborted preparations for a surprise attack on Poland on 1 June 1410 related to the recruitment of mercenaries by the Order. Ulrich von Jungingen had started a recruiting campaign, but in the light of the expected meeting at Thorn he stopped it immediately after receiving the message from Sigismund in Käsmark on 11 May.[21] The costs of keeping more than the 1,800 mercenaries who were already in Prussia were, under these new circumstances, clearly too high. Two weeks later, however,

---

[17]   *Codex diplomaticus Regni Poloniae et Magni Ducatus Litvaniae*, ed. M. Dogiel, 1 (Vilnius, 1758), no. 6, pp. 41–2; *Lites ac res gestae*, 2, Additamentum, no. 56, pp. 446–7.

[18]   See OBA, no. 1280, a letter from the councillors of Thorn to the grand master dated 18 May 1410.

[19]   Ekdahl, *Die Schlacht*, p. 277.

[20]   *Joannis Dlugossii Annales*, liber XI, p. 59.

[21]   OBA, no. 1276.

he ordered the Order's *Komtur* of Thorn, who was residing in Prague, to recruit 900 mercenaries (300 *Spiesse*) so that they would be in Prussia before the end of the truce on St John's Day.[22] In the meantime he was aware of the problematic situation for Prussia and the need for the Order to be prepared for war after the end of the truce despite the planned meeting in Thorn on 17 June. But not until June did he realize the alarming dimensions of the danger and order an unlimited recruiting campaign.

The two weeks in May with contradictory orders resulted in a catastrophe for the Knights. Many mercenaries, who had originally expected to be enrolled in the Order's army and had planned marching to Prussia, were so disappointed and angry at being refused that they then instead offered their service to the king of Poland.[23] He engaged them with generous promises to pay them also for any harm occurring during their march.[24] The total number of soldiers who changed sides is not known, but a letter to the grand master speaks of at least two big troops (*Haufen*) of Bohemian mercenaries.[25] There may have been more.[26] This unusual event has left many traces in contemporary sources and also in later narratives, some of which have added a legendary touch to it. Because the real facts that prompted the Bohemians to act as they did were soon forgotten, the chroniclers sought other explanations. One such story is told by Simon Grunau about a Bohemian knight *Medudius von Trauttenaw*, who was refused by the grand master as a 'traitor' and therefore turned instead to the king of Poland with 800 war-horses.[27]

Another consequence of the aborted recruiting campaign, which was resumed partly on 24 May and fully at the beginning of June, can be deduced from

---

[22]   His order of 24 May is mentioned in the reply dated Prague, 4 June 1410. OBA, no. 1294. The reply is published in *Jahrbücher Johannes Lindenblatts oder Chronik Johannes von der Pusilie, Officials zu Riesenburg*, eds J. Voigt and F.W. Schubert (Königsberg, 1823), p. 209 n. 3.

[23]   OBA, no. 1248 (mentioned above in n. 15).

[24]   'us unde in czu cziehen': OBA, no. 1248. For the Teutonic Order's or the grand master's contracts with mercenaries, see S. Ekdahl, 'Verträge des Deutschen Ordens mit Söldnerführern aus den ersten Jahrzehnten nach Grunwald', *Questiones Medii Aevi Novae*, 11 (Warsaw, 2006), 51–95.

[25]   OBA, no. 1248.

[26]   Two declarations of war on the grand master from Silesian mercenaries, dated 3 and 4 July 1410, are published in *Codex epistolaris Vitoldi, magni ducis Lithuaniae, 1376–1430*, ed. A. Prochaska (Cracoviae, 1882; repr. New York, 1965), nos. 350, 351, pp. 211–12. Also see OBA, no. 1471e, an undated declaration of war from Moravian mercenaries, vassals of the bishop of Olmütz/Olomouc. See S. Ekdahl, 'Polnische Söldnerwerbungen vor der Schlacht bei Tannenberg (Grunwald)', in O. Ławrynowicz, J. Maik and P.A. Nowakowski (eds), *Non sensistis gladios!* (Studies in Honour of M. Głosek) (Łódź, 2011), pp. 121–34.

[27]   *Simon Grunau's Preussische Chronik*, ed. M. Perlbach, Die Preußischen Geschichtschreiber des XVI. und XVII. Jahrhunderts, 1 (Leipzig, 1876), pp. 734–5.

information in the Teutonic Order's Payment Book (*Soldbuch*).[28] That important
source provides evidence that about 2,000 mercenaries arrived too late to join the
main army when it marched from Thorn at the beginning of July in order to take
up position to prevent the anticipated invasion. Only 3,700 of the mercenaries
mentioned in the Payment Book could be put into action at Tannenberg.[29] The
total number of the Order's mercenaries at Tannenberg was thus 6,400, a figure
gained by adding the 1,800 men (600 *Spieße*), who had been secretly recruited by
the Order for the planned but suspended attack on Poland on 1 June, as well as
the 900 men (300 *Spieße*) brought to Prussia by the *Komtur* of Thorn before St
John's Day.[30] The accounts for these men are not registered in the Payment Book
and have been lost.

Considering that the mercenaries had had to be recruited in their home countries
and in view of the time-consuming preparations for their long march to Prussia, it
is obvious that Ulrich von Jungingen had reactivated his recruiting campaign too
late. Even a well-known relative of the future grand master Heinrich von Plauen,
Heinrich the Elder lord of Plauen, arrived too late in Prussia with his mercenary
troops 'because it was the will of God'.[31] He and his men could only help in the
defence of the Order's main castle Marienburg (Polish: Malbork) after the battle.
Ulrich von Jungingen had relied in vain upon Sigismund's promises and then had
to face the fatal consequences of that policy.

III

We may conclude that in the space of a few weeks the mighty Teutonic Order
in Prussia was been cast down from the peak of its strength because its military
planning, preparations and possibilities were undermined in the political power-
struggles. Some thousand mercenaries had arrived too late or had offered their
service to the king of Poland instead of the grand master. This is surely one of
the most important facts when looking for the causes of the disastrous defeat
suffered by the Knights on 15 July 1410, even when considerations such as the

---

[28]   *Das Soldbuch des Deutschen Ordens 1410/11. Die Abrechnungen für die
Soldtruppen. Mit ergänzenden Quellen*, ed. S. Ekdahl, Veröffentlichungen aus den Archiven
Preußischer Kulturbesitz, 23/I–II (Cologne, Weimar and Vienna, 1988, 2010). For more
information about this source, see Ekdahl, 'The Teutonic Order's Mercenaries', pp. 348–9.

[29]   A thorough analysis of the Payment Book in 1968 revealed this evidence. S.
Ekdahl, 'Kilka uwag o księdze żołdu Zakonu Krzyżackiego z okresu "Wielkiej wojny"
1410–1411' (Some Remarks on the Payment Book of the Teutonic Order at the Time of the
'Great War' 1410–1411), *Zapiski Historyczne*, 33 (1968), 111–30.

[30]   The mercenaries who were brought by the *Komtur* of Thorn are mentioned
in a letter of the *Komtur* of Schlochau to the grand master of 25 June. See OBA,
no. 1316.

[31]   'wend her czu spete quam, als das got habin wolde': *SRP*, 3, pp. 318–19. Cf. *Das
Soldbuch*, p. 14.

SVEN EKDAHL

disadvantageous location of the battlefield, the unfavourable position of the sun (dazzling the troops of the Order)[32] or the feigned retreat of a part of the Lithuanian army are taken into account.[33] No wonder that the Order was to complain bitterly about King Sigismund's role during the months preceding the battle.[34]

It was also rumoured that the Hungarian unit in the Order's army had fled from the battlefield. According to chronicle of the Burgundian nobleman Enguerran de Monstrelet:

> Et, comme il fut commune renommée, la bataille fut perdu par la coulpe du grant connestable de Hongrye, lequel estoit en la seconde bataille des chrestiens et se parti, lui et ses Hongrois, sans cop férir.[35]

The 'grant connestable' was the palatine Nicolaus of Gara, commander of the Hungarian unit. Only the German knight Christopher von Gersdorff ('qui turpe putavit ex prelio fugere'), King Sigismund's councillor and emissary, refused to fly and was captured together with his men.[36] The rise of Poland–Lithuania to become the most important power in the east of Central Europe after 1410 was thus partly made possible by Hungary, Prussia's ally. Sometimes history goes in intricate ways.

---

[32] Ekdahl, 'Aufmarsch und Aufstellung'.

[33] S. Ekdahl, 'The Turning-point in the Battle of Tannenberg (Grunwald/Žalgiris) in 1410', in V. Kelertas (ed.), *Lituanus: The Lithuanian Quarterly Journal of Arts and Sciences*, 56.2 (Chicago, 2010), pp. 53–72.

[34] *Codex epistolaris Vitoldi*, no. 498, pp. 236–7. Also see OBA, nos. 2962 and 2970.

[35] *SRP*, 3, p. 455. Cf. Ekdahl, *Die Schlacht*, pp. 186–7.

[36] Quotation from Jan Długosz's *Banderia Prutenorum*. See S. Ekdahl, *Die 'Banderia Prutenorum' des Jan Długosz – eine Quelle zur Schlacht bei Tannenberg 1410*, Abhandlungen der Akademie der Wissenschaften in Göttingen, Phil.-Hist. Klasse, Dritte Folge, 104 (Göttingen, 1976), pp. 176–7 (= pp. 156–7 in the Lithuanian edition (Vilnius, 1992). (A Polish edition will be published by ElSet, Olsztyn, in 2012.) Christopher von Gersdorff commanded the *Reichsbanner* (*Sturmbanner*) of the Holy Roman Empire (red with a silver cross) during the battle, a most honourable task. For this man, also see *Das Soldbuch*, 2, no. 143.

Chapter 29

Power to the Educated? Priest-brethren and their Education, using Data from the Utrecht Bailiwick of the Teutonic Order (1350–1600)

Rombert Stapel

On 15 August 1561, immediately after the death of the Hospitaller bailiff of Utrecht, the convent there chose Hendrik Berck as the new head of the bailiwick. Berck was widely regarded as a capable administrator and he was praised for his exemplary behaviour *in religiosis*, but the hasty procedure did not meet with the approval of the provincial government. Supported by Margaret of Parma, governor of the Netherlands from 1559 to 1567, representatives of the Hof van Utrecht (Court of Utrecht) were soon sent to investigate the nomination. The documents that were subsequently produced describing what happened provide some interesting information about the education of Hendrik Berck and his fellow brethren.[1] They state that the new bailiff was of noble descent, originating from the town of Zutphen in the duchy of Guelders. According to a fellow student, he had received education at the prestigious Latin School in Deventer as far as the ultimate or penultimate level. Directly afterwards he had entered the Hospitaller Order as priest-brother, but without continuing his studies at a university.[2] This lack of a university education was one of the arguments used by the attorney-general of Utrecht to challenge his election. According to the attorney-general, 'He received little more education ... in spiritual or secular business other than by being overseer of the kitchens of the aforesaid convent for six months'. Although this may be a somewhat one-sided account of the affair, it indicates that a prelate was expected to be well educated.[3] Having had an appropriate academic training even seems to have been regarded a precondition for appointment to important

---

[1] A.H.L. Hensen, 'Hendrik Berck, de laatste baliër der Sint Jansheeren te Utrecht', *Archief voor de geschiedenis van het Aartsbisdom Utrecht*, 35 (1909), pp. 1-77 at pp. 44, 48, 53, 63; R.R. Post, *Kerkelijke verhoudingen in Nederland vóór de Reformatie van ±1500 tot ±1580* (Utrecht and Antwerp, 1954), p. 351.

[2] Hensen, pp. 23, 48.

[3] Ibid., p. 63.

offices in the bailiwick.[4] Furthermore, in Utrecht, studying did not end once a
priest-brother entered the Hospitaller Order. Hendrik Ruysch, the commander
of Ingen, for instance, was said to delve deeply into theological subjects every
day and gave sermons on a regular basis. The members of the Utrecht convent
received instruction on the psalms of David and the letters of Paul from the learned
Hieronymus Vairlenius, who was later to become vicar general of the bishop of
Haarlem.[5]

This paper focuses on the education of the brethren of the military orders,
choosing as a case study the priest-brethren in the Teutonic Order's bailiwick of
Utrecht, for whom a series of valuable sources are available.[6] Priest-brethren have
hitherto only received limited attention from students of the military orders. It
is the brother knights who usually take centre stage, since it was their duty to
perform the main task of the military orders: fighting the enemies of Christendom
in the Holy Land, the Iberian Peninsula and the Baltic region. However, in all
military orders the priest-brothers constituted a large group, and without them the
combination of a military and a religious way of life would have been impossible.
Unfortunately, we know less about their backgrounds than about those of the
brother knights.

To an extent, the study of the education of the brethren in general has been long
neglected.[7] This is probably because the normative sources, such as for instance the
statutes of the Teutonic Order, do not make any mention of requirements to study.
Quite the contrary, they give the impression that study was officially discouraged
among its members.[8] It might therefore be assumed that priest-brethren of the
military orders generally were unlearned men, whose only roles were the religious

---

[4]   Ibid., pp. 44, 48, 53, 63; Post, p. 351.

[5]   Hensen, pp. 53, 44.

[6]   A more detailed prosopographical analysis on the priest-brethren of the Utrecht
is provided in R.J. Stapel, '"Onder dese ridderen zijn oec papen": De priesterbroeders
in de balije Utrecht van de Duitse Orde (1350–1600)', *Jaarboek voor Middeleeuwse
Geschiedenis*, 11 (2008), 205–48, at http://depot.knaw.nl/5007/ (accessed 3 May 2011).
An English translation is currently in preparation.

[7]   Main exception being H. Boockmann, 'Die Rechtsstudenten des Deutschen Ordens.
Studium, Studienförderung und gelehrter Beruf im späteren Mittelalter', in T. Schieder *et
al.* (eds), *Festschrift für Hermann Heimpel zum 70. Geburtstag am 19. September 1971*, 2
(Göttingen, 1972), pp. 313–75. There are several studies that include data on attendance
at universities by brethren of the military orders: K. Górski, 'Das Kulmer Domkapitel in
den Zeiten des Deutschen Ordens. Zur Bedeutung der Priester im Deutschen Ordens', in
J. Fleckenstein and M. Hellmann (eds), *Die Geistlichen Ritterorden Europas* (Sigmaringen,
1980), pp. 329–37; M. Glauert, *Domkapitel von Pomesanien (1284–1527)* (Toruń, 2003),
pp. 283–8; R. Biskup, *Domkapitel von Samland (1285–1525)* (Toruń, 2007), pp. 293–8;
A.J.A. Bijsterveld, *Laverend tussen kerk en wereld. De pastoors in Noord-Brabant 1400–
1570* (Amsterdam, 1993).

[8]   Boockmann, p. 363; A. Mentzel-Reuters, *Arma spiritualia. Bibliotheken, Bucher,
und Bildung im Deutschen Orden* (Wiesbaden, 2003), pp. 43–8.

inspiration and the spiritual care of the warriors. Nonetheless, from the example shown above and other evidence, it is perfectly clear that, contrary to what the statutes might suggest, the orders had plenty of use for well-educated brethren. Especially for the priest-brothers, who were often active as parish clergy, some level of education at least was considered an advantage.[9] Because not all chaplains who are mentioned could boast an academic training, a university education cannot be regarded as a requirement for the whole group. However, the first impression is that, at least for the military orders in the Netherlands, the priest-brethren were as well trained as the secular and regular clergy in general. We know that in the later Middle Ages the percentage of academically trained priests rose, and it has been observed that around 1500 the majority of the clergy in the Low Countries had enjoyed some form of university education.[10]

Since matriculation registers for Latin Schools are lacking, with a few exceptions such as for example Hendrik Berck, we know almost nothing about the teaching priests had received before they attended university. Taking into account the number of academically trained priest-brethren, we can infer that attendance at one of the Latin Schools was considered fairly normal. The only data on education that can be quantified is from the universities. That said, it is clear that even that information is hard to retrieve. The enrolment registers sometimes contain serious gaps. In his study on the careers of parish priests in late medieval North Brabant, the Dutch scholar Arnoud-Jan Bijsterveld calculated that, of all the parish priests who had received an university education, around 22 per cent left no trace in the university archives that have been preserved.[11] With this information in mind, we shall now look at the data collected from the priest-brethren of the Utrecht bailiwick between 1350 and 1600.[12] To avoid possible confusion, I have chosen to leave out the brethren of the Frisian commanderies. Their position in the Utrecht bailiwick, as well as their education, is, as Hans Mol has pointed out in his study on the Frisian houses of the Teutonic Order, complex and cannot be considered to have been typical.[13]

---

[9]   For comparison: Bijsterveld, p. 135.

[10]   Post, pp. 158–61.

[11]   Bijsterveld, pp. 149–50, 154–5.

[12]   For this purpose I have studied the enrolment registers or lists of Dutch students at the universities of Cologne, Louvain, Paris, Rostock, Heidelberg, Orléans and the most important Italian universities (Bologna, Padua, Siena, Ferrara, Pisa/Florence, Rome, Pavia, Turin, Naples and Arezzo), provided in A.L. Tervoort, *The iter italicum and the Northern Netherlands: Dutch Students at Italian Universities and their Role in the Netherlands' Society (1426–1575)* (Leiden, 2005).

[13]   J.A. Mol, *De Friese huizen van de Duitse Orde. Nes, Steenkerk en Schoten en hun plaats in het Friese kloosterlandschap* (Leeuwarden, 1991).

Table 29.1   Number of academic brethren in the Utrecht bailiwick of the Teutonic Order

| | Priest-brethren | | | | | | Knight-brethren | | Unknown | | Total | |
	1350–1600		Frisian commanderies		before 1350							
A	116	56%	61	88%	22	92%	102	82%	169	88%	470	76%
B	67	32%	6	9%	2	8%	21	17%	19	10%	115	19%
C	25	12%	2	3%	0	0%	1	1%	3	2%	31	5%
Total	208	100%	69	100%	24	100%	124	100%	191	100%	616	100%
B +*C*	92	44%	8	12%	2	8%	22	18%	22	12%	146	24%

A: No identification in university enrolment lists.

B: One identification in university enrolment lists.

C: Two or more possible identifications in university enrolment lists.

First of all: not only priests of the Teutonic Order in the bailiwick of Utrecht could boast a university education. We have evidence as well that one-fifth of the brother knights attended a university at some time. A strikingly large proportion of these knights reached the position of land commander, and a majority of the land commanders in the fifteenth and sixteenth centuries seem to have been learned. Until recently this information was unknown, and it sheds a new light on that office.

When we turn once more to the priest-brethren and compare our data on them with what is known about the intellectual training of the parish priests of North Brabant mentioned earlier, we can find that both groups produce rather similar figures. The comparison does not hold for the higher ranks in the church. Members of cathedral chapters, for instance those of the chapter of St Salvator in Utrecht, usually show a higher proportion with an academic *curriculum vitae*.[14] Most chapters actually demanded that their members be learned. For the military orders this was not required, but at least half of the Teutonic brethren among the North Brabant parish priests in the sixteenth century are recorded as having studied at a university.[15]

As for the situation in Prussia, there are figures available for two major groups of priest-brethren. These are treated in studies on the Prussian chapters of Pomesania (in Marienwerder) and Sambia (in Königsberg).[16] Of the 150 priest-brethren of the Pomesania chapter, it has been shown that at least 25 (17 per cent) had been trained at a university. A peak of academically trained men is reached around the year 1400. For the Sambia-group a similar peak around 1400 can be found, but here the graph shows another peak in the late fifteenth century. All in all, 43 of the 131 Sambia chapter priests (33 per cent) had received a university education. It is likely, however, that the percentage of academically trained priest-brethren in the Prussian chapters, especially in the fifteenth and sixteenth centuries, was much higher. For the years 1442, 1497 and 1517 complete lists of members of the Sambia chapter have been preserved. In these three years the number of academically trained priests exceeded 50–70 per cent, while in the early fourteenth century a substantial number of the chapter members could not even write their own names. Thus, at some point before 1442, the level of education must have risen considerably. Unfortunately, it is impossible to analyse more in detail when this development took place. What this data does show is that the figures presented for the priest-brethren of the Utrecht bailiwick are not unique within the Teutonic Order.

---

[14] A.J. van den Hoven van Genderen, *Heren van de Kerk. De kanunniken van Oudmunster te Utrecht in de late middeleeuwen* (Zutphen, 1997), pp. 248, 250.

[15] Bijsterveld, p. 163.

[16] Glauert, pp. 283–8; Biskup, pp. 293–8.

Table 29.2   Number of academic priest-brethren per period of 25 years (not including the Frisian brethren)[17]

Period	Number of priest-brethren	Academic education		Parish priests in Northern Brabant with academic education	
		(one identi-fication)	*(one or more identi-fications)*		
	Total (n=208)	Total (n=67)	Total (n=92)	'Certain'	'Certain' and 'uncertain'
1351–1375	13	0%	0%		
1376–1400	15	13%	13%		
1401–1425	14	7%	7%	17%	20%
1426–1450	27	30%	52%	28%	37%
1451–1475	37	43%	54%	38%	49%
1476–1500	16	31%	44%	36%	49%
1501–1525	34	41%	71%	40%	51%
1526–1550	30	50%	63%	46%	52%
1551–1575	33	36%	36%	47%	56%
1576–1604	4	50%	50%		
1350–1600	*223*	*34%*	*45%*		
1401–1575	*191*	*37%*	*51%*	*36%*	*45%*

Analysis

The data on the Utrecht priest-brethren of the Teutonic Order can be differentiated in time, as can be observed in this chart. In the fourteenth and early fifteenth centuries, almost none of the priests attended a university before or after entry into the Order. However, after about 1425 a significant growth of the number of academically educated priest-brethren must have taken place. Their number continued increasing into the sixteenth century, although the last quarter of the fifteenth century shows a slight decline in their volume. Later in the sixteenth

---

[17]   Bijsterveld, p. 158. Note that the total figure is somewhat distorted by brethren that were active in more than one period. The period examined by Bijsterveld was 1400–1570 instead of 1401–1575.

century we see a relatively lower percentage of priests who had studied at a university. The slowly rising figures as we come into the sixteenth century are quite compatible with the general tendency in the Low Countries, as can be seen from the figures for the North Brabant parish priests. The increasing demand for 'quality' in the parish ministers will have been the underlying driving force behind this growth of academically educated priest-brethren in the Utrecht bailiwick. Nonetheless, there are also some striking differences to the general trend, especially in the first half of the fifteenth century.

Before we dig deeper into this matter, some remarks have to be made on the priest-brethren's choice of university and faculty. The regional universities Louvain and Cologne attracted about 90 per cent of all the Utrecht priest-brethren with an academic background. Other universities attended include Paris, Rostock, Heidelberg, Orléans and Bologna. Well over a third of the priest-brethren went to more than one university. These findings are fairly similar to those of Bijsterveld with regard to the parish priests in North Brabant. In Sambia and Pomesania, there seems to have been a preference for the more 'regional' universities as well. Leipzig, Prague and Vienna were the most attractive academic centres for the Prussian chapter priests.[18] As for the faculties that were chosen, by far the most popular was the Faculty of Arts. Only 10 per cent of the Utrecht brethren had studied law and – surprisingly enough – only one priest may have entered the faculty of theology. Only one in five had received a degree and just two priests had graduated in one of the higher faculties, in this case at the law faculty. By comparison with the parish priests and the Prussian chapter members, these figures are quite low. The studies on these groups produced much higher figures in terms of the numbers attending the higher faculties and the numbers graduating. On the other hand, the faculty of arts was in any case by far the most popular faculty for future clergy.[19]

How should we interpret these figures? First and foremost, we cannot but draw the general conclusion that the level of education of the priest-brethren was substantial, even though in the Utrecht bailiwick relatively few had attended a higher faculty or gained a degree. The percentage of priest-brethren with an academic background may have been lower than, for instance, that of the Utrecht secular canons. But the figures are very comparable with those for the parish priests of North Brabant. The priest-brethren of the Teutonic Order who were active in the parishes there – they belonged to the bailiwick of Alden-Biesen – are reported to have studied even somewhat more frequently than most of the other parish priests belonging to a monastic order (see Fig. 29.1)[20]

Also noteworthy are the changes in the numbers of students over time. In order to trace developments in the number of students, we can best arrange

---

[18] Bijsterveld, pp. 172–3. Biskup, p. 296. A case study of students coming from Leiden resulted in different specific characteristics, but the general trend is again comparable. Tervoort, p. 30.

[19] Bijsterveld, pp. 190, 198; Glauert, pp. 287–8; Biskup, pp. 296–7.

[20] Boockmann, pp. 320–61.

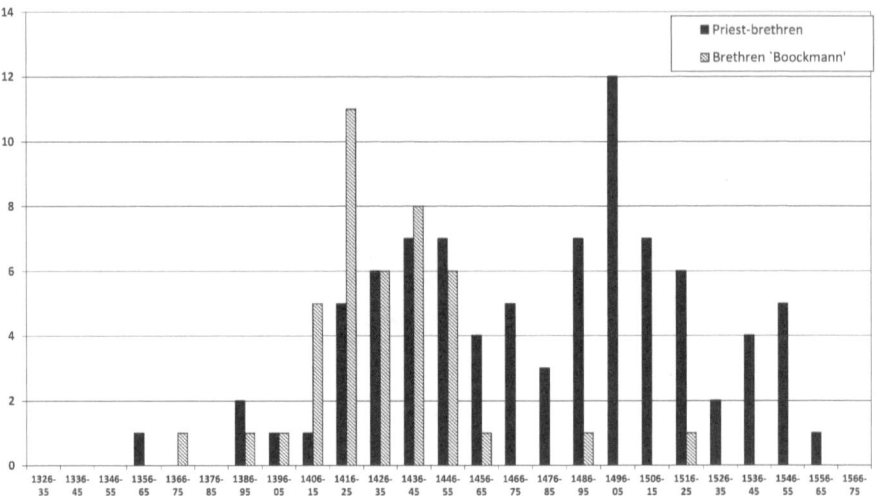

Figure 29.1   Number of matriculating (future) priest-brethren of the Utrecht
bailiwick and brethren in Boockmann, 'Rechtsstudenten'

them according to their first year of matriculation, as shown in the graph. This
graph gives not only the priest-brethren of the Utrecht bailiwick but also the
brethren, studied by Hartmut Boockmann, who applied for a scholarship from
the grand master of the Teutonic Order. Boockmann saw a distinct correlation
between these scholarship applications and the political and juridical affairs in
Prussia and Livonia in which the Order was entangled. He suspected that the grand
master recognized a desirability for brethren in diplomatic service who were well
educated in the field of law.[21] To demonstrate the need for law graduates in the
Order's diplomatic affairs, we can cite the example of the Prussian chapters, who
maintained close relations with the office of the grand master. Of the six priest-
brethren of the Sambia chapter who could call themselves doctors of law, four
were actively involved in the Order's diplomatic service.[22]

Indeed, when we take a closer look at Boockmann's data, it shows a notable
rise in the number of students around the year 1410 and a sharp decline at the
beginning of the Thirteen Years' War. In other branches of the Order, for instance
at the office of the German master or in the bailiwicks, Boockmann could not find
any evidence for similar scholarships, and that led him to conclude that, beyond
the office of the grand master, the need for academically trained brethren was
non-existent.[23] It will be clear that the newly gathered information on the Prussian
chapters and the Utrecht and Alden-Biesen bailiwicks does not support this view.

---

[21]   Ibid., pp. 372–3.
[22]   Biskup, pp. 296–7.
[23]   Boockmann, pp. 373–4.

The figures for the Utrecht priest-brethren show a rise of academic training for them after the Battle of Grunwald in 1410 as well. This increase pre-dates the founding of the University of Louvain in 1426, which triggered a substantial growth in the general number of students from the Low Countries.[24] Coinciding with the peak for Boockmann's brethren, a small peak 'of study' for the Utrecht priests occurs for the period just before the beginning of the Thirteen Years' War. It cannot be a coincidence that five of the seven priest-brethren who had studied law were active in the Order in these few decades.

We can thus offer the hypothesis that the situation in Prussia and Livonia in the first half of the fifteenth century was an important factor in the development of academic interest in the relatively peripheral Utrecht bailiwick. We know that the ties between the bailiwick on the one side and the Teutonic Order's lands in Prussia and Livonia on the other were still very strong in this period.[25] It looks as if in the first half of the fifteenth century there were attempts throughout the whole Order to bring about a growth in the number of academically trained brethren. It is doubtful whether these attempts were initiated or enforced by the highest authorities. But we cannot escape the impression that in these turbulent times a need for skilled lawyers was felt in different parts of the Order. It is difficult to say whether the Utrecht bailiwick really had the same need of such men for diplomatic purposes, as did the office of the grand master. Maybe we should assume that there were fears that the troubles in Prussia would eventually result in political and judicial difficulties in the bailiwick as well. Whether or not such fears existed, by the end of the fifteenth century they had receded: no more law students were recruited in the bailiwick. The growing number of educated priest-brethren in the Utrecht convent and its dependent commanderies comprised arts students. Arguably, for the priest-brethren with the spiritual charge of churches in the bailiwicks or even in Prussia, an education in law was not essential for their day-to-day tasks. Nonetheless, a certain level of education was considered an advantage for these priests. In these cases, attendance at one of Europe's faculties of arts was a tried-and-tested route, especially as the fifteenth and sixteenth centuries progressed.

This shift halfway through the fifteenth century seems to have been much more than just a change in the choice of faculty. From about 1450 onwards we notice a swing to more local factors influencing the number of priest-brethren attending a university, while possible correlations with the affairs in the Baltic region are getting more difficult to establish. The struggle for power in the Low Countries that started after the death of Duke Charles the Bold of Burgundy in 1477 resulted in a temporary standstill in the recruitment of priest-brethren with a university

---

[24]   Bijsterveld, pp. 146, 165.

[25]   As can be deduced by, for instance, the relatively large proportion of the brother knights in Livonia who originated from the Northern Netherlands in these years. J.A. Mol, 'Nederlandse ridderbroeders van de Duitse orde in Lijfland: herkomst, afkomst en carrières', *Bijdragen en mededelingen betreffende de geschiedenis der Nederlanden*, 111 (1996), 28.

education. The siege of the city of Utrecht in 1483 by Emperor Maximilian was undoubtedly a major factor in this. The low number of brethren that matriculated during the years 1526–35, reflected a general decline in the number of students at the universities of Louvain and Cologne who were coming from the Low Countries.[26]

In addition it has to be kept in mind that the vast majority of the academically educated priest-brethren in the Utrecht bailiwick had attended university before they entered the Order. It is therefore not surprising that the changes in their numbers ran parallel with the general pattern of development in the student population in the Low Countries, which was, after all, the group from which the priest-brethren were recruited. Because of this, I do not think it likely that the Order gave scholarships to its priests, especially as I have not found any documentary evidence to support such a claim.

Education and Career

What, then, can be said about the relationship between education and career prospects within the Teutonic Order? Did academically trained priest have better chances in the *cursus honorum*? The studies on the Prussian chapters do not discuss this issue at any length. The main impression to be gathered from their data, however, is that priest-brethren with an academic education were generally trusted and rewarded with the more important offices. As for the Utrecht bailiwick we have more detailed figures at our disposal, and so we can go a little further.

Table 29.3   Differences in careers between academic and non-academic priest-brethren

	Duration of career (years)		Number of offices	Number of commanderies	Avg. geograph. mobility (km)
	All priest-brethren	*More than once mentioned*			
Without academic education	10.9	16.5	1.95	1.30	49.3
With academic education	15.0	20.1	2.49	1.62	54.9
Total	12.3	17.7	2.15	1.42	51.9

---

[26]   Tervoort, p. 28.

When examining the quantitative aspects of the careers of the priest-brethren, we find some differences between the priest-brethren with a university education and those without. The academically trained priest-brethren usually resided and worked farther away from their place of origin. They also held more offices in separate commanderies. Most importantly, though, their careers lasted about four years longer than those of their brethren who had not attended university. As a rule of thumb it can be stated that the number of offices and commanderies with which a priest-brother was entrusted (thereby influencing his geographical mobility) grew roughly in accordance with the number of years in service he had in the Order.

But why did the careers of the academically educated priest-brethren cover a larger number of years? This phenomenon is closely connected to the fact that most brethren left no documentary evidence until they were appointed as a parish priest or took up the office of commander. The data we gathered on the careers of a dozen brethren of whom the exact year of entry is known, gives us reason to conclude that on average it took around eight years before these brethren reappear in the archival evidence, often as a parish priest or commander. Thus, the first decade of their service in the Order is largely unknown to modern researchers. As a consequence, the lengthy careers of the brethren with an academic background indicate that on average these men took up the most important offices for priests in the bailiwick – parish priest or commander – a couple of years earlier in their career. This impression is supported by a small group of priests for whom we know their date of birth. Six of them are known to have studied at university; on average they reached the position of parish priest or commander at the age of 32. Eight others did not attend university before their entry into the Order; they reached these positions four years later. Finally, there were four priest-brethren who never took up these offices. Coincidence or not, none of them had ever attended an academic institution.

We cannot, however, signal a clear difference in the high point that they attained in their careers. Priest-brethren who had studied and those who had not are to be found alongside each other anywhere in the bailiwick. The percentage of academically trained brethren could differ greatly from office to office, but without evidence that could disclose all the determining factors involved, it is hard to draw conclusions. The most important office for priest-brethren in the bailiwick for instance, the commandership of the prestigious *domus* in Leiden which included the position of rector of the prestigious city parish of St Peter, was assigned to only a few academically trained brethren. In contrast, its commanders could remarkably often boast noble descent, a rarity among the priest-brethren. In short, for priest-brethren there was no unambiguous correlation between a 'top career' and a university education. We can only conclude that a priest-brother who had studied at a university had a better chance of a flying start in the bailiwick as a parish priest or commander.

There was, however, one office for which a university education, along with a high social status, was very helpful, although in this case only knight-brethren could apply. As mentioned above, a majority of the Land Commanders in the Utrecht bailiwick in the fifteenth and sixteenth centuries could boast a university education. In contrast with the academically trained priest-brethren, all of whom had studied before they entered the Teutonic Order, most of them were already full members of the Order at the time when they attended university. Within the group of brother knights these future Land Commanders no doubt formed an exceptional group. It shows, though, that even in a military Order, an academic education was considered useful for brethren in leading positions.

Summary

We have seen that a considerable proportion of the priest-brethren of the Utrecht bailiwick had attended university: 44 per cent, rising to 70 per cent in certain periods. There is persuasive evidence that these figures cannot be considered unique within the Teutonic Order as a whole. After the turbulent events in Prussia in the first half of the fifteenth century, student numbers in the bailiwick rose swiftly; they seem to have been influenced by developments in the Baltic region in the following period as well. But from the second half of the fifteenth century onwards, local factors became more prominent in determining the number of educated priest-brethren.

When we look at career prospects for educated priest-brethren, the picture is not always as clear as we would like it to be. Academics had better chances than non-academics for being appointed as a parish priest or a commander early in their careers, but it is hard to prove that they also attained better positions in the end. We can conclude, however, that at least in the fifteenth and sixteenth centuries an academic education and membership of the military orders were not such an unusual combination as one might think.

Chapter 30

Hidden in the Bushes: The Teutonic Order of the Bailiwick of Utrecht in the 1780–1806 Revolutionary Period

Renger E. de Bruin

Twice a year, the members of the Knightly Teutonic Order of the Bailiwick of Utrecht meet in the house that was built for them over 650 years ago. They come together to discuss their centuries-old possessions and the charities they finance with the proceeds. The fact that the Teutonic Order still exists in Utrecht is more a miracle than the logical outcome of events. The Bailiwick of Utrecht barely managed to survive the period of the Reformation and the Dutch Revolt against Spain, but it did become separated from the central organization. Hans Mol reported on this in the proceedings of the 2004 conference on the military orders and the Reformation,[1] and the problem has been studied in further detail by Daniela Grögor-Schiemann in her 2009 dissertation.[2]

Slightly less than two centuries after surviving the Reformation, the Bailiwick was affected by the new storms that swept the European continent. The revolution that had started in North America brought fundamental social change to much of Europe. 'Freedom, equality and fraternity' became a rallying cry against privilege that was transmitted by birth, privilege that had hitherto been self-evident. Institutions that were based on such privilege, including the military orders, were under threat. The revolutionary leader who meted out the harshest treatment on such institutions was Napoleon Bonaparte: in February 1811, having already dispersed the Knights of Malta and dissolved the Teutonic Order in Germany, he disbanded the Bailiwick of Utrecht, a decision that was to be reversed by the Dutch king, William I, four-and-a-half years later. This article focuses on one of the key aspects of this period: the approach to revolutionary developments taken by the Bailiwick of Utrecht in the years before 1806 when Napoleon's brother Louis was crowned king of Holland. After briefly outlining the history of the Bailiwick until

---

[1]   J.A. Mol, 'Trying to Survive: The Military Orders in Utrecht, 1580–1620', in J.A. Mol, K. Millitzer and H.J. Nicholson (eds), *The Military Orders and the Reformation. Choices, State Building and the Weight of Tradition* (Hilversum, 2006), pp. 181–207.

[2]   D. Grögor-Schiemann, 'Die Deutschordensballei Utrecht während der Reform-ationszeit', diss., Ruhr-Universität Bochum, 2009.

the revolutionary period, I shall examine successive phases of these developments, and describe how the Bailiwick responded to them.

The Bailiwick of Utrecht until the Revolutionary Period

The Teutonic Order had gained property in the northern Netherlands within only a few decades of its foundation at the end of the twelfth century. Initially the Utrecht commandery was under the authority of the land commander of Alden Biesen; later it became an independent bailiwick. In 1346, the Utrecht land commandery received new quarters inside the city walls. There it remained down the centuries, and there it returned in 1995.

In the late Middle Ages, the Bailiwick was not fundamentally different from the Holy Roman Empire's eleven other bailiwicks. But the sixteenth century was to bring significant changes. Between 1516 and 1543 – the period in which the secularization of Prussia by Grand Master Albrecht of Brandenburg plunged the Teutonic Order into an unprecedented crisis – the emperor Charles V's territorial expansion in the Netherlands brought all the possessions of the Bailiwick of Utrecht under direct Habsburg control.

The rising that broke out some decades later against Charles's son Philip II isolated the Utrecht knights from the monarch, the Bailiwick's sovereign, increasing its separation from any central authority. After Philip had been abjured as sovereign in 1581, the States of Utrecht appointed themselves in his stead, and, following the example of the other rebellious provinces, forbade Catholic worship. Very much against this current, the Bailiwick of Utrecht strove to retain its Catholic identity, and also its links with the grand master of the Teutonic Order. This brought it into conflict with the new Dutch Republic in its struggle against the Habsburgs, and by 1620 the States of Utrecht had imposed a Protestant land commander.[3]

There were, however, still Catholic members, and ties with the grand master in Mergentheim had not yet been broken. Although the knights were still bound by the vow of celibacy, the situation changed in 1640 when the States of Utrecht granted them permission to marry.[4] This brought rupture with Mergentheim – permanently, as it proved; attempts to heal the rift continued until 1805, but were in vain.[5] Within the Dutch Republic, the Bailiwick of Utrecht thus developed into an organization of lay nobles. Its loyalty to the political order was guaranteed by the land commanders, who occupied central positions in the Dutch Republic. As

---

[3]   Mol, p. 202.

[4]   J.J. de Geer tot Oudegein, *Archieven der Ridderlijke Duitsche Orde, Balie van Utrecht sedert 1581* ('s-Gravenhage, 1871), 1, cxiii; H. Biesma, *Ridders in een klooster. Het Duitse Huis in Utrecht* (Utrecht, 1999), p. 32; J.H. de Vey Mestdagh, *De Utrechtse Balije der Duitse Orde. Ruim 750 jaar geschiedenis van de Orde in de Nederlanden* (Utrecht-Alden Biesen, 1988), p. 44.

[5]   Ibid.; U. Arnold, *Deutscher Orden 1190–2000* (Bad Mergentheim, 2004), p. 79.

members of the Provincial States or as army officers, the commanders – who now had to belong to the official Dutch Reformed Church – were also part of this order.

The procedures for admission to the Bailiwick showed that the earlier traditions had been adapted to the new reality. Candidates were put forward as children by the incumbent commanders. Upon reaching adulthood, they could, provided they fulfilled a number of conditions, be raised to the rank of *jonkheer* (esquire). One such condition was to have four noble quarters (i.e. four noble grandparents); another was to be a member of the Dutch Reformed Church. Whenever the death of a commander caused a shift in the chapter, the oldest *jonkheer* could accede. First, however, this move had to be affirmed in a meeting of the chapter. The assent of the States of Utrecht was also required.

Although a commander was supposed to live at his commandery, this requirement tended to happen less and less, and management of the commanderies was left to local officials. The buildings fell into disrepair, and many of them were eventually disposed of. In the absence of clear control, the Order's finances descended into chaos. The admission of new members also left much to be desired. Eventually the procedures for arranging a commander's succession were so poor that interregnums became inevitable. By the middle of the eighteenth century, the bastions of nobility were at risk.

A first move towards turning the tide came in 1740 with the adoption of rules for admission that stressed the conditions requiring that a candidate should have four legal noble quarters and be of the Dutch Reformed faith. Some ten years later the rules were tightened further in a thorough overhaul inspired by the coadjutor, Count Unico Wilhelm van Wassenaer van Twickel (1692–1766), who, during the meeting of the chapter in 1753, had analysed the problems facing the Order. He was charged with the search for solutions. His plan included far-reaching changes to the Order's administration, bookkeeping and property management. The latter would be put into the hands of a steward-general. According to their rank, which was to be determined on the basis of seniority, the commanders would henceforth receive an annual stipend, and, if there was any surplus, also a bonus. The count's committee also proposed how duties should be divided between the steward and the secretary, who would also be paid a stipend. To avoid interregnums, the land commander would automatically be succeeded by the coadjutor. At the following meeting of the chapter, the committee's proposals were accepted. They then had to be submitted to the sovereign, the States of Utrecht, which, after lengthy deliberations, eventually agreed.

The new structure was quick to deliver results. While the first annual accounts – for 1762 – showed a deficit of 1,396 guilders, those from 1763 onwards showed a surplus.[6] In the years that followed, this surplus increased rapidly, so much so that in 1783 the stipends could be raised. The growing surpluses were due primarily to rising incomes from leased land. The steward, Gijsbert Dirk Cazius (1722–1804),

---

    [6]   Archief Ridderlijke Duitsche Orde Balije van Utrecht (ARDOU-OA) 337 dl. 2 (1763) en dl. 3 (1764).

not only ensured that these were actually collected, but also that the rent rose when a lease was renewed.

The times were favourable. Grain and dairy prices rose in the years after 1765, and there was also a fall in the frequency of flooding, livestock epidemics and other disasters that had plagued the earlier part of the century.[7] Farmers thus had less difficulty in paying for higher leases. Cazius's proactive approach to rent collection and his careful administration of income and expenditure began to bear fruit. While the commanders' stipends were the greatest cost item, increasing sums could also be spent on the maintenance of buildings, thereby putting an end to their deterioration.

Revolution and Restoration in the Netherlands

In 1780 the Bailiwick of Utrecht was therefore in much better shape than it had been some decades earlier – and also much better prepared for the difficult times that lay ahead. The Dutch Republic was now afflicted by a revolution born out of dissatisfaction with the regime of Stadtholder William V and his aristocratic and patrician regents. Some of these – men such as Count Johan Walraat van Welderen (1725–1807) and Baron Joost Taets van Amerongen van Natewisch (1726–91) – were members of the Teutonic Order.

The fact that this dissatisfaction came to a head had much to do with the American struggle for independence, for which there was considerable sympathy in the Netherlands.[8] As arms deliveries to America were also a good source of income for Amsterdam's merchants, relations with Great Britain became strained. Despite attempts to avert it by Stadtholder William V and the Dutch envoy in Westminster, the Count van Welderen referred to above, war broke out in late 1780. For the Netherlands, the results were disastrous. The British, with their control of the seas, were able to close off trading routes with catastrophic consequences for an economy that was already sickly. Trade and industry slumped, unemployment rose rapidly, and war boosted food prices, which had been rising already. The result was widespread poverty.

The scapegoat for this misery was soon found: Stadtholder William V, whose position on foreign policy and whose family connections with George III made him appear pro-British, and thus a traitor. Nor could he be trusted in his capacity as Captain-General of the Dutch army, partly because of its many foreign mercenaries

---

[7]   R.J.N. Rommes, 'Werken op het platteland', in C. Dekker *et al.* (eds), *Geschiedenis van de provincie Utrecht* (Utrecht, 1997), 2, p. 186; J. de Vries and A. van der Woude, *The First Modern Economy: Success, Failure and Perseverance of the Dutch Economy, 1500–1800* (New York, 1996), pp. 270–74.

[8]   S. Schama, *Patriots and Liberators: Revolution in the Netherlands 1780–1813* (New York, 1977); J. Israel, *The Dutch Republic: Its Rise, Greatness and Fall, 1477–1806* (Oxford, 1995), p. 1210.

and officers. For this reason discontented citizens naming themselves Patriots after the American model insisted on being armed. The civic militias, which had fallen into disuse, were re-established. Citizens also drilled in associations known as Free Corps, the key element in this movement being the Free Corps of the city of Utrecht. It was not long before citizens made demands for a greater part in their government. The Utrecht Patriots demanded an elected city council, a demand that was repeated elsewhere.

After the civic militia had assumed power in August 1786, a democratic constitution was adopted in Utrecht. The States of Utrecht split into two: a revolutionary faction in Utrecht and a pro-stadtholder faction in nearby Amersfoort. The result in Utrecht was crucial to the Dutch Republic as a whole. Three provinces – Holland, Overijssel and Groningen – were dedicated to the revolution, while a majority in three others – Gelderland, Friesland and Zeeland – remained loyal to the stadtholder. Utrecht was divided.

Soon after power had been seized, the chapter of the Bailiwick of Utrecht met in the revolutionary city. In the only reference to the revolution contained in their resolutions, Prince Charles Louis of Anhalt-Bernburg-Schaumburg-Hoym (1723–1806), who had recently assumed the position of land commander, noted that, 'owing to the current situation, seven Lord Commanders were not present'.[9]

Politically, the chapter meeting took no position, although most of its members were inclined towards the stadtholder. As well as the members referred to above – van Welderen and Taets van Amerongen – four other commanders were confidants of the stadtholder: Arend Sloet tot Tweenijenhuizen (1722–86), Count Sigismund Vincent Lodewijk Gustaaf van Heiden-Hompesch (1731–90), Baron Arend van Raesfelt (1725–1807), and Baron Jacob Derk van Heeckeren van Kell (1730–95). Land Commander Van Anhalt even had family connections with William V, although he was no longer active in the Dutch Republic: after a career in the Dutch army, he had returned to Germany to succeed his father as prince of the family's mini-state. The pro-stadtholder majority in the chapter was not opposed by any real supporters of the revolution, although four members from Gelderland were undecided or, if anything, slightly inclined towards the Patriots.

Dutch politics were in stalemate. The irresolute stadtholder was unable to restore order. After consultations with the British ambassador, the stadtholder's wife, Wilhelmina of Prussia, wanted to lead a popular uprising in The Hague, but her way there was barred by a Free Corps. She was sent back to Nijmegen where the stadtholder's family was residing. She asked her brother, Frederick William II of Prussia, to avenge this disgrace. His ultimatum to the States of Holland was followed by feverish diplomacy. When it became apparent that Prussia and the Dutch stadtholder had the unconditional support of Great Britain, and that France was unable to come to the aid of the Patriots, Frederick William decided on

---

[9] 'door de omstandigheden des tijds zeven Heren Commanders niet present geweest zijn'. ARDOU-OA 11, Resolutiën van de landcommanderij van Utrecht (1561–1827), deel 3, fol. 212v.

armed action. In September 1787, he brought the revolution to an end. Thousands of Patriots fled abroad. Among them were two sons of a commander who had belonged to the group of doubters identified above and who, unlike their father, had declared their support for the revolution.

The Shadow of the Bastille

Most of the refugees eventually ended up in France, where, not long afterwards, the next phase of the transatlantic democratic revolution was to begin. After financial problems had caused Louis XVI to convoke the States General, the pace of events accelerated until they reached their dramatic climax in the storming of the Bastille on 14 July 1789.

The decisions to abolish feudal rights and to confiscate church property were of enormous importance. The latter was to prove a mortal threat to the military orders. First to be affected was the property of the Teutonic Order on French soil, in Alsace. Next came the Order of the Knights of St John in France. Then, when war was declared on Austria and Prussia, and the radicalizing revolution thus began to be exported, the property of the Teutonic Order in the Rhineland and the Austrian Netherlands was threatened. The definitive conquest of these regions in the second half of 1794 led to the confiscation of property of five bailiwicks: Alden Biesen, Alsace-Burgundy, Lorraine, Coblenz and Franconia. Grand Master Maximilian Francis, who was also archbishop of Cologne, had to flee.[10]

In the meantime, the French had declared war on the British king and the Dutch stadtholder, whose troops in the Austrian Netherlands initially stood firm. In the autumn of 1794, however, the French moved into the territory of the Dutch Republic. At first the region's great rivers seemed to present a barrier, but heavy frosts soon enabled the French troops to advance unhindered. Resistance collapsed quickly. On 18 January 1795, William V fled to England, followed by a small number of his supporters, including Johan Walraat van Welderen. Another member of the Utrecht chapter, Jan Carel van der Borch, was William's envoy in Sweden, where he sought asylum. Robespierre's reign of terror had ended only six months earlier, and the refugees feared they would suffer the fate of their French counterparts.

Some of those proclaiming the Batavian Revolution did indeed advocate severe measures against former rulers. But while the latter lost their positions in government, more moderate revolutionaries, supported by the French authorities, ensured that they lost neither their lives nor their freedom. In fact, the sole focus of revolutionary fury was the dead: in Utrecht, Leeuwarden and elsewhere, the graves

---

[10]   K. Oldenhagen, *Kurfürst-Erzherzog Maximilian Franz als Hoch- und Deutschmeister (1780–1801)* (Bad Godesberg, 1969).

of aristocrats were despoiled.[11] Among the bones that rolled through the churches were those of earlier land commanders of the Teutonic Order and their kin.

The authorities responded to plunderers with severity. The primary concerns of the French were to receive contributions to the cost of the war and to the care of their troops; the primary concern of the Batavian rulers was to build a new democratic government based on the principles of the revolution.[12] The remaining nobles and patricians kept a low profile; if those of them who were sympathetic to the revolution participated in the elected bodies, they did so only as individuals. As a class, the aristocracy was finished: governing bodies composed on the basis of class – such as the States of Utrecht, the sovereign of the Teutonic Order – were replaced by elected revolutionary bodies such as the Representatives of the Land of Utrecht.[13]

The Teutonic Order, too, lay low. The chapter no longer met; everything was left to the secretary and the steward-general. But at the end of the first year of the revolution, Utrecht's provincial government started to take an interest in it. On 19 December 1795, the government discussed a request by the province of Gelderland to 'examine the affairs of the so-called Teutonic Order'.[14]

Overall Utrecht's administration had little desire for the consultation requested by their counterparts in Gelderland. In this particular case, however, they took action. On 2 February 1796 they summoned the Order's secretary, Willem Jacob van Nes, and presented him with an order forbidding him to make further payments without authorization. Van Nes stated that, as it was not permitted to send letters abroad, where the majority of the chapter's most important members resided, he had been unable to consult the commanders. He was also of the opinion that this ban 'in no way applied to the Teutonic Order'.[15] He asked respectfully how he was to proceed, and was ordered to halt payments until a committee had advised on the matter.

Its answer came within ten days. The committee's members had examined 'whether and to what extent the Representatives can and may concern themselves with the management of private business',[16] and it was decided that the Bailiwick

---

[11]   R.E. de Bruin, *Burgers op het kussen. Volkssoevereiniteit en bestuurssamenstelling in de stad Utrecht 1795–1813* (Zutphen, 1986), p. 46; J. Kuiper, *Een revolutie ontrafeld. Politiek in Friesland 1795–1798* (Franeker, 2002), pp. 78–9.

[12]   Schama, *Patriots and Liberators*, p. 199.

[13]   R.E. de Bruin, *We and our Successors shall do Justice by All. Provincial Government in Utrecht: A Historical Perspective* (Utrecht, 2003), pp. 102–3.

[14]   'door de omstandigheden des tijds zeven Heren Commanders niet present geweest zijn': Het Utrechts Archief (HUA) 233, Staten van Utrecht 1581–1810, 1071–7, Lappen-notulen van Gedeputeerden van de Provisionele Representanten en de Representanten, 19-12-1795, bijlage 2 (brief 16-12-1795).

[15]   'de Duitsche Order geenzints specteerde': HUA 233, Staten van Utrecht 1581–1810, 1071–8, 2-2-1796, fol. 1.

[16]   'of en in hoeverre de Representanten zich met de domestieke huyshoudinge derzelve mogen en kunnen inlaten': ibid.

of Utrecht was a private organization whose properties had been acquired through purchase and donations, and that the province of Utrecht was not authorized to intervene in its affairs. In the context of the revolution, this was a crucial judgement. Private property was held in deep respect, and the position of the Order had now been considerably strengthened.

The recognition of a body that administered private assets could have several consequences, including fiscal ones. Due especially to the financial demands imposed by the French, the Batavian Republic was perpetually short of funds; to meet its shortfalls, it decided to levy a wealth tax.[17] While individuals were to be assessed for this, organizations such as orphanages and charitable hospitals, and, potentially, the Order were not – a state of affairs that some people thought unjust. In January 1797, a member of the newly elected Provincial Government asked whether such institutions should not also contribute. While no action was taken, the payment by the Bailiwick of Utrecht of a total of 12,000 guilders to the provinces of Utrecht and Gelderland nonetheless resembled some kind of settlement.

The Order would have had little difficulty in paying such a sum. In the years after 1795, steward-general Unico Willem Teutonicus Cazius (1766–1832), who had succeeded his father in 1789, pursued the same policies that had been successful in preceding years. Between 1795 and 1800, annual incomes rose from 45,117 to 48,090 guilders, while expenditure rose from 37,352 to only 38,504 guilders.[18] As well as the stipends, whose payment continued after some hesitation in early 1796, expenditures included land taxes and the maintenance of real estate. These applied not only to the farms under lease, but also to the *Duitse Huis* or 'Teutonic House'.

Living in the latter was not only the steward-general, but also *jonkheer* (esquire) Volkier Rudolf Bentinck van Schoonheten (1738–1820), who had asked for permission in 1792 to use one of its rooms as a *pied-à-terre*. Protection against any plans for the house's confiscation was provided by the fact that it was not only well maintained but also in use as a private residence. It is true that the Order had to agree to French troops being quartered there, but that was nothing unusual: the same applied to many other houses in Utrecht.

Back in the Open

Thanks to its low-profile policy, the Bailiwick of Utrecht did not even come to harm during the radical phase of the Batavian Revolution, which started in January 1798. After a *coup d'état*, the Batavia Republic was converted into a unitary state after the French model in which the historic provinces were replaced with departments

---

17   S. Schama, 'The Exigencies of War and the Politics of Taxation in the Netherlands, 1795–1810', in J.M. Winter (ed.), *War and Economic Development: Essays in Memory of David Joslin* (Cambridge, 1975), pp. 103–29.

18   ARDOU-OA 342 (1795–1800).

named after rivers. With this, the political environment in which the Teutonic Order was rooted disappeared. Quietly, the Order's members awaited better times.

These duly arrived. In 1801, after two more *coups d'état*, the revolution was partially reversed. The old provinces were reinstated, officially as 'departments', and supporters of the banished stadtholder were allowed to resume political activities. The Knights of the Teutonic Order also risked a new meeting. Together with the two *jonkheren* who had been appointed during the last meeting in 1791, the four surviving commanders were invited by Land Commander Charles Louis van Anhalt to convene in Utrecht in a meeting of the chapter on 16 August 1802. Two of the commanders did not attend – the coadjutor, Johan Walraat van Welderen, who was living in London, and the 83-year-old commander of Maasland, Wolter Godefried van Neukirchen genaamd Nyvemheim –, but the other two did: Baron Herman Willem Jan van Lynden and Baron Arend van Raesfelt. So, too, did the two *jonkheren* (Baron Volkier Rudolph Bentinck and Baron Gijsbert Jan van Pallandt), as well as the land commander himself.

To fill the vacancies in the chapter that had been created by decease, the two *jonkheren* were raised to commander. Four additional commanders were appointed, as well as two new *jonkheren*. Two of the former had served the Batavian Revolution, Baron Jan Arend de Vos van Steenwijk as a parliamentarian, and Baron Frederik Gijsbert van Dedem van de Gelder as ambassador in Constantinople. In sharp contrast, one other new member, Baron August Robbert van Heeckeren van Suyderas, had fought against the revolution while in exile, and, in 1799, had even led an unsuccessful invasion into the eastern Netherlands, when a British and Russian army threatened the Batavian Republic in the north-west. The nomination of another 'émigré', Baron Frederik Christiaan van Reede van Athlone, a member of the British House of Lords, was barred on the grounds of an appeal based on insufficient proof of ancestry.

With regard to financial matters, the meeting of the chapter in August 1802 had to deal with a considerable backlog of paperwork that had accumulated over the previous years. These were offset by considerable surpluses – good news after such worrying times. The chapter expressed 'the satisfaction of those present' ('het genoegen van dese vergadering') with the steward 'for his conduct, for the lengths to which he went to preserve this Order and to maintain its rights and interests in the recent difficult times, and for his faithful administration of its property'.[19] He was given a fine gift: a solid gold snuffbox worth 400 guilders. The secretary was offered the same, but asked to be given money instead.

As it was Bentinck who had to implement the decisions of the chapter, there can be little doubt that he was its key figure. In this sense it was therefore convenient that he still lived at the Duitse Huis. The building now assumed a certain permanence; Bentinck was even exempted from paying rent for it (which

---

[19] 'wegens zijn gedrag, en aangewende pogingen, tot conservatie van dese order, en het handhaven van deszelfs rechten en belangens in de laadst verlopen moeijelijke tijden, alsmede wegens zijne getrouwe administratie der goederen': ibid., 11-4, fol. 24.

in the past he had not in fact paid anyway). This formal exemption was given in recognition of 'the many services and wise precautions taken in recent years by the Most Noble Lord V.R. Bentinck for the general preservation of this Order, and particularly for the Duitse Huis; these provided protection against misfortunes whose consequences might have been of the utmost seriousness'.[20] As we have seen, it was Bentinck's residence of the Duitse Huis that had helped the Bailiwick of Utrecht survive the Batavian Revolution unscathed.

The chapter meeting of August 1802 also resolved a sensitive political problem. Before the revolution there had been intimate links with the House of Orange, which the revolution had then compromised. As part of the Bailiwick's low-profile policy, no mention had been made of these links. After 1801, a history of sympathy to the royal house was no longer a bar to political office, but, in political terms, the position of the royalty remained problematic: the state could offer no place to an exiled stadtholder. The fact that powers such as Great Britain and the Prussians continued to insist on his return hung like the sword of Damocles over the compromise of 1801.

William V freed his supporters from this predicament by granting them permission to participate in the new establishment. In this way he also benefited the Bailiwick of Utrecht. Land Commander van Anhalt announced that the Prince of Orange 'had requested him that no further cognizance of his Person should be taken with regard to the Teutonic Order of Bailiwicks in Utrecht'.[21] The members of the chapter accepted his help with gratitude: it freed them from accusations of disloyalty for failing to cold-shoulder the Batavian government – and, in its shadow, the French authorities.

Less than a year later, the Order's policy of maintaining a low-profile in revolutionary times was articulated in explicit terms by coadjutor Johan Walraat van Welderen, who still resided in London. It came in his response to a proposal for reunion with the grand master. Many years earlier, van Welderen had already advocated this. In 1775 and again in 1791 he had even initiated concrete action to this effect, but the disasters that had befallen the country since 1795 had caused dreams of reunion to evaporate. It had now become necessary to make the best of matters; he pointed out that steward-general Cazius had 'avec tout de prudence et de succès' been able to manage the Order without attracting the attention of the authorities, 'to some extent hiding us from Government scrutiny, in order to save

---

[20]   'de menigvuldige diensten en goede voorzorg, die door den Hoog Wel Geb. Heer V.R. Bentinck, zo tot conservatie van deze Order in het gemeen, als ook met betrekking tot het Duitschen Huijs in het biezonder gedurende de laatst verlopen jaren heeft genomen waardoor veele onaengenaamheden zijn verhoed, die van de nadeeligste gevolgen hadden kunnen zijn': ibid., 11-4, fol. 39.

[21]   'hem verzogt hadde, dat geene verdere notitie van zijn Persoon concerneerende de Duitsche Orden der Balije binnen Utrecht mogte genomen worden': ibid., 11-4, fol. 38.

us through the bushes, as one says'.[22] The best way of guaranteeing the survival of the Bailiwick, van Welderen maintained, was by continuing this policy. The idea of not doing so perturbed him greatly. In the end, however, nothing came of this last attempt to heal the 1640 rupture.

Conclusion

Van Welderen was right to observe that the best strategy for survival was by not attracting the attention of the revolutionary regimes. The philosophy of the revolution was diametrically opposed to that of the military orders, which was deeply rooted in the Ancien Régime and even in its most reprehensible feudal and ecclesiastical aspects.

As we have seen above, the French confiscated the property of the Teutonic Order and the Order of the Knights of St John, not only in France, but also in the lands they occupied between 1792 and 1795. Napoleon Bonaparte, who had determined the totality of French policy since 1799, bore a particular grudge against old institutions, which he was therefore glad to humiliate, strip of their assets, or even disband, two examples being the Venetian Republic and the Papal State. When dealing with Bonaparte, clumsiness could be fatal, as the Knights of Malta were to discover after the impulsive German Ferdinand von Hompesch had succeeded the circumspect Frenchman Emmanuel de Rohan in 1797. Bonaparte's conquest of the island took place in the run-up to his attack on Egypt. The policies of von Hompesch played into his hands, and in June 1798 he brought an end to the Hospitaller State in Malta.

Maximilian Francis and Charles of Austria, the grand masters of the Teutonic Order, were more careful, and managed to prevent their organization from being disbanded. With French support, the German monarchs sought to take church property – including that of the Order – in compensation for their losses on the left bank of the Rhine. However, when the matter of compensation was settled in 1803, the Teutonic Order came to no harm. Instead, the end came six years later in a brief order of the day from Napoleon.[23] In 1811, shortly after annexing the Netherlands, the emperor also abolished the Bailiwick of Utrecht. As the bailiwick had been able to survive until then, William I, the new Dutch king, was to reverse this decision when independence was re-established. Although this period is outside the scope of this paper, the survival strategies I have described had proved essential to the continued existence of the Teutonic Order of the Bailiwick of Utrecht.

---

[22]   'en nous dérobant en quelques sorte aux regards du Gouvernement, pour nous sauver à travers du brousailly, comme on dit': ibid., 131, stuk 65.

[23]   F. Täubl, *Der Deutsche Orden im Zeitalter Napoleons*, QuStDO 4 (Bonn, 1966), pp. 171–7.

PART 6
The Iberian Peninsula

Chapter 31

Troubles and Tensions before the Trial: The Last Years of the Castilian Templar Province*

Philippe Josserand

Et axi seria molt profitosa la vostra entrada de Castella e de Portugal, segons que ha fet saber per lurs letres alcuns frares daqueles partides.[1]

Thus wrote the Catalan Templar, Pere de Sant Just, in a letter to the grand master from Miravet dated 23 May 1307. He had probably gone there to visit his relative, Berenguer de Sant Just, who was now commander having previously, in 1292, been grand commander in Cyprus.[2] Like his kinsman, Pere de Sant Just – who was then commander of Alfambra – enjoyed close links with the Order's headquarters in the East,[3] where he had recently resided,[4] and he was trading on this association in order to persuade Jacques de Molay, who had arrived in the West at the end of 1306,[5] to take the opportunity to come in person to the Iberian Peninsula.[6] The affairs of the two Templar provinces, Aragon-Catalonia and Castile-Portugal,[7] were causing concern, and the master's involvement was clearly regarded as advantageous and conducive to finding solutions. That, at least, was the opinion

*   I am very grateful to my colleague Diana Maloyan for correcting my English text.

[1]   H. Finke, *Papsttum und Untergang des Templerordens* (Münster, 1907), 2, p. 37, no. 24.

[2]   A. Demurger, *Les Templiers. Une chevalerie chrétienne au Moyen Âge* (Paris, 2005), p. 162.

[3]   A. Forey, 'The Career of a Templar: Peter of St Just', in N. Housley (ed.), *Knighthoods of Christ: Essays on the History of the Crusades and the Knights, Presented to Malcolm Barber* (Aldershot, 2007), p. 188.

[4]   A. Forey, 'Letters of the Last Two Templar Masters', *Nottingham Medieval Studies*, 45 (2001), 166, no. 13.

[5]   Demurger, *Les Templiers*, p. 424.

[6]   A. Forey, *The Templars in the Corona de Aragón* (London, 1973), p. 383, n. 184; A. Demurger, *Jacques de Molay. Le crépuscule des Templiers* (Paris, 2002), p. 214.

[7]   At this time Castile and Portugal seem to have been separate. See J. Valente, 'Soldiers and Settlers: The Knights Templar in Portugal, 1128–1139', unpublished dissertation, Santa Barbara, University of California, 2002. This deserves further investigation.

of Pere de Sant Just and those Castilian and Portuguese brethren whose letters he had to hand.

But although that is what the letter says, it may nevertheless come as a surprise because in neither Castile nor Portugal is the position of the Templars in the years immediately before the trial considered to have been problematic. In contrast to the situation in the Crown of Aragon, there is an absence of studies on this topic, and what scholarship there has been in recent years does not amount to much.[8] The Temple is the only great military order in the Castilian kingdom that has not benefited from a recent dissertation, and, although I have tried to give the topic some attention,[9] a specialized monograph would certainly not go amiss. At present, for the whole of the Crown of Castile-León, we have only the works of Gonzalo Martínez Díez,[10] which, despite their undoubted merits, are rather static in their approach to the problems and, as so often in the Iberian Peninsula, are concerned mainly with estates.[11] Thus a lot of unanswered – and unasked – questions still remain relating to the trial and the immediately preceding period.

The final years of the Templar Order in Castile and León before the arrests is usually considered to have been very tranquil. According to Martínez Díez, the period passed 'con toda normalidad'.[12] Yet this normality is a myth, and the years – indeed the months – following the death of King Sancho IV and the accession of his young son, Fernando IV, in the spring of 1295 saw the Order sink into a crisis that until now has gone unnoticed but was, to say the least, serious. The troubles and tensions then occurring among the Templars find echoes elsewhere. At the turn of the thirteenth and fourteenth centuries, Calatravans and Hospitallers had experienced destructive schisms as a result of recurrent although unsuccessful royal pressure.[13] Examining whether the Temple was a victim of similar forces would certainly be worthwhile. Some points can be developed, and, despite the difficulties that surround them, the sources reveal the extreme instability of the Order's Castilian province during the decade before the start of the trial. It is

---

[8]   C. de Ayala Martínez and C. Barquero Goñi, 'Historiografía hispánica y órdenes militares en la Edad Media (1993–2003)', *Medievalismo*, 12 (2002), 116–18.

[9]   P. Josserand, *Église et pouvoir dans la péninsule Ibérique. Les ordres militaires dans le royaume de Castille (1252–1369)* (Madrid, 2004); idem, 'Entre Orient et Occident: l'ordre du Temple dans le contexte castillan du règne d'Alphonse X', *Alcanate. Revista de estudios alfonsíes*, 2 (2000–2001), 131–50; idem, '*Et succurere Terre sancte pro posse*. Les Templiers castillans et la défense de l'Orient latin au tournant des XIIIe et XIVe siècles', *Cahiers de Recherches Médiévales. A Journal of Medieval Studies*, 15 (2008), 217–35.

[10]   G. Martínez Díez, *Los Templarios en la Corona de Castilla* (Burgos, 1993); idem, *Los Templarios en los reinos de España* (Barcelona, 2001).

[11]   P. Josserand, 'L'historiographie des ordres militaires dans les royaumes de Castille et de León. Bilan et perspectives de la recherche en histoire médiévale', *Atalaya. Revue française d'études médiévales hispaniques*, 9 (1998), 25–8.

[12]   Martínez Díez, *Los Templarios en la Corona de Castilla*, p. 59.

[13]   Josserand, *Église et pouvoir*, pp. 531–7.

usually said that authority was in the hands of Gonzalo Yáñez and then Rodrigo Yáñez;[14] in fact four, or perhaps five, men were competing for control.

The Sources: Difficulty and Scarcity

Materials for the later history of the Templar Order in Castile and León are relatively scarce and particularly difficult. As elsewhere in Christendom, the testimonies taken at the trial are the most obvious sources. Traditional Spanish scholarship, however, has not focused on the interrogation of the Templars, but has tended to concentrate on the verdict of not guilty delivered by the Council of Salamanca on 21 October 1310.[15] Only three inquiries are preserved.[16] All are in a poor state of preservation. Only one, the shortest, has been fully published, and that does not deal directly with the brethren but with four clerics from outside the Order.[17] The other two are only known in part. Like the first, the principal one was conducted in Medina del Campo on 27 April 1310; it deals with thirty-three witnesses, among whom were thirty Templars.[18] In 1996 Josep Maria Sans i Travé published an excellent edition of this text.[19] Unfortunately the roll in the Archivio Segreto Vaticano is seriously damaged, and an important part of the information is lost, particularly from the first ten testimonies,[20] although at the beginning of the nineteenth century François Just-Marie Raynouard did transcribe a few sentences.[21] At the same point, although the precise date is unknown, another enquiry was held in Orense. Here there were thirty-six witnesses of whom twenty-eight were brethren. Great confusion prevails even among the most distinguished

---

[14]   Martínez Díez (*Los Templarios en los reinos de España*, p. 111) even mixed up the two successive masters.

[15]   J.M. Sans i Travé, 'L'inedito processo dei Templari in Castiglia (Medina del Campo, 27 aprile 1310)', in F. Tommasi (ed.), *Acri 1291. La fine della presenza degli ordini militari in Terra santa e i nuovi orientamenti nel secolo XIV* (Perugia, 1996), p. 234.

[16]   M. Barber, *The Trial of the Templars* (London, 1978), p. 213, noted only two inquiries. Yet C.G. von Murr, 'Anhänge zur Geschichte der Tempelherren', in *Über den wahren Ursprung der Rosenkreuzer und des Freymaurerordens* (Sulzbach, 1803), pp. 107–60, referred to all three.

[17]   F. Fita i Colomé, *Actas inéditas de siete concilios españoles celebrados desde el año 1282 hasta el de 1314* (Madrid, 1882), pp. 90–99.

[18]   Martínez Díez, *Los Templarios en la Corona de Castilla*, pp. 225–9.

[19]   Sans i Travé, 'L'inedito processo', pp. 249–64.

[20]   Archivio Segreto Vaticano (ASV), Castel Sant'Angelo, armario D-220.

[21]   F.J-M. Raynouard, *Monumens historiques relatifs à la condamnation des chevaliers et à l'abolition de l'ordre du Temple* (Paris, 1813), pp. 264–5.

scholars,[22] and the manuscript itself appears to be lost.[23] As for its contents, we have no more information than Raynouard, and, like him, we are indebted to late eighteenth-century Masonic erudition, notably that of Friedrich Münter and Christoph Gottlieb von Murr.[24]

Evidence for the Templar trial remains scarce in Castile and León and awaits further investigation. It could be thought that this difficulty would be offset by relying on other Templar documents from this final period. This, alas, is impossible. The central archives of the Temple have disappeared, and even the date and place of their loss are uncertain.[25] Castilian provincial documentation has not been preserved either. Sometimes a provincial archive was housed in a particular commandery, such as Miravet in the case of Aragon.[26] It is conceivable that in the late thirteenth century a similar arrangement existed for the Castilian province at Zamora, but even if it had – and the preservation of the not guilty verdict at Sta María de la Horta until the mid-sixteenth century proves nothing[27] – nothing remains. In some countries Templar commanderies compiled cartularies in order to preserve their memory locally, but in Castile and León, unlike Provence or Aragon, documents of this kind, if they did exist, do not survive. To overcome these difficulties the researcher has to turn to surviving materials from the individuals or institutions with which the Order had relations. For the later history of the Castilian Templars, for which sources are especially scarce, the archives of the cathedral of Zamora and the Dominican convent of Benavente are particularly useful.[28] So too is the rich collection of Teresa

---

[22]   Barber, p. 213, and Demurger, *Les Templiers*, p. 458, both placed Orense in Portugal.

[23]   I am not sure whether, at the end of the eighteenth century, the document was conserved in the Archivio Segreto Vaticano despite the opinions of Martínez Díez, *Los Templarios en la Corona de Castilla*, p. 229, and Sans i Travé, 'L'inedito processo', p. 236.

[24]   Von Murr, pp. 140–45. For comprehensive surveys, E. Rosenstrauch-Königsberg, *Freimaurer, Illuminat, Weltbürger, Friedrich Münters Reisen und Briefe in ihren europäischen Bezügen* (Berlin, 1984); J. Wojtowicz, 'Die Templertraditionen in den Vorstellungen der Aufklärung', in Z. Nowak (ed.), *Die Ritterorden zwischen Geistlicher und Weltlicher Macht im Mittelalter* (Toruń, 1990), pp. 163–5.

[25]   R. Hiestand, 'Zum Problem des Templerzentralarchivs', *Archivalische Zeitschrift*, 76 (1980), 17–37.

[26]   A. Forey, 'Sources for the History of the Templars in Aragon, Catalonia, and Valencia', *Archives*, 21 (1994), 16–17; idem, 'Notes on the Templar Personnel and Government at the turn of the Thirteenth and Fourteenth Centuries', *Journal of Medieval History*, 35 (2009), 159.

[27]   Fita i Colomé, pp. 105–7; Martínez Díez, *Los Templarios en la Corona de Castilla*, pp. 243–4.

[28]   Archivo Catedral de Zamora (ACZ), leg. 13, no. 7 and leg. 1, no. 16. J. C. de Lera Maíllo, *Catálogo de los documentos medievales de la catedral de Zamora* (Zamora, 1999), pp. 286–8, nos. 897–9, 901–2 and pp. 331–2 nos. 1055–7; Archivo Histórico Nacional (AHN), Clero, carp. 3525, no. 1, published in R. Fernández Ruiz, *Colección diplomática del monasterio de Santo Domingo de Benavente (1228–1390)* (Benavente, 2000), p. 65, no. 27.

Gil, presented to the monastery of Sancti Spiritus de Toro, where in 1307 she asked to be buried,[29] for it reveals the deep conflict between her and the brethren that had been heard twelve years earlier.[30] In the wider sphere documentary discoveries are also possible in the great archives around the Mediterranean such as those in Rome, Naples and, above all, Barcelona, where the rich Templar collections occasionally refer to Castile and León.[31]

In view of the poverty of the archives, it might be expected that royal documentation would supply a greater quantity of information with a greater degree of continuity. The chancery practice for registration in the kingdom of Castile remains relatively unknown and, unlike Aragon or even Portugal, limited records survive from before the late Middle Ages.[32] However, from the mid-thirteenth century it had been customary to include the main dignitaries of the military orders in Castile among those who confirmed the most solemn royal charters.[33] These privileges are called *rodados* because they show a wheel (*rueda*), which is the signature of the king, and on each side are listed the main representatives of political society: on the left, in two columns, the Castilian nobles and clerics, with, below, the master of Calatrava and, from the accession of Sancho IV, the prior of the Hospital; on the right, in a similar arrangement, the masters of Santiago, Alcántara and, from 1255, the Temple, after the Leonese ecclesiastics. For a long time now historians of the military orders have made use of these privileges that are especially helpful in institutional studies.[34] As will be seen, in the case of the Templars there is more to be made of the information from these documents. In April 1299 the name of the Castilian provincial master ceases to appear on the *rodados*, never again to be included.[35] Martínez Díez has drawn attention to this

---

[29]   A. Rucquoi, 'Le testament de *doña* Teresa Gil', in *Femmes, mariages, lignages (XIIe–XIVe siècles). Mélanges offerts à Georges Duby* (Brussels, 1992), pp. 305–23.

[30]   Archivo del Monasterio de Sancti Spiritus de Toro (AMSS), Doc. Particulares 57. P. Galindo Romeo, 'Catálogo del archivo del monasterio de Sancti Spiritus de Toro', *Archivos Leoneses*, 30 (1976), p. 225, no. 57.

[31]   Josserand, *Église et pouvoir*, pp. 24–6.

[32]   M.J. Sanz Fuentes, 'La influencia de la cancillería pontificia en las cancillerías reales castellano-leonesas', in S. Domínguez Sánchez and K. Herbers (eds), *Roma y la Península Ibérica en la Alta Media. La construcción de espacios, normas y redes de relación* (León, 2009), pp. 81–90.

[33]   P. Josserand, 'Les ordres militaires et le service curial dans le royaume de Castille (1252–1369)', in *Les Serviteurs de l'État au Moyen Âge. Actes du XXIXe Congrès de la Société des Médiévistes de l'Enseignement Supérieur* (Paris, 1999), p. 77; idem, *Église et pouvoir*, p. 547.

[34]   P. Rodríguez Campomanes, *Dissertaciones históricas del orden y cavallería de los Templarios* (Madrid, 1747), pp. 33–6.

[35]   The last *rodado* naming the Templar master dates to 11 April 1299. E. González Díez, *Colección diplomática del concejo de Burgos (884–1369)* (Burgos, 1984), pp. 258–62, no. 159.

absence, but without offering any explanation.[36] This change is not simply a matter of diplomatics: it has historical significance as well. The break stems directly from the deep crisis that affected the final years of the government of Gonzalo Yáñez, the last Castilian provincial master to have been named in the *rodados* as the main representative of his Order.

Gonzalo Yáñez: A Troubled Mastership

Gonzalo Yáñez is considered to have been provincial master in Castile and León continuously during the ten years from 1289 to 1299, and his authority is also said to have extended to Portugal.[37] This last point is extremely doubtful,[38] and the first, as I shall show, is questionable as well. It is impossible to say anything about his social and family background or about his career in the Temple before his election as provincial master. This election may have occurred in the second half of 1289. On 13 December Gonzalo Yáñez was in office,[39] and he had succeeded Gómez García, last attested on 9 June.[40] In the autumn of 1289 the Templars had taken part in the repression of the Bejaranos who had roused Badajoz against Sancho IV in the name of the *infantes* de la Cerda, the sons of Fernando, the king's deceased elder brother.[41] Did the brethren act under the aegis of Gonzalo Yáñez or Gómez García? We do not know.[42] Be that as it may, Gonzalo Yáñez followed the political stance of Gómez García, who had always served Sancho IV, supporting him while still *infante* in 1282 in the revolt against Alfonso X. This had involved Gómez García in a long conflict with the provincial master, the Portuguese João Fernandes, who remained faithful to the old king.[43] A split in the Castilian Templar province had followed and was not resolved until 1285.[44] In the end, despite the

---

[36]   Martínez Díez, *Los Templarios en la Corona de Castilla*, pp. 57–8.

[37]   Ibid., pp. 56–7, 66.

[38]   Valente, pp. 193–6, 283.

[39]   F. Pino Rebolledo, *Catálogo de los pergaminos de la Edad Media (1191–1393)* (Valladolid, 1988), pp. 98–102, no. 20.

[40]   J. Torres Fontes, *Documentos de Sancho IV*, Colección de Documentos para la Historia de Murcia 4 (Murcia, 1977), pp. 78–80, no. 88.

[41]   M. Gaibrois de Ballesteros, *Historia del reinado de Sancho IV de Castilla* (Madrid, 1922), 2, pp. 8–15.

[42]   The *Crónica del rey don Sancho el Bravo*, in *Crónicas de los reyes de España desde Alfonso el Sabio hasta los Reyes Católicos*, ed. C. Rosell (Madrid, 1953), 1, p. 82, only refers to the 'maestre del Temple'.

[43]   Josserand, *Église et pouvoir*, pp. 505–6.

[44]   Ibid., pp. 506–7.

difficulties,[45] Gómez García won, and João Fernandes was forced into exile.[46] In these troubles nothing is known of the role of Gonzalo Yáñez, who was certainly already a Templar; it is likely that he supported Gómez García, whose policies he continued after his nomination. As provincial master his concern for the Order was real. For example, in the first months of his magistry he received a sergeant into the Order at Seville in presence of important brethren such as Gómez Guerra, Martín Martínez and Lope Conchado.[47] Nevertheless problems arose, and on two occasions the Temple was obliged to pay heavy damages, first, in 1291, to the cathedral of Zamora and then, in 1295, to Teresa Gil.[48] Did these events undermine the master's credibility? They may well have done so, especially when we take into consideration the many problems he had had to face after the death of Sancho IV and the accession of Fernando IV to the Castilian throne in April 1295.

The minority of Fernando IV was an extremely troubled period in Castile. Against the young prince, who, in the eyes of the Church, was illegitimate, various pretenders emerged: first his cousin, Alfonso de la Cerda, and then an uncle, Juan, who managed for a short time to be recognized in León.[49] These conflicts were further complicated by armed intervention from both Portugal and Aragon, but the military orders generally maintained their loyalty to Fernando IV: this stance, however, was not absolute and was motivated in part by self-interest.[50] Despite the poor documentation, it would appear that the Temple acted similarly. As early as 1295, a brother named Martín Martínez is said to have admitted the *infante* Juan into Coria.[51] He is surely the same man who five years previously had received a sergeant at Seville in association with Gonzalo Yáñez, with whom he was evidently close. One might therefore infer that the provincial master had wavered in his loyalty to Fernando IV. He disappeared from the *rodados* twice: on 4 June 1296 and 26 March 1297.[52] The first of these dates coincides with the attack of King Jaume II of Aragon whereupon the master of Santiago, Juan Osórez, had

---

[45] The main difficulty concerned the castle of Bullas. Even in victory, Gómez García did not succeed in being recognized as a master, and only the title of *comendador mayor* was given to him in the *rodados* until to the end of his government.

[46] P. Josserand, 'João Fernandes', in N. Bériou and P. Josserand (eds), *Prier et combattre. Dictionnaire européen des ordres militaires au Moyen Âge* (Paris, 2009), p. 504.

[47] Sans i Travé, 'L'inedito processo', p. 259.

[48] ACZ, leg. 1, no. 16; AMSS, Doc. Particulares 57.

[49] C. González Mínguez, *Fernando IV de Castilla (1295–1312). La guerra civil y el predominio de la nobleza* (Vitoria, 1976), pp. 31–54; idem, *Fernando IV (1295–1312)* (Palencia, 1995), pp. 25–36.

[50] Josserand, 'Les ordres militaires', p. 79; idem, *Église et pouvoir*, pp. 550–55.

[51] *Crónica del rey don Fernando IV*, in C. Rosell (ed.), *Crónicas de los reyes de España desde Alfonso el Sabio hasta los Reyes Católicos* (Madrid, 1953), 1, p. 95.

[52] J.L. Martín Martín, *Documentación medieval de la iglesia catedral de Coria* (Salamanca, 1989), pp. 74–5, no. 37; L. Rubio García, *Mayoría de edad de don Juan Manuel. Consolidación aragonesa en Murcia (1297–1302)* (Murcia, 2001), pp. 115–18, no. 1.

also wavered;[53] the second corresponds to the aftermath of the siege of Paredes de Nava when the royalists failed to overcome the noble opposition.[54] On these two occasions the loyalty of Gonzalo Yáñez may have faltered. In any case there were some problems facing the Castilian Templar province. We do not know precisely how serious they were; perhaps they were less important than those arising from the return of João Fernandes to Iberia. The former provincial master had been driven out in 1285 or shortly after, but he had persevered in his claims and ambitions first in Naples, where, together with King Charles II, he had committed himself in favour of the *infantes* de la Cerda,[55] and then in Rome as *cubicularius* of Pope Boniface VIII.[56] On 15 August 1296, at the general chapter held in Arles, the grand master Jacques de Molay made him a life grant of the commanderies of Faro, Canabal, Neira and Ceinos de Campos; a papal confirmation of this award was requested and secured on 27 November.[57] It is likely that João Fernandes used this grant as the basis for trying to recover his authority from Gonzalo Yáñez, and this ambition provoked a deep crisis within the Castilian Templar province that is, surprisingly, unknown to scholars.

The crisis that occurred in 1297–98 was certainly as serious as that which had arisen in 1282–83. As then, a schism broke out among the Castilian brethren. Apparently Gonzalo Yáñez remained the sole provincial master, and on 24 February 1298 his name appeared in a royal charter issued during the Cortes of Valladolid.[58] Nevertheless, at this point the Order, like the realm, was divided. João Fernandes had returned to rival Gonzalo Yáñez. No document accords him the title of master, but two testimonies given during the trial of the Templars from opposite ends of the Mediterranean refer to him as asserting provincial authority and receiving brethren. On 28 May 1310 in Cyprus, the knight Martín Martínez declared that, almost twelve years earlier, on 24 June 1298, he had been received by João Fernandes, 'tunc preceptor balivie de Castella'.[59] A lacuna in the document prevents us from identifying the location of this event: the opening letters 'Alc..' might be transcribed as 'Alcanadre' or 'Alconétar',[60] but, as Pierre-Vincent

---

[53]   González Mínguez, *Fernando IV* , p. 40.

[54]   Ibid., pp. 41–3.

[55]   *I Registri della cancilleria angioina*, ed. S. Palmieri (Naples, 1991), 38, p. 291, no. 868.

[56]   Josserand, *Église et pouvoir*, p. 507, n. 279.

[57]   ASV, Reg. Vat. 48, fols 143v–144. *Les registres de Boniface VIII*, eds G. Digard *et al.* (Paris, 1884–1939), 1, no. 1508.

[58]   González Díez, pp. 253–7, no. 157.

[59]   K. Schottmüller, *Die Untergang des Templerordens mit urkundlichen und kritischen Beitragen* (Berlin, 1887), 2, p. 212; A. Gilmour-Bryson, *The Trial of the Templars in Cyprus: A Complete English Edition* (Leiden, 1998), p. 145.

[60]   Schottmüller, 2, p. 212 n. 3: 'Kann sich sowohl auf Alcanadre, wie Alconitar beziehen'; Gilmour-Bryson, p. 145, n. 373.

Claverie has recently explained, they denote Alcañices in León.[61] Unfortunately this author failed to realize that it was from this frontier castle, where the year before an important treaty had been signed between Castile and Portugal,[62] that João Fernandes was attempting to control a much larger area.[63] This report is confirmed by the other testimony I have referred to: on 27 April 1310, in Medina del Campo, the sergeant Pedro attested that 'receptus fuit apud Alcaniças, XII anni sunt elapsi, per fratrem dominum Iohannem Fernandi, tunc illius bayluie comendatorem'.[64] Confronting each other, João Fernandes and Gonzalo Yáñez could each claim to be legitimate. The impasse was obvious. That is most likely why a third person was invested as provincial master: first, possibly, Rodrigo Rodríguez, and then Gómez Pérez. The former is said to have been master in a single statement during the trial.[65] The latter was in office on 14 May 1298, as is revealed in an exchange between the Order and the Dominican convent of Benavente.[66] The document in question, which would really need a separate study, refers to Gómez Pérez as 'humilloso maestre de lo que ha la ordem del Tenple en Leom e en Castiela' and corroborates a piece of information from the parish archives of Villalpando.[67] In spite of his title, he never exercised authority over the whole province and in all likelihood had to resign because of the firm opposition of João Fernandes to whom he had once been close.[68] The compromise solution that Gómez Pérez represented had failed. Gonzalo Yáñez was reinvested, and he is named as master in April 1299 before disappearing from the sources.[69] Did he die? Perhaps, but it is doubtful, and he could have resigned again.[70] Be that as it

---

[61]   P-V. Claverie, *L'ordre du Temple en Terre sainte et à Chypre au XIIIe siècle* (Nicosia, 2005), 1, p. 209.

[62]   M. González Jiménez, 'Las relaciones entre Portugal y Castilla durante el siglo XIII', and M.Á. Ladero Quesada, 'Reconquista y definiciones de frontera', in L.A. da Fonseca (ed.), *As relações de fronteira no século de Alcanices. IV Jornadas Luso-Espanholas de História Medieval* (Porto, 1998), 1, pp. 1–24, 655–91.

[63]   Contrary to Claverie's opinion (1, p. 209), there are no grounds for believing that João Fernandes was wrongly designated commander of Castile. On the contrary, the title reveals his ambition.

[64]   Sans i Travé, 'L'inedito processo', p. 260.

[65]   Fita i Colomé, p. 96.

[66]   AHN, Clero, carp. 3525, no. 1.

[67]   Á. Vaca Lorenzo, *Documentación medieval del Archivo parroquial de Villalpando (Zamora)* (Salamanca, 1988), pp. 34–6, no. 18.

[68]   Sans i Travé, 'L'inedito processo', p. 260.

[69]   Present in the *rodados* on 2 April (J.A. Fernández Flórez, *Colección documental del monasterio de Sahagún (1200–1300)* (León, 1994), pp. 574–6, no. 1894) and on 11 April (González Díez, pp. 258–62, no. 159), Gonzalo Yáñez was absent on 15 April (ed. A. Floriano Cumbreño, *Documentación histórica del Archivo Municipal de Cáceres (1229–1471)* (Cáceres, 1987), pp. 29–31, no. 17).

[70]   Gonzalo Yáñez might still have been alive during the trial. He or his namesake was described as old and sick. Fita i Colomé, p. 89.

may, the crisis had seriously affected Gonzalo Yáñez's magistry, and at the end of the thirteenth century the situation of the Castilian Templar province was a cause for concern.

Rodrigo Yáñez: A Failed Solution

Under these circumstances, the Templar central authorities had to act. Even if Spanish scholars pass over the question,[71] the central convent's interventions in the Castilian province had increased since the second quarter of the thirteenth century.[72] In the time of Jacques de Molay they were well to the fore.[73] The first necessity was to nominate a new provincial master who would be both able and undisputed. Rodrigo Yáñez was their man. His position, poorly documented as it is, is fascinating. Alongside my own treatment of his career,[74] a Spanish writer, Jesús Fuentes Pastor, has provided him with some apocryphal memoirs.[75] According to this author, Rodrigo Yáñez is supposed to have been born in Zamora around 1257 into a family of the middling nobility.[76] This is not impossible, but absolutely unprovable. Indeed, information concerning the brother before his election as provincial master is scarce. He may have served in the East, for in 1281 a certain Rodrigo Yáñez assisted at the reception as sergeant of Tomás de Pamplona by the Castilian commander of Tripoli, Rui de Cuero.[77] If this were so, it would have been a significant source of prestige and an important factor in his later promotion.[78] A doubt, however, remains. During the trial a secular priest named Juan Fernández, who is the witness who attested Rodrigo Rodríguez as provincial master, had stated that Rodrigo Yáñez was received 'tempore suo', in other words after he himself had entered the Order in 1289–90.[79] If this testimony is reliable, he would only have become a Templar in the 1290s. Be that as it may, Rodrigo Yáñez had enjoyed genuine prestige among his brethren. Before the spring of 1298, he was in Zamora,

---

[71]   Martínez Díez, *Los Templarios en la Corona de Castilla*, pp. 61–2; C. de Ayala Martínez, 'Fernando III y las órdenes militares', in *Fernando III y su tiempo. VIII Congreso de Estudios Medievales* (Ávila, 2003), pp. 83–4.

[72]   Josserand, '*Et succurere Terre sancte*', pp. 225–7.

[73]   As early as the magistry of Thomas Bérard, the *fueros* granted to Castelo Branco in 1271 and to Valencia del Ventoso the following year were granted with the convent's consent. J.A. de Figueiredo, *Nova história da militar ordem de Malta é dos senhores grão-priores della em Portugal* (Lisbon, 1800), 2, p. 256; Rodríguez Campomanes, pp. 30–31.

[74]   Josserand, 'Rodrigo Yáñez', in Bériou and Josserand (eds), p. 801.

[75]   J. Fuentes Pastor, *Las memorias de Rodrigo Yáñez, último maestre del Temple* (Madrid, 2005).

[76]   Ibid., p. 17.

[77]   *Le Procès des Templiers*, ed. J. Michelet (Paris, 1841–51), 2, p. 16.

[78]   Josserand, '*Et succurere Terre sancte*', p. 224, n. 41.

[79]   Fita i Colomé, pp. 95–6.

probably the main house of the Order in Castile and León,[80] and there he received a knight named Juan Rodríguez in the presence of several important dignitaries.[81] At that time he was most likely commander, and shortly afterwards he was appointed provincial master. When did that occur? The date of 8 September 1299 is a mere supposition.[82] Once again the testimonies from the trial provide the only reliable evidence. On 27 April 1310 in Medina del Campo, Rodrigo Yáñez's chaplain, a secular priest by the name of Juan Yáñez, declared that he had served him for over nine years;[83] at the same time, the knight Rodrigo Fernández testified that he was received in Zamora by the provincial master before the spring of 1301.[84] So Rodrigo Yáñez had most likely been appointed in 1300 or very shortly before.[85]

In the Castilian province, where he was supposed to turn things round, Rodrigo Yáñez may have faced immediate and strong opposition. We do not know what it comprised, but the difficulties were sufficiently serious to bring about the end of his appointment. In April 1301, according to a letter he wrote to Pere de Sant Just on the way back from Cyprus, Berenguer de Cardona, the master of Aragon and visitor of Spain, had appointed Pere de Tous, an important Catalan brother,[86] as 'loch tinent de comanador de Castela'.[87] One year later, in May 1302, Berenguer de Cardona came in person to Castile:[88] from another letter he had written, we learn that he met Fernando IV, who was holding the Cortes in Medina del Campo, and was preparing a provincial chapter in Zamora 'per endreçar les cases e los logars del Temple, los quals, si a Deu que plau, tornaran en estament de be'.[89] In neither of these letters is the name Rodrigo Yáñez mentioned. In 1302 the presence of the Temple provincial master is attested at the Cortes of Medina del Campo,[90] but, contrary to the opinion of Martínez Díez,[91] it is most likely that the dignitary in question was another brother, even surely a competitor. By then Rodrigo Yáñez had withdrawn to the southern locality of Jerez de los Caballeros. This is apparent from the testimony of the sergeant Juan Matias given during the trial: on 27 April 1310, he declared that he was received there by 'frater Rodericus Iohannis, preceptor maior, tunc commendator d'Exeres, in dicta villa, Pacensis diocesis, VIII anni sunt elapsi'.[92] Clearly, at that time, before the spring of 1302, Rodrigo Yáñez was not acting as provincial master. What had happened? Nothing

---

[80] Josserand, *Église et pouvoir*, p. 576.
[81] Sans i Travé, 'L'inedito processo', p. 251.
[82] Fuentes Pastor, pp. 189–90.
[83] Sans i Travé, 'L'inedito processo', p. 263.
[84] Ibid., p. 252.
[85] Josserand, 'Rodrigo Yáñez', p. 801.
[86] J.M. Sans i Travé, *La fi dels Templers catalans* (Lleida, 2008), p. 260.
[87] Archivo de la Corona de Aragón (ACA), CRD, Templarios, no. 181.
[88] Josserand, '*Et succurere Terre sancte*', p. 227, n. 59.
[89] ACA, CRD, Templarios, no. 322.
[90] *Cortes de los antiguos reinos de León y Castilla* (Madrid, 1860), 2, p. 162.
[91] Martínez Díez, *Los Templarios en la Corona de Castilla*, p. 58.
[92] Sans i Travé, 'L'inedito processo', p. 254.

is certain, but it was probably royal authority that kept Rodrigo Yáñez out. This is also what happened in respect of Calatrava.[93] An anti-master emerged in the Temple as well. At the moment he is impossible to identify, but he may have been Gómez Pérez, who is known to have benefited from a large and peculiar degree of autonomy at the time of the trial,[94] or even João Fernandes, if indeed he was still alive. That Rodrigo Yáñez was deprived of the Templar Castilian magistry is, on the other hand, absolutely certain, and, after more than a year, it was only the authority of Berenguer de Cardona that allowed his complete restoration.

After that Rodrigo Yáñez did not cease his efforts to reform and to improve the state of the Templar province he had to govern. The situation was worrying enough. Indeed, the letter of May 1307 from Pere de Sant Just to Jacques de Molay quoted at the beginning does not say anything new. Six years earlier, the letter he had received from Berenguer de Cardona was similar in tone.[95] Faced with such problems, Rodrigo Yáñez seems to have reacted vigorously, and he took action on various fronts. He developed a noteworthy interest in the reception of brethren: each year from 1304 to 1307 he received at least one new member into the Order.[96] He was also preoccupied with issues to do with estates. For example, in the kingdom of Murcia, which had been long contested between Castile and Aragon, he attempted to recover and strengthen the Templar houses following the treaty of Torrellas of August 1304 which he had been called upon to attest in person.[97] That is why he granted Cehegín a municipal charter (*fuero*) in accordance with the decision of the provincial chapter held in Zamora in May 1307.[98] Rodrigo Yáñez did not neglect the eastern Mediterranean either, and he certainly sailed to Cyprus in 1305.[99] In May 1310, in the context of the trial, the sergeant Bertran de Brandio declared that he had been received in Marseilles some five years earlier, on 15 August 1305, by 'quidam, qui erat preceptor in Spania et in Castello, cuius nomen non cognovit'.[100] The title is rather confused, but it refers to a Castilian provincial master, who in my opinion was Rodrigo Yáñez, although we cannot entirely exclude the possibility that João Fernandes, if alive, was the unnamed dignitary.[101] If I am not mistaken, Rodrigo Yáñez's visit to Cyprus allowed him to reinforce his links with the Templar central convent. In order to reform the Castilian province,

[93]   Josserand, *Église et pouvoir*, pp. 531–4.
[94]   Fita i Colomé, pp. 84–6; Martínez Díez, *Los Templarios en la Corona de Castilla*, pp. 218–19.
[95]   ACA, CRD, Templarios, no. 181.
[96]   Sans i Travé, 'L'inedito processo', pp. 253, 256–7, 260.
[97]   A. Benavides, *Memorias de don Fernando IV de Castilla* (Madrid, 1860), 2, p. 416.
[98]   B. de Chaves, *Apuntamiento legal sobre el dominio solar que por expressas reales donaciones pertenece a la orden de Santiago en todos sus pueblos* (Madrid, 1740), fol. 47v.
[99]   Josserand, 'Entre Orient et Occident', p. 139; idem, *Église et pouvoir*, p. 600.
[100]   Schottmüller, 2, p. 208; Gilmour-Bryson, p. 138, n. 337 (correcting 'Brandisio' to 'Brandio').
[101]   The presence of Martín Martínez who João Fernandes received into the Templar Order might be an argument in that sense.

that was certainly necessary but not sufficient in itself. Since the reign of Alfonso X, the monarchy was prepared to exert influence on the military orders,[102] and the Temple was as much affected as any of the other orders.[103] The relations of Rodrigo Yáñez with the monarchy were therefore crucial for his success. The difficulties that he had experienced were a major problem. As I have said, Rodrigo Yáñez never appeared in the *rodados*, and that would suggest he never enjoyed full royal recognition. Under these circumstances the solution for the Castilian Templars as personified by Rodrigo Yáñez had no chance of success.

In the Castilian Templar province, there was no peace and quiet at the turn of the thirteenth and fourteenth centuries. In terms of institutional control, the weaknesses were obvious, and in the other areas of the Order's life, despite the silence of the sources, the problems would have been real too. In conclusion I should like to point out two features that certainly deserve some attention and might improve our understanding of the Castilian trial: the first is the curious autonomy of Gómez Pérez in relation to Rodrigo Yáñez;[104] the second, the title of 'preceptor mayor' given to the latter in the main enquiry, which never uses 'magister provincialis'.[105] Was that simply the choice of the scribe? It might have been, but that is doubtful. The two terms, in my opinion, then conveyed very different meanings,[106] and one might see in these features further evidence of the Temple's fragility in the Iberian West just before the trial, just as for Portugal the *Livro das Lezírias d'el-rei dom Dinis*, even if rarely utilized, also provides indicators on that subject.[107]

---

[102]   C. de Ayala Martínez, 'La monarquía y las órdenes militares durante el reinado de Alfonso X', *Hispania*, 51 (1991), 409–65; idem, 'Las órdenes militares y los procesos de afirmación monárquica en Castilla y Portugal (1250–1350)', in da Fonseca (ed.), 2, pp. 1289–312.

[103]   Josserand, *Église et pouvoir*, pp. 459–647.

[104]   Fita i Colomé, pp. 86–9; Martínez Díez, *Los Templarios en la Corona de Castilla*, pp. 219–20.

[105]   Sans i Travé, 'L'inedito processo', pp. 249–64.

[106]   Josserand, 'Entre Orient et Occident', p. 149; idem, *Église et pouvoir*, p. 506–7, as against Martínez Díez, *Los Templarios en la Corona de Castilla*, p. 55, and Forey, 'Notes on Templar Personnel and Government', p. 160, who consider Gómez García as a *lugarteniente* or a *deputy* of João Fernandes.

[107]   *Livro das Lezírias d'el-rei dom Dinis*, ed. B. de Sá Nogueira (Lisbon, 2003).

Chapter 32

The Relationship between the Crown and the Monastery of Santos during the Middle Ages

Joel Silva Ferreira Mata

The monastery of Santos was founded at the end of the twelfth century and given to the Order of Santiago by King Sancho I. Later, as it gained lands further south from the Muslims, the Order was able to occupy other castles, and Santos became a place of residence for women who were relatives of members of the Order. Its relationship with the crown is interesting, because both the kings and the queens had granted these women privileges, protecting them and their entourages in ways that can help explain the success of several families.

In Portugal, the Order of Santiago needed a convent for the safe housing of its women members. As we know, many of the brothers were married, and, as a consequence, in times of war it was necessary to protect the wives and daughters of men whose duties meant they were absent for long periods. The institutional protection of the *donas*,[1] the *comendadeira*,[2] and other persons who belonged to the monastic community was set out in the Order's foundation bull:[3] any man who dared touch a nun or harm her physically was to be punished with excommunication.[4]

---

[1]   The term *dona* is used in the documents to describe a religious woman who lived in the monastery. She would swear to obey the rules of the Order but enjoyed extensive freedoms as, for example, in her right to administer her own patrimony, nominate lawyers, administrators and controllers, and petition in the courts, and she was allowed to leave the monastery in order to marry. J.S.F. Mata, 'Around a Theme. The Female Community of the Order of St James in Portugal: A Journey from the Late 15th Century to the 16th Century', at   http://www.brown.edu/Departments/Portuguese_Brazilian_Studies/ejph/html/issue11/ pdf/jmata.pdf, 6,1 (2008), 2 (accessed 12 May 2011).

[2]   The word *comendadeira* has the same meaning as *prioresa* or *abadessa* – terms that appear occasionally in the documentation. From the beginning of the thirteenth century it was used to designate the religious woman who had charge of the administration of the monastery. She was chosen for her experience in leading the other religious women. In short, the smooth running of the *cenóbio* and her reputation depended on her sensitivity. Ibid., p. 3.

[3]   Alexander III confirmed the Order of Santiago in 1175 in the bull *Benedictus Dei*. University of Coimbra Library (CBG), R-31-20.

[4]   Instituto dos Arquivos Nacionais/Torre do Tombo (hereafter IAN/TT), *Colecção Especial*, cx.1, nº12, inserted in a later document dated 20 June 1396.

During the first quarter of the sixteenth century, however, King Manuel (1495–1521) took the protection of the nuns upon himself, when he decided what sanctions were to be applied if someone entered the monastery without permission, or if a nun was raped within the precincts or was abducted. The penalties to be exacted included whipping, monetary fines appropriate to the physical or moral injury, exile to Africa for a period of over two years, or expulsion from the kingdom to the colony of S. Tomé.[5] What we are seeing is a shift in the protection paradigm: the nuns' safety is guaranteed by royal authority, much as the king was responsible for law and order throughout his realm; excommunication, a spiritual penalty, has been replaced by physical punishment and ostracism – penalties that resembled those set out in legislation for the whole of Portuguese society.

From its beginning, the Order of Santiago developed in close association with royal power. The members of the Order identified their interests with those of the crown, as they shared a common interest in establishing the frontiers and so regaining areas occupied by the Muslims. The monarchy saw the Order as a powerful ally that cooperated closely in the definitive reconquest of territory south of the River Tejo. The nuns benefited from a statute of their own within the Order, a counterpart to the one conferred on the men. But other factors – royal authority, papal power, and, on a smaller scale, the rights of the bishop of Lisbon – impacted on their daily life, although with a more variable frequency. So, for example, the *comendadeira* had the right to appoint priests to Aveiras and Coina, and it was for the bishop or the chapter of Lisbon to confirm her nomination. However, it is in the realm of royal authority that we find a distinctive relationship between the crown and the *donas* of the veil of Santiago.

The scope of royal protection in relation to this community was complex, enshrined in letters of privilege, grace and mercy that, taken as a whole, gave protection in general terms. However, there were other public instruments that spelled out specific matters relating to the court of justice, to arbitration in disputes arising from conflicts of jurisdiction, and the delegation of legal authority in matters such as property restitution, the abrogation of measures taken by royal officers that injured the nuns' interests, as well as charitable donations intended for the support of the *comendadeira* or even monetary contributions for repairs to the monastery. The legal provision governing this wide range of actions was mostly implemented by letters of *guarda* and *encomenda* for the benefit of the monastery, their fundamental features being very similar. However, their contents reflected the petition that prompted the document concerned.

The first of these letters was granted by Afonso IV on 17 August 1333, and it was written by Pêro do Sem, controller of the chancellery;[6] in it the king clearly expressed his desire to defend the *comendadeira*, D. Joana Lourenço from Valadares, the monastery, the *donas*, the dependents, the cattle and the patrimony. As a preventive measure against possible abuse or confrontation, the monarch specified a penalty of

---

[5]   *Ordenações Manuelinas, Livro V* (Lisbon, 1984), pp. 71–2.

[6]   IAN/TT, *Mosteiro de Santos-o-Novo*, cx.1, m.1, nº3; cx.1, m1, nº12; cx.1, m2, nº4.

6,000 *pays*, besides a double indemnity for any loss or damage. Then in 1373, King Fernando (1367–83) conferred a similar letter, adding that the tenants and those belonging to the monastery's ecclesiastical court should not be compelled by the Lisbon council to render any form of service whether on land or sea. However, this measure did not include either military service or guard duties.[7]

Dependents of the monastery were well aware of royal protection, and, when appropriate, they would invoke it in the face of infringements by outsiders. So for example, Lourenço Domingues presented himself to the judge João Afonso Dinis in 1379 to ask for an authenticated copy of a privilege issued by King Fernando, as a safeguard in the event of threats from the municipal officers.[8] More significantly, in 1388 João Anes, a farmer who depended on the *comendadeira*, D. Leonor Gomes de Azevêdo, showed Vasco Vicente, controller of the repairs in Sintra, a letter that, among other things, included an exemption from rendering services to the city council, and argued that his property lay on the boundary of Sintra. Vasco Vicente had summoned him to work on the repairs that were taking place under the aegis of the city council, and he protested on the basis of a royal letter that exempted him for as long as he remained a dependent of the monastery of Santos. The controller insisted that he perform the disputed duties, and in the end the king, João I (1385–1433), had to intervene. The monarch decided that the royal charter protecting the Order of Santiago's farmer should be upheld.[9]

At times of political and economic instability, the lords or the royal officers would use their authority to collect levies in kind, for example cereals, wine and olive oil, far in excess of the amounts stipulated in tenancy agreements.[10] This practice was regularly denounced by various administrative bodies.[11] In these instances tenants would directly address the king, who, in most cases, answered reasonably quickly, placing the complainants under his protection. Thus Gomes Pires, procurator of the monastery warns Gil Vicente, treasurer of the royal household, at the boundary of Torres Vedras to fulfil and respect the rights recognized and guaranteed in a letter written by D. Leonor and dated 1381.[12] The requirements of the war with Castile, still alive in the collective memory, included recruitment for military purposes. In this connection, in 1399 a new complaint was presented to João I by the *comendadeira* D. Inês Pires,[13] denouncing the collusion

---

[7]   King Fernando ruled that the sum of 1,000 *soldos* was to be paid to the monastery as a fine. IAN/TT, *Mosteiro-Novo de Santos*, cx.1, m.1, n°12.

[8]   Ibid., cx.1, m.6, n°6.

[9]   Ibid., cx.1, m.3, n°4.

[10]   The sums demanded by landlords were related to the royal fee, known as *jugada*, a payment associated with the yoke of oxen used by a farmer in ploughing. I. Gonçalves, 'Jugada', *Dicionário de História de Portugal* (Porto, 1981), 3, p. 415.

[11]   A. de Sousa, *As cortes medievais portuguesas de 1385 a 1490* (Lisbon, 1990), 2, p. 317.

[12]   IAN/TT, *Mosteiro de Santos-o-Novo*, cx.1, m.6, n°16.

[13]   Inês Pires lived with João while master of the Order of Avis in the castle of Veiros. From this liaison two children were born: D. Afonso, count of Barcelos, and D. Beatriz.

between the officials of the local council in Aveiras de Cima and the king's officer in obstructing the privilege which exempted the local people from compulsory military service (*fossado*). The king annulled his officer's ruling, thus upholding the privileges granted to those living in that place.[14]

To be a dependent of the Order of Santiago in general and of the monastery of Santos in particular meant enjoying privileges conferring rights[15] that by their very nature could not be extended to everyone but benefited only a restricted circle. These rights might include exemption from paying various taxes[16] or performing other duties, although such measures only protected farmers with their own holdings who lived and worked on their own properties.[17] However, if a tax was raised purely for the defence of the village, nobody would be exempt.[18]

By the beginning of the second half of the fifteenth century, Afonso V (1438–81) confirmed these privileges but urged that everyone should contribute to the defence of the peasants' dwellings and to public works such as reconstructing walls, maintaining bridges, cleaning fountains and repairing paths and paved roads. This directive was transmitted to Diogo de Abreu and Álvaro Mendes Godinho, and to the vassals, judges, the *corregedores* of the council of the central and southern regions of Portugal, Estremadura and Entre-Tejo, and was also intended for all other court officials.[19] The measure is important because it covered the area that encompassed most of the monastery's agrarian patrimony, ranging from the small farms to the villages and communes belonging to it. Examples include Coina, Aveiras de Cima and Vale do Paraíso. The royal privileges still allowed for the exemption from the so-called *direitos de pousada*, the right to requisition bedclothes, straw, barley, hens or anything else against the owners' will.[20] Another privilege concerned exemptions from having to escort prisoners to prison or anywhere else, wherever justice was done.[21]

Royal intervention in the monastery's affairs acted as a buffer between monastic authority and those who illegally occupied monastic properties or even the illicit demands made by the kings' officials. For example the butcher Lourenço Pires was obliged to return a property with olive trees he had illicitly occupied

---

J.S.F. Mata, *A comunidade feminina da Ordem de Santiago: a comenda de Santos na Idade Média* (Porto, 1991), p. 233.

[14]   IAN/TT, *Mosteiro de Santos-o-Novo*, cx.1, m.2, n°16.

[15]   A.L. de C. Homem, *O Desembargo Régio (1320–1433)* (Lisbon, 1990), p. 64.

[16]   The exemptions were the *talhas*, *fintas*, *pedidos*, *empréstimos* and *peitas*. I. Gonçalves, 'Peita', *Dicionário de História de Portugal* (Porto, 1981), 5. p. 43.

[17]   *Cabeça de quintã* or executor of the joint estate is the title given to the *foreiro* responsible for delivering the income to the landlord by the terms agreed in the contract for renting out lands of a property. M.J. de B. de A. Costa, *Origem da enfiteuse no direito português* (Coimbra, 1957), p. 95.

[18]   IAN/TT, *Mosteiro de Santos-o-Novo*, cx.1, m.5, n°13.

[19]   Ibid., cx.1, m.3, n°6.

[20]   Ibid., cx.1, m.2, n°5.

[21]   Ibid., cx.1, m.u., n°23

in Lisbon. The order was issued by Fernando acting through Gomes Martins and João Pires, both officials at the *Casa da Suplicação*.

It often happened that within the borders of the royal estates there were small grants of land belonging to the Order of Santiago, and where the royal officials failed to respect the distinctive character of the property with regard to its privileges and rents that inevitably led to tensions between royal authority on the one hand, and the aggrieved party on the other. What happened with the property of Outorela is an example of this. The inhabitants became members of the monastery of Santos when D. Isabel Afonso died. The *comendadeira*, D. Inês Pires, had complained to João I that the treasurer of the royal palace, Vicente Vaz, had incorporated it into the royal land at Algés. Despite the legal position of the monastery, the king compelled the *donas* to pay a quarter of the produce annually, thus taking the view that his own officials were in the right because that was the tradition.

The Castilian sieges of Lisbon in the last quarter of the fourteenth century were responsible for damage to the monastery, where, in 1373, Henrique II of Castile even considered establishing his base. However, he then decided to make camp in the monastery of S Francisco. Then, in 1384, King João established his headquarters in Santos, considering it a suitable strategic point from which to launch an assault on the city.[22] Once the conflict between the Portuguese and Castilians came to an end, repairing the monastery was a matter of urgency. Responsibility for the expense fell on the king, who in 1395 set aside a considerable sum of money for the repair works on the monastic fabric.[23]

In the second half of the fifteenth century, in order to ensure regular maintenance, Afonso V exempted a mason and a carpenter who were to work continuously in the monastery from any services or responsibilities imposed by the Lisbon council, as well as from escorting prisoners. The king also excused these two workers from going to Ceuta and Alcácer Ceguer, as well as from having to participate in the army or the navy.[24]

The king's involvement in the women's community can also be observed in the subsidies granted to the *comendadeira*. She relied on them to maintain the dignity commensurate with her position. The first donation was offered by King Fernando in 1372,[25] but the crises of 1383–85, with all its associated economic and financial consequences, resulted in the king forgetting his obligations. Thus, the

---

[22]   The monastery of Santos was 'two shots of a crossbow' from the city walls. Fernão Lopes, *Crónica de D. João I* (Porto, 1990), 1, p. 219.

[23]   João I, in a letter dated 21 November 1395 countersigned by João Afonso, a scholar in laws and judge of the High Court in the place of Afonso Lourenço, ordered the judge Martim de Santarém to make available 7,000 pounds, a sum that was never fully paid. IAN/TT, *Mosteiro de Santos-o-Novo*, cx.1, m.1, n°13.

[24]   Ibid., cx.1, m.3, n°18.

[25]   In a letter written in Alenquer, Fernando ordered the treasurer of the royal household, Lopes Esteves, that 10 *soldos* should be paid daily to the *comendadeira* for her sustenance. Ibid., cx.1, m.6, n°2.

comendadeira reminded João I that he was obliged to honour the commitments of his predecessor. Accordingly, the king ordered Domingos Pires, the treasurer of the royal household, to put the payments, by then about four years in arrears, on a regular footing.[26]

In the area of justice, there is a royal charter in favour of the monastery issued in response to a petition presented by D Inês Pires, who had asked the king to choose a legally qualified official in Lisbon who could represent the monastery's interests in respect of the localities where the *comendadeira* retained jurisdiction over criminal and civil matters. However, it was mainly D. Beatriz, mother of King Manuel, who intervened with her son as a mediator to put an end to a dispute that had begun in 1376 and lasted until 1499.[27] The commune of Almada had been a commandery of the Order of Santiago until 1298, when it was exchanged for the small villages of Almodovar and Ourique and the castles of Monchique and Aljezur.[28] Before the end of the fifteenth century, Almada had been given to Infant D. Beatriz, the queen mother, who immediately started to intervene in a conflict over the border between the communes of Almada and Coina that had persisted for over a century without any sign of a peaceful and lasting solution. This dispute, sustained by the commune of Almada, demanded a well-thought-out solution, and the *comendadeira*, D. Violante de Nogueira, realized that it would be only through the king and by appealing to his justice that she would be able to attain her objectives.

It is in this context that the voice of the dissuader, D. Beatriz, appears. She undertook to put an end to this dispute, interposing between the vested interests and the politics that kept them alive. The *infant* was aware of the pitfalls in litigation and the uncertainty of the final decision, and, in the rhetoric style typical of the period, she conceded that religious people should not get involved in the courts and so waived the civil action that the Almada officials had brought against the monastery.[29] However, it is here that power-politics played its part. Against the seemingly intractable royal authority the monastery found a counterweight in the

---

[26]   The *comendadeira* D. Inês Pires sent João I his predecessor's letter asking him out of kindness to maintain the payment of the daily subsidy of 10 *soldos*. Ibid., cx.1, m.1, nº17.

[27]   D. Beatriz intervened in the negotiations with Castile in 1479, during the reign of D. Afonso V, having been with Isabel the Catholic in the Castilian village of Alcântara in March of that year. This was a crucial moment in the negotiations between her son-in-law, the duke of Bragança, and the catholic kings in 1480 that led to the agreement known as the 'Tratado das Terçarias de Moura' in which the catholic kings renounced the African coast beyond the Canaries. Rui de Pina, 'Crónica de D. Afonso V', in *Crónicas de Rui de Pina* (Porto, 1977), pp. 867–70. In Beja, D. Beatriz became the patron of the wool industry, and in 1490 she also obtained from João II a privilege for the *pisões*.

[28]   The deed of exchange, written in Santarém by Domingos Pires, is dated 1336 and not 1335 as previously understood. *Livro dos Copos* (Porto, 2006), 1, pp. 210–11; IAN/TT, *Chancelaria de D. Dinis*, Livro dos Privilégios, fl.10.

[29]   IAN/TT, *Mosteiro de Santos-o-Novo*, cx.16, m.1, nº5 (1110); *Leitura Nova – Estremadura*, Livro nº2, fl.180; *Leitura Nova – Guadiana*, Livro nº5, fl.99.

power and influence of the Queen Mother, who, in a letter written in Lisbon and dated 25 February 1499, informed King Manuel about the purpose of the litigation and her wish to withdraw the commune of Almada's longstanding complaints against the monastery of Santos in relation to its border with Coina.[30] By means of a donation, D. Beatriz was able to put an end to a problem that the courts had not been able to solve, thereby sensibly avoiding a judicial conflict that might not have come to a satisfactory conclusion and that might have dragged on for another century. D. Beatriz asked her son, King Manuel, to acquiesce in agreeing to confirm her donation in favour of the monastery. The king answered affirmatively, commending his mother's action.[31]

When faced with the prospect of litigation, the parties would sometimes try to avoid going to court and the associated costs and delays as well as the uncertainty of the outcome. The 'amicable agreement' was often preferable.[32] D. Beatriz opted not to go to law, and so, going against the wishes of the commune of Almada, she determined the limits of the communal territory. Her decision would have been to her own detriment, unless the commune of Coina's case was much stronger anyway. That could be the reason why Manuel did not object. Otherwise, as Almada belonged to the crown, it made no sense for its owner to relinquish territory to a third party. Manuel must have studied the records, and he probably concluded that the *comendadeira* and the *donas* were in the right.

All this helps us understand why what had proved impossible to resolve between two well-entrenched powers for approximately 123 years, was accomplished in only twenty days through what we might call 'feminine intervention' or 'indirect intervention' – something that might appear weaker but that proved effective with an efficiency that justified the line of action. The crown's procedures were normally protracted, but here a quick and decisive solution was found.

The relationship between the crown and the monastery of Santos allowed for a situation in which the roles were reversed. After the removal of the *donas* in 1490 to a new convent on the instructions of João II (1481–95), the monastery of Santos-o-Velho was not abandoned for any length of time. As with any other real estate, the property was subject to economic exploitation, due to the ever-increasing dereliction. The now-unoccupied convent was let at fee farm for the course of three generations, its first owner being Fernão Lourenço, a man of great prestige

---

[30]   D. Beatriz declared that she wished to renounce the demand that was against the monastery, the *comendadeira* and the *donas*. IAN/TT, *Mosteiro de Santos-o-Novo*, cx.14, m.10, n°27 (1305).

[31]   D. Manuel replied: 'mandamos que assy lha compram e que dem e façam d'aquy em diamte muy imteiramente comprir e guardar como nella he contheudo porquamto nos avemos por bem o que pela dita senhora asy he feyto como dito he'. Ibid., cx.16, m.1, n°5 (1110).

[32]   An amicable settlement was a legal device in which the litigants agreed to relinquish their claims and reach a mutually advantageous agreement when it was being said that no one could be sure about the judge's final decision.

and one of the most powerful of his day.[33] João II appointed him bailiff of Casa da Guiné in 1486, and in 1502 he became treasurer of the '*Trautos* (commerce) of Guiné and Casa da Mina in India'.[34] A wealthy man, Fernão Lourenço, completely transformed the old building into a modern palace that was both envied and frequented by senior court dignitaries, but the old king's courtier only enjoyed it for a short period. João II had already realized that the place provided an excellent occasional retreat from the routine of government, and it was there that in 1477 he had received the news that his father was handing over to him the regency of the realm.[35] After the departure of the *donas*, the king transformed this palace into a royal rural residence.[36]

King Manuel was also well acquainted with the palace of Santos. After his marriage to D. Isabel, he visited the widowed queen D. Leonor in Lavradio and lodged at the monastery, and it was there that the solemn entrance of the royal couple into Lisbon was organized.[37] In 1501 the building became the royal residence outside Lisbon, from where Manuel could sail along the River Tejo and engage in other forms of recreation.[38] The king made full use of this palace on Sundays and public holidays.[39] Duarte Forreiro was in charge of the upkeep of this summer retreat and was expected to provide fresh fruit, jam, cakes, wine and water, as well as musicians and young boys for rowing.[40] In 1502, when Manuel returned from his pilgrimage to Santiago de Compostela, he was received by the queen at the royal residence at Santos, and it was also in that year that the king received there a Venetian embassy that had come to Portugal to ask for help in the war against the Ottoman Turks. [41]

Under these circumstances, we can understand why Manuel should want this residence for his sole use, and so he persuaded Fernão Lourenço to exchange his title for Gestaçô and Penajoia.[42] This exchange meant that Fernão Lourenço formally relinquished his entitlement to Santos and its orchard although without altering what was a typical tenancy in fee contract. King Manuel then informed the *donas*, the other interested party, of his intentions by sending Francisco Pestana,

---

[33]   Norberto Araújo, *Peregrinações em Lisboa*, 2nd edn (Lisbon, 1992), 7, p. 14; Damião de Góis, *Crónica do Felicíssimo Rei D. Manuel* (Coimbra, 1926), I, 1, pp. 67–8.

[34]   M.E.C. Ferreira, 'Casa da Índia', *Dicionário de História de Portugal*, 3, p. 281 and 4, p. 301.

[35]   C. de Sabugosa, *A Rainha D. Leonor*, 2nd edn (Lisbon, 1979), p. 72.

[36]   T. d'A. M.D. Vilhena, 'O Paço de Santos-o-Velho', *O Instituto* (Imprensa da Universidade de Coimbra), 80/7 (1931), 8.

[37]   Ibid., p. 9.

[38]   Sabugosa, *A Rainha D. Leonor*, p. 407.

[39]   A.B. Coelho, *Quadros para uma viagem a Portugal no século XVI* (Lisbon, 1986), p. 32.

[40]   Damião de Góis, p. 92.

[41]   J. de Castilho, *A Ribeira de Lisboa: Descrição Histórica da margem do Tejo desde a Madre-de-Deus até Santos-o-Velho* (Lisbon,1983), p. 600; Damião de Góis, p. 105.

[42]   Castilho, p. 597.

bailiff of the 'Casa da Mina', to the monastery to negotiate his possession of the estate. [43] He carried a letter stating that the king had bought the title of the palace from the previous owner and that the *donas* were aware of what was happening.[44] For their part the *donas* were not against the change of ownership, but they imposed certain conditions on the king:[45]

1. that King Manuel and his wife should be mentioned first in the fee farm;
2. that their nomination of the second person should be by the king's wife;
3. that the fee farm could only last for three generations;
4. that Manuel, his wife and descendants would keep the place repaired in the event of fire, flood or earthquake, and improved;
5. that the owners should keep a silver chalice, priestly vestments, and all that might be necessary to say Mass in the monastery chapel, and at the end of the contract all the ecclesiastical ornaments should remain in the monastery without any further payment;
6. that the contract should stipulate that Fernão Vaz, a neighbour of the palace, should be allowed to use the well that was located in the farm yard, and that both usufructuaries were responsible for the cleaning the well.

Francisco Pestana, in the name of both the king and the queen, accepted the contract with these clauses, along with the properties and rents, as the only means of validating the transaction.

This study demonstrates how the central authority was always ready to intervene in the affairs of the monastery of Santos, beginning by implementing the system of protection that ended up being enshrined in the written law of the realm, and in the resolution of litigation, as well as by placing trusted people in charge of the convent, as is the case of the *comendadeira* Inês Pires, who had been the mistress of D. João, master of the Order of Aviz, before he became king.

---

[43] IAN/TT, *Mosteiro de Santos-o-Novo*, cx.5, m.1, n°19 (419).

[44] The document stipulates: 'per vosso prazer e conssentimento'. Ibid.

[45] Ibid.

Chapter 33

The Recruitment of the Portuguese Military Orders: A Sociological Profile (1385–1521)

António Pestana de Vasconcelos and Manuel Lamas de Mendonça

The role played by the military orders, almost always with the encouragement and support of the Holy See, in all the longstanding conflicts between Christian and Muslim powers scarcely needs highlighting. As in all prolonged wars, the antagonists were profoundly transformed, pacts and alliances were made and dissolved, and the political geography of the opposing sides underwent important modifications. Some of the most significant changes to occur came about as a consequence of the adaptation of crusading ideology to fit the realities of the twelfth-century Iberian Peninsula and the continued threat posed by the Islamic states. In this setting, frontier warfare, with its widespread if spasmodic and desultory conflicts, predominated. This situation, coupled with seasonal fighting, logistical difficulties, scarcity of trained men and precarious chains of command, to mention only a few factors, called for the introduction of the military orders, which had already proved their worth in the Holy Land. Similarly, the creation of new regional militias became imperative and unavoidable. The former had to adapt to the specific characteristics of this new theatre of operations, while the latter came into being in accord with local needs, resources and demography.

A glance at the social profile of the knights of the military orders in Portugal throughout the thirteenth and fourteenth centuries shows that they were mainly recruited from the urban population, 'with no shortage of sons of merchants, of learned men, of knights and landowning city-dwellers, or even relatives of notaries public or, indeed, of some well-off peasants'.[1] In actual fact, many were 'members of council militias (or yeomanry) or of the urban aristocracy'.[2]

Until around 1382, the main concern of orders was to maintain their manpower at all cost, regardless of the social origins of their brother-knights. There are therefore few references to nobles, who, although including some of the masters, before that date amounted to only a few dozen, and there is even evidence that the

---

[1] L.F. de Oliveira, *A Coroa, os Mestres e os Comendadores. As Ordens Militares de Avis e de Santiago (1330–1449)* (Faro, 2006), p. 450.

[2] J.A.S. Pizarro, 'The Participation of the Nobility in the Reconquest and in the Military Orders', *e-Journal of Portuguese History*, 4 (2006), 1–10.

nobility did not hold the military orders in high regard.[3] Two factors contributed significantly to this state of affairs. One was the fact that 'in Portugal second-born sons were not excluded from the paternal inheritance, with property being divided amongst all heirs until the introduction of entails in the late thirteenth and early fourteenth centuries'.[4] Secondly, the military orders did not make noble or gentry status mandatory for candidates aspiring to be brother-knights, a requirement that only emerged in the mid-thirteenth-century statutes of several orders including the Hospitallers,[5] Santiago,[6] and Avis,[7] and, later on, in the customs of the Order of Christ.[8]

---

[3]   A case in point is the example quoted by Oliveira on the attitude of D. Pedro, count of Barcelos, regarding Lourenço de Beja, 'freire' and later 'comendador-mor' of Santiago, accusing him of avarice and other villainies. Oliveira, *A Coroa*, p. 15. On the rivalry between the lesser nobility and the members of the orders, which is apparent in the *Livro de Linhagens*, see L. Kruz, *A concepção Nobiliárquica do Espaço Ibérico. Geografia dos Livros de Linhagens Medievais Portugueses (1280–1380)* (Lisbon, 1994), pp. 141–2, n. 303.

[4]   J.A.S. Pizarro, *Linhagens Medievais Portuguesas: genealogias e estratégias, 1279–1325* (Porto, 1999), 2, pp. 565–92. See also L. Ventura, *A nobreza de Corte de Afonso III* (Coimbra, 1992), pp. 187–8, 353, 381–2; B.V. Sousa, *Os Pimentéis. Percurso de uma Linhagem da Nobreza Medieval Portuguesa (Séculos XII–XIV)* (Lisbon, 2000), pp. 252–64.

[5]   Biblioteca da Ajuda: *Regra da Ordem de S. João de Jerusalém*, fol. 20. Hugh Revel, master 1258–77, forbade the entrance of all illegitimate sons, except the sons of counts and other nobles. *Regra da Ordem de S. João de Jerusalém*, fol. 18v.

[6]   Initially the Order of Santiago did not place any condition on potential recruits, but in 1249 the Order's regulations demanded that the entrant had to be *homen fidalgo que fuese cavalero*. (National Library (Madrid), ms. 8582, fol. 45v. I.M.L. Barbosa, 'A Ordem de Santiago em Portugal nos finais da Idade Média (Normativa e prática)', *Militarium Ordinum Analecta*, 2 (1998), 173. On the requisites for joining the Order of Santiago, see M.R.S. Cunha, *A Ordem Militar de Santiago (das origens a 1327)* (Porto, 1998), pp. 194–5. According to the legislation of 1509, no prospective knight could be, nor be descended from, a craftsman, farmer or disabled person, unless injured in war against the Moors, with the proviso 'unless the person was such that the Order would gain service from her': 'nem official macanico nem lavrador nem aleijado salvo se ha aleijam fosse avida em guerra de mouros ou ha pessoa for tal e de taes qualidades que ha ordem receba delle serviço'. *Regra, Statutos e Diffinções da Ordem de Santiago de 1509*, fol. 2.

[7]   Like the Order of Santiago, the Order of Avis placed no conditions on those who aspired to join as a brother-knight. M.C. Cunha, 'A Ordem de Avis', MA thesis, University of Porto, 1989, pp. 37–9. However, cap. 63 of the *Definiciones de la Orden de Calatrava* of 1468 regarded the status of *homem fidalgo* as a pre-requisite. J.F. O'Callaghan, *The Spanish Military Order of Calatrava and its Affiliates* (London, 1975), p. 264. This rule was repeated in *Definições de Avis de 1503* (TT (Torre do Tombo), *Livros do Convento de Avis*, 25, fol. 57v) and in the *Regra e Estatutos da Ordem de Avis de 1516* (National Library (Lisbon), Res. 3008 V, fol. 49).

[8]   This condition was mandatory for all prospective knights, something that the chronology and the circumstances of the foundation of this Order amply justify. *Regra e Definições de 1503*, TT., *Série Preta*, 1393, cap. VIII, fol. 18v.

The growing economic importance of the military orders, not the least because of the vast lands they owned, led to changes to this situation during the reign of King Dinis (1279–1325), which continued until the time of King Fernando (1357–83). These monarchs intervened in the orders' internal affairs, in particular in the appointment of the masters, and sought thereby to reinforce their influence and authority.[9] While the system of equal inheritance rights that predominated until the late fourteenth century made possible the distribution of estates among all family members, it also contributed to the fragmentation of ownership. This caused some to start looking to the military orders as a means of ensuring the maintenance of their status within the class into which they were born. The rise of the entail at the end of the thirteenth century, with the consequent indivisibility of estates and the exclusive succession by the eldest male, led many younger sons to opt for a career in the military orders.

The military orders thus underwent a gradual process of 'aristocratization', as masters enrolled those they deemed 'fitter',[10] many of them their relatives or vassals.

Table 33.1   Relationships among Masters and relatives

Lineage	Military Order	Relation to the Master
Vasconcelos	Santiago[1]	Sons
Sequeira	Avis[2]	Sons
Sousa	Christ[3]	Sons

With the rise of the Avis dynasty in 1385, the administration of the military orders was transferred to the *infantes*, the younger sons of João I (the first Portuguese princes to be known as such). The rise of the new dynasty also contributed to the 'aristocratization' of the military orders by opening them to more families. Entrance into the orders was indeed widened to embrace all sectors of the nobility, in particular those that had links to the crown and to the *infantes*. Furthermore, the military orders started to be regarded as a model to be followed. In the period between the accession of João I (1385) and the personal rule of Afonso V (1450), the social profile of the brother-knights underwent profound changes as witnessed by the growing presence of nobles in their ranks. Among the noble lineages represented in the military orders were members of the middle-ranking court nobility as well as individuals from families belonging to the upper and middle ranks of the provincial nobility and from those serving in the royal administration.

---

[9]   A.M.F.P. Vasconcelos, *Nobreza e Ordens Militares. Relações Sociais e de Poder (Séculos XIV a XVI)* (Porto, 2008), 1, pp. 42–66, 222.

[10]   Oliveira, *A Coroa*, p. 15.

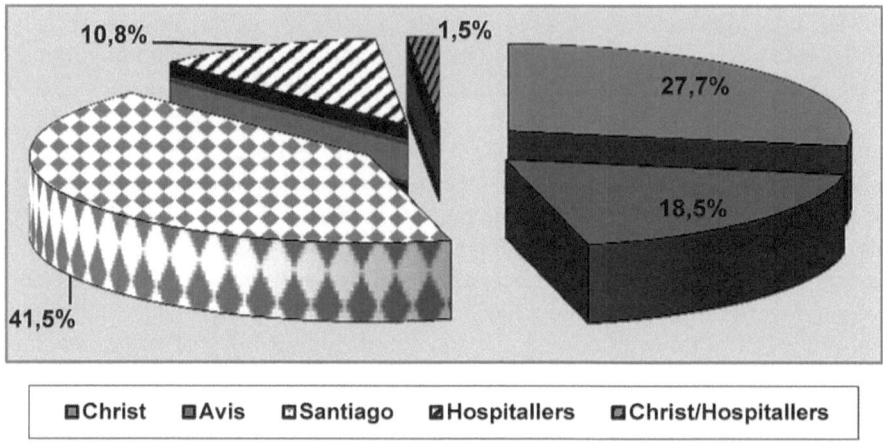

Figure 33.1   Noble Knights enrolled in the Military Orders (1385–1450) (graphics: A.P. de Vasconcelos and M.L. de Mendonça)

Source: The data is derived from the study of 33 noble lineages whose members were present in the military orders. Ibid., 2, p. 377.

This trend is observable in all the military orders, with the Order of Santiago standing out in this respect, followed by the Order of Christ, the Order of Avis and, finally, by the Knights Hospitaller (see Fig. 33.1).

The lead taken by the Order of Santiago can be explained by various factors. There were traditional links between some noble families and the Order; Santiago alone accepted married men as brothers, a state of affairs that favoured the recruitment of their offspring; its members enjoyed the right to administer the Order's lands – the so-called commanderies (*comendas*) – for their own benefit, and there was the possibility of bequeathing claims to *comendas* to relatives and children; and finally Santiago families had opportunities to operate in areas that since the *Reconquista* were closed to the nobility, such as the central and southern areas of Portugal, and where the military orders maintained a significant presence. The great political and economic influence that the Order of Santiago exerted should also be singled out. Its influence was even felt abroad, as witnessed by the intense and fruitful diplomatic activity of its master, Fernando Afonso de Albuquerque, in England during the succession of King Fernando and accession of João I. He first strove to obtain English support for the cause of the future King João, who was then great-master of Avis, and later, following the victory of Aljubarrota, he promoted an alliance between the two countries. These efforts culminated in the treaty of Windsor between England and Portugal, ratified on 9 May 1386.[11]

---

[11]   Visconde de Santarém, *Quadro Elementar das relações politicas e diplomaticas de Portugal com as diversas potencias do mundo desde o principio da monarchia portuguesa*

The interest of the nobility in this particular Order was perhaps also due to the fact that the next master was Mem Rodrigues de Vasconcelos, a nobleman who, as an active campaigner for his accession, enjoyed the favour of King João I. After his death, his successor, styled 'administrator and governor' of the Order, was for the first time a member of the royal family, the *Infante* João.[12] This backdrop makes the choice of this Order by noble lineages including the Abreu, Almeida, Barreto, Correia, Freire de Andrade, Furtado de Mendonça, Mascarenhas, Miranda, Moniz, Noronha and Vasconcelos more intelligible at this particular time.

The interest of the nobility in the military orders would increase considerably during the personal rule of Afonso V (1450–81) and his military and expansionist drive. Portuguese expansion made cooperation between the crown and the military orders increasingly important,[13] and so noble families regarded the enrolment of some of their members in the orders as a means of associating themselves with royal policy and thereby being in a position to benefit from it. It is therefore unsurprising to note the considerable increase (63 per cent) in the manpower of military orders during the reign of King Manuel I (1495–1521) by comparison with the years 1382–1495. The fact that most of the new recruits chose a different institution – the Order of Christ, an Order that until then had been least favoured – is also noteworthy.

Certain factors, some of which originated in the preceding period, contributed significantly to these changes. Two developments in particular should be emphasized. One was the effort to identify the Order of Christ with what would become the great project of the Avis dynasty: overseas expansion. This had started with the conquest of Ceuta in 1415, in which Dom Lopo Dias de Sousa, the master of the Order of Christ,[14] had participated, thereby associating his institution with expansionist ambitions of the crown. This association led noble families to look on this Order as a means of identifying themselves with royal policy. That in turn would allow them to pursue their own objectives: make a career, advance their status with privileges and benefits, and obtain financial rewards. Secondly, the appointment of Prince Henry as 'governor and administrator' of the Order of Christ in 1420 and his efforts to prepare it for the struggle against the Infidel encouraged some elements in the nobility to consider joining. Enrolment in the Order of Christ would allow them to benefit from the dynasty's expansionist policy into North Africa and along the West African coast.[15] We may note the presence of Prince

---

athe aos nossos dias (Paris, 1842–76), 2, pp. 96–7.

[12]   Papal document dated 8 October 1418. *Monumenta Henricina*, ed. A.J.D. Dinis (Coimbra, 1960–74), 2, pp. 303–5, doc. 148.

[13]   Vasconcelos, *Nobreza*, 1, p. 224.

[14]   I.M.S. Silva, 'A Ordem de Cristo durante o mestrado de D. Lopo Dias de Sousa', *Militarium Ordinum Analecta*, 1 (1997), 76.

[15]   For an example of a good relationship between the crown and the military orders, see the grants by Afonso V of 7 June 1454, whereby he gave Prince Henry the Navigator the temporal administration of everything he should collect from Cape Não to Guinea and

Henry at the head of the 1437 expedition to conquer Tangier,[16] when some of the Order's *comendadores* took part,[17] as they did at the conquest of Alcácer Ceguer in 1458,[18] and, after his death, at the conquest of Arzila in 1471.

The concentration of wealth and influence in the Order of Christ led King Afonso V[19] to consider becoming governor and administrator himself on the death in 1460 of his great-uncle, Prince Henry.[20] That, however, did not happen, and the administration of the Order passed to the king's brother, the *Infante* Fernando,[21] who was already the governor of the Order of Santiago.[22] After his death in

---

the Order of Christ the spiritual administration of the same places. *Monumenta Henricina*, 12, pp. 1–6, docs 1–2.

[16]   A.B. Freire, *Brasões da Sala de Sintra* (Lisbon, 1996), 3, p. 201; A.C. Sousa, *Provas de História Genealógica da Casa Real Portuguesa* (Coimbra, 1946–54), 12, p. 174; J.V.S. Guimarães, *Marrocos e os três mestres da Ordem de Cristo. Comemorações do V Centenário da tomada de Ceuta* (Coimbra, 1916), pp. 71–97.

[17]   Among others, the presence is documented of Gonçalo Vaz Coutinho, *comendador mor* of the Order of Christ, accompanied by twenty men on horse and thirty on foot (*Documentos das chancelarias reais anteriores a 1531 relativos a Marrocos*, ed. P. de Azevedo (Lisbon, 1915–34), 1, p. 205, doc. 168.); of Diogo Lopes de Faro, knight and *comendador* of Castro Marim (TT., *Chancelaria de D. Duarte*, Liv. 1, fols 230v); of Fernão Lopes de Azevedo, knight and *comendador* of Casével ('Crónicas. D. Duarte', ed. R. De Pina, in *Tesouros da Literatura e da História* (Porto, 1977), pp. 147–51; 'Crónica de D. Duarte', ed. D.N. de Leão, in *Tesouros da Literatura e da História* (Porto, 1975), p. 758; H.B. Moreno, *A Batalha de Alfarrobeira. Antecedentes e significado histórico* (Coimbra, 1980), pp. 563, 731–2); of Gonçalo Rodrigues de Sousa, *comendador* of Nisa, Montalvão, Alpalhão and Idanha, *capitão dos Ginetes* (cavalry captain), with 300 horsemen under his command (Crónica de D. Duarte', pp. 756, 758; 'Crónicas. D. Duarte', pp. 155, 160).

[18]   In particular the reform carried out by Dom João Vicente, bishop of Coimbra, at the request of Prince Henry the Navigator, and which received the pope's agreement. See the bull *Super gregem dominicum* of 22 November 1434. *Monumenta Henricina*, 5, pp. 113–15, doc. 49.

[19]   In a letter sent to Dom Fernando da Guerra, archbishop of Braga, Afonso V stated that, although his brother *Infante* Fernando had requested from him the office of Master of the Order of Christ, he 'felt that to God's and my good service I should request the Holy Father to grant it to me or to the prince my son'. A.J.D. Dinis, *Estudos Henriquinos* (Coimbra, 1960), pp. 248–9.

[20]   For the text of Prince Henry's will, see *Monumenta Henricina*, 14, pp. 25–33, doc. 11.

[21]   *Infante* Fernando was appointed administrator for life of the office of master of the Order of Christ by Pius II: *Repetentes animo* (11 July 1461). *Monumenta Henricina*, 14, pp. 158–62, doc. 57. In 1469 he took part in the conquest of Anafé. 'Crónicas. D. Afonso V', ed. R. de Pina, in *Tesouros da Literatura*, cap. 160.

[22]   On the conquest of Alcácer Ceguer and Prince Henry the Navigator's behaviour see ibid., caps 138, 142. The active participation of the Order and its governor throughout this process resulted in the royal grant to the Order of Christ of the right of patronage over the town in the same manner as it enjoyed at Tomar. *Monumenta Henricina*, 13, pp. 152–3, doc. 87.

September 1470,[23] Afonso V awarded this post the eldest son of Fernando, *Infante* Diogo.[24] Through this choice, the king kept control of the Order firmly in the hands of the dukes of Viseu, one of the leading houses in the kingdom.

The accession of João II (1481) marked a change in royal policy towards the military orders. The threat posed by the Viseu prompted João II to adopt an anti-feudal policy, which entailed curtailing the privileges of the higher nobility and centralizing royal power. Found guilty of leading a conspiracy against the king, the aforementioned Diogo, duke of Viseu and governor and administrator of the Order of Christ, was executed on 28 September 1483. As a result the government and administration of the Order of Christ passed to his younger brother and effective heir, Manuel duke of Beja.[25]

Irrespective of who its 'governor and administrator' was and whether the Order of Christ pursued the crown's expansionist goals or not, the nobility sought to enrol in it so as to secure new career opportunities and sources of income as well as the consolidation of its influence. A productive means of achieving these goals was by being appointed to the position of commander of a fleet or an overseas stronghold, or to a key offices in their administration, and in this respect, the Order of Christ became central, with all these posts being held by its members.

At a lower level, access to the mainland commanderies (*comendas*) of any of the military orders provided another incentive for nobles to join. These lands were mostly located in areas where the aristocracy, with the exception of the dukes, did not traditionally hold significant estates: the vast lands south of the Mondego, and especially the Tagus and Guadiana basins. These areas had been incorporated into the kingdom in the twelfth- and thirteenth century largely as a result of the initiative or cooperation of the orders, and there they retained a dominant position. Indeed, it was only from the reign of Afonso V that titles referring to estates in this region were conferred.[26] Access to these lands explains how many of the younger sons of leading noble families in the kingdom who joined the military orders came

---

[23]   'Crónicas. D. Afonso V', cap. 160.

[24]   The government of this Order was handed to the grand commander, Fr Gonçalo de Sousa, owing to the minority of Dom Diogo; indeed, according to the ordinance of 1326, it was the responsibility of the grand commander to replace the master in his absence. *Monumenta Henricina*, 1, pp. 152–5, doc. 64. After the death of this grand commander, Afonso V and Beatriz, as guardians of the duke of Viseu, successively handed the governance and administration of the Order of Christ to Dom Fr Pedro de Abreu, the acting-custodian of Tomar and the Isles and chaplain to Dom Diogo, and to Dom Fr Antão Gonçalves, castellan and commander of Tomar. IAN/TT., *Ordem de Cristo/Convento de Tomar*, Liv. 52, fols 25–6. See Silva, 'A Ordem de Cristo', p. 86.

[25]   'Crónicas. D. João II', ed. R. de Pina, in *Tesouros da Literatura e da História* (Porto, 1977), cap. 18; *Crónica de D. João II e miscelânea*, ed. G. de Resende (Lisbon, 1973), cap. 54. On this matter see also Silva, 'A Ordem de Cristo', p. 91, n. 339.

[26]   On the concession of titles by Afonso V, see Vasconcelos, *Nobreza*, 1, pp. 173–81.

to acquire estates in these areas belonging to the orders.[27] Although it meant a growing influence of the nobility over hitherto-inaccessible lands, noble control of commanderies was in fact less harmful to the crown than outright ownership. It should not be forgotten that the administration and governance of the Order of Christ remained firmly in the hands of members of the royal family. Moreover, the policies of the governor did not solely benefit his own affinity but also men whom the crown thought useful in promoting its own ends.

The death of Crown Prince Afonso in 1491 compelled João II to choose an heir to the throne and a successor for the governance and administration of the Orders of Avis and Santiago. The decision on the governance of the orders was duly taken, and the choice fell on Dom Jorge, the king's bastard son. Nonetheless, international pressure from Castile and Rome[28] led João II to name his brother-in-law Manuel, Duke of Beja, as both heir to the throne and governor and administrator of the Order of Christ. João's will stated that Manuel should relinquish these posts to Jorge as soon as he acceded to the throne.[29]

As explained above, the Order of Christ was understandably the order of choice for much of the nobility throughout this period, followed by the Orders of Santiago, Avis and the Knights Hospitaller (see Fig. 33.2).

A virtual stagnation in the number of enrolments into the Order of Santiago by comparison with the preceding period becomes apparent. Nevertheless, it should be noted that most of the families that had previously been involved with this Order maintained their links. Only the Almeida, Correia and Freire de Andrade

---

[27]   Mention can be made of presence of lineages including the Brito, holders of the commandery of Castelo Novo; the Castelo Branco, holders of the commandery of Pindo; the Castro, holders of the commandery of Segura and Cardiga; the Coutinho, holders of the commanderies of Trancoso, Almourol, Alpalhão, Portalegre, Anciães, Touro and Rosmaninhal; the Cunha, holders of the commanderies of Castelejo and Castelo Novo; the Freire de Andrade, holders of the commandery of Lousã; the Leitão, holders of the commandery of S. Vicente da Beira; the Meneses, holders of the commanderies of Mendo Marques and of Penamacor; the Miranda, holders of the commanderies of Torres Vedras and Sta Maria de Póvos; the Pereira, holders of the commandery of Casével; the Silva, holders of the commanderies of Ferreira, Soure, Marmeleiro and Reigada; the Sousa (Arronches branch), holders of the commanderies of Idanha, Niza and Soure; the Sousa (Prado branch), holders of the commanderies of Redinha, Segura, Lardosa, Sta Ovaia, Jejua, Salvaterra, Ega, Niza, Idanha, Rates and Arruda. Vasconcelos, *Nobreza*, 1, p. 227.

[28]   On this matter, see M. Mendonça, *D. João II: um percurso humano e político nas origens da modernidade em Portugal* (Lisbon, 1995), pp. 449–66.

[29]   'Item encomendo muito ao dito Duque meu Primo que suplique ao sancto Padre que proveja ao dito D. Jorge meu filho o Mestrado de Christo que elle dicto Duque agora them que ho possa ter com o Davjz e Sanctiago que tem.' ('I strongly commend the said duke my cousin to petition to the pope that he grants my son Jorge the mastership of Christ, which the said duke holds, together with those of Avis and Santiago'). Sousa, *Provas de História*, 2, p. 215.

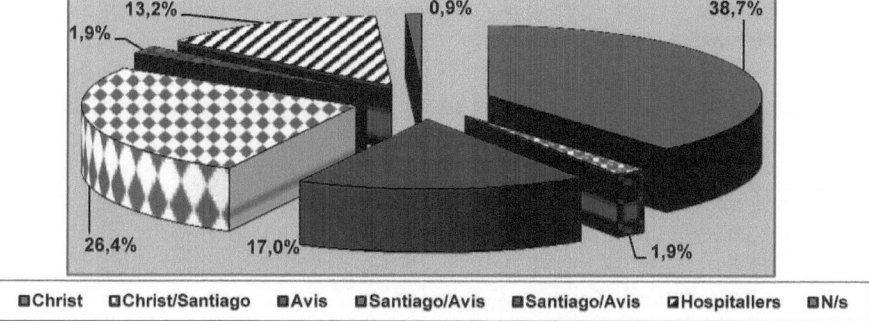

Figure 33.2   Distribution of enrolment among Military Orders (1450–1495) (graphics: A.P. de Vasconcelos and M.L. de Mendonça)

Source: The data is based upon the study of 33 noble families whose members were present in the military orders. Vasconcelos, Nobreza, 2, p. 377.

families opted for other institutions, while, on the other hand, it was chosen by lineages such as the Henriques, Pereira, Meneses and Sá for the first time.

Contrary to the terms of João II's will, Manuel, duke of Beja, did not on his accession to the throne pass the governance and administration of the Order of Christ to Dom Jorge. The new king was all too aware of the importance of the military orders and of the Order of Christ in particular. He sought to consolidate his political position and power with the support of this Order.[30] His decision reflected the views of Afonso V, who held that the Order of Christ should remain under the administration of the sovereign or his heir-apparent[31] as it pursued complementary goals to those of the crown, because of its vast estates, and because of its participation in the North African conquests and the exploration of the West African coast. Conscious of the importance and power of the Order in both economic and political terms, as well of its privileged relations with the Holy See, King Manuel enacted a series of innovative reforms with the purpose of converting this institution into a suitable instrument for the strategic objectives of the realm. Among these reforms were the following:

---

[30]   J.P.O. Costa, *D. Manuel I, 1469–1521* (Lisbon, 2005), p. 74.
[31]   See on this matter, the letter sent to Afonso V by Dom Fernando da Guerra, archbishop of Braga. See Dinis, *Estudos Henriquinos*, pp. 248–9.

1. The modification of the vow of chastity in Order to allow the knights of Christ and Avis to marry (1496)[32]
2. The creation of a limited number of commanderies (*comendas*) to be granted to knights who served for at least four years in the North African strongholds[33]
3. The creation of further thirty commanderies with an annual rent of 10,000 *reais*, paid out of the Order's own funds, which would be granted 'to those that serve in the wars against the infidels ... overseas in Africa ... and who ... live in said places and there reside and have their households and wives and some others who do not'.[34]
4. The creation of the so-called 'new' or '20,000 *reais* worth' commanderies endowed with ecclesiastical revenues that would be used to reward those who served the key strategic goals of the crown, for example, the search for possible allies in Africa, and the struggle against Islamic (Mamluk and Ottoman) commercial hegemony as well as against Ottoman advance in the East. These new commanderies accorded with the original crusading principles and goals of the military orders.[35]
5. The creation of yet more *comendas* using the income from fifty churches of which the king was patron and from a further fifteen whose patron was Jaime, duke of Braganza.[36]

---

[32]   *Romani pontificis sacri apostolatus* (20 June 1496). Sousa, *Provas de História*, 2 pp. 326–8.

[33]   This reform was decided in the general chapter of the Order held in 1503. Cf. IAN/TT., *Série Preta*, Cod. Nº 1393 – *Diffinçõoes do capitulo que el Rey nosso senhor governador do meestrado de Nosto Senhor Jhesu Christo fez no convento da villa de Thomar no mes de Dezenbro do anno de mill e quinhentos e tres*, cap. LI, fols 41v–43; A.M.F.P de Vasconcelos, 'A Ordem Militar de Cristo na Baixa Idade Média. Espiritualidade, Normativa e Prática', *Militarium Ordinum Analecta*, 2 (1998), 85–6.

[34]   IAN/TT., *Série Preta*, Cod. Nº 1393 – *Diffinçõoes do capitulo que el Rey nosso senhor governador do meestrado de Nosto Senhor Jhesu Christo fez no convento da villa de Thomar no mes de Dezenbro do anno de mill e quinhentos e tres*, Chap. LIIII fols 48–48v; Vasconcelos, 'A Ordem Militar de Cristo', p. 89.

[35]   Leo X authorized Manuel I to seize 20,000 *cruzados* from the annual dues of churches, monasteries and priories in order to endow newly created commanderies and to fund all those that went overseas to fight the infidel on land and sea (*Redemptor Noster Dominus Iesus Christus* (29 April 1514) (*As Gavetas da Torre do Tombo*, ed. A.S. Rego, 2 (Lisbon, 1962), pp. 472–8). On the institution of the *comendas novas*, see also I.M.S. Silva, 'As comendas novas da Ordem de Cristo no Entre Douro e Minho', in *Actas do I Congresso sobre a Diocese do Porto Tempos e Lugares de Memória* (Porto, 2002), 2, pp. 43–71.

[36]   The bull *Honestis votis* of Leo X (19 April 1517) authorized the creation of fifteen new commanderies whose holders would be appointed by the duke of Braganza rather than by the master of Christ. These commanderies were created in order to reward knights that followed the duke in the capture of Azemmour in 1513. Sousa, *Provas de História*, 4, pp. 63–8; M.S. da Cunha, *A Casa de Bragança 1560–1640. Práticas senhoriais e redes clientelares* (Lisbon, 2000), pp. 312ff.

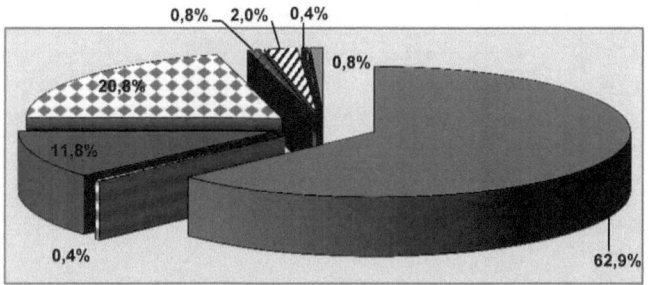

Figure 33.3   Distribution of enrolment among Military Orders (1495–1521)
              (graphics: A.P. de Vasconcelos and M.L. de Mendonça)

Source: The data is again based upon the study of thirty-three noble lineages whose members were present in the military orders. Vasconcelos, Nobreza, 2, p. 377.

By not relinquishing the governance and administration of the Order and by instituting the necessary internal reforms, King Manuel created the conditions necessary for diverting hitherto inaccessible revenues (notably from church lands and dues) to his patronage.[37] Indeed, during his reign, 1495–1521, the number of nobles enrolled in the orders underwent an increase of 131 per cent by comparison to the preceding period.

This increase was in evidence throughout this reign, but it was not, however, uniform in all the orders, nor did it follow the same logic of progression in those orders that were under the governance of Dom Jorge duke of Coimbra. Nevertheless, it is indisputable and significant that many of the nobles opted for the Order of Christ, followed by the Orders of Santiago, Avis and the Knights Hospitaller (see Fig. 33.3).

The substantial increase in the enrolment into the Order of Christ also testifies to a shift in the strategies hitherto adopted by some noble lineages: the abandonment of their traditional 'family order' in favour of the Order of Christ that, as it embodied royal policies, gave them better career and economic prospects, as well as other possible benefits. With this in mind, most of the families studied here elected to place as many as possible of their members in the Order of Christ from 1495 until the end of the reign. The Order eventually became the institution of choice for the large majority of families in stark contrast with the experiences of the Orders of Santiago and Avis (at least, as far as brother-knights were concerned). Among the lineages in this group were the Abreu, Ataíde, Azevedo, Brito, Castro/Eça, Coelho, Coutinho, Cunha/Albuquerque, Faria, Góis, Henriques, Melo, Meneses,

---

[37]   Silva, 'As comendas novas', p. 48.

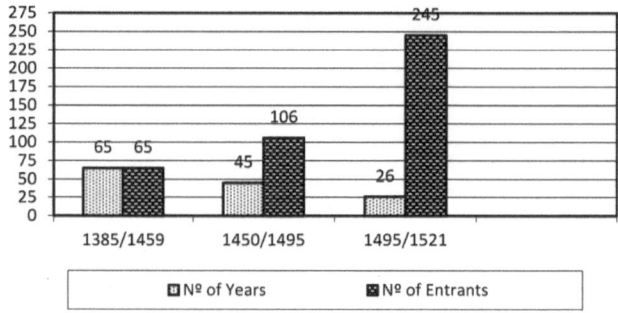

	Entrants	Entrants per annum
1385-1450	65	1
1450-1495	105	2.35
1495-1521	245	12.25

Figure 33.4   Number of entrants in all Military Orders (1495-1521) (graphics: A.P. de Vasconcelos and M.L. de Mendonça)

Moniz, Noronha, Pereira, Sá, Sequeira, Silva, Sousa (of Arronches and of Prado), Tavares, Távora and Vasconcelos.

It should, however, be noted that other lineages retained their preference for their traditional 'family order', even though, in adapting to the new circumstances, they simultaneously sought to enrol some of their members in the Order of Christ, as was the cases of the Almeida, Barreto, Furtado de Mendonça and Mascarenhas.

To sum up: the sociological profile of the entrants in the military orders underwent profound changes in the period running from the accession of the Avis dynasty to the end of the reign of Manuel I. Similarly, the appeal exerted by these institutions within the nobility also went through major shifts, the most important of which is shown by the substantial increase in enrolments into the orders, a trend that reached its peak during the reign of Manuel I, as is clear from the figure 33.4:

While between 1385 and 1495 all orders saw their ranks swell, in the following period the number of entrants into the Knights Hospitaller considerably decreased. This decline is exceptional, as the orders of Christ, Santiago and Avis all benefited from growing interest among the nobles. This was probably due to the fact that this Order was an international institution and was therefore excluded from the national expansionist policy of Manuel I. Similarly, it did not undergo the decisive reforms, such as those that allowed the brothers of Avis and Christ to marry. The recovery and growth of the Order of Christ clearly demonstrates its rising importance and influence within the realm, reaching its apogee from the moment it began to be governed and administered by the king himself. As thenceforward the rationale of the Order became fully in tune with the most important projects of the monarchy, membership conferred a clear advantage on all nobles who wished to take part in those projects and thereby gain access to royal patronage.

Chapter 34

The Portuguese Military Orders, the Royal Power and the Maritime Expansion (Fifteenth Century)

Luís Adão da Fonseca

There is a general belief among scholars that the military orders in Portugal, despite their religious inspiration and organization, had from the outset an intimate relationship with the Crown. I am referring, of course, to the Portuguese orders of Avis, Santiago and Christ; the Temple and the Hospital of St John, as international orders, lie outside the scope of this paper.

By the end of the fourteenth century, Portugal had three national orders, whose wealth, power and influence were closely linked to the monarchy. Their development during the second royal dynasty, which began in 1383, provides ample evidence for these links. At the time of the accession of the first member of the new dynasty, the masters of the three national orders, Fernão Rodrigues de Sequeira in Avis, Lopo Dias de Sousa in Christ, and Mem Rodrigues de Vasconcelos in Santiago, were all men who enjoyed the monarch's confidence.[1] The relationship became even more interdependent when these masters died and the king took the opportunity to replace them with his own sons.[2] Strictly speaking, the *infantes* were not the masters of the orders but their administrators, thereby embodying the orders' direct links with the Crown. Conscious of their importance, the monarch would invariably appoint members of the royal family to these positions, until in 1551 the orders were taken over by the Crown.[3] Thus, from 1418 onwards, all the governors of the Orders of Avis, Santiago and Christ were members of the Portuguese royal family.

---

[1] M.C.G. Pimenta, 'A Ordem Militar de Avis (durante o mestrado de D. Fernão Rodrigues de Sequeira)', *Militarium Ordinum Analecta*, 1 (1997), 127–242; I.M.S. Silva, 'A Ordem de Cristo durante o mestrado de D. Lopo Dias de Sousa (1373?–1417)', *Militarium Ordinum Analecta*, 1 (1997), 5–126; M.C.G. Pimenta, 'As Ordens de Avis e de Santiago na Baixa Idade Média: o governo de D. Jorge', *Militarium Ordinum Analecta*, 5 (2001), 39–42.

[2] *Monumenta Henricina*, ed. A.J.D. Dinis (Coimbra, 1960–74), 2, pp. 303–5, 367–9 and 5, pp. 70–72.

[3] *Gavetas [As] da Torre do Tombo*, ed. A.S. Rego, 2 (Lisbon, 1962), pp. 382–91, 402–7, 60–68, 392–9.

This policy had profound consequences. The military orders had, by the standards of that time, a fairly centralized organization and they possessed extensive estates that encompassed a significant proportion of the kingdom. They were therefore able to develop as institutions that during the later middle ages and the early modern period came to acquire a growing importance and relevance in Portuguese society as a whole. They acted in pursuit and defence of certain specific interests, frequently in keeping with clearly defined political and economic strategies. At the same time, throughout the fifteenth century, their links with the monarchy enabled them to acquire positions of great influence within the political life of the kingdom and act as channels for the various pressure groups and interests that were at stake.

It is impossible to understand Portuguese history in the fifteenth century without taking this fact into consideration. It explains why the new Portuguese dynasty consistently sought to involve the orders in their military initiatives outside the kingdom, thereby strengthening and extending a practice that dated back to the fourteenth century. For example, through the bull *Eximie Devocionis* (20 March 1411), the antipope John XXIII responded in positive terms to the king of Portugal's petition for the assistance of the military orders to be made available to him in any form of just war.[4] As mentioned already, from 1418 onwards the acquisition of control over the military orders by the king's children acquired an increasing significance.

This phenomenon had both political and sociological implications. It is therefore important to take into account the role of each order in the political and military history of the kingdom, their areas of interest, the weight and the social composition of their membership, and the similarity of their material and property interests.

In Portugal the military orders, whose establishment can broadly speaking be said to have occurred in the middle years of the twelfth century, soon acquired major importance. However, with the definitive conquest of the Algarve in the mid-thirteenth century, military activity gradually ceased to be their main concern. There is no doubt that they continued to guarantee the protection and maintenance of the castles and strongholds they had been given, many of which were situated in border areas, but it can be readily understood that war was becoming progressively less important for them. To some extent, this change in their social, political and military standing that had been particularly visible during the fourteenth century entailed a sharp decrease in crusading as their motivating force, something that until the thirteenth century had remained potent in the Iberian Peninsula. However, this traditional motivation never really disappeared. Adapting to the new constraints that were characterized by their involvement with the Portuguese monarchy in the later Middle Ages and early modern times was not a problem. In general, there were three main aspects of this process: the military orders' seigneurial attributes, the increasing secularization of their members' lifestyle, and, finally,

---

[4]   A.H.O. Marques, *A Expansão quatrocentista, Nova História da Expansão Portuguesa*, eds J. Serrão and A.H.O. Marques, 2 (Lisbon, 1998), p. 48.

their increasing dependency on royal power. The most significant moments in this process occurred between the early fifteenth and the mid-sixteenth centuries. In Portugal it was through the military orders that the impact of crusading as a military experience found expression, the most significant aspect being in what Norman Housley has described as the 'intangible field of social values'.[5]

To show how this process developed would be too ambitious for this paper, as I would have to reassess the whole of fifteenth-century Portuguese history. I shall therefore only draw attention to those issues I consider of the greatest interest.

To begin with the Order of Avis: the geographical focus of its endowments was in the northern Alentejo, to the south of the River Tagus. This geographical factor certainly helps explain the Order's special awareness of the importance of communications with the interior of the Iberian Peninsula, in other words, the political relations with the neighbouring kingdom of Castile. This can be seen, for example, in aspects of the 1383–85 crisis or at certain points in the government of Master D. Pedro in 1448–49 and 1463–66.[6]

In the case of the Order of Santiago, the situation is different. Here too the geographical spread of its properties was a determining factor, as were the diverse sources of their income. At the Order's headquarters and in some of the coastal commanderies, they were especially dependent on the sea, but in the commanderies that lay in the interior, they were more dependent on the cross-border movement of cattle to and from Castile. To give some examples: the Order of Santiago appears in association with the negotiations that led to the 1386 Treaty of Windsor;[7] it had important connections with the Mediterranean in the mid-fifteenth century;[8] at different times it assumed either a clearly anti-Castilian position,[9] or distanced itself from the Portuguese royal power,[10] and many of its members were in some

---

[5]   N. Housley, *Contesting the Crusades* (Malden, MA, 2006), p. 155.

[6]   L.A. Fonseca, *O Condestável D. Pedro de Portugal, a Ordem Militar de Avis e a Península Ibérica do seu tempo (1429–1466)* (Porto, 1982), pp. 48–63, 125–36; 'Algumas considerações a propósito da documentação existente em Barcelona respeitante à Ordem de Avis: sua contribuição para um melhor conhecimento dos grupos de pressão em Portugal em meados do século XV', idem. *Revista da Faculdade de Letras do Porto* (Porto, 1984), 19–56.

[7]   L.A. Fonseca, *O Essencial sobre o Tratado de Windsor* (Lisbon, 1986).

[8]   For example, we have to take into account the biography of Christopher Columbus. See L.A. Fonseca, 'Importanza e significato del soggiorno di Colombo in Portogallo', in G. Airaldi (ed.), *Cominciai a navigare in giovanissima età ... Genova e Cristoforo Colombo* (Genoa, 2004), pp. 41–54; idem, 'La colonia italiana in Portogallo, l'Ordine di Santiago e Colombo', in M. Macconi (ed.), *Genova Europa Mondo. Cristoforo Colombo cinque secoli dopo* (Genoa, 2006), pp. 53–73.

[9]   For example, the attitude of Prince D. João immediately after the death of King Duarte. Fonseca, *O Condestável*, p. 25.

[10]   For example, with King João II in 1484. L.A. Fonseca, *D. João II* (Lisbon, 2005), pp. 76–8.

way connected with maritime expansion. At different periods in the history of this Order, circumstances dictated widely varied political policies.[11]

What was important for the Order of Christ were the close links that, from the time of its foundation, bound it to the monarchy, rather than the location of its property. Even more potent was the constant updating of its vocation that associated crusade and maritime war. In the light of this triple profile (closeness to the king, crusading and the sea), the Order's role in Portuguese expansion is perfectly understandable.[12]

Indeed, the military order with the strongest participation in maritime expansion was undoubtedly the Order of Christ. One of the first aspects to be considered is the decision to bestow upon this Order spiritual jurisdiction over newly discovered lands. It is a problem that cannot be explained solely in the context of the military orders, but it is related to the broader cultural topic of the relationship between European expansion and the problems inherent in identifying the 'otherness' of conquered peoples. However, it undoubtedly shows the important role that the orders, in this case, the Order of Christ, had in conforming to the intellectual constructs that underlay that expansion.

The problem was already observable in the petitions Master D. Henrique addressed to Pope Eugenius IV in 1434. On 1 April of that year, the Portuguese prince presented the pope with a series of petitions, which were then granted. The first is very significant: it is a request for a confirmation of the foundation bull of the Order of Christ, with all its privileges and benefits.[13] The meaning seems obvious. Anticipating responsibilities in a new geographical and maritime realm,[14] the prince, as administrator of the Order of Christ, was seeking from the

---

[11]   L.A. Fonseca, 'As Ordens Militares e a Expansão', in J.O. Costa and V. Rodrigues (eds), *A Alta Nobreza e a Fundação do Estado da Índia* (Lisbon, 2004), pp. 321–47; idem, 'The Portuguese Military Orders and the Oceanic Navigations: From Piracy to Empire (Early Fifteenth to Sixteenth Centuries)', *MO* 4 (2008), pp. 63–73; M. Mendonça, 'As Ordens de Cristo e Santiago nos primórdios da Expansão Portuguesa (séculos XIV–XV)', in *Amar, sentir e viver a História. Estudos de homenagem a Joaquim Veríssimo Serrão* (Lisbon, 1995), pp. 859–84; M.C.G. Pimenta and I.M.S. Silva, 'As Ordens de Santiago e de Cristo e a fundação do Estado da Índia. Uma perspectiva de estudo', in Costa and Rodrigues (eds), pp. 349–86.

[12]   In addition to the studies by Fonseca cited in n. 11, see E.J.N.A. Jana, 'Considerandos sobre a presença da Ordem de Cristo no Ultramar português', in *Congresso Internacional de História Missionação Portuguesa e Encontro de Culturas*, 3 (Braga, 1993), pp. 423–39; also C.M. De Witte, 'Les bulles pontificales et l'Expansion Portugaise au XVe siècle', *Revue d'Histoire Écclesiastique*, 48 (1953), 683–719; 49 (1954), 438–61; 51 (1956), 413–53; 809–36; 53 (1958), 5–46; 443–71. For a collection of sources, see *Monumenta Henricina* (as n. 2).

[13]   Ibid., 4, pp. 335–8.

[14]   The first nautical explorations along the Moroccan coast started in 1419, but there is no doubt that it was from 1434 onwards that, due to the navigation of Cape Bojador, these expeditions were intensified. Marques, pp. 56, 62.

papacy official recognition for the activities he intended to develop in these new horizons. Other petitions relate to this one. I will mention only a few: the granting of indulgences to the faithful who visit the church of St Thomas of Canterbury in Tomar on the feast of St James;[15] the transfer of certain properties to the parish of St Marie of Africa, where the Order had built a church;[16] the plenary remission of sins for all Christians who, under the prince's command, fight against the enemies of the faith of Christ;[17] the appropriation by the Order of all churches built and endowed on the island of Madeira;[18] and also the appropriation of the hermitage of St Marie of Belém, where a priest appointed by the Order of Christ could hear confessions of people about to set sail.[19]

It seems clear that the purpose was to endow the Order of Christ with greater resources and more extensive rights. Accordingly, it is significant that on the same date (1 April 1434), the administrator, D. Henrique, requested the pope to confer the endowments belonging to the Castilian Order of Alcantara that were located in Portugal on the Order of Christ. During the Schism these endowments had been given to particular individuals; Henrique was arguing that Alcantara had made little progress in recovering them, but, in the hands of the Order of Christ, they would serve the cause of the Christian faith. The petition was granted.[20] It is interesting that, on the same day, the pope was asked about the possibility of revising the Order's rule,[21] and at the same time requested to confirm the Order's rights and freedoms in perpetuity.[22]

In January 1443, Pope Eugenius, through the bull *Etsi suscepti*, allowed Prince Henry to receive the habit and to profess in the Order of Christ, and at the same time he recognized the Order's spiritual rights over the Atlantic islands, and gave the church of St Marie of Africa of Ceuta parochial status. Later, on 8 January 1454, Pope Nicholas V confirmed a donation to the same Order by the king of Portugal, in which he granted the administration and spiritual jurisdiction over all coasts, islands and lands whether already conquered or yet to be conquered. This grant was to be confirmed in March 1456 in a bull of Calixtus III.[23] To quote a small part of the text:

---

[15] *Monumenta Henricina*, 4, pp. 343–4.

[16] Ibid, 4, pp. 345–6.

[17] Ibid, 4, pp. 347–9. For similar legislation, ibid, 7, pp. 336–7 (19 Dec. 1442); *Descobrimentos portugueses. Documentos para a sua história*, ed. J.M.S. Marques, 3 (Lisbon, 1971), pp. 242–3 (11 Sept. 1481).

[18] *Monumenta Henricina*, 4, pp. 354–5.

[19] Ibid, 4, pp. 357–8, 360, 361.

[20] Ibid., 4, pp. 352–3.

[21] Ibid., 4, p. 359.

[22] Ibid., 4, p. 361.

[23] Ibid., 8, pp. 2–4; *Descobrimentos portugueses. Documentos para a sua história*, ed. J.M.S. Marques, 1 (Lisbon, 1944), pp. 503–8.

> We perpetually decree, ordain and say that the spiritual and ordinary jurisdiction, the command and the authority, at least in spiritual matters, in these islands, towns, ports, lands and places acquired and to be acquired, beyond Cape Bojador and Cape No, and held throughout Guinea and as far as the southern shores of the Indies [...] be owned for ever by the Military Order (of Christ) [...] and we decree that those islands, lands and places are *nullius dioecesis*.[24]

The origin of this religious autonomy can be traced back to the decision of English pope Adrian IV in 1159 to exempt the Templar-owned church of St Marie of Olival from episcopal jurisdiction, thus bringing it into direct dependence on the Holy See.[25] This church is located in Tomar, the town that later became the headquarters of the Order of Christ. Probably as a consequence of the continuing restructuring of the Order of the Temple,[26] this vicariate would be later regulated – it is documented in early thirteenth-century sources – giving rise to the prelature *nullius dioecesis*.

This is a very important development, and for two reasons. Firstly, by the terms of the already quoted bull of 1456 and in the light of previous papal acts regulating the power of the vicar of St Marie of Olival, the Order of Christ had prerogatives that correspond to the ecclesiastical *padroado*, with the obligation of exercising pastoral care.[27] Secondly, these papal documents, especially the 1456 bull, define the rights and the places where they can be exercised. That in itself is not new: there was ample precedent for the establishment of a *padroado*. The great innovation in these fifteenth-century bulls is the enormous extension of the area in which it is to be applied: in place of a single church in Tomar, we move into the entire Atlantic. This bull therefore confirms to the king of Portugal and to the Order of Christ in the religious domain what, in the political domain, Nicholas V's *Romanus Pontifex* of 1454 had done in granting Portugal its achievements and discoveries, and now confirmed by Calixtus III in his bull of 1456.[28]

On the death of Prince D. Henrique in late October 1460, D. Fernando, his nephew and adopted son, who was also the brother of King Afonso V, succeeded. To judge by the letter *Dum tua* of 25 January 1461 by which the Order of Christ was delivered permanently to King Afonso, it would seem that the Portuguese monarchy's original idea was to bring the Order into direct dependence on the king

---

[24]   *Monumenta Henricina*, 12, pp. 286–8.

[25]   A.J.D. Dinis, 'A prelazia "nullius dioecesis" de Tomar e o Ultramar português até 1460', *Anais da Academia Portuguesa da História*, 20 (1971), 235–70. See also A.J.D. Dinis, 'A prelazia "nullius dioecesis" de Tomar e o Ultramar português na segunda metade do século XV', *Boletim Cultural da Guiné Portuguesa*, 27 (1972), 5–93; I.M.S. Silva, 'A Ordem de Cristo (1417–1521)', *Militarium Ordinum Analecta*, 6 (2002), 231–7.

[26]   Dinis, 'A prelazia' (1971), p. 243.

[27]   Ibid., p. 249.

[28]   Ibid., p. 263.

with immediate effect.[29] But this project was not implemented. Instead, on 11 July 1461, Pope Pius II, by the letter *Repetentes animo*, conferred the government of the Order on Prince D. Fernando.[30] Fernando adopted the same course of action, and henceforth control of the Order of Christ remained in his own family.

With King Manuel, during whose government the Portuguese arrived in India, the situation became extremely complicated, not least because of the stance taken by the neighbouring kingdom of Castile. Apparently, the diplomatic *status quo* agreed in the treaties of Alcáçovas-Toledo (1479–80) and Tordesillas (1494) remained in force. But with the arrival of the Portuguese in the Indian Ocean, a minimalist interpretation of these treaties in some Castilian quarters created difficulties. These groups advocated an interpretation that limited the application of the previous treaties to the Atlantic. They argued, therefore, that the voyage of Vasco da Gama would have meant annulling the Treaty of Tordesillas.[31] This is a well-known problem and explains several of the measures taken by the king of Portugal. For example, it enables us to understand the new title adopted by Portuguese monarchs 'Lord of the Conquest, Navigation and Commerce of Ethiopia, Arabia, Persia and India', announced in the early summer of 1499, immediately after Vasco da Gama's arrival in Lisbon.[32] This title is, in the words of the chronicler João de Barros, a 'sign and denotation of law and justice that each has over what he has'.[33]

Already in the 1480s, at the end of the reign of Afonso V and before the final attempts to navigate the southern extremity of Africa, the Portuguese monarchy had, through the bull *Aeterni Regis clementia* (21 June 1481), secured papal support. Confirming the bulls of the earlier popes, Nicholas V (in 1454) and Calixtus III (in 1456) – already quoted in this paper – Pope Sixtus IV accepted that they were commensurate with the Portuguese–Castilian demarcation treaty concerning their Atlantic territories as signed in 1479–80, and he incorporated a partial transcription of this treaty in this bull.[34] Accordingly, *Aeterni Regis clementia* is very important, again for two reasons: first, because it confirmed a series of papal documents, thereby demonstrating the continuity of papal policy, and secondly, because it was the pope himself who, in this sequence, expressed the donation of spiritual rights to the Order of Christ and their practical results in terms of a definition of sovereignty that had now attained diplomatic recognition.

---

[29]   *Monumenta Henricina*, 14, pp. 126–9.

[30]   Ibid., 14, pp. 158–62.

[31]   L.A. Fonseca, 'O regresso de Vasco da Gama e a definição da estratégia marítima portuguesa em finais do século XV', in *Sessão Solene do Lançamento do 3° vol. da História da Marinha Portuguesa. 19 de Maio de 1999* (Lisbon, 1999), pp. 13–20.

[32]   J.M. Garcia, 'Carta de D. Manuel a Maximiliano sobre o descobrimento do caminho marítimo para a Índia', *Oceanos*, 16 (1993), 28–32; J. Barros, *Ásia*, 1 (Lisbon, 1945), p. 174.

[33]   Ibid., pp. 227–8.

[34]   *Descobrimentos Portugueses*, 3, pp. 223–9, 230–38.

Later bulls that confirmed the Order of Christ's spiritual jurisdiction in lands beyond the sea reiterated the doctrine already set out in these fifteenth-century bulls. In my view, Pope Alexander VI's *Cum sicut nobis* of 23 August 1499 is a key document in this process. It grants in perpetuity to the king of Portugal, who was now also the governor of the Order of Christ, *padroado* rights in all African churches.[35]

As the end of the century approached, Alexander VI's main concern in his relations with the Iberian kingdoms was the war against the Turks. But aside from this issue, he was also concerned with problems arising from overseas expansion. On more than one occasion the pope responded in positive terms to the king of Portugal's requests relating to Lusitanian military campaigns in North Africa. It is interesting to note that in some cases, the pope expressed a concern not to differentiate between the two Iberian monarchies.[36]

The intricacies of papal policy are well illustrated in the situation that then developed. In the bull *Ineffabilis et summi Patris Providentia* (1 June 1497), Pope Alexander granted perpetual possession of the lands conquered from the infidels to King Manuel and his successors without prejudice to the Christian monarchs who are entitled to these lands.[37] Apparently, he was upholding the formal equality between Portugal and Castile as maintained since the beginning of his pontificate. But in contrast to what had happened in 1493,[38] now, in 1497, the practical effect of the papal letter was to benefit the Lusitanian side.[39]

This interpretation is clearly expressed in the three bulls of 26 March 1500. Through the first, *Cum sicut majesas tua*, King Manuel, as governor of the Order of Christ, was empowered to appoint an Apostolic Commissioner with ordinary powers for the Portuguese places in the Indian Ocean beyond the Cape of Good Hope.[40] Through the second, *Cum sicut nobis super majestas tua*, and for the same reason, the king was granted a third of the rights in all the lands already conquered and yet to be conquered in the Indian Ocean beyond the Cape of Good Hope.[41] Through the third, *Exponi nobis*, addressed to the bishops of Viseu, Guarda and Fez, the pope authorized, at King Manuel's request, the founding of monasteries and mendicant houses in places in the Indian Ocean beyond the Cape of Good Hope.[42]

---

[35]   Ibid., 3, p. 548.

[36]   L.A. Fonseca, 'Alexandre VI e os descobrimentos portugueses', in M. Chiabò, S. Maddalo, M. Miglio and A.M. Oliva (eds), *Roma di fronte all'Europa al tempo di Alessandro VI*, 1 (Rome, 2001), pp. 227–47.

[37]   *Descobrimentos Portugueses*, 3, pp. 479–80.

[38]   L.A. Fonseca, *O Tratado de Tordesilhas e a diplomacia luso-castelhana no século XV* (Lisbon, 1991), pp. 45–51.

[39]   Fonseca, 'Alexandre VI'.

[40]   *Descobrimentos Portugueses*, 3, pp. 591.

[41]   J.S. Abranches, *Fontes do direito eclesiástico português 1. Suma do bulário português* (Coimbra, 1895), p. 52.

[42]   F.F. Lopes, 'Fr Henrique de Coimbra. O missionário, O diplomata. O bispo', *Colectânea de estudos de História e Literatura*, 3 (Lisbon, 1997), pp. 425–6.

What is the importance of these three documents? They are all dated March 1500, just a few days after the departure for India of the fleet commanded by Pedro Álvares Cabral. Probably, they would have been requested in the summer of 1499, following the return to Lisbon of Vasco da Gama, the first Portuguese captain to sail to the East. These three papal documents recognize the legitimacy of Lusitanian actions in the area beyond the Cape, and so provide a clear expression of the pope's support for Portuguese expansion in the Indian Ocean. At the same time, they skirt around and implicitly deny the opposite position as taken in some quarters in Castile.

It should be emphasized that papal recognition was effected through the Order of Christ. And so it remained until 1551,[43] with the incorporation of the Orders into the Crown, ecclesiastical jurisdiction over the overseas territories was assigned to the monarch, which led to the gradual decline of this jurisdiction (a situation already evident from the beginning of the sixteenth century).

It should be added that this continuous policy of allocating such extensive spiritual powers to the Order of Christ and, later, to the monarchy has to be carefully understood. There were obvious religious and political motives. But it is essential to remember the need for conceptual and institutional formulae to legitimize the military and political presence of the Portuguese monarchy in distant lands. The Portuguese navigation of the coast of western Africa and then the south Atlantic and the Indian Ocean in turn forced them to confront a wide range of challenges – geographical, cultural, political, military and commercial – in which the only common denominator was that they were unexpected and new. The expeditions have traditionally been described as experiences of discovery, but in fact that was only true in some instances. Indeed, at that time, what the Portuguese did was to integrate these experiences in terms of their existing cultural and conceptual *schemas*. That was true of those who went to the various hitherto unknown places and also, subsequently, of those in power in Lisbon who had to understand the information they received. How these experiences were understood was not always adequate or consistent. Dealing with pirates on the coasts of Morocco and Mauritania, sailing the ocean in search of more appropriate routes (whether to the Atlantic islands or even Brazil), and arriving in India with all the attendant political and military problems, all entailed the clash of civilizations and simultaneous cultural misunderstanding, which could only be properly understood with the benefit of hindsight.

The problem belongs to the social history of intellectual understanding (if it makes sense to use this expression), and that makes it necessary to consider each historical situation separately, for each is different.

In this context, the role played by the Order of Christ has a special importance. Thanks to its ideological orientation and its traditional connection with the monarchy, it was employed as the institution that enabled the kings of Portugal in

---

[43]   I will only cite as examples the bulls of 7 June 1514 and 3 Nov. 1514. A.C. Sousa, *Provas da História Genealógica da Casa Real Portuguesa*, 2,1 (Coimbra, 1947), pp. 269–97.

a wide variety of situations to formulate and legitimize Portuguese intervention, both internally and within the diplomatic framework, especially with regard to Castile.

The privileges granted to this Military Order may therefore have the following interpretation: they gave them the institutional instrument for the monarchy to extend the traditional doctrine of *Padroado*, as defined since the time of Pope Nicholas II, to the overseas territories. They were, strictly speaking, not merely spiritual benefits, but material, linked to the spiritual. This is indeed the doctrine of St Thomas Aquinas:

> Quaedam autem sunt annexa spiritualibus inquantum ad spiritualia ordinantur; sicut jus patronatus, quod ordinatur ad praesentandum clericos ad ecclesiastica beneficia.[44]

In this case papal approval was a consequence of two ideas: the necessity of accepting the role of the Holy See for the validation of these privileges, and the then current doctrine that made the pope the universal lord of the land in Infidel possession. In this context, this Military Order was ideally equipped to facilitate the widening process mentioned earlier: it was legally an ecclesiastical institution mostly comprising lay people, first among them being the king himself.

The Order of Christ is thus the institution *par excellence* that, from the beginning of the fourteenth century, embodied the continuity of the crusading concept in combination with expertise in maritime navigation. It was also, since its creation, an obedient instrument in the hands of the monarchy. Its history illustrates the close connection in Portuguese history between the expansion of Christianity, political ambition and economic interests. Backed by a significant number of papal bulls, the Order of Christ serves as a good example of the perceptive comment by António Saldanha:

> Perhaps the place where the delicate balance between the ideas of trade, war and preaching is reflected with the greatest intensity is in the content of the papal bulls, especially in light of the way in which they were interpreted.[45]

It should be noted that these bulls were an important element in the legal basis of the Portuguese monarchy's overseas ambitions.

Finally, in sociological terms, the Order of Christ provided an appropriate tool for fostering loyalties adjusted to these overseas activities. For example, it is no coincidence that a significant element among those responsible for the first

---

[44]   Cited in A.S. Rego, *História das Missões do padroado português do Oriente*, 1 (Braga, 1993), p. 92.

[45]   A.V. Saldanha, 'Sobre o Officium Missionandi e a fundamentação jurídica da Expansão Portuguesa', in *Congresso Internacional de História Missionação Portuguesa e Encontro de Culturas*, 3 (Braga, 1993), p. 557.

voyages to India were knights of the military orders,[46] and that, in the reign of King Manuel, the pope at the monarch's request, authorized the transfer of knights from other orders to the Order of Christ. Originally such movements between orders had been very difficult to secure, but now it should be understood as arising from the kings' concern to broaden the 'universe of his support'. The first documentary reference where this concern is noticed dates to October 1501,[47] but it was only explicitly sanctioned some years later, on 24 January 1506, through Pope Julius II's bull *Sincerae devotionis*. There the pope authorized knights from other orders to transfer to the Order of Christ.[48] The significance of this provision is clear. Manuel was king and also governor of the Order. Accordingly, he was using the Order under his control as a vehicle for recruiting the military personnel he needed for his political programme.

In short, Norman Housley was absolutely right to remind us of the fact that, in the Iberian crusade of the fifteenth century, we witness the replacement of the notion of *recuperatio* by *dilatatio* as the justifying and legitimizing elements of the crusade.[49] But I believe that the implications of this conceptual evolution are even greater. As is the case with the Order of Christ, the concept of crusade acquires the function of legitimizing the actions of the monarchy and also of determining the area where it can act. In my view, this is extremely important.

In the context of Portuguese history, I have tried to define the circumstances and the chronology in which the notion of crusade lost its substantive value and became an appealing argument that won people over by persuasion. And I have also wanted to demonstrate that the Order of Christ represents a decisive element in this process that was, at the same time, institutional, political and intellectual.

---

[46] L.A. Fonseca, 'Os comandos da segunda armada de Vasco da Gama à Índia (1502–1503)', *Mare Liberum*, 16 (1998), 11–32; idem, 'O significado político em Portugal das duas primeiras viagens à India de Vasco da Gama', in *Conferência Internacional Vasco da Gama e a Índia*, 1 (Lisbon, 1999), pp. 69–100.

[47] L.A. Fonseca, 'Alessandro VI e l'espansione oceanica: una riflessione', in M. Chiabò, A. M. Oliva and O. Schena (eds), *Alessandro VI dal Mediterraneo all'Atlantico* (Rome, 2004), pp. 221–33.

[48] Ibid., p. 230.

[49] N. Housley, *The Later Crusades: From Lyons to Alcazar. 1274–1589* (Oxford, 1992), pp. 288, 309.

Chapter 35

The Port City of Setúbal (Portugal) under the Control of the Order of Santiago (1400–1550)

Ana Cláudia Silveira

The capture of Lisbon in 1147 was of major significance in the advance of the *Reconquista* in the Iberian Peninsula. In order to consolidate control over this important military gain, the defence of the Sado estuary was crucial for creating the necessary conditions for both the conquest of territories further south and the settlement of new communities in the region.

In view of these considerations, it is easy to understand the interest of the Portuguese kings in enlisting the support of institutions such as the military orders with their warrior vocation to consolidate their conquests, and these included the strategic territories around the Sado estuary and a number of maritime ports. Accordingly, as a result of its participation in the *Reconquista* in the course of the twelfth and thirteenth centuries, the Order of Santiago received large endowments and privileges from the Portuguese monarchy. With headquarters at Alcácer do Sal in the earlier stages and at Palmela in the fifteenth century,[1] the estates of this Order extended from Almada, on the south bank of the River Tagus, along the Alentejo's Atlantic coast to the Algarve.

In the course of the fourteenth century, as a consequence of the economic developments occurring in Europe since the eleventh century and also as a result of the political and military stabilization of this region, the port-settlement of Setúbal, located on the Sado estuary, was able to build up its position between the Order of Santiago's other urban spaces in this area, namely the fortified villages of Palmela and Alcácer do Sal. It was in this context that in the middle of the thirteenth century Paio Peres Correia, the grand master of Santiago, granted Setúbal a borough charter that offered a range of rights and privileges to potential settlers. The aim of this document was to further the consolidation of the settlement and encourage its economic development. Some years later the Order of Santiago was to receive from the crown the taxes levied on the fisheries in the Setúbal peninsula, and these would have been of considerable importance even in this early period. The royal grant mentions salt-production as well as the preparation and conservation of fish through the salting and seasoning processes. The Order

---

[1] I.C. Fernandes, *O Castelo de Palmela: do islâmico ao cristão* (Lisbon, 2004), p. 287.

therefore obtained an important income from the exploitation of salt marshes and fisheries around Setúbal and from the associated development of trade.[2] It seems that, thanks to the exceptional conditions afforded by the mouth of the River Sado, Setúbal was marked out for a unique position among the Order of Santiago's estates. As other studies of the possessions incorporated into the domains of the military orders in the Iberian Peninsula have shown,[3] the financial efforts the orders made to ensure the defence of their territories led to the development of strategies designed to raise the necessary income from them. Furthermore, we can be certain that there was a deliberate policy behind the interventions of the Order in its territorial administration, and it is in this context that we should interpret the concession of a borough charter to Setúbal in 1249,[4] the agreement between the master of Santiago and the king about the assessment and collection of taxes in the port of Setúbal,[5] the organization of the commanderies in the time of Master Pero Escacho (1327),[6] and the definition of the areas comprising the territory of Setúbal, which were transferred from the neighbouring municipalities of Palmela and Alcácer do Sal, a decision confirmed by King Afonso IV (1325–57) in 1343.[7]

Subsequently, the settlement of Setúbal became the main source of income for the *Mesa Mestral* of the Order of Santiago, and that explains the frequent presence of the master there. By the fourteenth century, the authority and power of the Order of Santiago were made visible in the urban fabric of Setúbal by the presence of remarkable buildings, notably the *paço da Ordem* (palace of the Order), first mentioned in 1339, which was where the master stayed when visiting Setúbal.[8] It was seen as a symbol of the Order's lordship, and it was also the place where the dues owed to the Order had to be paid.

However, the economic, military and political importance assumed by the Order of Santiago during the thirteenth and fourteenth centuries led the

---

[2]   A.A. Andrade and A.C. Silveira, 'Les aires portuaires de la péninsule de Setúbal à la fin du Moyen Âge: l'exemple du port de Setúbal', in M. Bochaca and J.-L. Sarrazin (eds), *Ports et littoraux de l'Europe atlantique. Transformations naturelles et aménagements humains (XIVe–XVIe siècles)* (Rennes, 2007), pp. 158–65.

[3]   Philippe Josserand, 'Nourrir la guerre. L'exploitation dominiale des ordres militaires en Castille aux XIIIe et XIVe siècles', in M. Bourin and S. Boisselier (eds), *L'Espace Rural au Moyen Age. Portugal, Espagne, France (XIIe–XIVe siècles). Mélanges en l'honneur de Robert Durand* (Rennes, 2002), pp. 167–92.

[4]   *Portugaliae Monumenta Histórica, Leges et Consuetudines*, 1 (Lisbon, 1856), p. 634.

[5]   *Descobrimentos Portugueses. Documentos para a sua História*, Suplemento ao vol. 1, ed. J. da Silva Marques (Lisbon, 1988), pp. 11–13.

[6]   I.M. de C.L. Barbosa, 'A Ordem de Santiago em Portugal nos Finais da Idade Média (normativa e prática)', *Militarium Ordinum Analecta*, 2 (1998), 231–6.

[7]   A. Pimentel, *Memória sobre a História e a Administração do Município de Setúbal*, 2nd edn (Setúbal, 1992), pp. 136–40.

[8]   Arquivo Histórico da Torre do Tombo (hereafter ANTT), Mosteiro da Santíssima Trindade de Lisboa, Livro 99, fols 229–31v.

Portuguese kings, who were still trying to consolidate their position after the end of the *Reconquista*, to develop strategies to bring under their control the most important territories, that is to say, those with military and economic value such as the port of Setúbal.[9] Accordingly, during the first quarter of the fifteenth century, King João I (1383–1433) adopted a policy of entrusting the government of the Portuguese military orders to his sons.[10] Thus in 1418 Prince João, the son of the King João I, was appointed administrator of the Order of Santiago, and his presence in Setúbal became frequent. After his death, his nephew D. Fernando, the brother of King Afonso V (1438–81) was designated his successor, but as he was still under the legal age, the administration of the Order was controlled for some years by the *Infante* D. Pedro, the regent for Afonso V.[11] The government of the Order was later granted to Prince João, the future King João II (1481–95), and to his illegitimate son, D. Jorge de Lencastre, who controlled the Order until 1550. After his death, the Order was finally annexed by the crown.

During this period, the grand masters or administrators of the Order, along with their families and officials, were a permanent reality in Setúbal. In fact, the attraction exerted by this port-village and its growing importance in a local, national and even international context can be confirmed by an analysis of the urban patriciate that included knights of the Order of Santiago as well as important members of the royal bureaucracy.

These developments prompt several questions. What consequences did the presence and administration of the Order of Santiago have for Setúbal? How did the interests of the Order of Santiago and the interests of the crown interact? Did the officers of the Order insinuate themselves into local institutions? What kind of relationship did the officers of the Order establish with local government? How did these issues impact on the organization of urban space and on local social, economic and political dynamics? The exploitation of the natural resources of the area around Setúbal provides a good perspective for analysing these questions.

As for the exploitation of the salt marshes, there is evidence from as early as the first quarter of the fourteenth century for construction of salt pans owned by the Order.[12] However, during the fifteenth century investment in this activity

---

[9]   For the study of the royal strategies for the Portuguese ports during this period, see A.A. Andrade, 'A estratégia régia em relação aos portos marítimos no Portugal medieval: o caso da fachada atlântica', in *Ciudades y Villas Portuárias del Atlântico en la Edad Media. Najera. Encuentros Internacionales del Medievo: 27–30 de Júlio 2004* (Logroño, 2005), pp. 57–89.

[10]   L.A. Fonseca, 'The Portuguese Military Orders and the Oceanic Navigations: From Piracy to Empire (Fifteenth to Early Sixteenth Centuries)', in *MO* 4, pp. 64–5.

[11]   ANTT, Chancelaria de D. Afonso V, Livro 24, fols 19v and 32.

[12]   Arquivo Distrital de Setúbal (hereafter ADS), CNSA/B/004/Lv1, Parte II, fols 172–172v.

was not promoted directly by the Order of Santiago, but by various private
individuals who nevertheless needed special authorization from the master
of the Order before they could begin any development on the salt marshes.
Analysis of the surviving documents, although relating to only a small part of
the properties, provides confirmation that there was a considerable expansion
of the area occupied by salt pans during the fifteenth and sixteenth centuries.
These salt pans became increasingly important as a result of the crises that
affected some of the French production areas during the fourteenth, fifteenth
and sixteenth centuries, and these have been linked to a wide range of factors
including the consequences of the Hundred Years War, the destruction of salt
pans by serious storms, the raising of tax assessment on salt production, with an
obvious effect on prices,[13] or the tying up of capital and workers in viticulture.[14]
This period also corresponds to an increase in the demand for this product by
the Dutch fleets.[15]

In fact, at the beginning of the sixteenth century, Setúbal was already
the principal salt production area in Portugal.[16] The development of the salt
production in the Sado estuary has to be explained as a result of the strategies
put in place by the Order. In 1422, a few years after his nomination as
administrator of the Order of Santiago, Prince João obtained royal permission
to introduce improvements into the mismanaged lands belonging to the Order.[17]
During the following years, major investments seem to have been made in some
salt marshes on the south bank of the Tagus estuary, [18] and in the Sado estuary.
Furthermore, the participation in these investments of knights and officers of

---

[13]   J.-C. Hocquet, *Le Sel et le Pouvoir. De l'An Mil à la Révolution Française* (Paris,
1985), p. 156; J.-L. Sarrazin, 'Les ports de la Baie à la fin du Moyen Âge: évolution des
rivages et problèmes d'accès' in Bochaca and Sarrazin (eds), pp. 45–9; A. Malpica Cuello,
"La sal del reino de Granada en el marco de las actividades salineras bajomedievales
(siglos XIII–XV)', in J.-C. Hocquet and J.-L. Sarrazin (eds), *Le Sel de la Baie. Histoire,
archéologie, ethnologie des sels atlantiques* (Rennes, 2006), pp. 296–7; P. Pourchasse, 'La
concurrence entre les sels ibériques, français et britaniques sur les marches du Nord au
XVIIIe siècle', in Hocquet and Sarrazin (eds), pp. 325–8.

[14]   M. Tranchant, 'La place du sel dans l'économie rochelaise de la fin du Moyen
Age', in Hocquet and Sarrazin (eds), pp. 223–7.

[15]   P.C. Emmer, 'The Dutch Salt and Sugar Trades, 1585–1650', in I. Amorim (ed.),
I Simpósio Internacional sobre o Sal Português (Porto, 2005), pp. 29–37; Malpica Cuello,
pp. 296–7; Pourchasse, pp. 325–8.

[16]   J.C. Pereira, 'Organização e administração alfandegária de Portugal no século XVI
(1521–1557)', in *Portugal na Era de Quinhentos. Estudos vários* (Cascais, 2003), p. 86.

[17]   A.C. Silveira, 'Novos contributos para o estudo dos moinhos de maré no Estuário
do Tejo: empreendimentos e protagonistas (séculos XIII–XVI)', in A.A. Andrade,
H. Fernandes and J.L. Fontes (eds), *Olhares sobre a História. Estudos oferecidos a Iria
Gonçalves* (Lisbon, 2009), pp. 581–610.

[18]   Silveira, 'Novos contributos', pp. 581–610; eadem, 'New Contributions to the Study
of Tide Mills of the Tagus Estuary: The Case of Seixal', in J.A. Miranda and M. Harverson

Santiago as individuals was crucial, although special permission from the Order was necessary. The Order also benefited from these investments thanks to the tax assessment and collection, receiving not only the tithe of the production but also the taxes relating to the salt transactions.[19]

As was true of several European salt production areas, this type of investment represented a common strategy followed by a large range of institutions and individuals. Among them we find several monasteries, business men, members of the traditional aristocracy and even the king himself.[20] In some cases, investment in salt marshes gave rise to an economic and social enterprise that led to political involvement in local government.[21] The study of Setúbal as a specific case, besides confirming the existence of a wide range of interests connected with salt production and the salt trade, makes possible the identification of several men, who, as investors in salt production during the fifteenth and sixteenth centuries, were either connected with the administration of the Order of Santiago or with local government or, sometimes, with both (see Table 35.1).

Exploitation of the salt marshes was also bound up with the construction of tide mills, a trend that features not only on the Sado estuary but also at other appropriate locations on the estates of the Order of Santiago including the estuaries of the Mira[22] and the Tagus.[23] These mills had an important role in supplying flour to a growing population and were also important for the circulation of salt water in the marshlands, essential for the management of the salt pans, as has been noted in respect of other salt production areas.[24] This kind of investment required a large amount of money, and, like the construction of salt pans, also needed the Order's permission, which those who were close to its administration found easier to obtain.

(eds), *Transactions 11th International Symposium of The International Molinological Society, Portugal, 25th September–2nd October 2004* (Belas, 2007), pp. 153–62.

[19]   ANTT, *Mesa da Consciência e Ordens, Ordem de Santiago / Convento de Palmela* (hereafter *MCO, OS/CP*), Livro 10, fol. 136v.

[20]   Hocquet, pp. 59, 112–20, 126–9.

[21]   J. Briand, 'L'exploitation de salines à la fin du Moyen Âge: les possessions des Blanchet dans la Baie de Bourgneuf', in Hocquet et Sarrazin (eds), pp. 125–36.

[22]   A.M. Quaresma, *Rio Mira – Moinhos de Maré* (Aljezur, 2000).

[23]   Silveira, 'Novos contributos', pp. 581–610; eadem, 'New Contributions', pp. 153–62.

[24]   For Venice, see Hocquet, p. 74; for France, see Sarrazin, pp. 47–50.

Table 35.1   Investments made by the knights, officers and tenants of the Order
of Santiago in the salt marshes of Setúbal (1439–1550)

Year	Knights and Officers of the Military Order of Santiago	Form of investment in the salt marshes of Setúbal
1439	Vasco Peres	Salt pan
1457	Gonçalo Pires de Andrade	Salt pan
1460	Diogo Dias	Salt pan
1468	João Pires	Salt pan
1482	Rui Mendes	Salt pan and Tide mill
1485	Diogo Aires	Salt pan
1485	Álvaro Esteves	Salt pan
1485	Álvaro de Vila Franca	Salt pan
1495	Vasco Queimado	Salt pan
1504	Fernão Rodrigues	Salt pan
1518	Nuno Fernandes	Tide mill
1522	Fernão de Reboredo	Salt pan
1522	João Pires Carvalho	Salt pan
1525	Bastião Antunes	Salt pan
1525	Nuno Vasques	Salt pan
1527 (before)	Jerónimo Pires	Salt pan
1527	Diogo Carvalho	Salt pan
1527	Fernão Velho	Salt pan
1527	Afonso Limão	Salt pan
1527	Nuno Casado	Salt pan
1527	Gonçalo de Ferreira	Salt pan
1527	João Vaz do Castelo	Salt pan
1529 (before)	Pero de Gouveia	Tide mill
1529	Pero Coelho	Salt pan and Tide mill
1529	Fernão de Miranda	Salt pan and Tide mill

Year	Knights and Officers of the Military Order of Santiago	Form of investment in the salt marshes of Setúbal
1530	Jorge Barreto	Salt pan
1530	João Rodrigues de Lucena	Salt pan and Tide mill
1530	Jorge de Quebedo	Salt pan
1530	Lisuarte de Lis	Salt pan
1533 (before)	João da Costa	Salt pan
1537	Luís Soares	Salt pan
1541	Tristão Delgado	Tide mill
1542	Francisco de Faria	Salt pan
1554 (before)	António de Gouveia	Tide mill

The participation of men who were close to the administration of the Order of Santiago in this sort of investment seems to demonstrate a strategy of active participation in the emergent opportunities. It should not be forgotten that during this period Portuguese fleets were involved in long distance expeditions to the Orient and Brazil, and that Portugal held several fortresses in Africa, activities that required a significant amount of biscuit for the sea voyages. The increasing demand for flour justified the construction of tide mills in the most active harbours, since these buildings were able to function all year and produce more than any other kind of mill.

If it is true that the intervention in the salt marshes required considerable investment, it is also true that the profits and consequent social reputation were substantial, as can be seen by an analysis of the later *cursus honorum* of some of the men, who were members of the urban aristocracy. This trend was identical everywhere. Economic power was in the hands of a leading group whose members had enriched themselves by a variety of means, but with the investment in rural and urban property being a common feature.[25]

It is possible to confirm the participation of some of the knights, officers and tenants of the Order of Santiago in the local government of the city of Setúbal, especially from the last quarter of the fifteenth century onwards (see Table 35.2). This may also have been true earlier, although the organization of the chancery of the Order of Santiago, which dates to the last quarter of the

---

[25]   F. Sabaté i Currull, 'Oligarchies and Social Fractures in the Cities of Late Medieval Catalonia', in M. Asenjo-González (ed.), *Studies in European Urban History (1100–1800)*, 19 – *Oligarchy and Patronage in Late Medieval Spanish Urban Society* (Turnhout, 2009), pp. 1–3.

fifteenth century, does not allow us to have a complete picture of the Order's social structure during the previous period.

Table 35.2   The participation of knights, officers and tenants of the Military Order of Santiago in the government of the city of Setúbal (1450–1550)

Year	Name
1451	João Fernandes Chanoca
1455	Fernão Rodrigues
1466	João Rodrigues Mousinho
1469	Pero de Vila Real
1477	Gomes da Serra
1481	Fernão Rodrigues
1482	Fernão Rodrigues
1485	Fernando Afonso de Aguiar
1489	Fernando Afonso de Aguiar
1490	Gonçalo de Freitas
1491	Fernando Afonso de Aguiar
1496	Gonçalo Eanes
1505	Pero Vaz do Castelo
1509	Duarte Teixeira
1510	Nuno Casado Gomes da Serra
1511	Fernão de Reboredo Gomes da Serra
1520	Gomes da Serra
1522	Nuno Casado Gonçalo Pereira
1523	Diogo de Velosa
1524	Diogo de Velosa Gomes da Serra
1527	Gonçalo Piteira Jorge Piteira
1528	Jorge Piteira Henrique Lobo
1529	António de Gouveia
1531	Henrique Lobo

Year	Name
1532	Nuno Álvares João Rombão Gomes da Serra
1533	Henrique Lobo António de Gouveia
1534	Brás Cordeiro António de Gouveia Gomes da Serra
1539	António de Gouveia Gomes da Serra
1540	João Rombão Gomes da Serra
1541	Brás Cordeiro
1542	António de Lucena Brás Cordeiro João Martins Gomes da Serra
1545	Martim Cordeiro Gomes da Serra António de Gouveia

Despite the conflicts between the Order of Santiago and the local government of the city of Setúbal that occurred during the fourteenth[26] and fifteenth centuries,[27] it is possible to demonstrate that relations became more peaceable from the second half of the fifteenth century onwards. This improvement can be explained by the control exerted by the Order over the local administration, either by the continuous presence of its own knights or officers in local government institutions or as a result of the convergence between the economic interests of the Order of Santiago and those of the local power elites. In fact in 1505 the Order's administrator chose two of the members of the local government from among the six candidates that were presented to him, and this was said to be an old practice that took place there. On the other hand, the participation of the senior officer of the Order of Santiago in the local assembly meetings was also a common practice during the first half of the sixteenth century.[28] Another process used for controlling the institutions of local government was the permission given by the Order in agreement with the king for a certain type of magistrate in the service of the Order of Santiago to be made aware of decisions taken during the meetings of the city council.[29]

---

[26]   *Chancelarias Portuguesas. D. Afonso IV*, 3 (1340–44) (Lisbon, 1992), pp. 154–63.

[27]   ANTT, *Chancelaria de D. Afonso V*, Livro 2, fols 5v–7v.

[28]   M.C.G. Pimenta, *As Ordens de Avis e de Santiago na Baixa Idade Média. O Governo de D. Jorge* (repr. from *Militarium Ordinum Anacleta*, 5) (Palmela, 2002), p. 432.

[29]   ANTT, *MCO, OS/CP*, Livro 272, fols 349v–350.

In addition, the participation of officers of the Order of Santiago in the royal administration at local level and its reality in Setúbal must also be considered (Table 35.3).

Table 35.3   The participation of officers and tenants of the Order of Santiago in the royal administration in Setúbal

Year	Name
1433	Diogo Gomes
1435	Diogo Gomes
1436	Diogo Gomes
1439	Diogo Gomes
1439	Fernão Rodrigues
1440	Fernão Rodrigues
1440	Estevão Rodrigues
1440	João de Coimbra
1440	Diogo Martins
1441	João de Elvas
1443	Fernando Afonso
1444	Diogo Vasques
1446	Fernão Rodrigues
1446	Estevão Rodrigues
1449	Fernão Rodrigues
1453	Gonçalo Eanes
1453	Fernão Lourenço
1454	João Pires
1455	Pero Lourenço
1457	Fernão Rodrigues
1469	Martim Calado
1471	João Rodrigues Mousinho
1473	João de Palmeira
1475	Diogo Fernandes

Year	Name
1478	Álvaro Dias
1482	João Pires
1482	Jerónimo Pires
1484	Jerónimo Pires
1486	Martim Calado
1496	Jerónimo Pires
1501	Pero Gonçalves Calado
1502	Duarte Teixeira
1504	Álvaro Monteiro
1508	Jerónimo Pires
1511	João Rodrigues
1511	Vicente Anes
1511	Fernão Rodrigues
1513	Jorge Cabedo
1513	João Vaz do Castelo
1521	Vicente Eanes
1539	Cristóvão Gomes
1539	Cristóvão Mouzinho
1539	Jerónimo Luís
1539	Manuel Fernandes Preto
1542	João de Lis
1542	Lisuarte de Lis
1543	Brás Cordeiro

The increasing complexity of administrative tasks required the employment of the best equipped, and those who made a career as officers of the Order of Santiago held an advantageous position. Bureaucratic specialization occurred in trading cities relatively early, and there was a need for officials with specific training either for particular tasks[30] or for local government. On the other hand, the central authority in a region needed figures of sufficient standing, as well as of assured

---

[30]   H. de Ridder-Symoens, 'Training and Professionalization', in W. Reinhard (ed.), *Power, Elites and State Building* (Oxford, 1996), pp. 149–55.

loyalty, for its policies to be implemented effectively at the local level. We must not forget that relatively small towns such as Setúbal did not possess many people with the appropriate qualifications for a number of administrative tasks, and that favoured the convergence of royal officers, the officers of the Order of Santiago and the local government officials.[31]

The tendency observed in other urban centres at the turn of the fifteenth and the sixteenth centuries for the power of the urban oligarchy to be reinforced can also be demonstrated at Setúbal. Furthermore, those who had a connection with the Order of Santiago, the king or to local assistance institutions could be favoured. In fact, it is possible to identify a connection between the majority of individuals who were involved in the economic development of Setúbal and in the local government. Besides their relationship with the Order, they were also involved with a local religious fraternity – the Fraternity of the Anunciada. The participation of the officers and knights of the Order of Santiago in this institution, in some cases with responsibilities in its administration, can be traced back at least to 1358,[32] and among its members it is possible to identify even one of the fifteenth-century administrators of the Order, *Infante* D. Fernando, as well as his son, King Manuel I (1495–1521).[33] This institution seems to have been very influential in the life of Setúbal, contributing to the development of fellowship between its members and probably to the establishment of common interests.[34] The coincidence between some of these spheres of action leads to the conclusion that during the fifteenth and the first half of the sixteenth centuries members of the Order of Santiago were involved in the development of Setúbal, either through the production of salt or the development of trade, activities that produced a considerable income for the Order and that had repercussion for urban management and improvement.

---

[31]   A.A. Andrade, 'Estado, territórios e "administração régia periférica"', in M.H. da Cruz Coelho and A.L. de Carvalho Homem (eds), *A Génese do Estado Moderno no Portugal Tardo-Medievo* (Lisbon, 1999), pp. 171–2.

[32]   ADS, *CNSA/B/004/Lv1*, Parte II, fols 66v–68v.

[33]   *Portugaliae Monumenta Africana*, 1, gen. ed. M.L.O. Esteves (Lisbon, 1993), pp. 420–22.

[34]   The influence of fraternities in urban life has been discussed for the Basque Country by E.G. Fernández, 'Las cofradías de mercaderes, mareantes y pescadores vascas en la Edad Media', in B.A. Bolumburu and J.Á. Solórzano Telechea (eds), *Ciudades y Villas Portuarias del Atlântico en la Edad Media. Nájera. Encuentros Internacionales del Medievo (Nájera, 27–30 de Julio de 2004)* (Logroño, 2005), pp. 277–94, and for Valencia by M.B. Bolorinos, 'Las cofradías en el reino de Valência. Análisis y claves interpretativas' *Anuario de Estudios Medievales*, 36 (2006), 569–77. For Portugal, see M.H. da Cruz Coelho, 'As confrarias medievais portuguesas: espaços de solidariedades na vida e na morte', in *Cofradías, grémios, solidariedades en la Europa Medieval. XIX Semana de Estudios Medievales. Estella '92* (Navarra, 1992), p. 161; L.A. Santos Nunes Mata, *Ser, Ter e Poder. O Hospital do Espírito Santo de Santarém nos Finais da Idade Média* (Marinha Grande, 2000), p. 179.

This individual participation and investment in emerging opportunities is well illustrated by some of the men, whose *cursus honorum* it is possible to construct. Fernão de Reboredo provides an example of an interesting career due to his social advance. Mentioned as a member of the Fraternity of the Anunciada in 1507[35] and as its notary in October 1511,[36] in 1507 he was also a town councillor.[37] In 1520 he was knight of the royal household and a notary with responsibilities for the receipt of a local tax.[38] A few years later he obtained leases on several properties that the Order of Santiago had in Setúbal.[39] In 1532 he is mentioned as a knight of the Order of Santiago[40] and as auditor of the Order's accounts,[41] an office that he carried out at least until 1536.[42] His connection with the Fraternity of the Anunciada is confirmed in 1537, when he is mentioned as one of the officers responsible for its administration.[43] Jerónimo Pires was also a member of the Fraternity of the Anunciada, whose participation at a meeting was referred to in 1493.[44] He is mentioned as a squire of the Order's administrator, *Infante* D. Fernando, and as the officer responsible for the receipt of some port taxes in 1482.[45] Some years later he was also a prosecutor in Setúbal,[46] and we know that he invested in the production of salt.[47] Another interesting example is Brás Cordeiro, who was developing his activities in Setúbal from at least as early as 1510 when he appears as a public notary appointed by the Order of Santiago.[48] He too was a member of the Fraternity of the Anunciada from at least 1522,[49] and in 1525 he is mentioned as one of the officers responsible for its administration.[50] In 1534 he had moved on to being a town councillor.[51] Between 1539 and 1543 he obtained leases on several properties in Setúbal owned by the Order of Santiago.[52]

All these points are intended to clarify some of the questions raised at the start of this paper, but a more thorough analysis is in preparation within the framework of an academic dissertation. A prosopographical survey and genealogical

---

[35]   ADS, *CNSA/B/004/Lv1*, fols 23v–24v.

[36]   Ibid., fol. 36v.

[37]   ANTT, *MCO, OS/CP*, Livro 233, fol. s/ n.

[38]   Ibid., Livro 11, fols 75–6v.

[39]   Ibid., Livro 13, fols 157v–159, 171–2v.

[40]   Ibid., Livro 14, fols 205v–206.

[41]   Ibid., Livro 15, fols 276v–277.

[42]   Ibid., Livro 36, fols 83–83v; ibid., Livro 42, fols 411v–412.

[43]   ADS, *CNSA/B/004/Lv1*, Parte II, fols 188v–189.

[44]   Ibid., fols 27–8, 49v–50v.

[45]   ANTT, *Chancelaria de D. João II*, Livro 6, fol. 47v.

[46]   ANTT, *MCO, OS/CP*, Livro 8, fol. 65.

[47]   Ibid., Livro 42, fols 142–142v.

[48]   Pimenta (as n. 28), p. 361.

[49]   ADS, *CNSA/B/004/Lv1*, Parte II, fols 125v–127.

[50]   Ibid., Parte II, fols 105–8.

[51]   Ibid., Parte II, fols 108–10.

[52]   ANTT, *MCO, OS/CP*, Livro 21, fols 205–6.

reconstructions will be useful in clarifying the relationship between the Order of Santiago and the local institutions in the port city of Setúbal, thereby allowing a better understanding of who made up the ranks of the Order at a local level and the sociological composition of this group, and also to shed light on who was admitted into the Order of Santiago, as it would seem that during the fifteenth and sixteenth centuries many of its members had an urban background.[53]

---

[53] The importance of the urban background of the members of the orders of Avis and Santiago has been recently pointed out by L.F. Oliveira, 'A Coroa, os Mestres e os Comendadores. As Ordens Militares de Avis e de Santiago (1330–1449)' (unpublished PhD thesis, Universidade do Algarve, 2009).

Chapter 36

Inquiring about Honour in the Portuguese Military Orders (Sixteenth to Eighteenth Centuries)[1]

Fernanda Olival

This essay aims to analyse a specific range of issues in the period between the time when the orders were brought under the tutelage of the Portuguese Crown and 1773, the date by which the distinction between New Christians and Old Christians had officially been brought to an end. With that in view, it is necessary to outline the prerequisites for admission into the Portuguese military orders and to scrutinize the details and meaning of these requirements in comparison with those of the Castilian orders. Finally, it is crucial to demonstrate the way these prerequisites were ascertained and set them against the procedures followed by other institutions that also had to establish eligibility for membership, such as the Holy Office (Inquisition) in Portugal. In tackling these questions, I shall take into account not only the theoretical framework, which underpins them, but also the actual institutional practices, which are essential for a proper understanding of what was at issue.

As early as the first half of the sixteenth century background investigations were carried out to ascertain whether candidates were fit for the military orders. The earliest ones on record date from the time of Dom Jorge, master of the orders of Avis and Santiago (†1550). A number of processes dating from the 1520s and later survive in the archives;[2] earlier ones have apparently been lost. At that time the requirements in force in the orders replicated those for the secular clergy. The process was essentially geared towards corroborating the candidate's honesty, ensuring that he was not part of any ongoing criminal investigation as well as establishing that he possessed material assets that would allow for a standard of living befitting the social status enjoyed by the institution.

[1]   Research work carried out within the scope of the PTDC/HAH/64160/2006 – FCT (= Portuguese Foundation for Science and Technology). The Author expresses her gratitude to Professor Francis A. Dutra for reviewing the English translation of this paper.
[2]   M.C.G. Pimenta, *As Ordens de Avis e de Santiago na Baixa Idade Média: o governo de D. Jorge* (Palmela, 2002), p. 230.

In the 1542 statutes of the Order of Santiago the estate of candidates aspiring to knighthood was set at 400,000 *reals* (*reais*) in capital value or 20,000 in income.[3] This was a large amount bearing in mind that at that time the financial requirements for qualification for secular clergy amounted to 30,000 *reals* (*reais*) in real estate[4] – that is to say, 13.3 times less. Although in particular cases one could invoke religious antecedents, as did Francisco Veloso, a scribe on the king's ships on the Guinea commercial route, when in 1538 he stressed 'I am an Old Christian and not of Jewish or Muslim extraction',[5] this information was not mandatory.

The 1542 statutes introduced more rigorous requirements for qualification. For the first time candidates were excluded if they, their parents or any of their four grandparents were Jews or Muslims, although converts were accepted: 'but if anyone blessed and enlightened by the grace of God should convert to our holy faith and is a person such as would serve or honour the Order, in such a case the master may welcome him into it'.[6] These statutes similarly excluded from the Order men such as 'mechanics' (manual workers) or persons of artisan background, farm labourers and the disabled, although in this last category exceptions might be made if the disabilities resulted from the war against the infidel or if 'the person be such and of such qualities that will benefit the Order'.[7]

After the annexation of the orders of Avis, Christ and Santiago to the Portuguese Crown in 1551, the profiles of these institutions changed substantially. The Crown wanted to set a value on what was distinctive about these institutions so as to make them socially more appealing. Hence it could pay for services rendered with habits (knighthoods) and commanderies in a more efficient manner. From the beginning of the sixteenth century onwards, as for example at the 1503 general chapter of the Order of Christ, the Crown had sought to cultivate the idea that a habit (knighthood) or a commandery should be obtained through services rendered and not solely by virtue of religious fervour or as a result of any system of hereditary transmission. After 1551, it did its best to enhance the honour inherent in membership. If one compares this development with the procedures followed within the Castilian orders, this is the only plausible explanation for the late introduction of the statutes concerning purity of blood and purity of occupation, which in Portugal only appeared in 1570. These demands resulted from a papal diploma, obtained as the result of royal initiative. In this context, it is worth highlighting that this same

---

[3]   *Regra et statutos da ordem de Santiago* (Lisbon, 1542), cap. IV.

[4]   F. Olival and N.G. Monteiro, 'Movilidad social en las carreras eclesiásticas en Portugal (1500–1820)', in F. Chacón Jiménez and N.G. Monteiro (eds), *Poder y movilidad social: cortesanos, religiosos y oligarquías en la Península Ibérica (siglos XV–XIX)* (Madrid, 2006), p. 105.

[5]   Arquivo Nacional da Torre do Tombo (hereafter ANTT), *Colecção Especial*, Cx. 75, Mç.1; Pimenta, p. 433.

[6]   *Regra et statutos da ordem de Santiago*, cap. IV.

[7]   Ibid.

papal bull (*Ad Regie*, 18 August 1570) also sanctioned military service in North Africa as a route towards proving oneself worthy of enrolment.

From this point onwards, being a knight in a military order in Portugal implied that one was noble, of pure blood and a good servant of the king, all of which were highly valued social assets. In practice the meaning attributed to the notion of 'services' varied immensely over time and was tailored to a variety of political interests. To mention three typical examples: (1) in the political conjuncture of 1580, allegiance to the new dynasty was often rewarded with habits – this was true of the procurators (*procuradores*) of the *Cortes* of Tomar in 1581; (2) in the mid-1630s some individuals received habits in recompense for providing soldiers destined for Pernambuco (Brazil), or even for providing money for that purpose; (3) although from 1706 onwards there was a limit to the type of services that could be rewarded by the Crown,[8] after 1750 the act of delivering a huge quantity (at least eight *arrobas*) of gold to the foundries in Brazil in the space of one year might be rewarded with a knighthood in one of the military orders.[9] This was a way of encouraging compliance with the new means for paying the royal fifth (*quintos*) due to the Crown.

After services had been rendered and once the grace of an insignia had been granted, even if 'graciously' (that is by virtue of the king's liberality rather than in exchange for services), the candidate could not yet wear his habit. The award of an insignia was merely the first step. Once it was achieved, another followed, one that could lead to enormous social embarrassment for those who did not fulfil the preconditions set out in the order's statutes and regulations. In addition to those already mentioned (purity of blood and occupation going back to one's grandparents), in Portugal from January 1570 onwards candidates also had to be between the ages of eighteen and forty-nine. This was a way of ensuring that, on the one hand, the future knight had had time to gain military experience and, on the other, he was physically able to continue to do so. The same reasoning lay behind the requirement that he should not have any physical defects. Furthermore, he should be a legitimate child and have no debts for which the order could become liable.

When seen in comparison with what was happening in Castile, the requirements outlined above appear similar but not identical. In Portugal, almost all of them were established for all the three orders at the same time, later than in Castile and through the Crown's initiative. This means that the majority of them had not been initiated in the general chapters, as had been the case with the Castilian orders. There were other differences too: in the Castilian orders, nobility of blood and not of privilege was demanded, whereas in Portugal purity of occupation – in other

---

[8]   This applied mostly to military services, services performed by members of the royal bureaucracy who had university training, diplomatic services and services carried out within the royal household.

[9]   Cap. IX charter law dated 3 Dec. 1750. Published in *Collecção das leis, decretos, e alvarás, que comprehende o feliz reinado del Rei Fidelissimo D.José o I. Nosso Senhor* (Lisbon, 1797), 1.

words, the fact that the candidate did not work with his hands – was enough. This aspect entailed a greater social openness, as well as an easier and more inclusive access, prompted by the fact that the Portuguese Crown, given its vast empire, needed a wide spectrum of servants and could ill afford a restrictive policy. Be that as it may, in Portugal the demand for purity of blood and of occupation were adopted simultaneously, whereas in Castile purity of blood was established first and it was only in the second half of the sixteenth century that there was a greater rigour in excluding men of 'mechanic' background. Not until 1563 did the Order of Calatrava begin to inquire as to the purity of occupation of candidates and of their fathers, and it was only as a result of the general chapter of 1600 that the same requirement was demanded of grandparents,[10] albeit not in the feminine line.[11] In the Order of Santiago, until the general chapter of 1653, the admission of wholesale merchants was allowed and the nobility of the maternal grandmother was not stipulated.[12] In 1600 and as late as 1776 in the Order of Alcântara, the purity of occupation was only investigated in the case of the candidates themselves and their fathers.

So far as legitimacy was concerned, only the Order of Santiago traced it back to the generation of the candidate's grandparents. The Order of Alcântara only went so far as the parents' generation, and in the Order of Calatrava the process was identical to that of the Portuguese orders: the candidate's legitimacy was deemed sufficient.

In the orders of Calatrava and Alcântara, ten was the minimum age for admission. In the Order of Santiago, at least from 1560 onwards, 7-year-olds were already eligible.[13] By comparison, the Portuguese minimum age limit, set at eighteen, was considered very high. In fact, it was even greater than that prescribed by the Council of Trent for admission into the regular orders. It is also worth pointing out that in the Castilian orders there was no maximum age limit – once beyond the minimum age, all were eligible. One should also note, however, that these attributes, particularly that of exceeding the upper age limit, were easily dispensed by the Portuguese monarchs. In fact, the profusion of dispensations on the part of the king was another defining trait of these institutions in Portugal, one that set them apart from those of Castile where a dispensation from any admission requirement for the military orders entailed the intervention of Rome. In Portugal,

---

[10]   F. Fernández Izquierdo, *La Orden Militar de Calatrava en el siglo XVI: infraestrutura institucional. Sociología y prosopografía de sus caballeros* (Madrid, 1992), pp. 92–100.

[11]   *Difiniciones de la Orden, y caballería de Calatrava, conforme al capitulo general celebrado en Madrid Año M.DC.LII.* (Madrid, 1661), tít. VI, cap. VI.

[12]   E. Postigo Castellanos, *Honor y privilegio en la Corona de Castilla: el Consejo de las Órdenes y los caballeros de hábito en el s. XVII* (Soria, 1988), pp. 135–6.

[13]   F. de Vergara y Alaba, *Regla, y establecimientos, de la Orden y Cavalleria, del gloriosso Apostol Santiago, Patron de las Spañas, con la Historia del origen y principio deella* (Madrid, 1655), tít. I, cap. VII.

this was true only in cases of Jewish or Muslim blood. Furthermore, only in investigations carried out by the Portuguese orders into a candidate's background was the exclusion of native blood (from Africa, Asia or America) traced back to the grandparents' generation; by contrast, in the Castilian orders this issue was not directly raised.

Another special aspect of the Portuguese investigation procedures was the fact that heresy and lese-majesty were amalgamated in the same question: 'If you are a son or a grandson of a heretic or of one who has committed a crime of lese-majesty ...'. This provision seems to imply that transgressing against God and divine doctrine was comparable to transgressing against the monarch, and reflects the degree of power the Crown exercised over the Portuguese orders.

In the Order of Santiago of Castile it was asked whether 'the said candidate knows how to ride a horse, can do so, and possesses one'. In the other Castilian orders the only question raised was whether the candidate was sound or had any disease that would prevent him from serving in the cavalry. The Portuguese Order of Avis, however, in the wake of the 1619 general chapter, also established the link between being healthy and the use of weapons and horsemanship. As late as 1640, inquiries were made as to whether 'there is any knowledge of the candidate having a reputation for cowardice, and uselessness in warfare'.[14] In the Order of Christ, the most important of the three Portuguese orders in the period under examination, the form of inquiry remained unaltered from c.1628 until 1773 in line with the earlier statutes.[15] Printed copies were made available and sent to the commissaries.[16] From the mid-seventeenth century onwards these copies began to be used by the other military orders. When illnesses or deformities were checked, such inquiries were aimed at knowing whether they constituted an obstacle to serve the 'order' rather than the cavalry. This was a subtle change, for not all candidates were involved in military service.

It should be noted that the process of uniformity that followed from the moment the three Portuguese masterships came under the tutelage of the Crown was less conspicuous in Castile. The questions asked during the investigations into the candidates' background was not the same in all orders. Throughout the sixteenth and seventeenth centuries the Castilian orders were on paper more restrictive, but the qualification rules had lacunae, in particular with regard to the feminine lineage and the grandparents' background. Rules were more stringent when it came to issues of nobility and purity of blood, as well as in the proscription of heretics. The Order of Alcântara went so far as to trace the candidate's lineage back four generations and to demand that the grandparents' coat of arms be offered as proof. However, important though it may be to analyse the demands outlined

---

[14] ANTT, *Habilitação da Ordem de Avis*, Letra L, Mç. 1, doc. 39.

[15] *Definições e estatutos dos cavaleiros & freires da Ordem de N. S. Iesu Christo, com a historia da origem, & principio della* (Lisbon, 1628), 1st part, tít. XIX, pp. 89–90.

[16] The earliest document we know of is dated 1633 (ANTT, *Habilitação da Ordem de Cristo*, Letra M, Mç. 3, doc. 15), but it is assumed that there were earlier ones.

above, it is even more pivotal to bring to light the process through which they were verified, and to ascertain who conducted the questioning and what type of preparations were involved. The rigour of the background check (*habilitações*) hinged on all these factors.

In the first half of the sixteenth century, judging by the procedures of the orders of Avis and Santiago, the background check consisted of no more than a hearing of a series of witnesses, conducted more often than not by the Judge of the Order. The witnesses were not numerous – three or four were enough in some cases – but were called by the interested party, which obviously indicates a lack of rigour.

After the annexation of the Orders to the Portuguese Crown, these institutions came to be administered by the *Mesa da Consciência*, which had been created in 1532 to deal with the pious obligations of the Crown. From 1564 onwards, background investigations began to be conducted under its supervision, and regulations to oversee the process were elaborated.[17] In Castile, the same mission was in the hands of the Council of Orders, an institution largely controlled by the knights of the military orders, something that was not true of the *Mesa da Consciência*. Until the regency of Dom Pedro, the latter was dominated by clergy,[18] a situation of which the knights strongly disapproved and against which they protested on more than one occasion. The Crown could thus extend its control over the institution.

The period between the 1619 general chapters and 1627–28 can be characterized as a time of reorganizing the formalities of the background check. According to the regulations of the mid-1560s, background investigations usually took place in Lisbon. Only in cases where no one could account for the candidate's ancestry was it common practice to write to the *juiz de fora*, to the *corregedor da comarca* or to any other person that could hear the testimony of witnesses, following a form of inquiry that was sent to them. However, in standard cases, it was the General Judge of the military orders in the *Mesa da Consciência* (Lisbon) who was responsible for selecting, summoning and questioning the witnesses who would number between two and four. They were summoned in secrecy by the *Mesa da Consciência* doorman.

It was only in 1597 that Philip II made it compulsory for background checks to be conducted in the birthplaces of the candidate, his parents and both sets of grandparents.[19] They were to be carried out by a brother priest of the respective order, who was sent to the place in question with the form of inquiry form and instructions so that he could undertake the necessary formalities in the name of the king as the governor of the order. It fell to this cleric to make a record of the

---

[17]   Biblioteca Nacional de Portugal (hereafter BNP), Cód. 10887, fols 65, 433–4; Biblioteca Geral Universidade de Coimbra (hereafter BGUC), Cód. 479, fols 19–21.

[18]   BNP, *Manuscritos Avulsos*, Cx. 91, nº 5, fols 1–2v; L.P. de Carvalho, *Enucleationes Ordinum Militarium* (Ulyssipone, 1693), 1, *Enucl.* I.

[19]   ANTT, *Mesa da Consciência – Ordens Militares – Papéis Diversos*, Mç. 22, doc. 126.

proceedings, while the interrogation itself was conducted by the *corregedor da comarca*, a judge with university training who was the king's representative in that jurisdiction. However, if the latter or his wife did not have the requisite of purity of blood, the inquiry was to be conducted by the *provedor* and, as a last resort, by the *juiz de fora* or the *corregedor da comarca* closest at hand. In places that lay outside of the jurisdiction of the *corregedor*, he was to be accompanied by the *provedor*. Having concluded the interrogations, the cleric returned to Lisbon with the documents he had drawn up and was not to leave any copy behind.

Only investigations into the background of people born and residing in Lisbon continued to be conducted by the Judge of the Order in the *Mesa da Consciência*, where witnesses were called by the doorman. When the candidates originated from any of the Atlantic islands or from a part of the Portuguese empire, the brother priest would not go there, and the person responsible for hearing and recording the testimonies (the *ouvidor*) was assisted by the local clergy and parish priests who served as scribes. However, if the candidate or his ancestors originated from the Castilian peninsula, the brother priest responsible for making the background checks had to go to Castile with a written request to the local authorities to have the necessary investigations made. If the candidates were from the Castilian Indies, Italy or elsewhere, it fell to the *Mesa da Consciência* and the King to determine the necessary procedures.[20] Until 1614, the brother priest would go to the *Mesa da Consciência* beforehand to swear that he would attend to his duties with the utmost zeal, but after that date he was relieved of the obligation of swearing before the Court of the Orders, and the local authorities could take his oath.[21] At this stage, the initial recommendation was for a transcript of the statements of three or four witnesses, who should be chosen from among the oldest Old Christians and who were not partial to the individuals under investigation.[22] However, at least as early as 1613 it was being recommended that the testimony of five or six should be heard.[23]

Later, in the wake of the *definitoria* that emerged from the general chapter initiated in 1619, the figure of the commissaries was introduced in line with standard procedures in Castile and in the Holy Office in Portugal.[24] In a particular chapter of that *definitorium*, dated 1620, it was determined that if an inquiry had been negligent or inadequate it should be conducted again, at the commissaries' expense.[25] The Court eventually set up a network of commissaries throughout the entire kingdom and empire. It was composed of knights from the different localities and who were to operate with the assistance of a brother priest of the

---

[20] Ibid., L° 310, fols 62–4.

[21] BGUC, Cód. 479, fol. 25v.

[22] ANTT, *Habilitações da Ordem de Cristo*, Letra P, Mç. 3, doc. 5, fol. 1.

[23] Ibid., Letra M, Mç. 2, doc. 7.

[24] ANTT, *Mesa da Consciência – Ordens Militares – Papéis Diversos*, Mç. 22, doc. 126.

[25] BNP, *Col. Pomb.* 156, fol. 7.

same order. Only the investigation procedures relating to candidates in the North African fortresses were to be conducted in Lisbon, summoning people from the candidate's place of origin, who were frequently individuals with the status of *merceeiros* of *Belém*. In the remaining cases, when faced with the places of birth included in each application, the *Mesa* would write to a local commissary, in most instances nominating the scribe to record the statements. Before every inquiry began, the officials took an oath in each other's presence. This network thus differed from the Castilian system; there the commissary always proceeded from the Court of the Orders to the place where the inquiry was to be conducted. In this respect, the procedure of the *Mesa da Consciência* was similar to that of the commissaries of the Holy Office in Portugal, who also had agents spread out throughout the territory.

Both the post of commissary and that of scribe were highly sought after. These posts provided some monetary remuneration, but above all the holders enjoyed significant power in a local context. The commissaries were true guardians of the honour of their localities, a function that in and of itself conferred a certain social status. Some of these individuals would even inspire fear, given the power they could yield. However, it should be noted that the individuals performing these functions had no specific training and acquired the necessary skills through practice and through their exchanges with the *Mesa de Consciência*.

In order to become part of this network it was sufficient to petition for this office a few years after having become a knight in one of the orders. As a rule, in their petitions applicants would allude to the fact that they were wealthy – perhaps as a way of stressing their impartiality – and in a few cases they noted that they had not benefited from dispensations when admitted to their order, thus emphasizing further their social standing.[26] We are thus led to believe that the posts had come into the hands of the local elites, the most important families in each region, especially from the end of the seventeenth century onwards.

The case of the Holy Office had slightly different contours, although it too did not involve any training period. Only ordained clergy were eligible for the post of commissary, although in fact – particularly from the early seventeenth century onwards – it would be hard for a mere priest to obtain it. Instead, he would most likely attain the office of notary of the Holy Office. A candidate for the office of commissary was required to have a superior rank or another distinctive sign of social status, such as a degree in the liberal arts, or a regular income or a canonry. All of these factors were dependent on the hierarchy of clergy in the diocese or the region. The application to the Holy Office could also be submitted for a specific post, although there are records of notaries who became commissaries a few years after being admitted. The background investigations that allowed them access to these offices were not in themselves different from the provisions for the admission as *familiares* to the Holy Office, except that its scrutiny by the commissaries in

---

[26]   ANTT, *Mesa da Consciência – Ordens Militares – Papéis Diversos*, Mç.1, docs. 3, 7, 12–13, 19, 21, 79.

charge could bring the specific post into the equation that was applied for; the same could be said of members of the General Council (*Conselho Geral*) of the Holy Office when they gave their approval to an applicant. If someone was a familiar or a notary and wanted to become a commissary, it was necessary to carry out a new background investigation.

Once the office of commissary was attained, the methods and procedures adopted during interrogations in the military orders also deviated somewhat from those of the Holy Office, even though the commissaries of both institutions went to the birthplaces of the candidates and selected witnesses in secrecy. In the Inquisition there were either *extra-judicial diligences* or information collected previously, although systematic records of these only began in the early years of the eighteenth century. In the orders these procedures were extremely rare. As a rule the interrogations by the Inquisition involved the testimony of twelve witnesses, as against the six normally present in those of military orders.

The Holy Office was one of the first institutions to accept background checks previously carried out by the Inquisition on one or more of a person's ancestors when checking that person's qualifications for admission. It was already accepting their checks in the 1690s, whereas the military orders only began to do so, as far as we know, from about 1745. Given the reputation for thoroughness and the power enjoyed by the Holy Office in Portugal, it is only to be expected that it was the first to adopt this practice. It was easy to recognize the validity of this procedure.

In both institutions the rigour of the background investigations was indeed reinforced during the first decades of the eighteenth century, a fact that may strike us as somewhat paradoxical given that the opposite trend was to take place from the mid-century onwards. This late tightening of the rules can be explained by, on the one hand, the pervading puritan atmosphere in Portugal at that time, and, on the other, by the power and prominence that both institutions enjoyed in Portuguese society. Everyone aspired to possessing the habit of the Order of Christ, or to being a *familiar* of the Holy Office, or both, regardless of how commonplace such honours had become. However, the *Mesa da Consciência*, when compared to the Inquisition, was generally relegated to second place, despite the power that the Crown enjoyed over the Court of Orders. This aspect is put into sharper focus when compared with the Castilian orders. It was common knowledge that the king was benevolent and granted dispensations liberally, which also contributed to the image of the Court. The reputation for rigour stuck firmly to the Holy Office – it regularly convoked *autos-da-fé* – even if every once in a while it let its standards slip and discreetly allowed the admission into its ranks of persons whose genealogy was impure. By comparison also with the Castilian orders in the sixteenth and seventeenth centuries, the rigour of the Portuguese military orders came a clear second. The Castilians were not satisfied with the testimony of witnesses alone, and the commissary was required to gather endless pieces of material evidence concerning the candidate's status.

Despite all this, gaining admission to one of Portuguese orders conferred status. Even more than ending the statutes requiring purity of blood, it was when the political centre started to grant habits without any investigation into the candidate's background that in the 1790s the value of the habit fell beyond repair. The background investigations into candidates for the military orders were the best means of recognizing nobility in a kingdom which in early modern times was characterized by far-reaching social mobility. The Inquisition only proved purity of blood.

PART 7
Templar Mythology

Chapter 37

'From the Holy Grail and the Ark of the Covenant to Freemasonry and the Priory of Sion': An Introduction to the 'After-History' of the Templars

John Walker

Few medieval institutions have attracted as much interest as the Knights Templar, whether from academic historians, pseudo-historians or the general public. This Military Order was founded after the First Crusade and played a hugely significant role in the fortunes of the crusader states. After the fall of Acre in 1291 it eventually fell foul of the machinations of King Philip IV of France and was suppressed in the early fourteenth century after a dramatic series of accusations, arrests, trials and confessions.[1] However, although academic historians have, for the most part, confined their attention to the history of the Order up to this point, many pseudo-historians have focused particularly on its 'after-history'. This term is used here to include the myths that surround the Order either during its documented existence from the twelfth to the fourteenth centuries or its supposed existence to the present day. This after-history ranges from the barely credible to the downright absurd and encompasses a vast range of subjects that will be discussed in this paper. However, it should be emphasized that this study is not intended to be an exhaustive survey of the literature and the myths they present, as that would be impossible in an article of this length. Instead, it aims to provide an introduction to the subject, showing the types of myths that are associated with the Order before considering why these myths appear in such great variety and why it is important to challenge the publication of such material.

There is a wealth of material concerning the after-history of the Templars and, indeed, the amount that is now available is so vast that one can be overwhelmed by the range of the studies that have been made. It includes *The Holy Blood and the Holy Grail* by Michael Baigent, Richard Leigh and Henry Lincoln, published in 1982, which weaves the Templars into an elaborate plot encompassing the Cathars, Rennes-le-Château, art, theology and the bloodline of Christ, and *The*

---

[1] For standard academic treatments of Templar history in English, see M. Barber, *The New Knighthood* (Cambridge, 1994); P. Partner, *The Knights Templar and Their Myth* (Rochester, 1987); H. Nicholson, *The Knights Templar: A New History* (Stroud, 2001).

Templar Revelation by Lynn Picknett and Clive Prince, published in 1997, which follows a similar pattern.[2] There are also works that appear more specific to the Templars, although they often go far beyond the narrow subject of the Order.[3] In addition to these studies the Templars have also featured in numerous works of fiction including *The Templar Legacy*, *The Last Templar*, and most (in)famously *The Da Vinci Code*.[4] Although these are clearly works of fiction (despite the publisher's note at the beginning of *The Da Vinci Code*),[5] the myths that they propound relating to the Order are not necessarily seen in that light by many of their readers. Further material concerning Templar after-history can be found on the internet, where a good starting point for viewing Templar speculation is www.templarhistory.com. However, in many ways, the advent of the internet has made the study of Templar after-history far more difficult because of the huge amount of material that has been published through that medium.

In terms of content, there are a number of significant themes relating to Templar after-history that keep cropping up in this vast literature. For the purposes of this paper they can be categorized as those concerning specific events in the Order's history; particular places with which the Order has been associated; and the wealth, power and influence of the Order. In terms of specific events, there are two main points in Templar history that have attracted a great deal of speculation: the origins and foundation of the Order in the twelfth century, and the suppression of the Order in the early fourteenth-. The origins and foundation have attracted considerable interest from pseudo-historians who have a particular penchant for conspiracy theories. Some of these writers have suggested an earlier date for the foundation of the Templars than is accepted by academic historians;[6] what most interests these writers is the assumption that the Order was founded for reasons other than is suggested in the standard academic texts. In recent decades the idea of the Templars being involved in something other than the protection of pilgrims has developed further and has been brought to a much wider audience, particularly since the publication of *The Holy Blood and the Holy Grail* in 1982 and the fictional novel *The Da Vinci Code* in 2003. The alternative foundation myth put forwards in these and other works suggests that the Order was created by a larger but hidden organization known as the Priory of Sion to excavate under Solomon's Temple in

---

[2]   M. Baigent, R. Leigh and H. Lincoln, *The Holy Blood and the Holy Grail* (London, 1982); L. Picknett and C. Prince, *The Templar Revelation* (London, 1997).

[3]   For example, A. Sinclair, *The Sword and the Grail* (New York, 1992); S. Sora, *The Lost Treasure of the Knights Templar: Solving the Oak Island Mystery* (Rochester, 1999); E. Haagensen and H. Lincoln, *The Templars' Secret Island: The Knights, the Priest and the Treasure* (Moreton-in-Marsh, 2000); J. Markale, *Templar Treasure at Gisors* (Rochester, 2003).

[4]   S. Berry, *The Templar Legacy* (London, 2006); R. Khoury, *The Last Templar* (London, 2005); D. Brown, *The Da Vinci Code* (London, 2003).

[5]   Ibid., p. 15.

[6]   Baigent, Leigh and Lincoln, pp. 83–6.

Jerusalem to try to find documents proving, among other things, the marriage of Mary Magdalene and Jesus Christ and the fact that their bloodline continued into the crusader era and beyond.[7] There are many variants on this theme, particularly when authors speculate about who created the Order and what they were looking for, but the general pattern of alternative foundation motives tends to be the same.[8]

The suppression of the Templars at the beginning of the fourteenth century has also attracted a great deal of speculation from pseudo-historians. This surrounds the reasons behind the suppression itself and the possible survival of the Order beyond 1312. The question of whether the Templars were guilty as charged has continued to be a source of fascination for academic and pseudo-historians alike. For the most part, they have tended to suggest that the charges were the imaginative work of Philip IV's administration, although quite recently this viewpoint has been questioned up to a point by Jonathan Riley-Smith.[9] However, among pseudo-historical writings there is a widely held belief that, whether guilty or innocent, the Order was able to engineer an escape from France – La Rochelle is the place often mentioned – with huge amounts of treasure. The suggestion has been made that there was a tip-off before the arrests on Friday 13 October 1307, and this gave the leadership the opportunity to spirit away some of its members. In *The Holy Blood and the Holy Grail*, the authors state that 'persistent but unsubstantiated rumours speak of the [Templar] treasure being smuggled by night from the Paris preceptory, shortly before the arrests. According to these rumours, it was transported by wagons to the coast … loaded into eighteen galleys, which were never heard of again',[10] while Lynn Picknett and Clive Prince argue that 'there is slender evidence that the Knights Templar were ever effectively killed off … It is very likely that knights … went off and founded their own underground movements'.[11] The speculation about the escape of the Templars then goes on to consider where the survivors went and what they did when they got there. Although numerous places have been mentioned in this context, the clear favourite among pseudo-historians appears to be Scotland, at least initially.[12]

A second feature of the content of Templar after-history is the link that has been made between the Order and specific places. Some of these include cities like

---

[7]   See for example, ibid., pp. 83–107; Brown, pp. 216–19.

[8]   For a critique of these myths see, for example, J. Wood, *Eternal Chalice* (London, 2008), pp. 139–46.

[9]   J. Riley-Smith, 'Were the Templars Guilty?', in S.J. Ridyard (ed.), *The Medieval Crusade* (Woodbridge, 2004), pp. 107–24.

[10]   Baigent, Leigh and Lincoln, p. 72.

[11]   Picknett and Prince, p. 163.

[12]   J. Walker, '"The Templars are Everywhere". An Examination of the Myths behind Templar Survival after 1307', in J. Burgtorf, P.F. Crawford and H.J. Nicholson (eds), *The Debate on the Trial of the Templars* (Farnham, 2010), pp. 347–57. For works that link the Templars to a Scottish escape, see for example M. Baigent and R. Leigh, *The Temple and the Lodge* (London, 1989), pp. 102–15; T. Wallace-Murphy, *The Knights of the Holy Grail* (London, 2007), pp. 185–7.

Jerusalem and countries like Scotland where the Order had a presence in reality. However, there are plenty of examples of associations that have been made between the Order and places for which there is no proof of a connection whatsoever, or where the connection has been greatly exaggerated. These include such diverse locations as North/Central America, Scotland and Scandinavia, which will now be examined in turn to illustrate the varied nature of these associations.[13]

As mentioned earlier, the starting point for much Templar speculation is Scotland, with a number of writers suggesting that members of the Order escaped persecution in France, England and other countries, by fleeing there. Two of the authors of *The Holy Blood and the Holy Grail*, Michael Baigent and Richard Leigh, argued that the fugitives probably landed in Argyll and may have settled there, citing evidence from gravestones in several churchyards in the region. Following this settlement, Archie McKerracher, and others, believe that Templar knights were the decisive factor at the Battle of Bannockburn (1314), during which members of the Order appeared over a hill in a cavalry charge, throwing the superior English forces into a panic and sending Robert the Bruce to victory. The Templars are also alleged to have been heavily involved in the foundation of Rosslyn Chapel, eight miles to the south of Edinburgh, in 1446, and a large number of studies have suggested strong Templar connections with this building. These include the supposed relationship between the Order and the Sinclair family, descended from the founders of the Chapel; a bewildering array of supposedly Templar carvings and images, including a knight on horseback and a collection of stars, crosses and doves; and the idea that Rosslyn Chapel was used by the Templars to hide their wealth, brought over from France in the early fourteenth century.[14]

The Scottish connection with the Templars can be found in many pseudo-historical works. Another popular association made in connection with the Order is North and Central America. In a series of books on the subject, Michael Bradley suggested that they may have had in their possession the *mappamundi*, a late-medieval world map that showed land on the west of the Atlantic. He also speculated that the Order could have possessed *portolans*, mariners' charts from the fourteenth and fifteenth centuries, which had coastlines, harbours, rivers and other features noted from personal experience. Bradley argued that the Templars may have brought this knowledge and equipment to Portugal after the suppression that then enabled the Portuguese to lead the way in sea exploration in the fifteenth century. Following this line of thought, the Order can then be credited with being forerunners and instigators of the exploration of North America by later European

---

[13]   For Scotland, see below. For Switzerland, see A. Butler and S. Dafoe, *The Warriors and the Bankers* (Hinckley, 1998).

[14]   Baigent and Leigh, pp. 24–35; A. McKerracher, 'Bruce's Secret Weapon', *The Scots Magazine*, 135, no. 3 (June 1991), 261–8; Sinclair, pp. 44–7; Picknett and Prince, pp. 167–8; K. Ralls, *The Templars and the Grail* (Wheaton, IL, 2003), pp. 175, 187–90; Sora, pp. 116–19. A number of these myths are discussed and debunked in M. Oxbrow and I. Robertson, *Rosslyn and the Grail* (Edinburgh, 2005), pp. 115–36, and Walker, *passim*.

explorers including Columbus whose ships were said to have born insignia that were very similar to the Templar cross.[15] The idea that the Templars may have been involved in early exploration across the Atlantic is a theme taken up by a number of other writers including Louis Charpentier. In *The Mysteries of Chartres Cathedral*, published in 1975, he argued that the Order may have found large deposits of gold while exploring Mexico that they shipped back to Europe and used to start the construction of Gothic cathedrals.[16] Further north, writers including Lionel and Patricia Fanthorpe and Stephen Sora have suggested an early Templar presence in connection with Oak Island, a small uninhabited island in Mahone Bay, Nova Scotia. In Sora's book, *The Lost Treasure of the Knights Templar*, published in 1999, the author argued that the Order first concealed their huge wealth beneath Rosslyn Chapel and then transported it across the Atlantic to Oak Island. They then hid it in a deep pit, complete with booby-traps. The Money Pit, as it has become known, was discovered by three boys in 1795. However, despite huge efforts, great expense and some fatalities, the secrets of the pit have never come to light.[17]

A less popular, but equally fascinating, association made between the Templars and a particular location is that put forwards by Henry Lincoln and Eric Haagensen in their book *The Templars' Secret Island*, published in 2002. In this complex work they suggested a link between the tiny Danish island of Bornholm, lying about twenty-five miles from the coast of Sweden, the hilltop village of Rennes-le-Château in southern France, and the city of Jerusalem.[18] Their hypothesis suggested that the Templars, with the support of St Bernard of Clairvaux, Bertrand de Blanchefort (a Templar grand master) and Archbishop Eskil of Lund, built a series of churches, including the four round churches of Osterlars, Olsker, Nyker and Nylars, on the island of Bornholm. The significance of these constructions was that they were built in very specific locations using the Templars' knowledge of sacred geometry. Lincoln and Haagensen alleged that 'the churches on Bornholm are locked together in an intricate geometric design'[19] that incorporated three-, four-, five-, six- and seven-sided figures with the largest church at Osterlars forming the centre of this geometrical wonder. The authors also suggested that the Bornholm landscape was mirrored in the landscape of Rennes-le-Château that, according to Lincoln, contained its own pentacle of mountains.[20]

---

[15]   M. Bradley, *Holy Grail across the Atlantic* (Toronto, 1988); idem, *Grail Knights of North America* (Toronto, 1998); idem, *Swords at Sunset: Last Stand of North America's Grail Knights* (Ancaster, ON, 2005). See also Ralls, pp. 167–70.

[16]   L. Charpentier, *The Mysteries of Chartres Cathedral* (New York, 1975), 173–5. For a survey of the subject, see Ralls, pp. 171–3.

[17]   Sora, *passim*; L. and P. Fanthorpe, *The Oak Island Mystery* (Toronto, 1995).

[18]   Haagensen and Lincoln, *passim*.

[19]   Ibid., p. 41.

[20]   H. Lincoln, *The Holy Place* (Moreton-in-Marsh, 2005).

The final feature of Templar after-history to be considered here is the suggestions that have been made relating to the immense wealth, power and influence of the Order. There is no denying the fact that it possessed huge amounts of wealth in terms of land, churches and other possessions. However, pseudo-historians have gone much further than this and made suggestions about the Order's wealth that have concerned not just vast sums of money but treasures of a fantastic nature. The most obvious example of this is the way in which the Templars have been closely linked to the Holy Grail, and Helen Nicholson has made the point that in a lot of pseudo-historical works the link between the Templars and the Grail is assumed to be so obvious that it requires no real justification.[21] Links between the Templars and the Holy Grail have come under increasing scrutiny following the publication of *The Holy Blood and the Holy Grail*, which challenged accepted views about the nature of the grail itself. The book's authors suggested that, rather than being a physical object like a chalice or plate, the grail was in fact the bloodline of Christ that the Templars were created to protect. Whatever writers on the grail conclude about its nature, its connection with the Templars and the power and influence it brought them is never seriously questioned.

It is not just the Holy Grail that has been linked to the Templars, and the work of three other writers will serve to illustrate further the nature of Templar myths concerning their wealth. Keith Laidler in *The Head of God*, published in 1998, suggested that the Templars received 'the head of God' from the Cathars, who had in turn received it from Mary Magdalene. It was the embalmed head of Jesus Christ and proved that God was a Man – hence it gave the Templars great power.[22] Patrick Byrne, *Templar Gold: Discovering the Ark of the Covenant*, published in 2001, speculated that the Ark of the Covenant had come into the possession of the Order and was kept either in Cyprus or at the Paris Temple. He suggested that Philip IV may have been aware of the Templars' possession of the Ark because he was married to a daughter of the count of Champagne, and that may have prompted his desire to destroy the Templars.[23] Finally Jean Markale, in the *Templar Treasure at Gisors*, published in 2003, argued that there was a secret order within the Templars that possessed esoteric secrets concealed from the ordinary knights. He examined the idea that, hidden beneath the castle and town of Gisors in Normandy, which had been held briefly by the Templars in the twelfth century, there are many secret passageways in which lies a treasure (physical or spiritual) that could transform the fortunes of whoever discovers it.[24]

The Templars have also been associated with numerous groups that are deemed to have power and influence beyond the normal bounds of possibility, including

---

[21]   H. Nicholson, *Love, War and the Grail* (Leiden, 2004), p. 102; Baigent, Leigh and Lincoln, *passim*; Picknett and Prince, p. 156.

[22]   K. Laidler, *The Head of God: The Lost Treasure of the Templars* (London, 1998).

[23]   P. Byrne, *Templar Gold: Discovering the Ark of the Covenant* (Nevada City, 2001). See also G. Phillips, *The Templars and the Ark of the Covenant* (Rochester, IL, 2004).

[24]   Markale, *passim*.

most recently the Priory of Sion. However, in many people's minds it is with freemasonry that the Templars are still most closely linked, and the connection between these groups is widely accepted without question. Interestingly, this is exactly the outcome that many freemasons in the eighteenth and nineteenth centuries were hoping for, as Masonic historians tried to link their particular organizations to the medieval past, presumably to give themselves greater credibility. The starting point for this process was on 26 December 1736 when Andrew Michael Ramsey, a Scottish freemason and chancellor of the Grand Lodge of France, gave a speech to French masons (repeated on 20 March 1737) in which he presented an outline of Masonic history. In this speech he was keen to stress the knightly antecedents of the freemasons who were descended from former crusaders. Ramsey did not make any direct link between the Masons and the Templars, but it was not long before others, particularly German masons, drew the Templars into their histories.[25] One of the most intriguing groups to make links with the medieval Order was that associated with Baron Karl Gotthelf von Hund (d. 1776), a substantial landowner from north-eastern Saxony. He claimed that he was introduced to a branch of 'Templar Freemasonry' known as Strict Observance while in France. He argued that the Templar link to his Order was created when Pierre d'Aumont, preceptor of Auvergne, and nine associates fled persecution in France around 1310, concealing themselves as stone masons, and ending up on the island of Mull off the west coast of Scotland. D'Aumont was supposed to have been elected grand master of this group on St John the Baptist's Day in 1313.[26] The links that freemasons tried to make between themselves and the Templars have been developed by modern writers including Michael Baigent and Richard Leigh in *The Temple and the Lodge*, published in 1998, in which the authors trace the development of freemasonry out of the establishment of Templar groups in Scotland after the arrests of 1307. Another writer to examine the origins of freemasonry is John Robinson, whose book, *Born in Blood*, includes suggestions that the Templars and, by implication, the freemasons, were linked to historical events not normally associated with either group, most notably the English Peasants' Revolt of 1381.[27] Whatever the nature of the links that are made between freemasonry and the Templars, they serve to create an image of the Order as being in receipt of power and influence, which enables it to manipulate events from behind the scenes, either in the past, present or future.

As was stated in the introduction, this survey of the content of Templar after-history is hardly exhaustive. Any search under the heading 'Templars' on Google and Amazon.co.uk, or even in the Mind, Body, Spirit section of a high street bookstore will reveal that the myths referred to here are really only the tip of the iceberg and that the variety in these myths is quite astounding. Malcolm Barber has suggested that one of the reasons for the flexibility of the myths that creates such

---

[25]   Partner, pp. 100–114.
[26]   See ibid., pp. 117 22; Baigent and Leigh, p. 112 .
[27]   J.J. Robinson, *Born in the Blood* (London, 1989).

variety is the fact that the Order attracted quite disparate groups of people who used it for their own ends. He pointed out that radicals, conservatives, medieval romantics, freemasons and charlatans have all used the Templars in one form or another and in the last couple of decades the Order has continued particularly to serve the needs of anyone interested in conspiracy theories of any kind. Indeed, the explosion in investigations in this area has encouraged greater interest in, and almost an obsession with, the Order itself. Unfortunately, as Barber suggests, it is the after-history in all its different strands, rather than the documented history of the Order, that tends to cement itself in the minds of many people.[28]

In the preceding paragraphs the elements of Templar after-history have been outlined without contradiction, and it is therefore worth emphasizing clearly at this point that none of these myths have any supporting evidence that would be taken seriously by a credible historian. As has been stated elsewhere, the writers of this type of literature do not follow traditional academic 'rules' in the sense that contemporary source material is often ignored when looking for evidence. Instead, anecdotes, false assumptions and other pseudo-histories themselves are regarded as having real authority. Even the lack of any concrete evidence is deemed to be irrelevant as this proves that something has been hidden by conspirators who do not want us to know the real truth.[29]

Although there is no evidence that any of these myths can really be taken seriously, it is worth pointing out that there is an element of truth at the starting point of some of these ideas. For instance, it is accepted by historians that a small number of Templars did escape the arrests in France in 1307,[30] there is no doubt that the Templars did own property and had two preceptories in Scotland,[31] and no one would doubt that the Order was in receipt of large sums of money from land rents, financial dealings and other sources, and also had an influential position in the crusader states and Europe from the twelfth to the early fourteenth centuries.[32] However, pseudo-historians have gone far beyond these basic elements of truth to create something that has no basis in reality: so for example, the number of Templars to escape is exaggerated, their role in Scottish history is over-developed, and their wealth and power transformed into something quite unrealistic. Although these myths have been challenged to a greater or lesser extent in a number of academic works on the Templars, they are, however, still accepted by many people outside the academic community. The reasons why so many people are prepared

---

[28]   Barber, *New Knighthood*, p. 334.
[29]   E. Lord, *The Knights Templar in Britain* (Harlow, 2002), pp. 280–81; Walker, *passim*.
[30]   M. Barber, *The Trial of the Templars* (Cambridge, 1978), pp. 46–7.
[31]   R. Aitken, 'The Knights Templar in Scotland', *The Scottish Review*, 32 (1898), 1–36; J. Edwards, 'The Templars in Scotland in the Thirteenth Century', *Scottish Historical Review*, 5 (1908), 13–25; I. Cowan, P. Mackay and A. Macquarrie, *The Knights of St John of Jerusalem in Scotland* (Edinburgh, 1983), pp. xviii–xxvi.
[32]   Barber, *New Knighthood*, pp. 64–178.

to accept these myths, apparently without question, have been considered in an another article by this author and have a lot to do with the gaps in our knowledge of the Order's documented history, the dramatic nature of its end, a growing interest in conspiracy theories, and the literary style of the writing of pseudo-historians.[33]

Of course, many historians do not appear to see this as a significant problem and would argue that there is no need to be concerned about the after-history of the Templars. At a conference in 2007 a participant suggested to this author that there was no point in looking into this area at all, and others would agree that pseudo-historians and their audience should be left on their own to pursue their interests in peace with little interference from the academic world. However, this seems to be a rather short-sighted viewpoint. In Michael Shermer and Alex Grobman's book, *Denying History*, in which they challenge those who deny the Holocaust, the authors make an important point that is relevant to our subject. They state that, 'Once we allow the distortion of one segment of history without making an appropriate response, we risk the possible distortion of historical events'.[34] Obviously, they are dealing with a much more significant subject, but the same principal applies to Templar after-history. If we are happy to allow the acceptance of published material that is a distortion of the truth to go unchallenged, then where do we draw the line?

The after-history of the Templars is a subject that deserves further investigation and a strong challenge. The variety of subject material that is covered, the use or abuse of evidence that it reveals, and the conclusions that are drawn by pseudo-historians make for fascinating, if at times, intensely irritating reading. The notions of secret activity beneath the city of Jerusalem, flight from France to Scotland and beyond, sacred geometry, immense wealth and secret societies, all serve to draw the general reader into the subject matter and it is hardly surprising that many people do not seek, or are unable, to separate fact from fiction. Despite the valid criticisms that have been made of this body of literature, there is no doubt that its authors will continue to produce more material and will continue to deny the accusations of falsehood and fiction that are levelled at them. It is, therefore, the responsibility of historians to make sure that these writings are not left unchallenged so that 'the modern image of the real Templars'[35] is not lost beneath the weight of Templar after-history but is based on the documented history of the Order from the twelfth to the early fourteenth centuries.

---

[33]   See for example, ibid., pp. 314–34; Nicholson, *Knights Templar*, pp. 238–46; Oxbrow and Robertson, pp. 115–36; M. Haag, *The Templars: Fact and Fiction* (London, 2008); Walker, *passim*.

[34]   M. Shermer and A. Grobman, *Denying History* (Berkeley, 2000), p. 16.

[35]   Barber, *New Knighthood*, p. 334.

Chapter 38

The Myth of Secret History, or 'It's not just the Templars involved in absolutely everything'

Juliette Wood

One of the joys of *Foucault's Pendulum* is that one can recognize the central premise that 'the Templars are involved in absolutely everything' and be able to appreciate the irony of the postmodern twist in which the conspirators become engulfed by their own creation. Even Umberto Eco himself has been inserted into the Templar myth, despite the fact that the novel satirizes just that popular, if somewhat simplistic, model in which notions of Templar secrets and world conspiracies have been advanced as serious, albeit speculative, revisions of history.[1]

The notion of a malevolent conspiracy intent on destabilizing society has a considerable history of its own. Recent scholarship has identified similar attitudes towards perceived threats in the rhetoric of anti-Semitism as well as witchcraft and heresy. Often the violation of taboos, such as incest, infanticide or cannibalism, heightens the perceived malevolence of the conspirators. Sometimes the performance of taboo activity is presented as ritualized cult behaviour under the control of a cabal, like the blood-libel magic attributed to an unspecified Jewish intelligentsia or the preternatural powers of the *Illuminati*.[2] Ironically, such cabals can be at the centre of a conspiracy that threatens to destroy the world, or they can be the core of a secret society whose knowledge will save us. The Templars have fulfilled both roles. A striking feature of alternative Templar material is that similar assumptions about history and the nature of tradition produce such apparently contradictory results. This paper intends to examine the cultural model which allows Templars to be both villains and heroes and to suggest reasons why this perception of them is so tenacious.

---

[1]   U. Eco, *Foucault's Pendulum*, trans. William Weaver (New York, 1989); J. Wood, *Eternal Chalice: The Enduring Legend of the Holy Grail* (London, 2008), p. 175.

[2]   *Conspiracies and Conspiracy Theory in Early Modern Europe: From the Waldensians to the French Revolution*, eds B. Coward and J. Swann (Aldershot and Burlington, VT, 2004).

Bad Templars, Good Templars

The suggestion that magic, rather than just blasphemy, had been part of their secret practices encouraged the perception of the Templars as more than just a military order. Cornelius Agrippa von Nettlesheim (1486–1535) included Gnostic magicians, pagan worshippers of Priapus and Pan, and the Templars in his discussion of the misuse of magical knowledge. It was a prescient observation, and today Templars, Gnostics and the adherents of ancient nature paganism are largely interchangeable. Standard fantasies of sexual perversion and corruption, the stock in trade of descriptions of otherness, were applied to the Templars. Charges of sodomy were embroidered with tales of orgies, and they became part of a conspiracy to undermine moral and political stability. By the time Agrippa's work was translated into French, as part of a nineteenth-century revival of interest in the occult, the Templars were not only rehabilitated, but also endowed with elaborate doctrines and a complex and secretive internal structure.[3]

In many popular treatments, the Templars guard an ancient secret, hints of which appear in later sources clothed in symbolic language. Such assumptions remain rooted in an older academic discourse, especially the notion of cultural survival used by nineteenth-century anthropologists. At that time a renewed appreciation of the rich oral folk heritage of Europe was strengthened by developments in the fields of anthropology, archaeology and the new discipline of folklore. This provided a mechanism by which intellectual innovations, like Darwin's theory of evolution, could be applied to sources such as ancient texts and oral tradition. This approach underpinned the work of E.B. Tylor and J.G. Frazer. Tylor defined 'survivals' as outdated practices and customs that were retained as primitive societies evolved and progressed. The continued existence of such practices was proof of an older stage of culture and could be used to reconstruct the past.[4]

Frazer's magisterial examination of the cultural basis of religion, *The Golden Bough*, sought the common basis of all belief systems and traced its influence on organized religions like Christianity.[5] For him, all religions originated as fertility cults that evolved into more rational forms of belief. The concepts of rationality and cultural survival were important for Frazer's theory. He used comparative and historical analysis as a way of demystifying religion. Linear progressive evolution was a given, and the concept of cultural survival provided a way to explain how modern rationality originated in primitive belief. As the mythic world of primitive man gave way to rationality, remnants of earlier rituals and practices lingered on as folk beliefs and customs among the rural peasantry. Like E.B. Tylor before him, the practices of primitive (i.e. less technologically advanced) societies, ancient peoples and European peasants were seen in terms of distinct and progressive

---

[3]   P. Partner, *The Knights Templar and Their Myth* (Rochester, 1990), pp. 92–5, 100–101, 110–12, 120–23, 128–9; Wood, *Eternal Chalice*, pp. 122–31.

[4]   R.M. Dorson, *The British Folklorists* (Chicago, 1968), pp. 187–201.

[5]   J.G. Frazer, *The Golden Bough*, 3rd edn (London, 1906–1915).

stages of culture. Thus the origins of contemporary social practices could be retrospectively linked to both primitive and ancient cultures. Anthropologists and folklorists have pointed out that such conclusions depend on selective and over-interpreted sources, and that analogues are not necessarily the same as origins. Nonetheless, this universal system harmonized with the Victorian worldview. What seemed more shocking then was that Frazer applied his theories about myth and ritual to Christianity, thereby undermining its authority, and it is this seemingly subversive element that may account for his continuing popularity with alternative writers.[6]

If Tylor and Frazer represent a positivist anthropological tradition in the way they linked the past and the present, then the French historian Jules Michelet embodies the romantic approach to the past. In his writings, Michelet reassessed the history of witchcraft and magic as well as the Templar trial records. The influence of the past on the present was essential to Michelet, a position influenced by Claude Fauriel, Herder and ultimately Giambattista Vico. In contrast to the anthropological school, Michelet had a much darker view of superstition and the way it had been used by authorities, principally the Church, to repress the people. For him what survived from the past was not an outmoded remnant, but the spontaneous collective wisdom of the people and an expression of national character. Michelet saw in the practices of the past, not irrational myth that would disappear with progress, but an effort to express natural libertarian sentiments and fight against repression.[7]

The notion of survival provided a mechanism for the transformation of primitive belief. Like a game of Chinese whispers, the belief changed as it was passed on until the original meaning became distorted. However, a connection between past and present was never lost completely and could always be reconstructed through its surviving remnants. In contemporary alternative writing, ancient beliefs are presented as the essential elements of true religion, and the process of change, which Frazer and Tylor regarded as the inevitable effect of social evolution, becomes a sinister conspiracy of repression. Goodness is shifted to an ideal past, not a progressive future. Unfortunately the process is very subjective and superficial similarities between ancient and modern sources are used to explain each other. This explains why sketchy sources can produce a negative view of the Templars comparable to the irrational survivals discussed by Tylor and Frazer, or a positive view, such as the possession of a wonderful secret, comparable to the romantic religion of Michelet.

---

[6]   R. Fraser, *The Making of The Golden Bough: The Origins and Growth of an Argument* (London, 2001).

[7]   C. Rearick, *Beyond the Enlightenment. Historians and Folklore in Nineteenth-Century France* (Bloomington, IN, and London, 1974), pp. 62–80, 82–102; J. Michelet, *La sorcière* (Paris, 1862; new edn, 1878); *Le procès des templiers*, ed. J. Michelet (Paris, 1841–51; repr. Paris, 1987).

Thus the same model produces a mythical Templar organization that is either a threat to or a salvation for human society.

In alternative Templar history, the attitude to the Order has moved generally from negative to positive, and several writers stand out for the way they synthesized various strands of Templar mythology. The Abbé Augustin Barruel (1797) depicted Templars and freemasons as negative forces out to destabilize the French monarchy and the Catholic Church. For Barruel, everything was connected in a vast historical conspiracy of evil beginning with the Manicheans and embracing Cathars, the Assassins, Oliver Cromwell, the Templars, the *Illuminati* and freemasons. Although the Templars were not yet involved in 'absolutely everything', Barruel's worldview emphasized the idea of a Templar conspiracy.[8]

Freemasonry epitomized a number of Enlightenment values such as a strong work ethic, social tolerance and rationality, but it also provided a context for a nostalgic view of the past, specifically that medieval institutions had been a repository of secret wisdom. Enlightened rationalism provoked a counter-movement that sought deeper realities. It looked for hidden connections, apparent only to the initiated, that would unify the world and transform society. Freemasonry invoked legends about early masons and the cosmic architecture of Solomon's Temple as metaphors for its origin and organization. The more speculative strain within this movement historicized such motifs to create a secret Masonic history. The official name of the Templars, the Order of the Knights of the Temple, and their association with the Temple Mount in Jerusalem, the traditional location of Solomon's building, provided an opportunity to combine the mythic history of the two organizations. Among the first to make such a connection was the self-styled Chevalier Andrew Michael Ramsey (c.1687–1743), who identified the military orders with freemasonry. Ramsey, a Catholic convert, stressed the Christian nature of the enterprise and differentiated it from 'a revival of the Bacchanals'. In other words, it was not pagan and had nothing to do with the worshippers of Pan, one of the groups cited in Agrippa's critique of the misuse of magic.[9]

The Templars figure prominently in the work of Austrian-born oriental specialist, Joseph von Hammer-Purgstall. He too believed that everything in history was connected by conspiracy and secrets. In *The Mystery of Baphomet Revealed* (1818), Hammer-Purgstall linked Templars with Gnostic sects whose members practised phallus worship and denounced Christ. An androgynous deity named Bahomet presided over this phallic cult. Wolfram's *templeisen* become the Knights Templar, and the Holy Grail becomes a gnostic vessel with no Christian links, a secret symbol appearing in all manner of ancient iconographies unrecorded outside Hammer-Purgstall's own work. With such dubious archaeological finds and medieval objects, he transformed accounts of

---

[8]   Abbé Augustin Barruel, *Application of Barruel's Memoirs of Jacobinism to the Secret Societies of Ireland and Great Britain*, trans. [Hon. Robert Clifford] (London, 1798).

[9]   Andrew Michael Ramsay, *The Travels of Cyrus to which is annex'd a discourse upon the theology and mythology of the ancients* (Dublin, 1728).

medieval heresy trials into a treatise on comparative religion, a topic of great interest to early nineteenth-century scholars. The images that he perceived on ancient architectural monuments set a precedent of reading secret meaning into iconography that would be taken up with enthusiasm.[10]

Both Templars and Cathars were powerful forces in southern France, and the area has been a fruitful arena for speculations about secret societies. The fact of persecution allowed both groups to be seen as martyrs. Their disappearance from historical records enabled them to be more easily recreated as secret societies, and today Cathars and Templars have been absorbed into modern myths. Stated thus, it seems a very naïve project, but the idea of a secret system waiting to be rediscovered has become a powerful tool of alternative history. Even good scholars get caught up. Sir Steven Runciman suggested that the Cathars practised unorthodox rituals with dualistic overtones.[11] These ideas do not fit comfortably with the ascetic nature of Catharism or with the surviving records. Although mysteries about this sect are no doubt more apparent than real, they provide great potential for romantic reconstructions.

Deodat Roché was among those who saw the Cathars as a vehicle for mysticism. In the 1920s he met Rudolph Steiner, whose beliefs, that one could regain lost spiritual faculties by developing the intellect, struck a chord with Roché's own spiritual interests. He was himself a mason and believed that Catharism was rooted in Gnosticism. In 1950 he founded a neo-Cathar society not unlike the later Priory of Sion. For Roché, and many others, the culture of the Languedoc region embodied a spirit of independent resistance to outside pressure even after the defeat of the Cathars.[12]

Nostalgia for a world of chivalry made it fashionable to theorize about deep philosophical roots in medieval romance literature. Kyot, a fictional source for Wolfram von Eschenbach's *Parzifal*, was identified with a real troubadour poet, Guot de Provence. The suggestion that Wolfram's masterpiece could be connected to historical events opened new possibilities for speculation. At the beginning of the twentieth century, Joséphin Péladan identified the Cathar stronghold of Montségur with Montsalvache, the grail castle in Wolfram's poem.[13] Péladan, the novelist Maurice Magre and Deodat Roché helped create the idea that Cathars and Templars were linked as custodians of an esoteric secret and that this secret was entwined in some mystical way with the national character of France. Such romantic ideas, rather than ancient documents, formed the background to the

---

[10] Wood, *Eternal Chalice*, pp. 73–7, 101–2.

[11] S. Runciman, *The Medieval Manichee* (Cambridge, 1947), Appendix.

[12] D. Roché, *Le catharisme: son dévelopement dans le Midi de la France et les croisades contre les albigeois*, 3rd edn (Narbonne, 1973); idem, *Contes et légendes du catharisme*, 4th edn (Arques, 1971); idem, *L'Église romaine et les cathares albigeois* (Arques, 1957).

[13] J. Péladan, *Constitutions de la Rose-Croix, le Temple et le Graal* (Paris, 1893).

creation of the Priory of Sion, and by extension, to the further elaborations in other alternative histories.[14]

By the time the young Otto Rahn came on the scene, Cathar mythology was well established and had merged with ideas about the Templars. Rahn drew on French writers like Péladan, and his own imagination coloured by Teutonic nationalism. For him, Catharism had been embedded in the Languedoc since the time of the Celts and the Visigoths. They, in their turn, preserved the sacred Persian wisdom of Zoroaster. In *Luzifers Hofgesind* (1937) Rahn envisaged a cosmology in which Lucifer was the central deity of an ancient untainted religion pitted against the Judaic god of the Old Testament who was later exploited by a corrupt Roman Catholic Church.[15] Rahn's association with the SS, his homosexuality and his suicide has attracted conspiracy legends of its own, and a whole stream of writing that seeks to explain Nazism as an occult aberration.[16]

Some revisions of Templar history focused on a tradition that before his execution Jacques de Molay hid a secret treasure in a hollow pillar at a Templar site. Hidden Templar treasure is another tenacious motif. The Masonic myth of Hiram the first mason, the murdered architect of Solomon's temple, became fused with the execution of de Molay and influenced traditions associated with the Apprentice Pillar at Rosslyn.[17] In Germany, the Templar revival was often linked to the occult, as well as to lost treasure and the idea that shadowy grand masters were behind complex economic and political plots. This created a link between the Templars and the disgraced *Illuminati*, a group of radical intellectuals based in Bohemia during the latter part of the eighteenth century, which gave these supposed conspiracies even greater sinister impact.

Fiction and the Templars

As neo-templar and masonic organizations took root and their history was reassessed, a more positive view of the Order began emerge. As early as 1808 a group led by Bernard Raymond Fabré-Palaprat celebrated the anniversary of the death of Jacques de Molay with all the pomp of modern re-enactors. This was but one indication that what had been regarded as a conspiracy to undermine the

---

[14]   Wood, *Eternal Chalice*, pp. 125–31; M. Magre, *Lucifer: roman moderne* (Paris, 1929); idem, *Le sang de Toulouse, Histoire albigeoise du XIIIe siècle* (Paris, 1931; repr. Paris, 1972); idem, *Le trésor des Albigeois: roman du XVIe siècle à Toulouse* (Paris, 1938); idem, *La clef des choses caches* (Paris, 1935); idem, *Magiciens et Illuminés* (Paris, 1930); idem, *The Return of the Magi*, trans. R. Merton (London, 1931).

[15]   Rahn's *Kreuzzug gegen den Gral* and *Luzifers Hofgesind* are reprinted in *Otto Rahn: Leben und Werk*, ed. H.-J. Lange (Engerda, 1995); Otto Rahn, *La croisade contre le Graal*, trans. R. Pitrou (Paris, 1934; repr. Paris, 1974).

[16]   C. Bernadac, *Montségur et le Graal: le mystère Otto Rhan* (Paris, 1994).

[17]   Wood, *Eternal Chalice*, pp. 131–6.

forces of society by Gnostics, Cathars, Templars and freemasons could be turned on its head to become a conspiracy against these groups, one aimed at suppressing the secret knowledge passed on by a gallant band of brothers struggling against the oppression of the Church and the tyranny of repressive government.

Nostalgia for a chivalric world that never existed was expressed in fictional re-workings of chivalric adventures. This too was an important factor in the transformation of the Templars into romantic outsiders and carriers of secret doctrines. The Chevalier Ramsey's book *The Travels of Cyrus* (1727) teems with secret chambers under Solomon's Temple and parchments in hollow pillars. These motifs are crucial to legends associated with Rosslyn Chapel and the Priory of Sion, but they also express the spirit of the gothic that became popular during the eighteenth century. A complete analysis of the characterization of Templars in literature and the elements they absorbed from the different strains of gothic and romantic literary imagination is beyond the scope of this paper. However, changing attitudes to the Templars in the realm of alternative history echo the emergence of 'the Fatal Man' of romantic fiction as identified by Mario Praz. This figure developed in the context of the romantic reinterpretation of Milton's Satan from demon to subversive rebel and was influenced by the eighteenth-century taste for gothic literature. Sinister Jesuits and inquisitors in literature and art prefigure the appearance of literary Templars, as do gothic outsider heroes, often with religious overtones, such as Mrs Radcliffe's Schedoni and Matthew Lewis's monk, Ambrosio.[18] The line between fiction and fact, or rather between fact and alternative history, is easily blurred. The fictional characters of the French novelist, Eugene Sue, are sinister Jesuits not Templars, but they become incorporated into an influential piece of alternative propaganda, the Protocols of Sion, which does figure in the modern Templar myth.

The fascination and revulsion evoked by the occult is ultimately if somewhat tortuously situated in a rational tradition going back to writers who saw the suppression of groups such as the Templars and the Cathars as an unjust and unnecessary cruelty on the part of corrupt government. The fact that these persecutions had taken place in enlightened, liberal France, rather than in some remote barbarian realm, suggested to some writers that dark forces must have been at work. As a result conspiracy theories with satanic overtones became a propagandist weapon in both fiction and political pamphleteering. A mixture of fiction, pseudo-history, ideology, popular culture and rumour is characteristic of such conspiracy legends. These themes work well in popular novels where the gothic atmosphere of black magic is mixed with the excitement of a spy thriller. The doyen of the occult thriller, Dennis Wheatley did just this in his novel, *They Used Dark Forces* (1964).[19] One of the fountainheads for this type of fiction was Sir Edward Bulwer-Lytton, a considerable force in the occult revival of the nineteenth century who incorporated esoteric ideas into his own novels. A secret

---

[18] M. Praz, *The Romantic Agony* (Oxford, 1933; 2nd edn Oxford, 1979), pp. 59–80.

[19] D. Wheatley, *They Used Dark Forces* (London, 1964).

theocracy ensures the success of *Vril, The Power of The Coming Race* (1871), while *Zanoni, a Rosicrucian Tale* (1861) is full of rituals and secret power.[20] Both find their way back into alternative works of history as well as into later fiction.

Myth and Fiction in Modern Templar Legend

Once a new Templar myth had become established, it could be applied to other contexts. For example, the garden of the Shugborough estate in Staffordshire contains an eighteenth-century monument with a carved relief of Nicolas Poussin's depiction of three Arcadian shepherds gazing at a tomb. The tomb is inscribed with the words 'Et in Arcadia Ego'. At least three designers, working under the direction Thomas Anson, an important figure in the revival of classical art and culture at the time, contributed to the Shugborough monument. The architect, Thomas Wright of Durham, created the basic design. James Stuart, a prime mover in the field of classical design and romantic atmosphere in landscaping provided additions, and the sculptor, Peter Scheemakers, executed the carved relief, based on a reversed print of Poussin's painting.[21] Beneath the relief is a series of letters D.O.U.O.S.V.A.V.V.M. It is this inscription, and the fact that the carved relief is a mirror image of the original painting, that has attracted the attention of modern Templar enthusiasts. Poussin's painting is a central feature of the Priory of Sion whose members used it to create an imaginative history for their society. However, the authors of *Holy Blood and Holy Grail* added Shugborough to the mix.[22] Since then a location with no prior links to the Templars has been absorbed into a dynamic modern legend and has attracted new motifs of its own.

The Templar solution to the cipher has been attributed to an unnamed code-breaker in an unspecified intelligence network. In actual fact, the solution, *Jesus H Defy,* makes little sense as it stands. So, the undaunted code-breaker identifies H with the Greek letter X (chi χ), then equates it with '*messiah*' or '*Christ*' and translates it as 'deity'. Since 'H', the Greek letter eta, refers to the letter 'e' in the name Jesus, and *messiah* and *Christ* mean 'the anointed one' not 'deity', this new phrase, *Jesus (the Deity) Defy* is still unclear. Once again the code-breaker adds more speculative history about Templars and their supposed secret doctrine that Jesus was human, not divine. The result has no more inherent sense than the original series of letters and owes more to the popularity of romanticized

---

[20]   A. Owen, *The Place of Enchantment: British Occultism and the Culture of the Modern* (Chicago and London, 2004), pp. 133–4.

[21]   Wood, *The Eternal Chalice*, pp. 159–64; A. Baker, 'The Shepherdess's Secret', Staffordshire Record Office, unpublished MS; *Arbours and Grottos: A Facsimile of the Two Parts of Universal Architecture (1755 and 1758)*, ed. E. Harris (London, 1979); James Stuart, *The Antiquities of Athens*, 1 (London, 1762).

[22]   M. Baigent, R. Leigh and H. Lincoln, *The Holy Blood and The Holy Grail* (London, 1984; revised edn London, 1996), pp. 190–91.

Templars that any convincing solution to the cipher. Indeed no code-breaker need look further than the book, *Holy Blood*, where a reference to the 'denial of Jesus' divinity' occurs immediately after the discussion of Shugborough Hall's so-called code.[23] However, code-breaking grids, an unknown code breaker in a mysterious intelligence network, the Templars, and the Priory of Sion are just the elements that make this legend so compelling.[24]

There is, in fact, considerable information about the Anson family, their classical interests and the monument. Thomas Pennant, an eighteenth-century traveller and a close friend of the Ansons, hinted that the monument was a memorial to someone close to the family.[25] The letters D.M., *Diis Manibus* ('To the Souls of the Departed'), are a standard abbreviation on Roman tombs, and the remaining letters might stand for the Latin phrase, '*Optimae Uxoris, Optimae Sororis, Viduus Amantissimus Vovit Virtutibus*'. A possible translation would be, 'Best wife, best sister, the most loving widower dedicates [this] to [your] virtue'. This would provide a clear solution to the cipher, and one consistent with the taste for the poetic ambiguities of Arcadian symbolism and stoic philosophy that were popular among the Anson family and their circle.[26]

A fourteenth-century mazer bowl made of wych-elm kept at Nanteos House near Aberystwyth in Wales is frequently presented as a candidate for the real Holy Grail. The cup was probably found during excavations of the ruins at Strata Florida in the late nineteenth century. The object, known as the Nanteos cup, was owned by the Powell family, who exhibited it at a meeting of the Cambrian Archaeological Society in the late nineteenth century. A drawing and a description also appeared in the journal *Archaeologia Cambrensis*. The drawing is contemporary with the Strata Florida excavations and shows a damaged object held together with metal staples. The description, however, contains some traditional elements and archaeological speculations, not unlike modern Templar genre pot-boilers. The object had a reputation for healing even before it was put on display, but the Holy Grail element dates only to the beginning of the twentieth century. According to this legend, several monks escaped to Strata Florida just before the king's minions entered Glastonbury Abbey. The monks took shelter with the Powell family, and, when the last one died, the cup was given into the family's keeping. It is this version that appears most frequently in newspapers or on internet sites.[27]

There is, however, a more localized version of the Nanteos legend in which the grail belongs to the Cistercian Abbey at Strata Florida, and this one involves crusading Templars. There are no records of pilgrims or relics at Strata Florida

---

[23]   Ibid., p. 192.

[24]   J. Wood, 'The Templars, the Grail and just about everything else: Contemporary Legends in the Media', *FLS News*, 45/2 (2005), 13–14.

[25]   Thomas Pennant, *Journey from Chester to London* (Chester, 1782).

[26]   Wood, *Eternal Chalice*, pp. 159–64.

[27]   J. Wood, 'Nibbling Pilgrims and the Nanteos Cup: A Cardiganshire legend', in G. Morgan (ed.), *Nanteos A Welsh House and its Families* (Llandysul, 2001), pp. 137–50.

during the medieval period. The Stedman family, not the Powells, acquired Strata Florida after it ceased to be an Abbey and they built a house on site. After they married into the Powell family, they moved to a new mansion at Nanteos. The circumstances by which the Stedman family acquired Strata Florida are embellished with a charming nineteenth-century story that claimed the first Stedman was a Templar crusader, the son of the duke of Arabia, a friend to Richard the Lionheart who was banished to Wales.[28] The murky period of the crusades provides an ideal setting for the transfer of relics to Europe, and this dramatic cliché is a mainstay of many relic legends. Local historians, keen to give their homegrown worthies the right pedigree, turned the most tenuous traditions into elaborate crusader histories and often simply equated crusaders with Templars. There is an echo of this in Nanteos traditions about a lathe-turned, olive wood cup (actually the object is wych-elm) made in Syria at the dawn of the Christian era.[29]

An alabaster cup associated with the fantasy gardens of Hawkstone Hall in Shropshire echoes the folklore surrounding the Nanteos cup. Both discoveries took place in mysterious circumstances, and both have links to once-prominent local families. The Nanteos cup was known locally as a healing object before becoming attached to the Holy Grail legend. By contrast, the alabaster cup, most likely a Roman make-up pot, supposedly found in a statue had no particular associations until it was identified as the grail in the 1990s. The Nanteos cup was, and still is, rooted in local tradition, while the modern genre of conspiracy history is the major influence on the Hawkstone object. Such theories have been influenced by mass media phenomena like *Holy Blood and Holy Grail* and *The Da Vinci Code* and are full of conspiracies, codes, sacred geography and odd sources.[30]

At Hawkstone, a local family guarded the object and passed its secrets on to their descendants. In the nineteenth century, the antiquary Thomas Wright concealed it in a statue. An obscure numerical code in one of his books leads, via Victorian stained-glass windows in the medieval church at Hodnot in Shropshire, to the object and its secret. The history presented here is typical of the chimerical arguments that crop up regularly in such studies. Thomas Wright's numerous publications on medieval literature, philology, archaeology and folklore contain nothing to link him to Hodnot church or Hawkstone Park, and crucial details are mysteriously absent from library catalogues and published works. Wright (1810–77) was a well-known Shropshire antiquary from a Methodist, not an Anglican, family. Far from finding secret codes in a romance about a local bandit, Fulk Fitzwarine, Wright noted his story combined historical matter enlarged by folk tradition. 'It is surprising how much of this class of legendary matter has crept into history itself and how difficult it is to eradicate it ... The practice of localizing legends seems so inseparable from the popular mind that we actually find it going

---

[28]   S. Wright, *Up the Claerwen* (Birmingham, 1948), p. 61; S.M. Powell, 'Pilgrim Routes to Strata Florida', *Cardiganshire Antiquarian Society Transactions*, 8 (1931), 9–24.

[29]   Wood, *Eternal Chalice*, pp. 60–75.

[30]   G. Phillips, *The Search for the Grail* (London, 1995).

on at the present day; and it is quite remarkable how rapidly a legend is sometimes formed.'[31] This is a good description of the process by which legends are being created, and expanded with Templar motifs, at places such as Hawkstone, Nanteos and Shugborough.

Wright also touched on an important theme in modern Templar myth, namely Templars and pre-Christian wisdom tradition. An ardent Francophile, Wright drew on the revisionist views of the Templar trial put forward by the French historian, Jules Michelet. This approach, not easily available to an English reading public at that time, stressed the use of falsified evidence and torture in achieving the political agenda of the French king. Wright regarded the trials as a 'use which might be made of popular superstition as a means of oppression and vengeance'. He was critical of speculative suggestions about Templar beliefs in the work of the German scholar, Hammer-Purgstall, who proposed that medieval monuments revealed heretical Gnostic beliefs held by the Templars. Wright is unambiguous in his dismissal. 'Von Hammer Purgstall ... attempted to show from medieval monuments that the order of the Templars was infested with Gnosticism ... In fact, Von Hammer totally misunderstood the character of the monuments on which he built his theory.' This is hardly the language of a man guarding a Templar secret. Nevertheless, in his commentary on Richard Payne Knight's treatise on *Priapus*, Wright did consider the possibility of a phallic cult among the medieval Templars. 'Whatever degree of truth there may have been in this story [regarding the Templars] ... the conviction of the existence of secret societies of this character during the middle ages appears to have been so strong and so generally held, that we must hesitate in rejecting it'.[32] This is not a high-minded pagan philosophy or liberal alternative to Christianity; it was an ancient superstition, but, influenced no doubt by Jules Michelet, Wright also believed that the medieval ('romish') Church had exploited such practices in order to discredit opposing factions.

Conclusion

If the popularity of Dan Brown's thriller has done nothing else, it has provoked an examination of the underlying assumptions about secret Templar societies.[33] The modern Templar myth is defined by a particular view of the past, either as a march of progressive evolution or as a coded secret for a pristine society untainted by

---

[31]   T. Wright, 'The Local Legends of Shropshire', *Collectanea Archaeologica, communications made to the British Archaeological Association*, 1 (1862), 50–66.

[32]   Thomas Wright, *Narratives of Sorcery and Magic from the most Authentic Sources* (London, 1851), 1 p. 41; *A discourse of the Worship of Priapus by Richard Payne Knight, Essay on the Generative Powers during the Middle Ages*, ed. Thomas Wright (London, 1865); Partner, pp. 166–7.

[33]   B. Putnam and J.E. Wood, *The Treasure of Rennes-le-Chateau: A Mystery Solved* (Stroud, 2005).

the corruptions of present life. Currently the latter, romanticized, view is popular. Both are highly selective readings of the past, rooted in theories of anthropological and comparative religious history that have been largely superseded in more recent critical thinking. Despite this, ideas about Templar secrets remain curiously impervious to criticism. The emergence of this view of the Templars coincided with the rise of the popular novel in the 1790s. In this context it is striking that several figures influential in the creation of Templar mythology in the eighteenth and nineteenth centuries also dabbled in fiction as well.

Initially the novel was not greeted with universal approval, especially when it incorporated gothic or fantasy plot lines. Critics feared that the intense emotion and sensationalism of these novels would disturb the moral propriety of society and draw readers into the 'labyrinths of politics' (i.e. revolution).[34] The labyrinth metaphor was often used to express the ambiguities of gothic and fantastic fiction. It encapsulated anxieties that such narrative forms would produce a real effect, and their horrors would be intensified by being associated with political terrors. The ability of the gothic form to challenge accepted categories is exactly the feature that attracts literary critics today.[35] Gothic and fantasy novels, the genre to which so many works influenced by the Templar myth belong, have been described as a literature of subversion that seeks to undermine accepted philosophical and epistemological orders by suggesting new possibilities and alternatives.[36] Some of the critical problems posed by these texts, such as the interplay of the imaginary, the symbolic and the real could equally apply to works of alternative history. This writing blurs the distinctions between fictional worlds and the realm of experience and allows novels to read like non-fiction books and non-fiction books to have the drive and narrative coherence of novels. Indeed, in an early interview about *The Da Vinci Code*, Dan Brown seemed to suggest that the novel's reference to symbolism and secret societies was accurate historically.[37]

The most influential modern Templar legend began as serious television documentaries. The subsequent best-selling book, *The Holy Blood and the Holy Grail*, has influenced both the conspiracy genre of alternative history and the popularity of conspiracy-thriller fiction. It has also influenced historical novels like Bernard Cornwell's *Grail Chronicles*, interactive video games like the Gabriel Knight fantasy series, female fiction such as Kate Mosse's *Labyrinth* and the sophisticated satire of Umberto Eco's *Foucault's Pendulum*. It seems that the Templars are likely to be involved in absolutely everything for some time to come.

---

[34]   F. Botting, *Gothic* (London, 1996), pp. 80–82.

[35]   Ibid., p. 168.

[36]   R. Jackson, *Fantasy: The Literature of Subversion* (London, 1995), pp. 171–80.

[37]   'The Real Da Vinci Code', Wildfire Television, broadcast February 2006, Channel 4.

Index